Heat Transfer

3rd UK National Conference
Incorporating
1st European Conference on
Thermal Sciences

Volume 2

Institution of Chemical Engineers, Rugby, UK

Hemisphere Publishing Corporation

A Member of the Taylor & Francis Group

New York   Philadelphia   London

i

Heat Transfer
3rd National Conference incorporating 1st European Conference on Thermal Sciences

Members of the Institution of Chemical Engineers should order as follows:

Worldwide	Institution of Chemical Engineers, Davis Building, 165–171 Railway Terrace, RUGBY, Warwickshire CV21 3HQ, UK

Non-members' orders should be directed as follows:

UK, Eire and Australia	Institution of Chemical Engineers, Davis Building, 165–171 Railway Terrace, RUGBY, Warwickshire CV21 3HQ, UK
or	Taylor & Francis Ltd, Rankine Road, BASINGSTOKE, Hampshire RG24 0PR, UK
USA	Taylor & Francis Inc., 1900 Frost Road, Suite 101, Bristol, PA 19007, USA
Rest of the World	Taylor & Francis Ltd, Rankine Road, BASINGSTOKE, Hampshire RG24 0PR, UK

Library of Congress Cataloging-in-Publication Data

U.K. National Conference on Heat Transfer (3rd: 1992: University of Birmingham)
Heat transfer: 3rd UK National Conference incorporating 1st European Conference on Thermal Sciences: a three-day symposium organized by the Institution of Chemical Engineers, in association with the Institute of Mechanical Engineers, the Heat Transfer Society and EUROTHERM, held at the University of Birmingham, 16–18 September 1992/ organizing committee H.C. Simpson...[et al.].
p. cm. — (Institution of Chemical Engineers symposium series no. 129) (EFCE publication no. 192)
"EFCE event no. 460."
ISBN 1–56032–258–6 (cloth): $350.00
1. Heat—Transmission—Congresses. I. Simpson, H.C. II. Institution of Chemical Engineers (Great Britain) III. Conference on Thermal Sciences (1st: 1992: University of Birmingham) IV. Title. V. Series: Symposium series (Institution of Chemical Engineers (Great Britain)); no. 129. VI. Series: EFCE publication series; no. 92.
TJ260.U2 1992
621.402′2—dc20

92—33782
CIP

Heat Transfer
3rd UK National Conference
Incorporating
1st European Conference on
Thermal Sciences

A three-day symposium organised by the Institution of Chemical Engineers, in association with the Institution of Mechanical Engineers, the Heat Transfer Society and EUROTHERM, held at the University of Birmingham, 16–18 September 1992.

Organising Committee

H.C. Simpson (Chairman)	University of Stràthclyde
G.F. Hewitt	Imperial College of Science, Technology and Medicine
J.P. Bardon	ISITEM, France
R.J. Berryman	Harwell Laboratory
T.R. Bott	University of Birmingham
G.P. Celata	ENEA, Italy
K. Cornwell	Heriot-Watt University
C. Grant	University of Strathclyde
P.J. Heggs	University of Bradford
Y.R. Mayhew	University of Bristol
A.A. Nicol	Consultant
D.B. Spalding	CHAM Ltd
D. Webb	UMIST
P. Whalley	University of Oxford
D. Wilkie	Consultant
P. Pilavachi	Commission of the European Communities

INSTITUTION OF CHEMICAL ENGINEERS

SYMPOSIUM SERIES No. 129
EFCE Event No. 460
EFCE Publication No. 92
ISBN 0 85295 282 1

Printed by J.W. Arrowsmith Ltd, Winterstoke Road, Bristol BS3 2NT

Preface

In 1984, the introduction to the proceedings of the 1st UK
National Heat Transfer Conference looked forward to a
series of meetings taking place at four yearly intervals, at
which delegates from the UK and around the world
could gather to review and discuss the latest work and
developments in this important field. Since then, the
event has grown and matured through the successful
1988 conference at Strathclyde to this, the 3rd conference,
which incorporates the 1st European Conference on
Thermal Sciences, itself the first in what is intended to be
a series of major conferences hosted by the countries of
Europe.

The value of the forum is exemplified by the number of
papers submitted — some 150 — which make up a high
quality programme covering a wide spectrum of the
field. The level and quality of contributions demonstrate
that Heat Transfer maintains its position as a subject of
central importance both in traditional and newly
developing areas.

This conference is the result of a collaboration between
The Institutions of Chemical and Mechanical Engineers,
The Heat Transfer Society and EUROTHERM. Thanks
are due to all the members of the Organising Committee
and to the Commission for the European Communities
who have generously sponsored the event. It need not be
said that without the support of all concerned, the
conference would not have happened.

Errata

Paper 44 (page 377):Prediction of the lumped heat transfer coefficient in regenerators (P.J. Heggs and M. Sadrameli)
(i) Page 378: equation (5) should read
$U_1 = \dot{m}c_gP/\rho_s (1-p)c_sL$, utilisation factor
(ii) Page 380: lines 20 and 21 should read:
The obtained values of Λ from each model were denoted Λ_N and Λ_L, respectively, and for all cases, the values of Λ_I were greater than Λ_N...
and line 28 should read:
The evaluated heat transfer coefficients, α_N and α_L, for each flowrate investigated vary...
(iii) Page 385: the caption for Table 3 should read:
Experimental data and evaluated results for AL2

Paper 113 (page 907): Calculation of heat transfer on a flat plate for high Prandtl and Reynolds numbers (P. Hrycak)
The sentence following equation (1) should read:
Here, the form of the right-hand side is taken over from the pipe flow, as τ_w, the wall shear stress, represented as a function of two parameters p and n, that are themselves related to the flow Reynolds number.

Contents

Keynote Lectures

Boiling and Condensation

Heat Exchangers

Refrigeration and Air Conditioning

Natural Convection

Process Safety and Nuclear Reactors

Two-Phase Flow

Post Dry-Out

Combustion, Radiation and Chemical Reaction

Convective Heat Transfer

Fouling

Heat Transfer To and From Solids

ASSESSMENT OF HEAT TRANSFER PERFORMANCE OF RECTANGULAR CHANNEL GEOMETRIES : IMPLICATIONS ON REFRIGERANT EVAPORATOR AND CONDENSER DESIGN

F. Chopard, Ch. Marvillet, J. Pantaloni[*]
GRETh,Groupement pour la Recherche sur les Echangeurs Thermiques
CENG, 8SX, 38041, Grenoble Cedex

This paper presents experimental work on three different geometries of rectangular channel (smooth, studded with in-line array, studded with staggered array) during forced convection boiling and condensation of refrigerant R22. The heat transfer coefficient correlations have been established in wide operation conditions (mass velocity, vapor quality, heat flux and saturated pressure) and the comparisons have been made between the plates performances.

INTRODUCTION

The subject of enhanced heat transfer has developped to the point that it is of serious interest for heat exchanger application in *heat pumps, air conditionning systems and refrigeration units*. Enhanced heat transfer surfaces are used for different reasons:
- *to obtain less expensive* heat exchangers in reducing the number of heat exchangers or plates, the shell diameter and, on the whole, the heat exchanger size.
- *to obtain more compact heat exchangers*, anyone designing heat pumps or refrigeration units may try and keep the amount of refrigerant as small as possible, the more so as an "ozone friendly" substitute is not yet commercially available.
- *to reduce the mean temperature difference* for the heat exchangers and to provide increased thermodynamic process efficiency which yields saving in operating costs.

The plate heat exchangers are made of high performance heat transfer surfaces: the corrugated or the studded plates (**figure 1**) may dramatically increase the heat transfer coefficient with single phase fluids. Nowadays the plate heat evaporators are increasingly used in refrigeration plants. The welded or brazed plate heat exchangers may partially replace the conventional bundle heat exchangers designed for evaporation or condensation of refrigerant because of their greater compactness and smaller refrigerant capacity.

The design, the conception and the optimization of plate evaporators and condensers require precise, reliable and, if possible, complete thermohydraulic data and correlations. Available information on convective boiling and on condensation characteristics is very limited in the case of studded geometries. This paper presents experimental work on three different plate geometries during forced convection boiling and condensation of refrigerant R22. The heat transfer coefficient and pressure drop correlations have been established in wide operation conditions (mass velocity, vapor quality, heat flux and saturated pressure) and the comparisons have been made between the plates performances.

[*] IUSTI - Centre Saint-Jerome - MARSEILLE (FRANCE)

2/METHODOLOGY

To evaluate the performances of the different geometries, stainless steel test sections with studded or smooth have been built (**figure 2a**). **Table 1** shows the main features of the plates: test section 1 is made of smooth plate, test section 2 is made of studded plate with in-line array, test section 3 is made of studded plate with staggered array. The hydraulic diameter Dh is defined as a volumetric one and the heat transfer surface as a projected one. These test sections (in vertical position) are composed of three parallel channels and, of course, of four welded plates (**figure 2b**): inside the central channel, the refrigerant R22 flows upwards and is partially evaporated or condensed, inside the two lateral channels water flows downwards and is cooled or heated.

The objective of the present work is to evaluate the heat transfer coefficient between the wall of the internal plates and the saturated refrigerant fluid flowing upwards. The main parameters having a significant influence on this heat transfer coefficient are:

- the refrigerant *mass velocity* G (kg/m^{2}s), defined as the ratio of refrigerant mass flowrate to the **mean flow area** (in the cases of studded plates channels, local flow area does obviously not have a constant value)

- the *mean thermodynamic vapor quality* x (%). At the inlet of the central channel of the test section, the vapor quality of the refrigerant flow may be imposed from 10% up to 80% when the vapor quality does not change by more than 15% from the inlet to the outlet of the central channel.

- the *heat flux* $\dot{q}$ (W/m2) defined with projected heat transfer surface (and not the developed heat transfer surface).

- the *saturated pressure* P_{sat} (or temperature T_{sat}) of the refrigerant.

The test loop, whose flow sheet is represented in figure 3, provides tests in a wide range of conditions: **table 2** summarizes the explored test conditions of the different geometries during **R22** evaporation or condensation.The test loop is made of four main circuits:

-the **refrigerant circuit**'s components are: a variable speed circulation pump used to control the refrigerant mass flowrate, two volumetric flowrate sensors BOPP&REUTER, a system of three coaxial preevaporators to control the vapor quality at the inlet of the test section, the four welded plates test section, a liquid/vapor separator, and a refrigerant horizontal condenser.

-the **water circuit** on the test section whose components are a regulated electrical heater (maximal output: 24kW) (for evaporation tests), a cooler (for condensation tests), a recirculation pump and an electromagnetic massflowrate sensor (WAFERMAG BROOKS).

-the **warm water circuit** on the system of preevaporators whose components are a regulated electrical heater (maximal output:100 kW), a recirculation pump and an electromagnetic massflowrate sensor (WAFERMAG BROOKS).

- the **cold water circuit** on the loop condenser whose components are a refrigeration unit CIAT TK600, a recirculation pump, a heat exchanger and an electrical heater both hold to control the water temperature and refrigerant saturated pressure.

The inlet and outlet temperatures of fluids flowing through the test section are measured with 100 Ohm platinum resistances, the pressure drop on refrigerant flow is measured with a differential manometer ROSEMOUNT 1151 DP. The inlet and outlet vapor quality and the global heat transfer coefficient U are calculated from temperatures and massflowrate measurements.

From this experimental data, is deduced the mean heat transfer coefficient (αdp) between the wall and the saturated fluid R22. From a previous test, the mean heat transfer (αw) between water flow and internal plate is evaluated by the Wilson plot method. αdp is deduced from the following relation (e and λ are the thickness and thermal conductivity of plates) :

$$\alpha dp = \cfrac{1}{\cfrac{1}{U} - \cfrac{e}{\lambda} - \cfrac{1}{\alpha w}} \qquad (1)$$

The mean value of precision for the evaluation of the heat transfer coefficient αdp is about 15%.

3/EXPERIMENTAL RESULTS DURING R22 EVAPORATION

More than 100 different experimental tests on each test sections has been realized in a wide range of operating conditions (see table 2). Analysis of experimental data with refrigerant R22 evaporation shows:

- a sharp influence of heat flux on the heat transfer coefficient αdp between wall and saturated refrigerant that means the predominant heat transfer regime is nucleate boiling regime.

- no significative influence of vapor quality and mass velocity on heat transfer αdp: these results are in accordance with vertical intube evaporation results (STEINER (1)).

To correlate this experimental results, the factor R is defined as the ratio of the measured evaporation heat transfer coefficient αdp and the calculated pool boiling heat transfer coefficient αpool :

$$R = \frac{\alpha dp}{\alpha pool} \qquad (2)$$

The coefficient αpool is calculated with the following expression from VDI-Wärmeatlas (2):

$$\alpha pool_{(W/m^2)} = 2200 * (\dot{q}/20000)^{(0.9 - 0.3 * Pr^{0.3})}$$
$$* (2.1 * Pr^{0.27} + Pr * (4.4 + \frac{1.8}{1-Pr}))\qquad (3)$$

where Pr isthe refrigerant reducedpressure
$\dot{q}$ isthe heat flux (W/m2)

This factor R is represented in (figure 4) against the product of the two classical nondimensional numbers Bo (Boiling Number) et X_{tt} (Lockart-Martinelli Parameter) whose definitions are:

$$Bo = \dot{q} / (G.\Delta h_{LV}) \qquad (4)$$

$$X_{tt} = \left[\frac{1-x}{x}\right]^{0.9} * \left[\frac{\mu_L}{\mu_V}\right]^{0.1} * \left[\frac{\rho_V}{\rho_L}\right]^{0.5} \qquad (5)$$

The physical parameters $\Delta h_{LV}, \mu_V$, μ_L, ρ_V, ρ_L are the latent heat, the vapor and liquid viscosity, the vapor and liquid density.

As can be seen in (figure 4), it may be concluded for the three test sections that:

- when the product Xtt.Bo is larger than 0.00015, the factor R is constant and near the value 1: nucleate boiling appears to be the predominant heat transfer regime and heat transfer coefficient αdp may be evaluated from classical pool boiling correlation (equation 3).

- when the product Xtt.Bo is less than 0.00015, the factor R is higher than 1, and increases with the decrease of product Xtt.Bo (i.e with the increase of the vapor quality x, the mass velocity G and the decrease of the heat flux $\dot{q}$): two phase forced convection begin to have a more significant weight on heat transfer regime.

Concerning the comparaisons of thermal performance of the different geometries, we may conclude, from this experimental work, that the three different geometrical plate evaporators have very similar heat transfer coefficients in the operating conditions explored during this study and no heat significative transfer intensification has been caused by studs.

4/EXPERIMENTAL RESULTS DURING R22 CONDENSATION

More than 100 different experimental tests on each test sections has also been realized in a wide range of operating conditions (see table 2). Analysis of experimental data with refrigerant R22 condensation shows :

- a sharp influence of mass velocity G and vapor quality on the heat transfer coefficient α'dp between wall and saturated refrigerant can be seen on figure 5 for the three different geometries: α'dp is increasing with vapor quality and mass velocity.

- no significative influence of heat flux is noticed on heat transfer coefficient α'dp.

- a great influence of geometry of test sections on heat transfer coefficients : as can be seen on figure 5, studded plate heat transfer coefficients may be more than two times the smooth plate heat transfer coefficient and in-line array plate shows a stronger enhancement than staggered array plate.

To correlate experimental data on heat transfer coefficient during R22 condensation, we calculate the ratio M which is defined as:

$$M = \frac{\alpha' dp}{\alpha l} \tag{6}$$

where αl is the convective heat transfer coefficient for liquid phase flowing in the channel of the different geometry. This coefficient αl is calculated from the classical expressions:

$$\alpha l = a.\lambda_L / Dh . \left(\frac{G.Dh.(1-x)}{\mu_L} \right)^n . Pr_L^{0.33} \tag{7}$$

where λ_L and Pr_L are the thermal conductivity and Prandtl number of the saturated liquide phase.

where a and n are constants which depend of geometrical parameter of the channel and of the plates; the values of these constants have been evaluated from previous tests:

for smooth plates,	a = 0.0328	and	n = 0.8
for in line array studded plates,	a = 0.1447	and	n = 0.7076
for staggered array studded plates,	a = 0.1447	and	n = 0.7076

As can be seen on figure 6, the ratio M shows a strong dependance with the Lockart Martinelli number X_{tt} at low X_{tt} ($X_{tt}<0.5$) while at higher value of X_{tt}, the ration M remains constant for all the three tested geometries.

Experimental data are correlated in the following form:

$$M = a + \frac{b}{X_{tt}} + \frac{c}{X_{tt}^2} \qquad (8)$$

Where a, b and c are constants dependant of the plate geometry:

for smooth plates,	a = 0.744	b = 2.632	c = - 0.256
for in line array studded plates,	a = 1.344	b = 4.139	c = - 0.144
for staggered array studded plates,	a = 4.767	b = 4.106	c = + 0.094

5/CONCLUSION

In this paper, have been presented experimental results on three different geometries of rectangular channel (smooth, studded with in-line array, studded with staggered array) during forced convection boiling and condensation of refrigerant R22.

We notice a strong difference of thermal performance of plate geometries between evaporation and condensation:

- during R22 evaporation, the heat transfer coefficient (of the different geometries) is essentially dependant of heat flux and may be calculated by the means of classical nucleate boiling correlation (equation (3))

- during R22 condensation, mass velocity and vapor quality have a great influence on heat transfer coefficient :it has to be calculated with convective type equations (6) and (7).

No significative heat transfer enhancement is shown during R22 evaporation with studded plates. On the contrary, during R22 condensation, an important increase of heat transfer coefficient is shown with studded plates, in particular with in line array studded plate.

REFERENCES

1- STEINER.D, Flow boiling heat transfer in vertical tubes correlated by an asymptotic model, Heat transfer Engineering, Vol.13, n°2 (1992)

2- VDI Wärmeatlas, Behältersieden. Hal-Ha22 (1984).

Table 1

	Plate n°1	Plate n°2	Plate n°3
Geometry	smooth	studded	studded
length l	1284 mm	1284 mm	1284 mm
width w	146 mm	146 mm	146 mm
hydraulic diameter	7.6 mm	8.35 mm	9.03 mm
pitch between studs	-	30 mm	42 mm
heat transfer surface	0.194 m^2	0.250 m^2	0.250 m^2
geometrical parameters of plate test sections			

Table 2

	x (%)	G(kg/m2.s)	Tsat(°C)	ϕ(kW/m2)
Evaporation·	10-90	200-650	5-20	10-60
Condensation	20-90	150-700	30-55	10-60
Test conditions with R22 evaporation and condensation				

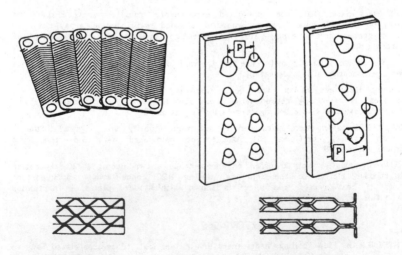

Figure 1: Corrugated and studded plates

SMOOTH CHANNEL

STAGGERED ARRAY

IN-LINE ARRAY

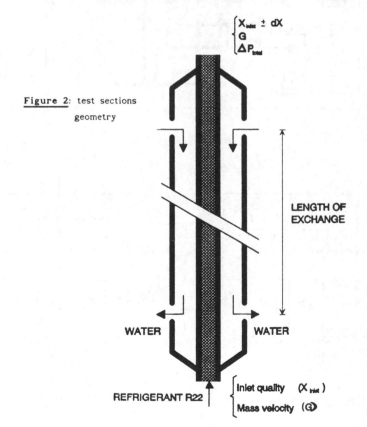

$$\begin{cases} X_{inlet} \pm dX \\ G \\ \Delta P_{total} \end{cases}$$

Figure 2: test sections geometry

LENGTH OF EXCHANGE

WATER   WATER

REFRIGERANT R22

$$\begin{cases} \text{Inlet quality} \quad (X_{inlet}) \\ \text{Mass velocity} \quad (G) \end{cases}$$

Figure 3: Flow sheet of evaporation test loop

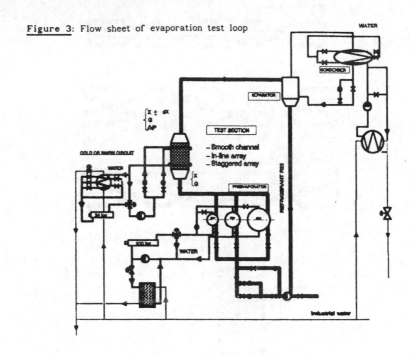

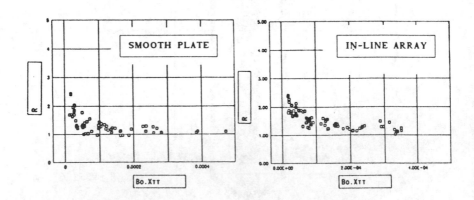

figure 4: Relation between R factor and the product Bo.Xtt
during R22 evaporation

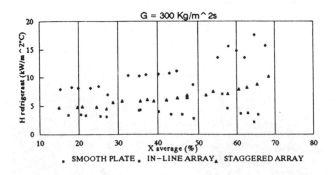

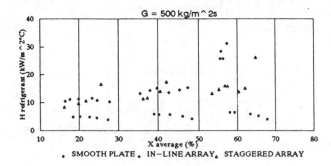

figure 5: Influence of vapor quality on heat transfer coefficient during R22 condensation

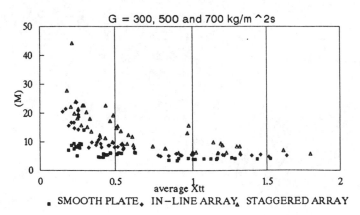

figure 6: Relation between M factor and the Lockart Martinelli number Xtt

AUGMENTATION OF FORCED CONVECTIVE HEAT TRANSFER IN TUBES WITH THREE
DIMENSIONAL EXTENDED SURFACE

Q. Liao and M. D. Xin
Institute of Eng. Thermophysics, Chongqing University, Chongqing 630044, China

The results are presented from experimental investigations of heat
transfer and flow friction behavior in thirteen copper tubes (ID 13.5
mm) with three dimensional internal fins varying in axial pitch,
circumferential pitch, height, width and arrangement form, with flow
of water in the Reynolds number range 7,000~40,000. By using a
heat-momentum transfer analogy, correlations were obtained that can
be used in practical designs.

INTRODUCTION

This paper presents heat transfer performance in the tubes with three dimensional extended
surface manufactured by a Chinese patent technics (1). The structure of the three
dimensional extended surface is shown in Fig. 1. G. Y. Liao et al (2, 3) have tested heat
transfer performance in several tubes with the three dimensional extended surface with flow
of air. No paper for experiments on heat transfer performance in this kind of tube has been
found with flow of water, except the author's paper (4) in which only a structure of the
three dimensional extended surface was reported. And the effects of various structure
parameters of the three dimensional extended surface on heat transfer performances in the
tubes haven't been studied systematically. In our investigation, the 13 tubes (ID 13.5mm,
OD 20mm) have a considerable range of configuration for three dimensional extended surface
as shown in Table 1. The tube inside diameter is defined as the diameter at the bottom of
the fin. The effects of strcture parameters on heat transfer performance are individaully
discussed in this paper.

EXPERIMENTAL APPARATUS AND TEST PROCEDURE

Fig. 2 shows a schematic drawing of the experimental aparatus. Water pumped from the
reservoir tank passes through a inlet mixing chamber, a 0.65m long smooth entrance section,
a 1m long test section with three dimensional extended surface, and an outlet mixing chamber,
and then drains out. Heat was supplied by electrical resistance heaters made of 0.8mm
diameter nickel chrome wires. The heating wires were uniformly over the test section.
Thermal insulation for the test section was provided by a 35mm thick glass wool cover. Wall
temperatures were measured by 9 couples of 0.2mm dia. nickel chrome and nickel silcon
thermocouples embeded uniformly on the outside surface of the test section. The water inlet
temperature at the end of the smooth entrace section was measured by 3 thermo-couples

inserted inside the tube. The outlet temperature was measured by 3 thermocouples in the outlet mixing chamber. Measurement of coolant flow rate was made with the weighing method. The pressure drop along the test section was measured by a U-tube manometer which was connected to two pressure measurement stations located on the inlet and outlet of the test section. At each station, 8 static pressure taps (hole dia. =2.0mm) were located uniformly around the tube wall.

TABLE 1 Structure geometries of the tubes with three dimensional extended surface

Tube Number	e/D_i	P_a/e	W/P_a	P_c/W	Fins arrangement form
No. 1	0.025	7.0	0.14	4.0	Staggered arrangement
No. 2	0.040	4.0	0.14	4.0	Staggered arrangement
No. 3	0.070	7.0	0.24	4.0	Staggered arrangement
No. 4	0.040	7.0	0.14	4.0	Staggered arrangement
No. 5	0.040	7.0	0.14	4.0	In-line arrangement
No. 6	0.040	8.0	0.14	4.0	Staggered arrangement
No. 7	0.040	9.5	0.14	4.0	Staggered arrangement
No. 8	0.055	7.0	0.14	2.5	Staggered arrangement
No. 9	0.055	7.0	0.14	4.0	Staggered arrangement
No. 10	0.055	7.0	0.14	5.5	Staggered arrangement
No. 11	0.055	7.0	0.14	7.5	Staggered arrangement
No. 12	0.070	7.0	0.10	4.0	Staggered arrangement
No. 13	0.070	7.0	0.14	4.0	Staggered arrangement

EXPERIMENTAL RESULTS AND DISCUSSION

Fig. 3 shows the experimental results of transport performance in a smooth tube (ID 0.013m, OD 0.02mm). Discrepancies between experimental data and Dittus-Boalter's correlation (Nu= 0.023 (Re)$^{0.8}$(Pr)$^{0.4}$) are in range of ±6%. As compared with Blassius' correlation (f= 0.3164 (Re)$^{-0.25}$), the maxium error scope of experimental data for flow resistance is ±8%.

Effects of structure parameters and arrangement forms of three dimensional extended surface on heat transfer and friction behavior

Effect of the fin axial pitch to height ratio (P_a/e) Four tubes with different values of P_a/e were tested. The experimental results are shown in Fig. 3. It can be seen that P_a/e has a more considerable effect on the heat transfer performance. If the P_a/e were overly large, fluid could not be re-disturbed timely, and the overly small p_a/e would create a stagnant wake at the back of fins. It indicates that there is an optimum of the P_a/e, approximately being 0.8 in our experiment.

Effect of the fin height-to-ID ratio (e/D_i) Turbulent flow in tube consists of the viscous sublayer, buffer zone, and fully turbulent core zone. Each zone has thermal resistance of its own. The total thermal resistance of turbulent flow in tube is the sum of the thermal resistances of three zones. So the proportion that the thermal resistance of each zone accounts for in the total thermal resistance varies with the physical properties of media, such as Prandtl number. For different medium, the optimum of the e/D_i is different . In Fig. 4, it is seen that the flow resistances decrease with the fin heigh increasing. It is probably by reason that a curve surface at the top of heigher fin, produced by the machining technology, leads to the decrease of the form drag. The similiar result can be seen in paper (2). The optimum of e/D_i is approximately 0.025 in our experiment.

Effect of the fin width-to-ID ratio (W/D_i)   The W/D_i has an effect on the density of Kármán vortex street inside the tubes as wall as the strength of Kármán vortex behind each fin. The strength of Kármán vortex increases with the W/D_i increasing, but the density of Kármán vortex street decreases at the same time. So there exists an optimum of the W/D_i. In our experiment, the optimum is in the range from 0.049 to 0.0686. The experimental results are shown in Fig. 5.

Effect of the fin circumferential pitch to width ratio (P_c/W)   The P_c/W has an effect not only on the distribution density of the sudden enlargement channel formed of the fins, but also on the strength of fluid disturbance caused by the each sudden enlargement channel. The experimental results are shown in Fig. 6. The P_c/W of No. 8 tube is so small that its internal surface approaches to the two dimensional rib surface, thus its flow friction is largest in the tubes of the figure and the heat transfer increases no more greatly.   The form drag of fin (i. e. the flow resistance) decreases with the P_c/W increasing, such as No. 9 and No. 10 tubes. But their P_c/W are small still, the heat transfer in the tubes changes no greatly due to small strength of fluid disturbance produced by the sudden enlargement channel. While the P_c/W increases to some value, the disturbance produced by the sudden enlargement channel intensifies. Both the flow resistance and heat transfer increases with the P_c/W decreasing.

Effect of the fin arrangement form   Fig. 7 shows the flow resistance and heat transfer characteristics in No. 4 tube (staggered arrangement) and No. 5 tube (in-line arrangement) with the three dimensional extended surface having the same fin sizes.   Both the flow resistance and heat transfer in No. 5 tube are larger than that in No. 4 tube,   which is different from the transport performance of the fluid flowing across the bundle of tubes. We consider it is by the reason that the fluid disturbance inside the tubes with the three dimensional extended surface is three dimensional disturbance, and the fluid disturbance across the bundle of tubes is two dimensional disturbance.

In our experiment, No. 6 tube has the most suprior heat transfer performance in the other tubes. The heat transfer coefficient in No. 6 tube can be increased by an average of about 3 times as compared with that in the smooth tube. And its friction factor can be 4 times as much as that in the smooth tube.

The friction factor and heat transfer correlations

The friction factor correlation   Nikuradse (5) developed the friction similarity law for sand-grain roughness surface. His data, covering a wide range of e/D_i, was correlated by equation (1), where e^+ is the roughness Reynolds number ($e^+ = (e/D_i) (Re) (f/2)^{1/2}$).

$$Re^+ (e^+) = (2/f)^{1/2} + 2.5 \ln (2e/D_i) + 3.75 \qquad (1)$$

The roughness function Re^+ is a general function determined empirically for each type of geometrically similiar roughness. In our investigation, based on 'the friction similarity law', the friction behavior for turbulent flow in tube with the three dimensional extended surface is correlated with the experimental data by means of the dimensionless parameters P_a/e, P_c/W, and W/D_i, as in Fig. 8. That is

$$Re^+ = 10.5 \, (P_m/e)^{-0.252} \, (P_c/W)^{0.685} \, (W/D_1)^{0.878} \, (e^+)^{0.136} \tag{2}$$

The friction factor can be found by

$$(2/f)^{1/2} = 10.5 \, (P_m/e)^{-0.252} \, (P_c/W)^{0.685} \, (W/D_1)^{0.878} \, (e^+)^{0.136} - 2.5 \ln(2e/D_1) - 3.75$$
$$\text{for } 40 < e^+ < 500 \tag{3}$$

The experimental data fit in with the correlation in the range of $\pm 22\%$.

The heat transfer correlation   Dipprey and Sabersky (6) developed a heat transfer similarity law, which is complementary to Nikuradse's friction similarity law. Based on the heat-momentum transfer analogy applied to a two-region flow modle,   they arrived at the functional statement

$$He^+ (e^+, Pr) = [f/(2St) - 1]/(f/2)^{1/2} + Re^+ \tag{4}$$

If He^+, Re^+ and the friction factor are known, then the Stanton number can be found as

$$St = f/[(He^+ - Re^+)(2f)^{1/2} + 2] \tag{5}$$

Equation (5) is the "heat transfer similarity law". He^+ is the heat-transfer function. They used the equation to correlate their experimental data in turbulent flow inside the tube with the sand-grain roughness and found that He^+ equals $5.19 \, (e^+)^{0.2} \, (Pr)^{0.44}$.   In our paper, based on the "heat transfer similarity law", the experimental data in turbulent flow inside the tubes with the three dimensional extended surface are correlated by means of the dimensionless parameters P_m/e, P_c/W, W/D_1, and e^+ as shown in Fig. 9. That is

$$He^+ = 4.48 \, (P_m/e)^{0.013} \, (P_c/W)^{0.0559} \, (W/D_1)^{-0.0977} \, (e^+)^{0.298} Pr^{0.5} \quad \text{for } 60 < e^+ < 500 \tag{6}$$

In the absence of a wider variation of Pr (i. e. $Pr \approx 5.5$) in our experiment, $Pr^{0.5}$ is used in the correlation according to the Withers (7). The St can be found by substituting Eq. (2), (3), and Eq. (6) into Eq. (5). The experimental data fit in with the correlation in range of $\pm 15\%$.

Performance comparison

We consider a shell-and-tube condenser having the following specifications. Saturation R. 11 vapor (the saturation vapor temperature 308k and pressure 1.5 kgf/cm^2)   condenses outside the copper tubes (OD 0.019m and ID 0.016m). The external condenser surface of the tubes are all the thermoexcel C (8).   Inside the tubes are various internal enhanced surfaces and a smooth surface. Inlet temperature of water coolant equals 288k.   Reynolds number of water flowing in the condenser tube with the smooth internal surface equals 35000. The ratio of entropy production for the tube with internal smooth surface to that for the tubes with various internal enhanced surfaces is calculated under the condition of constant heat flux and pump power by means of the method provided by the authors' paper (4) .   The results are shown in Fig. 10. For comparison with R. L . Webb's performance evaluation criteria (9) derived from the first law of thermodynamics, the ratio of heat transfer area in the tube with internal smooth surface to that in the tubes with various internal enhanced surface is also calculated under the same condition, and the results are shown in Fig. 11. In Fig. 10 and Fig. 11, it can be seen that heat transfer performance of the condenser tube with the three dimensional internal extended surface of No. 6 tube is superior to those of the tubes with the other internal enhanced surfaces.

CONCLUSIONS

1. The fin axial pitch, circumferential pitch, height and width of fins for the three dimensional extended surface have an effect on heat transfer and flow friction behavior. Every structure parameter has optimum of its own. In our experiment, the heat transfer coefficient of the most superior tube (i. e. No. 6 tube) can be increased by an average of about 3 times as compared with that of the smooth tube. And the relevant friction factor can be 4 times in No. 6 tube as much as that in the smooth tube.

2. The heat transfer and flow resistance of No. 5 tube (in-line arrangement) are larger than that of No. 4 tube (staggered arrangement) with the same fin sizes.

3. The heat transfer and friction factor correlations obtained in this paper can be used in practical designs.

NOMENCLATURE

A   Surface area of condenser tube at the bottom of internal fins (m^2)
A_s Surface area in the condenser tube with smooth internal surface (m^2)
D_i Inside diameter of tube at the bottom of internal fins (m)
e   Height of fins (m)
e^+ The roughness Reynolds number $(=(e/D_i) (Re) (f/2)^{1/2})$
f   Friction factor $(=\triangle p/ (L/D_i)/ (\rho u^2/2))$
h   Average convective heat transfer coefficient $(W/m^2 k)$
He^+ The heat transfer function
k   Thermal conductivity of coolant (W/mk)
Nu   Nusselt number $(=hD_i/k)$
P_a The axial pitch of fins (m)
P_c The circumferential pitch of fins (m)
Pr   Prandtl number
ΔP Pressure drop of coolant (Pa)
q   Heat flux (W/m^2)
Re   Reynolds number of the coolant flow $(=uD_i/ \nu)$
Re^+ The roughness function
Sp   Entropy production (W/K)
$(Sp)_s$ Entroy production of condenser tube with smooth internal surface
u   Average velocity of coolant (m/s)
ρ Coolant density (Kg/m^3)
ν Kinematic viscosity of coolant (m^2/s)

REFERENCES

1. G. Y. Liao, 1988 Chinese patent 88102575. 5.
2. G. Y. Liao et al, 1990 J. Eng. Thermophysics 11 422.
3. G. Y. Liao et al, 1990 Proc. of Chinese Acad Eng. Thermophysics Conf. 325.
4. Q. Liao, M. D. Xin and C. M. Shi, 1991 Preprints for Proc. of ISTP-IV 2 812.
5. J. Nikuradse, 1965 NACA TM-1292.
6. Dipprey, D. F. and Sabersky, R. H., 1963 Int. J. Heat and Mass Transfer 6 329.
7. Withers, J. G., 1980 Heat Transfer Engineering 2 48.
8. HITACHI, Copper and Copper-Alloy Products CAT. No. Ai-518d 4.
9. Webb, R. L., 1981 Int. J. Heat Transfer 24 715.
10. Bergles, A. E. and Jansen, M. K., 1977 Proc. of 4th Annual Conf. on Ocean Thermal Energy Conversion 6 41.
11. Kenji Takahashi et al, 1988 Heat Transfer Japan Research 17 12.

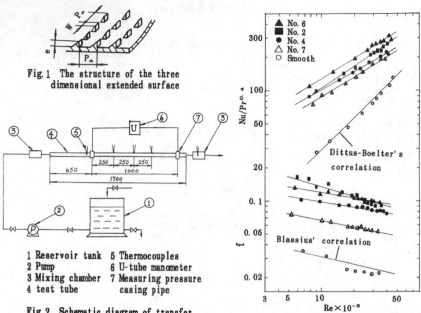

Fig. 1 The structure of the three dimensional extended surface

1 Reservoir tank   5 Thermocouples
2 Pump   6 U-tube manometer
3 Mixing chamber   7 Measuring pressure
4 test tube   casing pipe

Fig. 2 Schematic diagram of transfer experimental apparatus

Fig. 3 Effect of P_m/e on Nu and f

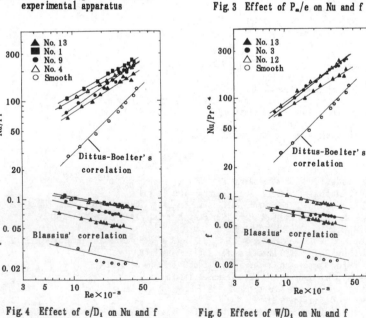

Fig. 4 Effect of e/D_1 on Nu and f

Fig. 5 Effect of W/D_1 on Nu and f

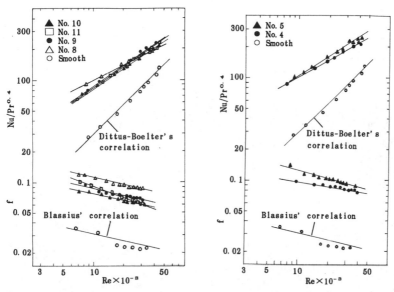

Fig. 6 Effect of P_c/W on Nu and f Fig. 7 Effect of fins arrangement form on Nu and f

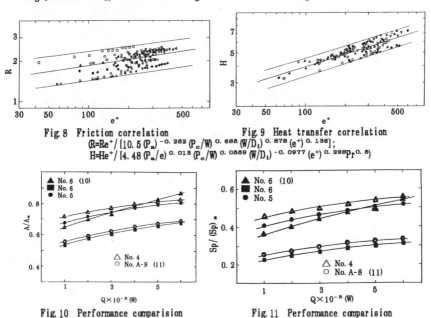

Fig. 8 Friction correlation Fig. 9 Heat transfer correlation

$$(R=Re^+/[10.5\,(P_a)^{-0.252}\,(P_c/W)^{0.685}\,(W/D_i)^{0.878}\,(e^+)^{0.136}];$$
$$H=He^+/[4.48\,(P_a/e)^{0.013}\,(P_c/W)^{0.0559}\,(W/D_i)^{-0.0977}\,(e^+)^{0.298}Pr^{0.5})$$

Fig. 10 Performance comparision   Fig. 11 Performance comparision

ENHANCEMENT OF HEAT TRANSFER IN SINGLE-PHASE HEAT EXCHANGERS

A.Žukauskas
Lithuanian Academy of Sciences, Gedimino 3, Vilnius

The paper deals with the possibilities of augmentation of convective heat transfer in single-phase flow of different liquids and low-temperature gases, as well as problems concerning the efficiency of heat transfer augmentation, on methods-and development of single-phase heat exchangers. The paper includes new experimental results. The different methods of enhancing heat transfer are developed with the sole purpose of designing small-size, metal-saving apparatus.

INTRODUCTION

Augmentation of heat transfer and improvement of the powerefficiency of devices in which processess occur are the principle goals in the development of convective heat exchangers. The solution of such problems is particularly timely for gas heat exchangers, where the heat transfer is typically low.

At present more and more attention is paid to this problem. The number of publications on heat transfer enhancement has increased tenfold over the last two decades. Here we consider only the possibilities of augmentation of convective heat transfer in single-phase flow of different liquids and low-temperature gases, as well as problems concerning the efficiency of heat transfer augmentation methods.

Enhancement of heat transfer leads to a decrease of the overall dimensions and weight of heat exchangers, of the temperature differences as compared with their magnitude under ordinary conditions. If increasing the velocity within permissible limits under practical conditions do not ensure obtaining the necessary dimensions of heat exchangers or the specified wall temperature, it is necessary to enhance heat transfer by reduction of the overall dimensions while causing only a moderate increase in the overall power penalty for pumping working fluids through heat exshangers.

Heat exchangers to which the selected method of heat transfer enhancement is applied should be suitable for serial production and sufficiently reliable and efficient in operation. When a hot solid surface interacts with the flow over it, a boundary layer is formed, which exerts the main resistance to heat transfer. The thicker the boudary layer and the lower the thermal conductivity of the working fluid, the lower the heat transfer coefficient.

LIQUIDS AND FLOW PATTERN

The amount of heat transfer can be increased by various means; the first is a proper selection of the working gas or liquid. Since $Nu \sim Pr^n$ the heat transfer in liquids flows is several taimes higher than in air flow, because at $t = 20^{\circ}C$ for air $Pr_1 = 0.703$, water $Pr = 7.02$ and for transformer oil $Pr_2 = 370$. For laminar flow power exponent usually is $n = 0.33$ and for turbulent flow $n = 0.43$.

Analitical for turbulent flow, when $n = 0.43$, $Nu_2/Nu_1 = 14.78$. So we see, that heat transfer in liquids flows is several times higher than in air flow.

After the working fluid is selected on the basis of its thermophysical properties, it is possible to consider enhancement of heat transfer by selecting the proper flow pattern. From the point of view of heat transfer the best flow pattern is one with turbulent or transition flow in the boundary layer however, natural development of turbulence start at a relatively high flow velocity, and consequently, a significant hydraulic drag. Since $Nu \sim U^m$, the greatest gain in the amount of heat transferred can be attained by raising the flow velocity. At the same time, high coolant velocities are obtained at the penalty of increasing pumping power. Data on local heat transfer from tubes show that the heat transfer coefficient over the tube perimeter increases significantly with increasing flow velocity and Re. In the case of crossflow over a tube, a laminar boundary layer, whose thick ness increases in the downstream direction, usually forms on the front past of a cylinder [1]. In a study performed over a wide range of critical and superecritical values of Re we noted, that the start of transition in the laminar boundary layer and the formation of the turbulent thermal boudary layer are a function of Re [2]. When these increase, the transition to a turbulent thermal boundary layer is steeply shifted toward the leading stagnation point. This phenomenon causes a significant increase in the heat transfer coefficient. Thus, the transition to the supercriticaal flow regime over a tube opens vast possibilities for the enhancement of heat transfer from tubes in crossflow.

FREE STREAM TURBULIZATION

Artificial flow turbulization is one of the effective means of heat transfer enhancement. Main flow turbulization has a significant effect on heat transfer in a laminar boundary layer.

It has been established (Fig. 1) that an increased flow turbulence exerts a strong effect on the hydrodynamics and heat transfer even in the case of a developed turbulent flow in the boundary layer on the plate.

An analysis of these data shows that this influence is exerted throught the disturbance of the outer zone of the boundary layer while the wall region remains uneffected [1,3].

A similar effect of free stream turbulence on heat transfer is obserwed in channels, over flat surfaces etc.

Efficient means of convective heat transfer enhancement is the swirling of flow in pipes, which increases local near-wall velocities and disturbs the whole free stream. For this purpose, strip and tape swirlers are used as helical tubes and tubes with inner spiral fins. Swirling must be maintained throughout the flow to provide a constant ratio between the axial and tangential velocities.

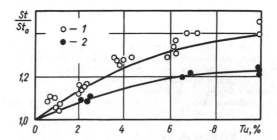

Fig. 1. The effect of free stream turbulence on the heat transfer for a turbulent boundary layer on a plate. 1 - Pr=0.7, 2 - Pr=5.4

Fig. 2 shows a substantial increase of the heat transfer in tubes with strip helical swirlers as compared to heat transfer of a uniform flow in a smooth tube. A decrease of the helix pitch s/d by a factor of two results in a three-fold increase of heat transfer. Benefits of using those enhancement measures are particularly large when the flow is laminar and the fluid has a high Prandtl number [1,4].

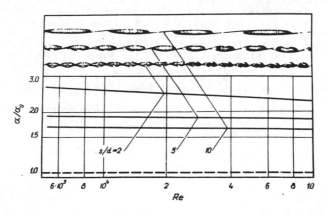

Fig. 2. The effect of a helix pitch on the heat transfer in a pipe flow of water

The average heat transfer of a tube in crossflow can be increased by up to about 55% by turbulization of the laminar boundary layer developing over the leading part of the tube.

Heat exchangers' tube bundles also act as turbulators. Therefore heat transfer from inner tubes is considerably higher than from those in leading row. In some highly compact bundles the heat transfer from inner tubes may be twice that from leading row tubes.

A high free stream turbulence aids the formation of the turbulent boundary layer on the tube surface, thus enhancing the heat transfer. At Re = 6.8×10^4, in the flow of water, an increase of turbulence intensity from 1 to 6.8% is accompanied not only by an increase of the heat transfer at the front part of the cylinder, but the laminar boundary layer, instead of separating at φ = 90° becomes turbulent and separates at φ = 140°; this also adds to the heat transfer augmentation [1,2].

At the same time, the onset of turbulent flow in the boundary layer reduces the size of the wake and the coefficient of hydraulic drag decreases.

DISTURBANCE OF THE BOUNDARY LAYER

The main thermal resistance in a solid-fluid interaction comes through the formation of a boundary layer, and efforts towards enhancing heat transfer must be directed to artificially destroy or disturb the boundary layer. In this respect, the most advantageous case is that of a turbulent boundary layer.

Measurements of velocity profiles across the turbulent boundary layer of a plate in different fluids revealed the universal nature of the wall law, including the logarithmic equation which is valid for different fluids.

The main difficulty in evaluating the heat transfer is presented by the molecular and turbulent transfer contributions with the growth of **Pr**.

At high **Pr** the relative contribution of turbulent transfer increases: at a fixed **Re** and increasing **Pr**, the heat transmitted by conduction goes to zero and the turbulent transfer may become much larger than the molecular heat transfer. Thus with an increase of **Pr** the viscous sublayer shows an increasing resistance to heat transfer. Analysis of the relation between the temperature gradient in the viscous sublayer and the value of **Pr** shows that the viscous sublayer is responsible for 25% of the total fluid-to-wall temperature difference in air, but 90% in transformer oil with Pr = 55.

Thus in air the main contribution to thermal resistance is in the turbulent boundary layer, but in liquids with Pr>1, it is mainly in the viscous sublayer. Therefore, heat transfer should be enhanced by the action of surface roughness in the wall layers. We encounter a new problem here - that of choosing the suitable height, shape and location of the surface roughness elements for each fluid in relation to its type and physical properties. Numerical calculations [5] of the thermal turbulent boundary layer resulted in ratios between the effective heat conduction λ_{ef} and the molecular ones λ for different positions $y^+ = y u_* / \nu$ in the boundary layer.

As shown in Table 1, the ratio λ_{ef}/λ equals 18 in turbulent cores of gases, but reaches the value of 742 in the buffer layer of liquids with Pr = 50, and increases to 2570 in the turbulent core. Thus, in order to reach the necessary heat transfer enhancement in gases, the whole wall layer must be disturbed. But in liquids with high **Pr** only the viscous sublayer should be

disturbed. The higher the **Pr** number, the shorter are the surface
elements needed for a significant heat transfer anhancement. One
only has to bear in mind, that in liquids it is a transition
region of roughness that gives most intensive heat transfer
rates.

Table 1. Ratio of effective and molecular heat conduction

Pr	y^+	λ_{ef}/λ	y^+	λ_{ef}/λ	y^+	λ_{ef}/λ
0.5	4.7	1.037	28.3	4.74	188	18
5	4.8	1.53	28.3	37.3	137	263
50	4.9	6.3	29.3	394	131	2570

The dimensionless height k^+ is introduced to describe the
roughness regime

$$k^+ = \frac{ku_*}{\nu} \qquad (1)$$

where k - height of roughness.
Measurements suggest that $k^+ < 5$ implies surface elements
submerged in the viscous sublayer and having no influence on the
heat transfer rate. At $k^+ > 70$ a full roughness effect is
observed. The range of k^+ from 5 to 70 covers transitional region
of its effect.
Heat transfer curves obtained with rough-surface plates
suggest different influence of k^+ on heat transfer with a
variation of **Pr**. In air a continuous increase of the heat
transfer is observed with a growth of k^+.
In transformer oil at Pr = 102 and in water flow at Pr = 5.4
a maximum increase of the heat transfer coefficient, as described
by the Stanton number, is observed at a transition region of
roughness effect at $k^+ < 70$. A further increase of k^+ results in
a decrease of St [1,6].
With closely spaced surface elements, stagnation regions are
formed in their interspaces, and the value of k cannot be treated
as a determining geometrical factor.
The process of heat transfer enhancement inside rough-
surface pipes, especially in the entrance region, is similar to
the above mentioned case of rough-surface plates.
The heat transfer is also similar in external flows over
rough tubes. Suitable roughness elements may provide significant
heat transfer enhancement from tubes in crossflow. Different
measurements [1,2] suggest that increasing the hight of roughness
elements on tubes in crossflow causes the laminar-turbulent
transition in the boundary layers to occur at lower values of **Re**,
while the average heat transfer increases up to 50%.
At high **Re** numbers a two- and three- fold increase of local
heat transfer is observed on rough surfaces, as compared to
smooth ones, Fig. 3. The presence of roughness introduces a sharp
change of the hydrodynamic pattern over the surface.

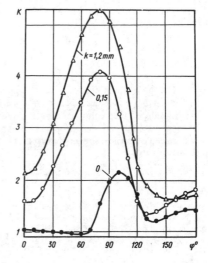

Fig. 3. Local heat transfer from a rough-surface cylinder at Re = 1x10^6 in water flow [2].

$$K = Nu\ Re^{-0.5}\ Pr^{-0.37}$$

CONCLUSIONS

The efficiency of a heat transfer augmentation technique depends on the flow mode, the type of the fluid, and the thermal conditions. The choice of a mode and a specific technique of heat transfer augmentation is a combined problem of enhancing heat transfer while reducing pressure drop.

References

1. Žukauskas A. (1989). High-performance single-phase heat exchangers, Hemisphere Publishing Corporation, New York, 515 p., 1989.
2. Dauyotas P., Žiugžda J. and Žukauskas A. (1976). Int. Chem. Engn., Vol. 16, No 3, p.p. 476-479.
3. Žukauskas A., Slančiauskas A., Pedišius A. (1982). Heat Transfer 1982, Vol. 3, p.p. 217-227.
4. Drizhyus, M.-R. M., Shkema, R. K. and Slančiauskas, A. A. Heat transfer in a twisted stream of water in a tube. Int. Chem. Engn., 1980, Vol 20, No 3, p.p. 486-489.
5. Vaitiekunas, P. P., Shlanchyauskas, A. A., Žukauskas, A. A. Int. Chem. Engn., Vol. 16, No 3, p.p. 403-412.
6. Slančiauskas, A. A. and Drižius, M.-R. M. Calculation of heat transfer on rough surfaces. Heat transfer - Soviet Research, 1977, Vol. 9, No. 4, p.p. 40-47.

OPTIMIZATION OF THE GEOMETRY OF A CORRUGATED CHANNEL

B.THONON*,M.FEIDT**,R.VIDIL*

The thermal and hydraulic performances of a corrugated channel are studied. To take account for both heat duty and pressure losses, we introduce the entropic efficiency of the heat exchanger. An optimization of the geometry is proposed and allows reduced thermodynamical irreversibilities and increased overall performances.

INTRODUCTION

In corrugated channels used in plate heat exchangers, the enhancement of heat transfer generally involves higher pressure losses, so the optimal pattern of the channel is a compromise between the different design constraints: heat duty, pumping power and heat transfer area [1]. In a rectangular channel, we can define the aspect ratio as the length of the channel reported to the width $(Ar = L / B)$; thus for a given heat transfer area, there is an infinite number of geometries possible.

Previous works have shown that the major geometric parameter of a corrugated channel is the angle between the corrugation and the main flow direction (α). And available data exist for the heat transfer and the pressure drop [2] [3], in form of relations between the Nusselt number and the friction factor versus the Reynolds number, for several angles of corrugation. In regard to the performance objectives of the plate heat exchanger, we study the influence of the different geometric parameters (Table 1) for a fixed hydraulic diameter and input conditions $(\mathring{M}, T_{in}$, physical properties). For the pressure drop:

$$dP = 4f \frac{\mathring{M}^2}{\rho A_c^2} \frac{L}{D} \tag{1}$$

$$dP = Bh \ f \ A^{-1} Ar^{3/2} \tag{2}$$

* GRETh   CEN Grenoble BP 85X 38041 GRENOBLE CEDEX FRANCE
** LEMTA   Université de NANCY I URA-CNRS 875
            2 av de la Forêt de Haye 54504 Vandoeuvre Les Nancy CEDEX FRANCE

For the energy balance:

$$\frac{D}{2} \rho \, Cp \, u \frac{dT}{L} = 2 \, U \quad dTlog \tag{3}$$

$$NUT = Bt \, St \, (Ar \, A)^{0.5} \tag{4}$$

Where the friction factor and the Stanton number are function of the angle of corrugation [3].

The optimization of the aspect ratio based on thermal and hydraulic performances (figure1) leads to opposite conclusions; the maximum heat duty is for high aspect ratios and the minimal pressure drop is for low aspect ratios.

ENTROPY GENERATION

To take into account both heat duty and pressure losses, we introduce an entropic efficiency (η) based on the thermodynamic irreversibilities of the system. BEJAN [4] has introduced the concept of irreversibility for the design of heat exchangers and numbers of authors have used this method for various types and configurations of heat exchangers [5] [6].

The local entropic balance of the channel per unit of width is:

$$d\mathcal{S} = \frac{\mathring{a} \, \Delta T(wall/bulk)}{T^2} - \frac{\mathring{M}}{\rho \, T} \, dP \tag{5}$$

In the case of the plate heat exchanger channel, the knowledge of the heat transfer and pressure drop laws allows us to establish a general relation:

$$\mathcal{S} = \frac{\mathring{Q}^2}{T^2} \frac{e}{L^2} \frac{1}{\lambda} \frac{1}{Nu} + \frac{\mu^3}{\rho^2 \, T} \frac{L \, B}{8 \, e^3} f Re^3 \tag{6}$$

The thermodynamical performance of the channel can be characterized by the dissipated energy; to keep a non dimensional form it is reported to the actual heat rate:

$$\eta = \frac{T \, \mathcal{S}}{\mathring{Q}} \tag{7}$$

Using eq (6) we obtain:

$$\eta = \frac{\mathring{Q} \, e}{T \, L^2 \, \lambda} \left[\frac{1}{Nu} + Bs \, f Re^3 \right] \tag{8}$$

With:

$$Bs = \frac{\mu^3 \, L^3 \, B \, T \, \lambda}{8 \, \rho^2 \, e^4 \, \mathring{Q}^2} \tag{9}$$

In function of the inlet conditions ($\mathring{M}, \mathring{Q}, T$) there is an optimal geometry which minimizes the entropic efficiency.

We express the constitutive laws of the channel:

$$Nu = a \, Re^b \, Pr^c \tag{10}$$

$$f = \alpha \, Re^{-\beta} \tag{11}$$

Where the coefficients (a,b,c,α,β) are chosen in the literature [2] [3].

With : $0.6 < b < 0.8$ and $0.3 < c < 0.4$ and $0.15 < \beta < 0.3$

If we consider the entropic efficiency as a function of the Reynolds number, we can search the minimum by derivating eq (8):

$$\eta = \frac{\dot{Q} e}{T L^2 \lambda} \left[\frac{Re^{-b}}{a Pr^c} + Bs \, \alpha \, Re^{3-\beta} \right] \tag{12}$$

$$\frac{d\eta}{dRe} = \frac{\dot{Q} e}{T L^2 \lambda} \left[\frac{-b}{a Pr^c} Re^{-(b+1)} + Bs \, \alpha \, (3-\beta) \, Re^{2-\beta} \right] \tag{13}$$

We finally get the optimal Reynolds number Re^*:

$$Re^* = \left(\frac{b}{a \, Pr^c \, Bs \, \alpha \, (3-\beta)} \right)^{\frac{1}{3+b-\beta}} \tag{14}$$

Replacing in eq (12), we express the variation of the entropic efficiency :

$$\frac{\eta}{\eta^*} = \frac{3 - \beta}{3 - \beta + b} \left(\frac{Re}{Re^*} \right)^{-b} + \frac{b}{3 - \beta + b} \left(\frac{Re}{Re^*} \right)^{3-\beta} \tag{15}$$

The coefficient b and β are known for each angle of corrugation, so we can study the influence of the Reynolds number on the entropic efficiency (figure 2). On the left side, the thermal irreversibilities are preponderant, the mechanical irreversibilities are on the right side.

OPTIMIZATION OF THE GEOMETRY

A general way to evaluate the thermal and hydraulic performances of an enhanced channel is to compare them to those of a smooth channel at the same nominal conditions, thus we introduce three performance evaluation criteria :

$$Kt = \frac{\dot{Q}a}{\dot{Q}o} \tag{16}$$

An efficient geometry needs to have Kt as high as possible.

$$Kh = \frac{d\dot{P}a}{d\dot{P}o} \tag{17}$$

Good performances involve Kh near 1.

$$Ks = \frac{\eta a}{\eta o} \tag{18}$$

The enhanced channel is performant if Ks is beneath 1.

Where subscript a is for the enhanced channel, and subscript o for the smooth channel.

The numerical study is performed for a range of aspect ratio from 0.6 to 4.0, two different corrugation angles ($\alpha=30°;\alpha=60°$), and two inlet mass flow rate ($\dot{M}1 = 0.33$ kg/s , $\dot{M}2 = 0.66$ kg/s)

It appears that increasing the aspect ratio leads to lower values of the thermal criteria (Kt) (figure 3), while the hydraulic criteria (Kh) is more constant (figure 4), this for both angles of corrugation. From this point of view, the enhanced channel becomes less efficient for high values of the aspect ratio.

The influence of the aspect ratio on the entropic efficiency (η) outlines two facts (figure 5):
- When the angle of corrugation is low ($\alpha=30°$) or for the smooth channel, the entropic efficiency (η) decrease with the aspect ratio; so it is interesting for these geometries to have high aspect ratios.
- When the angle of corrugation is high ($\alpha=60°$), the entropic efficiency reaches a minimum.

For the entropic criteria (Ks), we have to consider two cases: the influence of the Reynolds number and the influence of the angle of corrugation.
- For high values of the Reynolds number (figure 6 curves A-B) the performances of the channel decrease (Ks increasing), thus high aspect ratios are not significant. But, for lower values of the Reynolds number (figure 6 curves C-D) the entropic criteria reach a minima, which proves the existence of an optimal geometry.
- An efficient geometry deteriorates less energy than a smooth channel; therefore, low aspect ratios are interesting when the angle of corrugation is high ($\alpha=60°$), especially for a low Reynolds number; but for high aspect ratios ($Ar> 3.$) it is better to use a low angle of corrugation ($\alpha=30°$).

CONCLUSION

To optimize the geometry of a corrugated channel, performance evaluation criteria based on the heat duty and the pressure drop are not sufficient.

The study of the thermodynamic irreversibilities of the systems allows us to find an optimal geometry based on the inlet conditions. For high angle of corrugation (f and Nu high) it is interesting to have low aspect ratios, and for low angle of corrugation higher values of the aspect ratio are preferable.

Furthermore, when the angle of corrugation is low, problems of flow maldistribution might occur [7], and to avoid this the channel needs to have a high aspect ratio; at the opposite when the angle of corrugation is high, flow is uniformly distributed over the width of the channel, even for low aspect ratios.

References

1 WEBB R.L., 1981, Int Journal of Heat & Mass Transfer, Vol 24, pp 715-726

2 FOCKE W.W., ZACHARIADES, J., OLIVIER, I., 1985, Int Journal of Heat & Mass Transfer, Vol 28, pp 1469-1479

3 HUGONNOT P., 1989, PH. D. Thesis, "Etude Locale de l'Ecoulement et Performances Thermohydrauliques à Faibles Nombres de Reynolds dans un Canal Plan Corrugué", Chapter 6, pp 163-201, University of NANCY, FRANCE

4 BEJAN A., 1977, Trans of the ASME, Journal of Heat Transfer, Vol 99, pp 374-380

5 SEKULIC D.P., 1990, Trans of the ASME, Journal of Heat Transfer, Vol 112, pp 295-300

6 FEIDT M., 1987, "Thermodynamique et Optimisation Energétique des Systèmes et Procédés", Part 3, pp 297-314, Technique et Documentation LAVOISIER, FRANCE

7 THONON B., 1991, Ph. D. Thesis, "Etude et Optimisation de la Distribution du Fluide dans un Echangeur de Chaleur à Plaques", University of NANCY, FRANCE

Nomenclature

A	Heat transfer area	m^2	α	Angle of corrugation		
Ac	Cross area	m^2	λ	Thermal conductivity	$W / m\,K$	
Ar	Aspect ratio		μ	Dynamic viscosity	$kg / m\,s$	
B	Width of the channel	m	η	Entropic efficiency		
Cp	Heat capacity	$J / kg\,K$	ρ	Mass density	kg / m^3	
D	Hydraulic diameter	m				
e	Channel height	m				
f	Friction factor		Bh	Hydraulic parameter		
L	Lenght of the channel	m	Bs	Entropic parameter		
$\dot{M}$	Mass flow rate	kg / s	Bt	Thermal parameter		
$\dot{Q}$	Heat rate	W				
P	Pressure	N / m^2	Nu	Nusselt number		
$\dot{S}$	Specific entropy	$J / K\,s$	Pr	Prandtl number		
T	Temperature	K	Re	Reynolds number		
u	Mean velocity	m / s	Re^*	Optimal Reynolds number		
U	Heat transfer coefficient	$W / m^2\,K$	St	Stanton number		

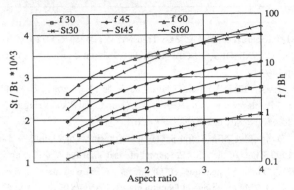

Figure 1 : Overall Performances

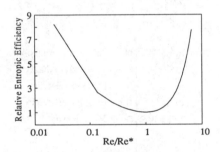

Figure 2 : Entropy Generation

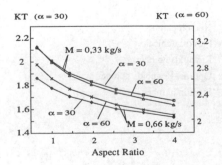

Figure 3 : Thermal Performances

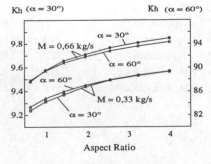

Figure 4 : Hydraulic Perfomances

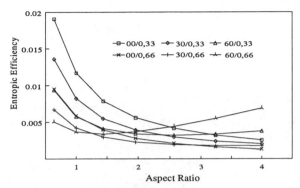

Figure 5 : Entropic Efficiency

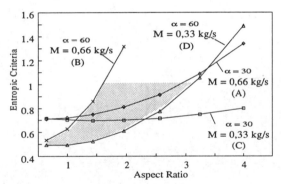

Figure 6 : Entropic Criteria

L ↗	dP ↗	A ↗	Q ↗
B ↗	dP ↘	A ↗	Q ↗
α ↗	dP ↗	A →	Q ↗

Table 1 : Influence of the geometric parameters
on overall performances

HEAT TRANSFER AND DRAFT LOSS PERFORMANCE OF AIR FLOWING ACROSS STAGGERED BANKS OF SPIRALLY CORRUGATED TUBES

J. F. Li
Shanghai Institute of Electric Power , Shanghai 200090
Q. T. Zhou , S. Y. Ye , N. Z. Gu
Southeast University , Nanjing 210018

Abstract

Experiments have been performed to determine the detailed average heat transfer characteristics of staggered tube banks in cross flow. The influence of H , P , S_1 , S_2 on heat transfer and pressure drop coefficients have been studied systematically, and the results have been correlated as follows:

$$Nuo = 0.2705 \; Reo^{0.5984} \left(\frac{P}{H}\right)^{0.1259} \left(\frac{H}{Do}\right)^{-0.007} \left(\frac{S_1 - Do}{Do}\right)^{0.1492}$$

$$\left(\frac{S_1 - Do}{S_2' - Do}\right)^{0.0665} Pr^{-1/3} \qquad \text{---------- (1)}$$

Reynolds number ranged from 3800 to 58000.

$$Euo = 3.5485 \; Reo^{-0.269} \left(\frac{P}{H}\right)^{0.1733} \left(\frac{H}{Do}\right)^{0.0863} \left(\frac{S_1 - Do}{Do}\right)^{-0.0435}$$

$$\left(\frac{S_1 - Do}{S_2' - Do}\right)^{-0.1241} \qquad \text{---------- (2)}$$

Reynolds number ranged from 6300 to 84000.
The tubes bank can get better benefits when H is about 1.3 mm.

Keywords : spirally corrugated tube, staggered bank, Nusselt number Euler number

Introduction

The goal of this paper is to determine experimentally the influence of P H, S_1, S_2 on the heat transfer and pressure drop characteristics of staggered banks of spirally corrugated tubes, having single start and 9 kinds of various geometrical-aspect ratios as shown in fig. 1. Though this kind of spirally corrugated tube has a rough surface, it can effectively prevent ash deposition when the working fluid is ash-laden flue gas. It can also increase the tube wall temperature, or reduce the number of tube to be used(compareed with the bare tube). So this kind of spirally corrugated tube is expected to be widely used in the air- preheater of boilers and various kinds of industical kilns or hot-blast stoves.

With wide-spread use of spirally corrugated tube bundles[1] [2], it is surprising that so little experimental results about their heat

transfer and pressure drop characteristics has been published(Although
plenty of experimental studies relate to inside research on individual
tubes[3]). Recently, more and more industrial heat exchangers had to be
designed referring to the results of single spirally corrugated tube and
plain tube banks. As such the data was not accurate enough[4].

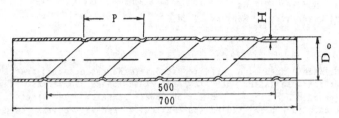

Fig. 1 Test specimen

The convective heat transfer coefficient of the spirally corrugated tube
banks here was obtained by applying the energy balance. The advantage of
this method are its higher accuracy and simplicity of the experimental
apparatus.

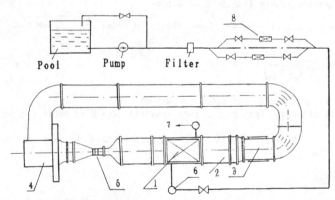

Fig. 2 Schematic diagram of experimental rig

1. test section passage	2. front measured passage
3. electric heater	4. induced draft fan
5. measured velocity passage	6. header
7. sewer	8. flow meter

Experiment apparatus

Experimental investigation had been carried out in a closed windtunnel
. The test rig was a water-to-gas heat exchanger as shown in Fig. 2, the
cooling water flowing inside the tubes of the test section with
electricaly heated air flowing on the outside impinging perpendicularly on

them. The test section was composed of a rectangular duct 1 with cross sectional dimensions of 400x500x680 mm, a contructed section 5 for measuring flow velocity. All test tubes were made of carbon steel with a value of the geometrical-aspect as follow : $\Phi 40 \times 1.5$ mm.

The spirally corrugated tubes were seated in blind holes on the floor of the test section. In order to facilitate the frequent access to the test section for installing or removing the experimental tubes, the side wall of the test section was made removable. The flow velocity of the air was measured by a pitot tube with blockage ratio less than 1.5%. Putty served to seal the test section during experimental runs.

Fig. 3 is the side view of the test section, within the streamwise direction ten rows of spirally corrugated tubes and in the spanwise direction five rows of spirally corrugated tubes with spacing $S_1/Do = 1.54$ and 1.825, $S_2/Do = 1$ and 1.75. The test spirrally corrugated tubes had values of P= 8, 12, 16, 20 and 28 mm , H=0.8, 1.3 and 1.8 mm.

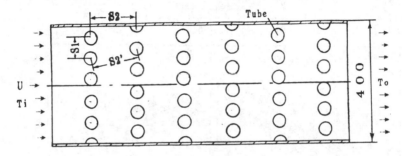

Fig. 3 Test section

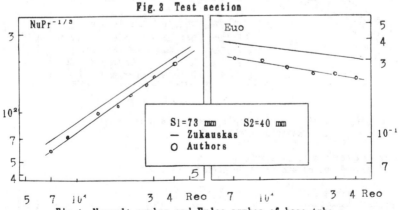

Fig. 4 Nusselt number and Euler number of bare tube
(fully developed)

Experimental procedure

The reliability of our experimental technique and apparatus was cofirmed in a preliminary experiment of a bare tube bank with spacing S_1 =73 mm, S_2= 40 mm. Agreement of the data, when compared with the well-known Zukauskas correlation[5], was within 7.3% as shown in fig.4. The reason for taking Zukauskas correlation as a comparison basis is its higher accurracy . The remakable agreement of present result with that of Zukauskas shows that the apparatus and methods of present experiment are reliable.

Data reduction

Equation :
$$\frac{1}{Ko} = \frac{Fo}{\alpha i \, Fi} + Rs + Rf + \frac{1}{\alpha o} \qquad \text{----------} \quad (3)$$

The inside heat-transfer coefficient was computed using the equation[6] :

$$Nui = 165\left(\frac{H}{Di}\right)^{1/3}\left(\frac{P}{Di}\right)^{-1/2}\left(\frac{Rei - 2000}{10000}\right)^{(0.8-3.5H/Di)} \text{--} (4)$$

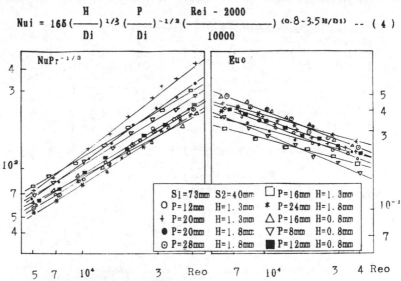

Fig. 5 Variation of Nusselt number and Euler number
with Reynolds number

Reynolds number ranged from 2000 to 8000.
The air-side Reynolds number is defined as follow :

$$Reo = \frac{\rho \, Umax \, Do}{\mu} \qquad \text{----------} \quad (5)$$

The friction factor is defined as follow :

$$Euo = \frac{2}{Z} \times \frac{\Delta P}{\rho \, Umax^2} \qquad \text{----------} \quad (6)$$

Heat transfer characteristics

The heat transfer characteristics of the spirally corrugated tube bank has been measured and the parameters P/H, H/Do and Reynolds number have been varied in the range of P/H = 9.23 ~ 20, H/Do = 0.02 ~ 0.045, Reo = 3800 to 58000. The fully developed heat transfer data of spirally corrugated tube bank which falls in the range of frequently used size, can be seen from fig.5 to fig.7 . All physical properties are evaluated at the average air temperature. The maximum deviation of experimental data from this correlation (1) is 11.1% as shown in fig.5 to fig.7.

The Nusselt number decrease with the increase of H, S_2 and increases with the increase of P, S_1. This can be explained as follows : (1) The gap between the two adjacent tubes will get larger with the increase of tube groove depth H . At the same time , this will deteriorate the flow of the air in the groove and hence reduce its heat transfer coefficient. (2) The spirally angle of tube groove. When P varied in experimental extent, The gas flow in the groove enhanced the heat tranbsfer .

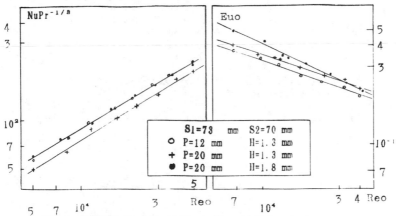

Fig. 6 Variation of Nusselt number and Euler number
with Reynolds number

Pressure drop characterstics

The pressure drop of spirally corrugated tube banks will increase with the P and S_2 , decrease slightly with H and S_1 , as shown in fig. 5 to fig.7 This is because the flow in the tube groove is enhanced with the increase of P , therefore increasing the surface drag. However bigger H will weaken the mass transfer in the groove with the free flow. Variation of Euo with the S_1, S_2 is similar with the bare tube bank. The maximum deviation of experimental data from the correlation (2) is 10.12% as shown in fig.5 to fig.7 .

Discussion

The spirally corrugated tube will tend to bare tube characteristics when $P \to 0$ or $P \to \infty$, and $H \to 0$, but Hmax will be confined by the effect of inside tube heat transfer and pressure drop . Thus the variation of Nuo and Euo with the tube parameter can be shown as fig. 8 and so an optimum P and H exists. Therefore , further experiments will be valube.

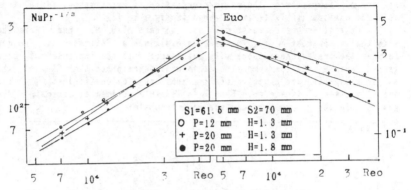

Fig. 7 Variation of Nusselt number and Euler number
with Reynolds number

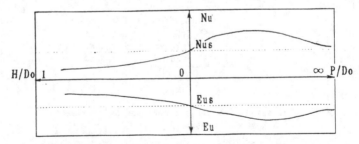

Fig. 8 Variation of Nusselt number and Euler number
with the tube parameter

Acknowledgement

This study was supported in part by a grant from the DongFeng Boiler Factory and NanJing Xiaguan Power Plant. The author also gratefully acknowledge the help rendered by Mr. Q. S. Si during the completion of this paper.

References

1) Bergles, A. E., Webb, R. L., Junkhan, G. H., 'Energy Conservation via Heat Transfer Enhancement', _Energy_, vol. 4, 1979, pp. 193 - 200

2) Cannavos, T. C., 'Some Recent Development in Augmented Heat Exchange Elements ', _Heat Exchanger: Design and Theory Sourcebook_, 1974

3) Bergles, A. E., Webb, R. L., Junkhan, G. H. and Jensen, M. K., 'Bibliography on Augmentation of Convective Heat and Mass Transfer', Report HTL-19, Engineering Research Institute Iowa State University, May 1979

4) Q. T. Zhou, _Heat Transfer Augmentation Boiler and Heat Exchanger_ Southeast University, 1989

5) A. A. Zukauskas _Convective Heat Transfer of Heat Exchanger_, Academic Pr., Beijing, 1986

6) Yosimi Eiaki, 'Heat Transfer and Pressure Drop of Spirally Corrugated Tube, ' _Nuclear and Thermal Electric Power_ Vol. 27, 1976

7) J. F. Li, 'Heat Transfer and Pressure Drop of Staggered Spirally Corrugated Tube Bank' , Master's Degree Thesis , Southeast University, 1991 (in chinese)

Nomenclature

D	Tube diameter, m
Eu	Euler number , $\triangle P / \frac{1}{2} \rho U^2$
F	Total heat transfer area, m^2
H	Tube groove depth, m
Ko	Heat transfer coefficient, $W/m^2 K$
Nu	Nusselt number
P	Tube pitch , m
Pr	Prandtl number
Re	Reynolds number
Rf	Fouling thermal resistance, $m^2 °C / W$
Rs	Wall thermal resistance, $m^2 °C / W$
S_1	Transverse pitch, m
S_2	Longitudinal pitch, m
T	Air temperature, °C
Umax	Maximum velocity of tube bank, m/S
Z	Row number
α	Coefficient of convective heat transfer , W/m^2
$\triangle P$	Differential pressure drop , N/m^2
ρ	Density of fluid, Kg/m^2
μ	Viscosity of fluid, $Kg/m \cdot s$

Subscripts :

i	Inside tube or inlet
o	Outside tube or outlet
s	Smooth tube bank

EXPERIMENTAL AND NUMERICAL ANALYSIS OF A TURBULENT SWIRLING FLOW IN A TUBE

W.F.J. Sampers, A.P.G.G. Lamers and A.A. van Steenhoven
Department of Mechanical Engineering, University of Technology Eindhoven

One of the problems of modelling a turbulent swirling flow in a tube is the choice of the turbulent closure model. To evaluate which model performs best, local velocity measurements of the flow field were carried out in a test-rig, consisting of a circular pipe in which a swirl is generated. The measurement method used is hot-wire anemometry in combination with visualization experiments. The measurements are compared with the results of a numerical analysis. In the numerical analysis two turbulent models are used: the k-ε model and the Algebraic Stress Model (ASM). In the experiment a symmetrical swirl was observed. Comparing the experimental data with the numerical results yields that the ASM represents the experiment quite well, whereas the standard k-ε model fails.

INTRODUCTION

The turbulent mass and heat transfer processes are often enhanced and stabilized by a swirl. For numerical analysis of this type of flow two turbulence closure models are commonly used: the k-ε model and the Algebraic Stress Model (ASM). The k-ε model is expected to describe a swirling flow accurate in a simple turbulent flow situation (Khalil (1)) but according to Boysan (2) and Kim (3) the ASM performs best in more complex flow situations . To evaluate the turbulence models, numerical results have to be compared with experimental data.
In a tubular test-rig, in which a swirl is induced, local velocity components and their fluctuations are measured and compared with the results of numerical calculations. Similar experiments have been conducted by Kitoh (4). He used air as working fluid. The swirl was generated with radial guidevanes at the entrance of the tube similar to the method used in this investigation. However Kitoh did not compare the experimental data with numerical predictions.

EXPERIMENTAL SET-UP

The swirling flow in the test-rig (Fig. 1) has a Reynolds number of 10,000 based on the bulk-velocity u_m measured by means of an orifice. The swirl-generator consists of twelve equally spaced vanes welded on an axis with a diameter of 63 mm. The total diameter of the swirl-generator equals the inner diameter of the tube being 192 mm. The airflow has an average bulk velocity u_m of 7.5 m/s.

The cross-sections where the velocity profiles are measured are located downstream from the swirlgenerator at a distance of 9, 10 and 18 pipe diameters. At those positions a hot-wire sensor can be inserted. The flow parameters to be measured are the mean velocity components in axial, radial and tangential direction, and their fluctuations. From the axial and tangential velocity distribution in a cross-section the swirl-number Ω, being a dimensionless measure of the intensity of the swirl, is calculated as:

$$\Omega = 2\pi\rho \int_0^{r_0} \frac{r^2 u_z u_\phi dr}{\pi\rho r_0^3 u_z^2} \tag{1}$$

Here u_z and u_ϕ are the instantaneous axial and tangential velocities, ρ the density of the fluid and r_0 the radius of the tube.

As mentioned the measurement method used is hot-wire anemometry. The sensor consists of two perpendicular wires. Those platinized tungsten wires having a diameter of 5 μm and a selective length of 2 mm, giving an aspect ratio l/d of 400. This means that Jörgenson's relation (5) can be used to convert the signals from the anemometers into the velocity components. In this manner the magnitude and the direction of the velocity vector are determined. The sign, however, is not defined. For each measuring point the sign is determined by means of two simple visualization techniques using smoke and tufts.

The turbulent kinetic energy k given by expression (2) is directly calculated from the fluctuations:

$$k = \frac{1}{2} (u_z'^2 + u_r'^2 + u_\phi'^2) \tag{2}$$

where u'_z, u'_r and u'_ϕ are the fluctuations of the velocity components. The anemometer used is a DISA Type 55D00 Universal Anemometer, whose signals are converted by a Tecmar Labmaster A/D converter. The data are stored in a PC-AT using the software package ASYST. At each measurement point 100,000 samples were taken with a frequency of 20 kHz.

NUMERICAL SIMULATION

To describe the turbulent flow phenomena the k-ε model and the ASM is used. The k-ε model (6) consists of the mean averaged Navier-Stokes equations combined with two additional differential equations for the turbulent kinetic energy (k) and the dissipation rate of turbulent kinetic energy (ε) to model the Reynolds stresses. However the k-ε model considers the turbulence to be isotropic. In a turbulent swirling flow the turbulence phenomena are expected to be anisotropic.

The Algebraic Stress Model is developed from the Reynolds Stress transport equations in order to provide a simpler model accounting for anisotropic effects. The terms containing gradients of the Reynolds stresses in the RSM are replaced by algebraic relations. Besides the three algebraic relations, the equations for the turbulent kinetic energy k and the dissipation rate ε need to be solved. A complete set of equations expressed in cylindrical coordinates can be found in Kim and Chung (3). Although the ASM is rather complex and some problems with the stability of the numerical solution may occur, its main advantage is that it accounts for anisotropy and hence is able to cope with more complex flows, like swirls, than the k-ε model.

BOUNDARY CONDITIONS AND SOLUTION PROCEDURE

The k-ε model and the ASM are only valid for fully turbulent flows. Close to solid walls there are inevitable regions where the local Reynolds number is so small that the viscous effects predominate over turbulent ones. To be able to account for the viscous effects and for the large gradients of the variables near the walls, the so-called function method of Launder and Spalding (6) is used.

The inlet boundary conditions for velocity and turbulent energy are derived from the measurements. The dissipation rate is coupled to the turbulent energy according to $\varepsilon_{in}=k^{1.5}/(0.5 \cdot r_0)$. At the outlet the normal gradients are set to zero.

The numerical experiments are carried out in a 2-D axisymmetrical and cylindrical domain with cyclic boundary conditions in tangential direction.

The numerical analysis is performed with both the k-ε model and the ASM model. The equations are solved by means of the finite volume package FLUENT. The variables calculated are the mean velocity components in z, r and φ direction, the pressure field, the turbulent kinetic energy and the dissipation rate of turbulent energy. The calculations are performed with a uniform grid of 18 cells in axial and 28 cells in radial direction. First the solution is derived with the k-ε model, taking about 700 iterations. Secondly this result is used as an initial flow field for the calculations with the ASM which needed 10,000 - 15,000 iterations on a micro VAX machine, taking a calculation time of about one day.

COMPARISON BETWEEN CALCULATIONS AND MEASUREMENTS

The velocity and turbulent energy profiles are measured at 6 cross-sections in total but only the results at 9, 10 and 18 pipe diameters from the swirl-generator are presented in this paper. The first measured position is taken at 9 pipe diamaters (9-D) where we expected that most of the local disturbances due to the swirl generator would be damped. The inlet boundary conditions are derived from the experimental data of the first measurement position (Fig. 2). The estimated relative errors of the measurements for u, k and Ω are 12, 24 and 23% resp..

In Fig. 3a measured and calculated axial velocity profiles at 10 and 18-D from the swirl generator are given. At 10-D both models predict the measurements quite well. At 18-D the k-ε model predicts an almost fully developed turbulent flow profile. The ASM approximates the measured profile more accurate. At the wall and the axis the differences are still large.

The calculated radial velocities at 10 and 18-D is almost zero for both models (Fig. 3b) whereas the measurements show a level between 1.5 and 2.5 m/s. The reason for the discrepancy between calculation and measurement is probably that the calculation is performed in a 2-D domain whereas the flowfield, due to the geometry of the swirl geometry is still essentially 3-D.

The tangential velocity profiles are shown in Fig. 4a. The result of the k-ε model shows that the swirl is dissipated fast; 25% of the initial value at 10-D and 80% at 18-D. The measurements show that in reality the swirl is not dissipated so fast. The ASM fairly approximates the measurement. At the wall however the difference with the measured profile is about 5% at 10-D and 20% at 18-D. Hence the modelling of the wall region is critical.

The profiles of the turbulent kinetic energy show large differences between measurement and calculation (Fig. 4b). In the calculations the turbulent kinetic energy is clearly dissipated while in the measurement production takes place locally. The exact reason for this difference is not known. A possible explanation is that the dissipation rate of turbulent kinetic energy is not formulated correctly in the used turbulence models (Bardina (7)). Another explanation is that the production of kinetic energy in the experiment is mainly due to the radial velocity component (3-D effects).

In Fig. 5 the measured and calculated swirl numbers along the test-rig are compared. In case of the k-ε model the swirl number is much lower than measured. The ASM predicts the same decay of the swirl number as observed in the experiment.

CONCLUSIONS

In three cross-sections of a measurement tube at 9, 10 and 18-D downstream of a swirl generator respectively, the mean and fluctuating velocity components are measured with hot-wire anemometers. The measured profiles at 9-D were taken as inlet conditions for the numerical calculations with a standard k-ε model and an ASM. The mean axial and tangential velocity components are calculated more accurate with the ASM as with the k-ε model. The discrepancy between the measured mean radial velocity profile and the calculated profiles for both models is probably due to 3-D effects in the flow whereas the calculations are performed 2-D. Further experiments to evaluate this effect are necessary. The dissipation rate of turbulent kinetic energy shows large differences for both models whereas the decay of the swirl-number is predicted correctly with ASM. In general it is clear that the ASM approximates the measured data the closest.

REFERENCES

1. Khalil E.E., Assaf H.M.W., 1981, Numerical Methods in Laminar and Turbulent Flow, pp. 363-376, Pineridge Press, Swansea.

2. Boysan F., Swithenbank J., 1981, Numerical Methods in Laminar and Turbulent Flow, pp. 425-438, Pineridge Press, Swansea.

3. Kim K. Y., Chung M. K., 1988, Int. J. Heat and Fluid Flow, vol. 9, nr 1, pp. 62 - 68.

4. Kitoh O., 1991, J. Fluid Mech., vol. 225, pp. 445 - 479.

5. Jörgenson F.E., 1971, DISA info, nr. 11, pp. 31 - 37.

6. Launder B.E., Spalding D.B., 1972, Mathematical Models of Turbulence, Academic Press, London.

7. Bardina J., Ferziger J.H., Rogallo R.S., 1985, J. Fluid Mech., vol. 154, pp. 321 - 336.

Nomenclature

D	inner pipe diameter	m
k	turbulent kinetic energy	m^2s^{-2}
L	pipe length	m
r	radial distance	m
r_0	radius of the tube	m
u_m	bulk velocity	ms^{-1}
u_z, u_r, u_ϕ	instanteneous velocity component in axial,	
	radial and tangential direction	ms^{-1}
ε	turbulent energy dissipation rate	m^2s^{-3}
Ω	swirl number	---
ρ	density	kgm^{-3}

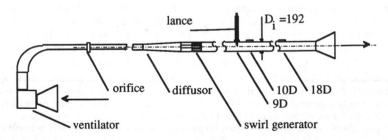

Fig. 1 Experimental test-rig. The inner diameter D_i of the test section is given in mm.

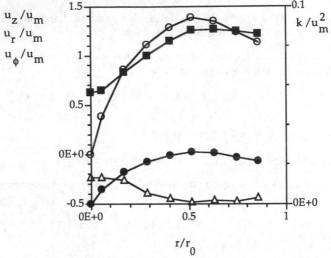

Fig. 2 The measured inlet profiles of the test section as function of the radius at 9D from the swirl generator. Normalised axial velocity (■), normalised radial velocity (Δ), normalised tangential velocity (O) and normalised turbulent kinetic energy (●).

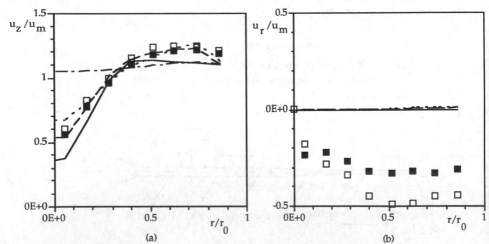

Fig. 3 The measured and calculated axial (a) and radial (b) velocity components as function of the radius.

□ measured at 10D, - - - - - k-ε at 10D, — — — ASM at 10D

■ measured at 18D, — - - — k-ε at 18D, —— ASM at 18D

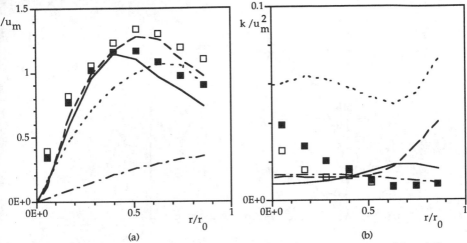

Fig. 4 The measured and calculated tangential velocity components (a) and the turbulent kinetic energy (b) as function of the radius.

□ measured at 10D, - - - - - k-ε at 10D, — — — ASM at 10D
■ measured at 18D, — - — k-ε at 18D, ——— ASM at 18D

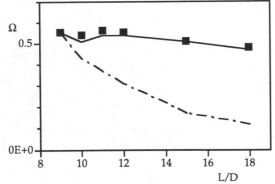

Fig. 5 The swirl number along the test-rig.

■ measurements, — - — k-ε model, ——— ASM

FIN EFFICIENCY OF ROUND TUBE AND PLATE FIN SURFACES WITH AND WITHOUT LOUVRES

LCA Yu, MR Heikal & TA Cowell
Dept of Mechanical Engineering, University of Brighton, Brighton BN2 4GJ.

The fin efficiency of heat transfer surfaces consisting of round tube and continuous plate fins cannot be arrived at analytically. Numerical results have been presented in the literature for surfaces with tubes arranged in a regular 60 degree layout and approximating equations have been proposed for layouts other than 60 degree. The first part of this paper presents numerical results for non-60 degree configurations which demonstrate that the approximating equations are perfectly adequate for engineering applications in these cases. The second part of the paper presents numerical results for surfaces in which louvres have been cut for heat transfer enhancement. The results demonstrate that cutting of the fin material to form the louvres results in a significant reduction in fin efficiency, which must be taken into account in assessing the enhancement effect of the louvres.

INTRODUCTION

The concept of fin efficiency goes back a long way, and analytic solutions for the simpler geometries have been available for many years, eg Harper and Brown (1922). These geometries include straight fins, spine fins and radial fins for a number of different types of cross-section, (Kern and Kraus, 1972). However, there exists a group of widely used fin geometries that do not lend themselves to straight forward analytic derivation of fin efficiency relationships. These are the surfaces made up of an array of round or oval tubes that pass through a stack of parallel, spaced-apart fins. Here numerical or approximate analytic methods must be used, and results for a limited number of these geometries have been presented in the literature. In the general case, methods for approximating the surfaces to the nearest equivalent radial fin have been proposed.

The work described in the first part of this paper uses numerical methods to derive fin efficiency figures for further tube layout configurations that are in quite common use. The results are compared with those of the available general approximating relationships.

It is becoming increasingly common for louvres to be cut into the surface of these fins in order to enhance the heat transfer performance. In the second part of this paper, attention is given to the effects of louvring on fin efficiency. Again numerical methods are used, and it is not surprising that the louvres give rise to significant modification of the fin efficiency values when compared with plain fin, since the louvres provide significant disturbance of the heat flow paths in the fin.

LITERATURE

The two basic types of round tube array, in-line and staggered, are shown in Fig. 1. As far as fin efficiency is concerned, both types of configuration can be specified by the non dimensional parameters (p_t/d) and (p_l/d), where p_t and p_l are transverse and longitudinal tube pitch respectively, and d is tube outer diameter. An alternative for the staggered array is to use (p_t/d) and the angle β. It is indicated in the diagram how each tube can be considered to have an elemental fin attached to it. Carrier and Anderson (1944) and Schmidt (1945/46) have both shown that a reasonable approximation for the fin efficiency can be given by treating the fin as radial with uniform fin height. Schmidt (1949) recognised that the use of equivalent area to calculate the relevant fin height leads to slight overestimate of fin efficiency, and he presents equations that claim to calculate the best value of fin tip radius for both in-line and staggered arrangements. However, some ambiguity in the nomenclature and lack of information about the origin of the equations limit their value. Carrier and Anderson checked out the accuracy of the assumption of equal fin area in order to calculate the relevant fin tip radius for the in-line tube layout. They split the fin area into radial wedge-shaped segments and used their radial fin equations separately for each of the segments to develop an efficiency for the complete fin. They evaluate two particular cases and showed that the equal area assumption gave an error of only 1.5% for their square array fin, but this went up to 4% for their fin with transverse to longitudinal pitch ratio of 1.7. They state that an equivalent assessment could be made for any staggered array, as long as the angle β (see Fig. 1) is 60 degrees. They recognised that it would not be valid for other values of β, since in that case the elemental fin boundary is no longer isothermal. Schmidt (1949) had not recognised this constraint.

Zabronsky (1955) presented analytic results for the square array, and although his solution met the adiabatic fin tip condition exactly, the isothermal fin/tube interface was only approximated. Sparrow and Lin (1964) took a different approach and considered both square and 60° hexagonal arrays. In contrast to Zabronsky, whilst their method met the isothermal fin/tube interface condition exactly, the adiabatic fin tip condition was only approximated, although the degree of accuracy could be freely selected. They presented fin efficiency curves (with calculation accuracy to 0.1%), for values of (p/d) from 1 to about 3.5 for both hexagonal and square arrays. They further showed that the equal area approximation can be very inaccurate for very small values of (p/d) - up to 18% in the extreme.

Fabris (1975) used finite element methods to derive fin efficiency values for ranges of hexagonal, square and rectangular tube arrays. His results covered ranges of values of (p_t/d):- 1.08 to 5.2 for hexagonal, 1.15 to 5 for square and 1.25 to 7.5 for rectangular layout. In the last case, the results were given for (p_t/p_l) values of 2 and 3.

Kuan et al (1984), used numerical methods to determine the fin efficiency of several different tube layouts. They presented their results in the form of percentage differences between these values and the values that result from the equal area approximation. For the hexagonal fin, the accuracy is better than 1% for (p/d) values between 1.15 and 5.8, and for square layout the accuracy is better than 1.5% for (p/d) between 1.5 and 10. Values for rectangular layout arrays were also quoted for the two cases (p_t/p_l) equal 2 and 5. For the first case, error could go up to 17% in the range of (pt/d) values 1.1 to 10, and in the second case the possible error rose to 20% with (pt/d) values ranging between 1.5 and 8.

It is clear that for realistic hexagonal and square arrays the equal area approximation gives adequate accuracy. However, for rectangular arrays significant error can result. No results are available for staggered layouts with β-angle values other than 60 degrees.

FIN MODELLING

The geometries considered in this paper are three different staggered arrays that were used in an experimental study of the surface performance (Yu, 1992). One array was regular hexagonal, whilst the other two were not. In one case the analysis was performed both for plain fin and for the case in which the fin had louvres cut into the surface. The relevant parameters are given in Table 1.

The fins were modelled using the general purpose finite element package PAFEC, which was originally developed for stress analysis. The basic element is indicated in Fig. 2. It can be seen that it is necessary to model a region containing parts of two adjacent tubes. This need only arises because of the non-equilateral nature of the staggered arrays, which means that the isotherm

Model Number	1	2	3	4
Fin Type	flat	flat	flat	louvred
Tube diameter- mm 8	10	10	10	
Transverse Pitch - mm	25.4	19	25.4	25.4
Longitudinal Pitch - mm	22	16	16	16
Angle β - degrees	60	61.4	76.9	76.9

Table 1. Fin Geometric Parameters.

dividing this region in two does not consist of easily identifiable straight lines, as is the case for the fully hexagonal array. A two dimensional mesh was created with smaller elements in the regions of higher temperature gradient near the tube walls. A sweep command then generated a three dimensional mesh from the two dimensional original. The finished model needed to represent only half of the fin thickness as a result of the symmetry. For the solution the following boundary conditions were applied:-

i. uniform heat transfer coefficient, ranging from 10 to 250 W/(m^{2}K), applied to surface of fin

ii. isothermal condition at fin base
iii. a temperature difference of 50K between the fin base and the convecting fluid
iv insulated surfaces elsewhere on model.

The term (kt) , ie fin conductivity x fin thickness, was set equal to 0.0265 W/K, the appropriate value for the fins under study.

In Fig. 3 it is indicated how the mesh was created in order to evaluate fin efficiencies for for both plain and louvred fin. The narrow rectangular regions represent the cuts in the fin material between adjacent louvres. For the first attempts the conductivity in these regions were simply set equal to zero. However it became apparent that this procedure effectively removes so much of the surface area from the heat transfer process as to distort the resulting fin efficiency values. The software did not allow the slits to be made any narrower. The solution was to make use of the software facility for generating orthotropic elements. With this, the conductivities in the narrow rectangular regions were set equal to zero in the plane of the fin perpendicular to the narrow regions, and equal to the fin material conductivity in the other two directions. This effectively allows convection to occur over the whole of the surface and allows heat to flow along each louvre over its whole width, but does not allow heat to flow across the gap between louvres. This mesh was also used for the plain fin calculations, only with the conductivity set to be the same in all directions. Typically about 1200 nodes and 650 elements were involved in a fin model.

A radial fin model was built with a similar mesh spacing in order to obtain a measure for the accuracy of the mesh. Agreement with the analytic solutions for this similar, but slightly simpler geometry were always better than 0.6 %.

RESULTS

In each case the fin model was run on PAFEC for the complete range of values of heat transfer coefficient. A heat balance calculated from the computed values of heat flow into and out of the model was always better than 0.01%.

To calculate the fin efficiency, the nodal heat flows from each of the fin base elements were first added together to give the total heat transferred by the fin. This was then divided by the heat that would have been transferred if the fin temperature had been everywhere equal to the fin base temperature, to give the fin efficiency. In all cases the results are presented as fin efficiency, h expressed as a function of a fin Biot number, here defined as:-

$$Bi = m(r_0 - r_i) \qquad \text{eq. (1)}$$

where r_i is the inner radius of the fin, ie the tube outer radius; r_o is the effective 'equal area' fin tip radius given by:-

$$r_o = (p_t \times p_l/\pi)^{0.5} \qquad \text{eq. (2)}$$

and the parameter m is given by:-

$$m = [\ 2\alpha/(k\ t)]^{0.5} \qquad \text{eq. (3)}$$

where α is the heat transfer coefficient.

The first configuration for which fin efficiency values were determined was a regular hexagonal array, Model 1. The values were compared with the curves for this configuration presented by Sparrow and Lin (1964) and Fabris (1975). The values agreed with both within the accuracy to which the values could be interpolated from the curves - better than 1 %. This provided further confirmation for the accuracy of the finite element procedure used here. The values were also compared with the values calculated using the Bessel function relationships for radial fin and the assumption of an effective 'equal area' fin tip radius. Again the agreement amongst the four sets of values was better than 1%. This is not surprising, since the regular hexagon array is the closest approximation to the radial fin, and in this case the tube pitch to diameter ratio, at 3.18, is large which further tends to increase the closeness of the approximation.

Model 2 had a β value of 61.4 degrees and a tube pitch to diameter ratio of 1.90. Despite the slight variation from the regular hexagonal array and the smaller value of tube pitch to diameter ratio, the values were found once again to agree with the radial fin approximation to within 1%. The curves are plotted in Fig. 4 as they represent radius ratio values that have not been plotted in the existing literature.

Model 3 was significantly non-equilateral triangular in layout with a β value of 76.9 degrees and a pitch to diameter ratio of 2.54. The curve for fin efficiency is shown in Fig. 5. Again, perhaps surprisingly this time, the curves remain within 1% of the values given by the radial fin approximation. However, now the values tended to lie around 0.5% below the radial fin values rather than 0.5% above as had been the case with the other fin models. This suggests that the deviation from the radial fin configuration is just beginning to show itself for this fin, but for engineering design purposes the difference is sufficiently small to be ignored.

Typical temperature profiles for fin model 3 are shown in Fig. 6 Alongside them are the equivalent temperature profiles for model 4 which is the louvred version of fin 3. For the plain fin it can be seen that the heat flow is largely radial, but in the louvred fin, the profiles are significantly distorted, as they must be, by the cuts in the metal. The expected pattern is seen, in which the louvre temperature gradients lie along their lengths, as heat has to flow to their ends before passing to the tube wall. The fin efficiency curve for the louvred fin is plotted together with that of the plain fin equivalent in Fig. 5. As expected the louvred fin shows a lower fin efficiency with the difference going up to around 3.5% for the higher Biot numbers. This degree of difference from the plain fin values and from the radial fin approximation is now sufficiently great to need to be accounted for in engineering design.

CONCLUSIONS

The results presented have shown that the equivalent area radial fin approximation for the determination of fin efficiency is sufficiently accurate for design purposes for the three hexagonal array configurations studied here. The method even proves to be valid for an array with the relatively large β angle of 76.9 degrees, although the results suggest that the effects of the approximation are just beginning to be seen. Since this sample also has relatively large tube pitch to diameter ratio, it is suggested that the errors may become relevant at this β value as the pitch to diameter ratio is reduced. Another area in which data is lacking is that of rectangular array fins. The isolated information in the literature shows that the radial fin approximation can lead to very significant errors, and a comprehensive study of the rectangular array fin remains to be done.

It has been shown that the louvring of the fin surface makes a significant difference to the fin efficiency, in the one configuration tested. Further configurations need to be studied; in particular the effect of extending the louvres to nearer the tube walls needs to be considered. Whilst it is known that this is likely to enhance the heat transfer coefficients, the effect on fin efficiency still needs to be determined.

ACKNOWLEDGEMENTS

The authors would like to thank ACR Heat Transfer for their support for the research programme of which the work presented here forms a part.

REFERENCES

Carrier WH and Anderson SW, 1944
Heating, Piping and Air Conditioning, May, pp 304-320.

Fabris O, 1975
Proc. 14th Int. Congress of Refrigeration, v. 2, pp 822-832.

Harper DR and Brown WB, 1922
NACA Report 158.

Kern DQ and Kraus AD, 1972
Extended Surface Heat Transfer
McGraw-Hill, Inc.

Kuan D-Y, Aris R and Davis HT, 1984
Int. J. Heat and Mass Transfer, v. 27, no. 1, pp 148-151.

Schmidt TE, 1945/46
Bull. Int. Inst. Refrigeration, Annex G-5.

Schmidt TE, 1949
J. ASRE, April, pp 351-357.

Sheffield JW, Wood RA and Sauer HJ Jr, (1989
Experimental Thermal and Fluid Science, v. 2, pp 107-121.

Sparrow EM and Lin SH, 1964
Int. J. Heat and Mass Transfer, v. 7, pp 951-953.

Yu LCA, 1992
Performance characteristics of round tube and plate fin heat transfer surfaces
PhD Thesis, Brighton Polytechnic, CNAA.

Zabronsky K, 1955
Trans. ASME, J. Applied Mechanics, March, p 119.

NOMENCLATURE

Symbol	Meaning	Units
Bi	Biot number, eq (1)	-
d	tube outer diameter	m
k	fin thermal conductivity	W/(mK)
m	fin efficiency parameter, eq (3)	m^{-1}
p_l	longitudinal tube pitch	m
p_t	transverse tube pitch	m
r_i	inner radius of fin	m
r_o	effective outer radius of fin, eq (2)	m
t	fin thickness	m
α	heat transfer coefficient	$W/(m^2K)$
β	staggered tube array angle	degrees

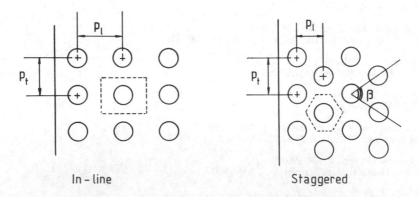

In - line   Staggered

Fig. 1 Tube layouts in tube and plate fin surfaces.

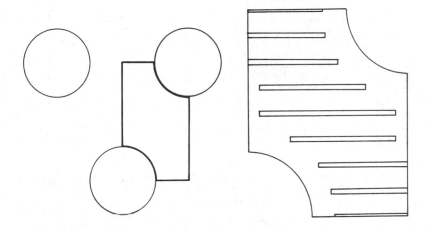

Fig. 2 Modelled fin element.   Fig. 3 Model basis for mesh generation.

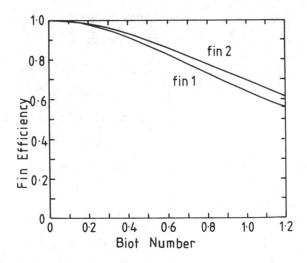

Fig. 4 Fin efficiency curves for two staggered arrays.
Fin 1:- $\beta = 60^o$, $r_o/r_i = 3.33$ and Fin 2:- $\beta = 61.4^o$, $r_o/r_i = 1.97$.

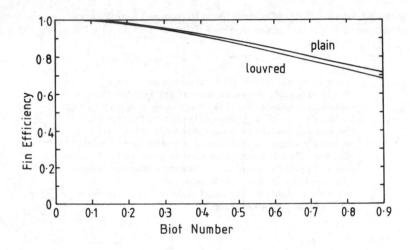

Fig. 5   Fin efficiency curves for plain and louvred fins, $\beta = 76.9^{\circ}$, $r_o/r_i = 2.28$.

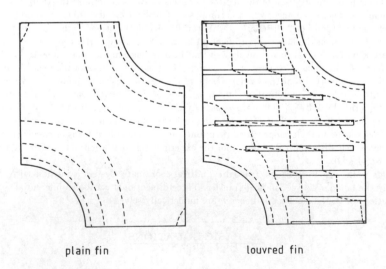

plain fin   louvred fin

Fig. 6 Temperature contours in plain and louvred fin.

A NUMERICAL ANALYSIS OF LAMINAR HEAT TRANSFER IN A STRAIGHT TUBE CONTAINING A TWISTED-TAPE SWIRLER

Yoshiyuki Aoyama, Koichi Mizukami, Kunio Hijikata and Kozo Futagami
Department of Mechanical Engineering, Ehime University, 3 Bunkyo-cho, Matsuyama, 790 Japan

A non-orthogonal twisted cylindrical coordinate system is introduced to study the laminar flow through a straight tube containing a twisted-tape and the heat transfer from the tube wall to the fluid. The effect of torsion of the tape on the axial flow, the secondary flow and the temperature profile is investigated by numerically solving the three-dimensional parabolic differential equations. The calculated Nusselt numbers coincide with the values from the experimental correlation obtained by other researchers. The trajectories are drawn on the surface of the axial velocity versus the axial pressure gradient, and the chaos-like fluid motion is found when the torsion is large.

1. INTRODUCTION

A technique of the insertion of twisted-tape swirlers is available to raise the thermal efficiency of a heat exchanger which is composed of straight circular tubes. Several experimental studies [1, 2, 3] reported heat transfer enhancement and an increase of the pressure loss when the flow was turbulent. Hong and Bergles [4] performed their experiments for laminar flows and showed that heat transfer in the tube is greatly enhanced. Nakayama [5] reviewed the researches on the performance of the twisted-tape swirlers and reported that the swirler was more effective for viscous liquids. However, there seems to be only two analytical investigations; one was reported by Date and Singham [6], and the other by Date [7]. They are essentially the same, and both provided not only the laminar heat transfer coefficient but also the turbulent one, which were numerically obtained on the fully developed condition. But either laminar heat transfer coefficient does not agree well with that calculated from the experimental correlation obtained by Hong and Bergles [4].

In this work, a non-orthogonal twisted cylindrical coordinates system is introduced to study the laminar convection heat transfer. Three dimensional parabolic differential equations of the momentum and energy are numerically solved.

2. BASIC EQUATIONS

Figure 1 shows a non-orthogonal twisted cylindrical coordinate system (x, ϕ, η), where x, ϕ and η are the longitudinal, the peripheral and the radial coordinates, respectively, (x', ϕ', η') is the ordinary cylindrical coordinate system and H is the pitch for 180° twist of the tape. The curvilinear coordinate ϕ is taken from the surface of the twisted-tape on every section of the tube and is not orthogonal to x unless $\eta = 0$. The coordinate system is the same as that Date [7] used. But our basic equations derived below differ from those by Date in usage of the contravariant components of velocity.

The coordinates x, ϕ and η are related to the ordinary cylindrical coordinates x', ϕ' and η' by the following equations, respectively.

$$\eta = \eta', \phi = \phi' - \tau^* x', x = x', \tag{1}$$

where τ^* is defined as $\tau^* = \pi/H$. The covariant metric tensors g_{ij} and the contravariant ones g^{ij} are obtained as follows,

$$g_{11} = g^{11} = 1, \quad g_{22} = \eta^2, \quad g^{22} = (1 + \tau^{*2}\eta^2)/\eta^2, \quad g_{23} = \tau^* \eta^2, \quad g^{23} = -\tau^*,$$

$$g_{33} = 1 + \tau^{*2}\eta^2, \quad g^{33} = 1, \quad g_{12} = g^{12} = g_{13} = g^{13} = 0. \tag{2}$$

The non-zero Christoffel symbols of the second kind can be expressed as

$$\Gamma_{22}^1 = -\eta, \quad \Gamma_{23}^1 = -\tau^*\eta, \quad \Gamma_{33}^1 = -\tau^{*2}\eta, \quad \Gamma_{12}^2 = 1/\eta, \quad \Gamma_{13}^2 = \tau^*/\eta. \tag{3}$$

In the derivation of the basic equations, we assume; (1) the fluid is incompressible; (2) the flow can be characterized as a three-dimensional parabolic flow and the longitudinal pressure gradient is uniform in the cross-section of the tube, that is, independent of both η and ϕ; (3) the viscous dissipation is negligible in the energy equation; (4) the buoyancy force is negligible.

The continuity equation and the momentum equation in tensorial form are

$$u^i|_i = 0 \tag{4}$$

$$u^j \frac{\partial u^i}{\partial x^j} + u^j \Gamma_{kj}^i u^k = -\frac{g^{ij}}{\rho} \frac{\partial p}{\partial x^j} + \frac{\mu}{\rho} \sigma^{ij}|_j, \tag{5}$$

where ρ and p are the density of fluid and the pressure, respectively, and σ^{ij} denotes the contravariant stress tensor that is expressed in terms of the velocity gradients and the fluid viscosity μ as

$$\sigma^{ij} = \mu(u^{j|i} + u^{i|j}). \tag{6}$$

The contravariant derivative $u^{i|j}$ and the covariant derivative for σ^{ij} in the above equations are expressed as

$$u^{i|j} = g^{jk}\left(\frac{\partial u^i}{\partial x^k} + u^l \Gamma_{kl}^i\right), \quad \sigma^{ij}|_j = \frac{\partial \sigma^{ij}}{\partial x^j} + \sigma^{lj}\Gamma_{jl}^i + \sigma^{il}\Gamma_{jl}^j. \tag{7}$$

The energy equation is

$$\rho c_p \frac{\partial \theta}{\partial x^i} u^i = \frac{\lambda}{\sqrt{g}} \frac{\partial}{\partial x^i}\left(\sqrt{g}\ g^{ij} \frac{\partial \theta}{\partial x^j}\right), \tag{8}$$

where g is the determinant of g_{ij}, and θ and c_p are the temperature of fluid and the specific heat at constant pressure, respectively.

In order to derive the basic equations in terms of the physical velocity and the notation of the coordinates (η, ϕ, x) from the above tensorial equations, it is necessary to put $x^1 = \eta, x^2 = \phi, x^3 = x$ and to transform the physical velocity components (u_η, u_ϕ, u_x) by $u_\eta = u^1, u_\phi = u^2, u_x = u^3$. Further the following dimensionless variables are introduced;

$$r = \eta/a, X = x/a, \tau = a\tau^*, U_i = u_i/u_m, T = \rho c_p u_m \theta/q_w$$

$$P = p/(\rho u_{m^2}), Re = 2au_m/\nu, Pr = \nu/\kappa, \tag{9}$$

where a is the radius of tube, u_m is the averaged axial velocity with respect to the cross-section of tube, q_w is the wall heat flux (it is assumed to be constant), and ν and κ are the kinematic viscosity and the thermal diffusivity, respectively.

Upon neglecting the X derivatives in the diffusion term according to the second assumption, the dimensionless continuity equation, momentum equations and energy equation can be obtained as follows.

Continuity equation:

$$\frac{1}{r}\frac{\partial(rU_r)}{\partial r} + \frac{1}{r}\frac{\partial U_\phi}{\partial \phi} + \frac{\partial U_{x'}}{\partial X} = 0 \tag{10}$$

Radial momentum equation:

$$\frac{1}{r}\frac{\partial(rU_rU_r)}{\partial r} + \frac{1}{r}\frac{\partial(U_\phi U_r)}{\partial \phi} + \frac{\partial(U_{x'}U_r)}{\partial X}$$

$$= -\frac{\partial P}{\partial r} + \frac{(U_\phi + \tau r U_{x'})^2}{r} + \frac{2}{Re}(\Delta U_r - \frac{U_r}{r^2} - \frac{2}{r^2}\frac{\partial U_\phi}{\partial \phi} - \frac{2\tau}{r}\frac{\partial U_{x'}}{\partial \phi}) \tag{11}$$

Momentum equation in ϕ:

$$\frac{1}{r}\frac{\partial(rU_\phi U_r)}{\partial r} + \frac{1}{r}\frac{\partial(U_\phi U_\phi)}{\partial \phi} + \frac{\partial(U_{x'}U_\phi)}{\partial X}$$

$$= -\frac{G}{r}\frac{\partial P}{\partial \phi} + \tau r\frac{dP_m}{dX} - \frac{U_r U_\phi}{r} - 2\tau U_r U_{x'} + \frac{2}{Re}(\Delta U_\phi - \frac{U_\phi}{r^2} + \frac{2}{r^2}\frac{\partial U_r}{\partial \phi} - 2\tau\frac{\partial U_{x'}}{\partial r}) \tag{12}$$

Momentum equation in x:

$$\frac{1}{r}\frac{\partial(rU_{x'}U_r)}{\partial r} + \frac{1}{r}\frac{\partial(U_{x'}U_\phi)}{\partial \phi} + \frac{\partial(U_{x'}U_{x'})}{\partial X}$$

$$= -\frac{dP_m}{dX} + \tau\frac{\partial p}{\partial \phi} - \frac{U_r U_\phi}{r} - 2\tau U_r U_{x'} + \frac{2}{Re}\Delta U_{x'} \tag{13}$$

Energy equation:

$$\frac{1}{r}\frac{\partial(rTU_r)}{\partial r} + \frac{1}{r}\frac{\partial(TU_\phi)}{\partial \phi} + \frac{\partial(TU_{x'})}{\partial X} = \frac{2}{Re\,Pr}\Delta T \tag{14}$$

In the above equations, Δ is the operator

$$\Delta = \frac{1}{r}\frac{\partial}{\partial r}(r\frac{\partial}{\partial r}) + \frac{G}{r^2}\frac{\partial^2}{\partial \phi^2}, \tag{15}$$

where $G = 1 + \tau^2 r^2$, and $U_{x'}$ denotes $Ux/G^{0.5}$, that is equivalent to the x' component of velocity.

The analysis domain is limited to a half space of the tube divided by the twisted tape because of symmetry. The boundary condition for the momentum equation is that the velocities on the all boundaries are zero. It is assumed that at the inlet the flow has the same velocity distribution as the fully developed flow in a tube with an untwisted tape, that the tube wall is uniformly heated, but that the twisted tape is thermally insulated. The boundary conditions for temperature are $(\partial T/\partial r)_{r=1} = RePr/2$ on tube wall and $(\partial T/\partial \phi)_{tape} = 0$ on the twisted tape.

The above differential equations are discretized into the finite difference equations by using the control volume method. The momentum and energy equations are marchingly solved in x direction, while the distributions of velocity and temperature are implicitly obtained from the iterative calculation in the cross-sectional area of the tube at every marching step. The pressure distribution coupled with the secondary flow velocities, U_r and U_ϕ, is determined according to the so-called SIMPLE algorithm [8] so that the local continuity condition of eq. (10) can be satisfied. The longitudinal pressure gradient dP_m/dX is obtained from the continuity of bulk flow.

3. NUMERICAL RESULTS AND DISCUSSION

Numerical analyses are performed for Pr=1. Figure 2 shows a result for the fully developed region, where $Re = 1000$ and the parameter of the torsion τ is 0.06. The left figure (a) shows the contours of the dimensionless axial velocity $U_{x'}$ and the central figure (b) shows the contours of the temperature, which is the normalized temperature difference $(T_{w(\phi=0)} - T)/(T_{w(\phi=0)} - T_m)$, where T_m and $T_{w(\phi=0)}$ are the bulk temperature and the wall temperature at $\phi = 0$, respectively. The right figure (c) shows the velocity field of the secondary flow. The counter-clockwise peripheral arrows indicate the direction of twist of the tape which locates on the horizontal diameter. In Fig. 2 the torsion of tape is so small that the distributions of velocity and temperature are slightly modified from those for an untwisted tape ($\tau = 0$) which are not shown here. In figure (c), there is only one clockwise vortex, and its strength is quite weak. The maximum velocity of the secondary flow is about 4% of the averaged axial velocity. As τ increases, the location of the maximum of $U_{x'}$ rotates clockwise towards $\phi = 0$, and the location of the maximum temperature rotates similarly. Figure 3 shows a numerical result for $\tau = 0.3$, that is, the torsion of tape is larger than that of Fig. 2. Two vortices are found in the cross-section of the tube, and separated near the line of $\phi = 2\pi/3$, as shown in figure (c). The second term of the right-hand side of Eq. (11) represents the contribution of the centrifugal force owing to the torsion of the twisted tape. The centrifugal force generates the radial flow from tube center towards the tube wall, and hence raise the pressure at the tube wall. But this pressure rise becomes less toward the twisted tape. Namely, the fluid is shot away towards the tube wall and turned back through the region close to the twisted tape. These typical circulations produce a pair of vortices.

Figure 4 shows the numerical results for $Re = 1000$ and $\tau = 0.7$ at $X = 600$. The location of the maximum velocity is nearer to $\phi = 0$ than that in the case of $\tau = 0.3$. Near the tube wall the profile of the contours of velocity is quite different from that in Fig. 3. There are the portions where the contours push out into the inner area, and the fluid drifts away from the tube wall towards the inner area. This deformation of the contours means that small vortices appear near the tube wall. As a result, a main pair of vortices induced due to the centrifugal force shown in Fig. 3 seems to be accompanied with some additional small vortices (see Fig. 9).

The velocities and the temperature shown in Fig. 4 are not fully developed yet. For $\tau = 0.7$ the velocity and the temperature fields are no longer invariant with X although numerical analysis is made to sufficient downstream of X=1000.

Figure 5 shows the variation of the peripherally averaged Nusselt number Nu against the reduced distance of $X(= X/(2RePr))$ with some values of τ as a parameter, where Re=1000 and Pr=1. The averaged Nusselt number is defined based on the peripherally mean wall temperature. For small τ values ($\tau < 0.4$), the Nusselt number decreases as X increases and becomes independent of X in the range of $X/(2RePr) > 0.1$. However, Nu varies with X unperiodically for $\tau \geq 0.4$ in the same range. Nu irregularly oscillates seriously for $\tau = 0.7$ since the temperature as well as the velocity fluctuate. In the case where τ is large, the fully developed state cannot be found, but the state in the range of $X/(2RePr) > 0.1$ is called a quasi-developed one since Nu can be regarded to fluctuate around an constant value.

The calculated Nusselt numbers in the fully developed or the quasi-developed region are plotted against a parameter $2\tau Re/\pi$ in Fig. 6. The parameter was introduced by Hong and Bergles [4] to correlate their experimental data. They proposed the following empirical correlation, and reported that the standard deviation of the data was 16.4 percent.

$$Nu = 5.172[1 + 5.484 \times 10^{-3} Pr^{0.7}(2\tau Re/\pi)^{1.25}]^{0.5} \qquad (16)$$

The solid line in Fig. 6 indicates the above equation. The dotted line shows the numerical result obtained by Date. In the range where $2\tau Re/\pi$ is large, that is, the quasi-developed state is established, the amplitude of the oscillation of Nu is shown by error bars. Our numerical results, though restricted to Pr=1, are in good agreement with the empirical correlation. It is not easy to point out the reasons why our numerical results have been different from those by Date. But one reason may be as follows. In his basic equations the velocity components were not described as the components of the twisted cylindrical coordinates but as those measured in the ordinary cylindrical coordinate system where the torsion did not exist. Accordingly, their peripheral convection terms were expressed by using $U_\phi + \tau r U_{x'}$, while U_ϕ alone appears in our terms as shown in eqs. (11) through (13). Since $\tau r U_{x'}$ is much larger than U_ϕ, the momentum transport in which U_ϕ concerns will have been larger in their equations than that in ours. It seems that their analyses were more unstable since the nonlinear effect of the convection term was larger.

In the quasi-developed flow region the axial pressure gradient dP_m/dX varies with X in a manner similar to Nu does. Figure 7 shows the unperiodical variation of dP_m/dX in the range of $100 \leq X \leq 1000$ for $\tau = 0.4$. In order to find the relation between the velocity variation and that of the pressure gradient in this range, the pairs of these two values are plotted in Fig. (8) to draw a trajectory on the two-phase surface. These variations are investigated at the position of $r = 0.81, \phi = \pi/2$ in the cross-section of the tube. The trajectory is consist of a series of the closed cycles that are similar each other but never the same. When τ is less than 0.38, the trajectory is attracted by only one point, that is, the point attractor since dP_m/dX and $U_{x'}$ are independent of X. For $\tau = 0.4$, the trajectory is never the point attractor nor the cyclic one, but seems to be a strange attractor. It is found that the fluctuation caused by the velocity variation is chaotic for $\tau \geq 0.4$.

4. CONCLUSION

(1) Heat transfer to the laminar flow through the straight tube containing a twisted-tape swirler is analyzed by numerically solving the basic equations which are described in the twisted cylindrical coordinate system. The result is in good agreement with the experimental correlation by other researchers.

(2) When the torsion of the tape is large, the heat transfer coefficient fluctuates unperiodically along the tube axis to the sufficient downstream. It is found by examining the trajectory of the velocity versus the pressure gradient that the fluctuation is induced due to the chaotic fluid motion.

References

[1] Smithberg, E. and Landis, F., 1964, Trans. of ASME, Ser. C, 86-1, pp. 39-49.

[2] Thorsen, R. and Landis, F., 1968, Trans. of ASME, Ser. C, 90-1, pp. 87-97, .

[3] Lopina, R.F. and Bergles, A.E., 1969, Trans. ASME, Ser. C, 91-3, pp. 434-442.

[4] Hong, S. W. and Bergles, A. E., 1976, Trans. of ASME, J. Heat Trans.,
pp. 251-256.

[5] Nakayama, W., 1982, Proc. 7th Int. Heat Trans. Conf. München, vol. 1,
pp.223-240.

[6] Date, A. W. and Singham, J. R., 1972, ASME Paper No. 72-HT-17.

[7] Date, A. W., 1974, Int. J. Heat Mass Trans., vol. 17, pp. 845-859.

[8] Patanker, S. V., 1980, Numerical Heat Transfer and Fluid Flow,
McGraw-Hill Publishing, pp. 126-129.

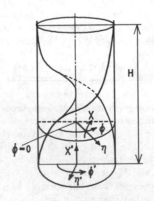

Fig. 1 The co-ordinate system

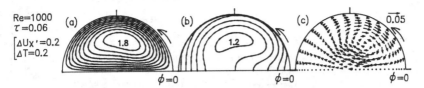

Fig. 2 (a) Contours of axial velocity, (b) Contours of temperature and
(c) Velocity of secondary flow

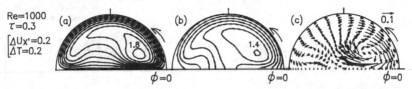

Fig. 3 (a) Contours of axial velocity, (b) Contours of temperature and
(c) Velocity of secondary flow

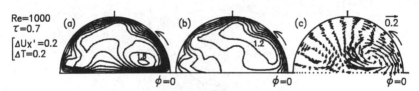

Fig. 4 (a) Contours of axial velocity, (b) Contours of temperature and
(c) Velocity of secondary flow (refer to Fig. 9)

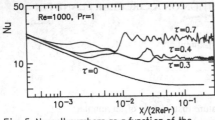

Fig. 5 Nusselt numbers as a function of the reduced axial length.

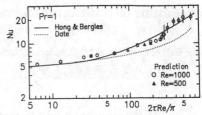

Fig. 6 A comparison of present predictions with the existing correlation

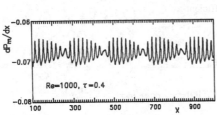

Fig. 7 An unperiodic variation of the axial pressure gradient

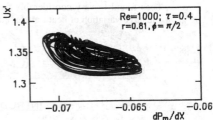

Fig. 8 A chaos-like trajectory on the two phase surface of the axial velocity versus the axial pressure gradient

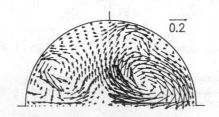

Fig. 9 A more detail of Fig. 4 (c)

LOCAL STUDY OF TEMPERATURE DISTRIBUTION ON A CIRCULAR PLANE FIN

R. Le Gall *, C. Bougriou*, J.C. Solecki* , and M. Laurent **

Measurements of the local fin temperature distribution of an annular finned tube in crossflow are described. We used the infrared thermography technique which proved to be useful for the investigation of the main factors influencing heat transfer. The scanned fin could be placed at the 1st , 5th , or 9th row for either in-line or staggered tubes bundle arrangements. Local temperatures have been found to vary considerably over the fin surface, resulting from the large variation of heat transfer coefficient.

INTRODUCTION

Due to the complex fluid flow pattern over the external surface of a typical finned tube heat exchanger, the analytical solution of the temperature distribution based upon a uniform heat transfer coefficient is often inadequate. The aim of this study is to obtain some detailed information on the actual temperature distribution for different gas-flow conditions. The development of a more efficient secondary surface could result in a considerable reduction in heat exchanger size and cost.

EXPERIMENTAL LOOP

The test section was incorporated in an open wind tunnel facility using air at atmospheric pressure and room temperature (see fig.1). Air is drawn through the working section by a fan. The air velocity is measured with a mass flowmeter.

(*)   GRETh   CEN Grenoble BP 85 X 38041 GRENOBLE Cedex FRANCE
(**) LPI   INSA Lyon 69621 VILLEURBANNE Cedex FRANCE

Test section

We used a model bank of unheated annular finned tubes to perform the flow pattern conditions. This consisted of nine transverse rows which could reproduce either a staggered or an in-line bundle (see fig.2). Each row had four complete finned tubes (or three and one pair of half-tubes). The finned tube geometry is described in the following table :

Base Tube Diameter	25 mm
Fin Tip Diameter	45 mm
Height of Fin	10 mm
Fin Thickness	1.5 mm
Width between Fins	8.5 mm
Number of Fins per meter	100
Fin Tube Height	51.5 mm
Transverse Pitch	50 mm
Longitudinal Pitch	50 mm

The air velocity in the minimal transverse free-flow area ranged from about 1 to 18 ms^{-1} , which yields a Reynolds number (based upon base tube diameter) between 2000 and 30000. The infrared visualization is made through one window of fluorspar (Caf2), facing the test tube, which is transparent to infrared radiation.

Test tube

The heated tube (see fig.3) was made up of three machined stainless steel fins and two fluorspar (Caf2) fins. This allowed us to perform our thermographic measurements on the middle fin, through the duct's window and the two fluorspar fins. Since these two fins reproduce exactly the local hydraulic conditions, the temperature field of the middle integral fin is representative of those encountered in common industrial heat exchangers. For the heat supply, we used a hot water circulation (about 60°c) inside the finned tube. Unfortunately the temperature difference between the entry and the outlet was too small to allow us some accurate heat transfer balances. The heated tube could be located in the middle place of either the 1st , 5th , or 9th row for the investigation of the row number effect.

Infrared measuring technic

We used a AGA SW 782 infrared camera which is sensitive to short wave length between 3 and 5.6 µm. We calibrated our camera with a laboratory black body in exactly the same measuring conditions (distance, window and fins of fluorspar). All infrared radiation densities were corrected by the fin surface emissivity of 0.96 (black paint of constant emissivity). All infrared pictures (100*280 pixels) were directly transfered to the associated micro-computer.

With our measuring conditions, we obtained more than 2 temperature readings per mm² of the scanned surface.

EXPERIMENTAL RESULTS

For the direct comparison of our different experimental results, and for the possible extrapolation to an outer temperature range, we defined a dimensionless temperature as :

$$\theta = \frac{T - T_{air}}{T_{wat} - T_{air}}$$ with :

T	Measured temperature
T_{air}	Air temperature
T_{wat}	Water temperature

Each point on the fin is precisely located with its associated radius position R and angular position α (see fig.4). A direct interpretation of the thermographic pictures gave us the dimensionless temperature distribution of the fin. Figures 5 and 6 show typical dimensionless temperature profiles for the same case at respectively, constant angular positions, and radius positions. From considerations of the radial temperature distribution (see fig.5), it is obvious that a single temperature distribution can not represent the performance of the fin. There is an important drop of temperature from root to tip of the fin, showing that the secondary surface was not running very efficiently. Figure 6 shows that the heat did not flow only radially and that there was a general migration of heat from the upstream portions to the rear portions of the fin. This tendency seemed to be valid throughout our wide range of Reynolds numbers.

Flow field considerations

The interaction between the developing flat plate boundary layer upon the fin and the stagnation flow on the circular cylinder leads to a system of horseshoe vortices described in (3), (4), (5), and (7). The main factors influencing the shape of these vortices are :
- the Reynolds number,
- the upstream turbulence,
- the fin spacing and shape,
- the bundle geometry.

The region affected by the horseshoe vortex is a region of enhanced heat transfer by comparison with the wake region. The fluid-dynamic features of these vortices is so complex, especially for a bundle of tubes, that we chose to perform a parametric study of some of these main factors.

Reynolds number effect

As the Reynolds number increases, the fin gradually cool down as a result of the augmentation of local heat transfer coefficient. This point was obvious on thermographic pictures, as were the representations of the fin minimal, average, and maximal dimensionless temperatures versus Reynolds number (see fig.7 & 8). There is also an increasing difference between the upper curve (maximal temperature) and the middle one (average temperature) showing that the area of poor heat transfer in the wake of the tube tends to reduce (in percentage of the total fin area). This is due to a general rise in the level of turbulence and to some associated modifications of the flow pattern (1) & (7).

Bundle geometry effect

The thermographic pictures showed some important differences between staggered and in-line bundles. For the first row, the in-line geometry seemed to be slightly more effective (lower fin average dimensionless temperature). It proved that the downstream bundle geometry could affect the first row performance. For inner rows, this gradually tends to reverse and finally for the 9th row the staggered bundle was more effective than the in-line. This better suits the litterature's data (2) & (6) that attribute higher heat transfer results to the staggered bundles. The in-line geometry leads to an important by-pass flow between the tubes whereas the staggered arrangement provides spanwise mixing that help to reduce the low-velocity wake region.

Row number effect

Two facts immediatly pointed out of our experimental results :
- the row number effect was markedly higher for the staggered than for the in-line configuration (see fig.11 & 12),
- the staggered bundle became more effective for the downstream rows (see fig.11).

As it is described in (2), the enhancement caused by the horseshoe vortex became significant only for higher Reynolds numbers. There is no common agreement in literature upon the number of rows that is necessary to reach the asymptotic value of a bank (2) & (6). This could be an effect of the upstream turbulence and some different bundle geometries (transverse and longitudinal tube pitches) used by the authors.

Heat transfer coefficient

The wide fin temperature variations are due to the associated variation of the heat transfer rate round the fin. However, isolated fin temperature readings are not by themselves sufficient to estimate precisely the local heat transfer coefficients. The use of such a thermographic technique associated with some transient temperature measurements during a step change of, either gas or water, temperature would lead to quantitative values of the local heat transfer coefficients (3) & (5). Most previous works involved with local heat transfer distributions used mass transfer analogy (4) & (7).

CONCLUSION AND DEVELOPMENTS

The high resolution temperature fields obtained are very useful for optimization of finned tube heat exchangers. Our experimental apparatus could apply to the investigation of some other important parameters : fin shape, fin pitch, fin surface modifications (turbulators), ...
The development of this sort of parametric study in terms of local heat transfer and fin's efficiency is clearly important and should be found in some further studies.

Acknowledgements

This research is financed in part by the Commmission of the European Communities within the frame of the Rational Use of Energy category of the JOULE R&D Programme.

All tests were carried out in the laboratories of GRETh (Groupement pour la Recherche sur les Echangeurs Thermiques) which we gratefully acknowledge.

References

(1) Neal, S.B.H.C. & Hitchcook, J.A., 1966, Proc. 3rd IHTC., CHICAGO, Vol 3, p 290.
(2) Webb, R.L., 1980, Heat Transfer Enginneering, Vol.1, n°3, p 33.
(3) Jones, T.V. & Russel, C.M.B., 1981, ASME Wint. Ann. Mtg., WASHINGTON DC, HTD-Vol 21, p 17.
(4) Goldstein, R.J. & Karni, J., 1984, ASME J.H.T., Vol 106, p 260.
(5) Ireland, P.T. & Jones, T.V., 1986, Proc. 8th IHTC., SAN FRANCISCO, Vol 3, p 975.
(6) Stasiulevicius, J. & Skrinska, A., 1988, Heat Transfer of finned tube bundles in crossflow, BERLIN, Springer-Verlag, 224 p.
(7) Schutz, G. & Kottke, V., 1989, Eurotherm Sem. n°9, HT in Single Phase Flows, p 17.

Nomenclature

T	Temperature	(°c)
θ	Dimensionless temperature	
R	Radius position	(mm)
α	Angular position	(°)
Re	Reynolds number (based upon base tube diameter)	

Subscripts

min	minimum
avg	average
max	maximum

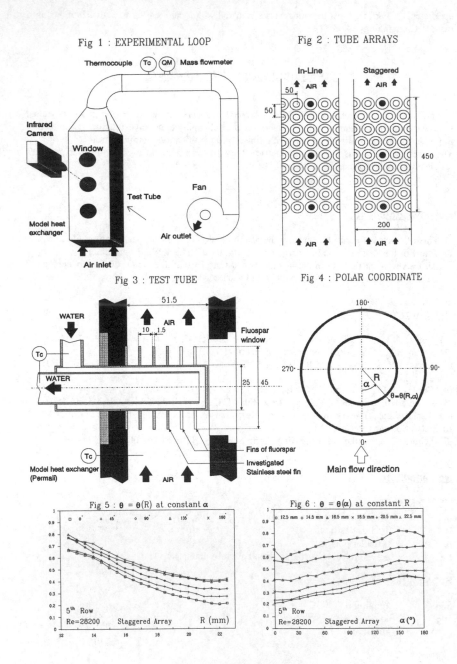

Fig 1 : EXPERIMENTAL LOOP

Fig 2 : TUBE ARRAYS

Fig 3 : TEST TUBE

Fig 4 : POLAR COORDINATE

Fig 5 : $\theta = \theta(R)$ at constant α

Fig 6 : $\theta = \theta(\alpha)$ at constant R

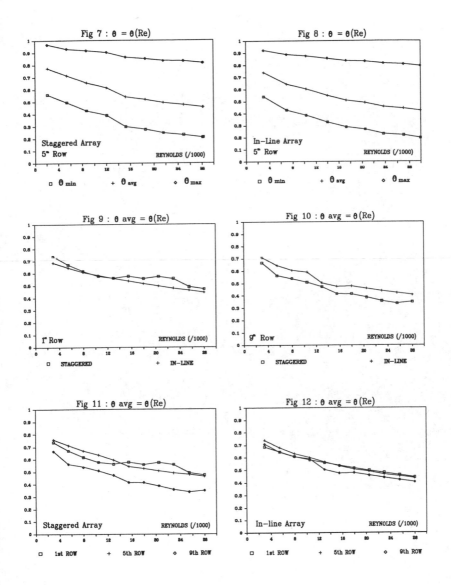

NUMERICAL ANALYSIS OF LAMINAR HEAT TRANSFER AROUND A SURFACE STEP

T. Ota * and Y. Kaga**
*Dept. of Machine Intelligence and Systems Eng.,Tohoku Univ., Sendai 980, Japan
**Sumitomo Metal Industries, Ltd., Kashima, Ibaraki 314, Japan

Two-dimensional laminar heat transfer and fluid flow around a surface step located in a channel with two parallel walls are numerically analyzed by employing the finite difference method to Navier-Stokes equations and energy one. Calculations were conducted on a wide range of parameters such as the step height–width ratio, the step height–channel height ratio, Reynolds number and Prandtl number. Clarified are the details of the local heat transfer characteristics along with the flow behaviors accompanying the separation and reattachment of flow. It is found that the local heat transfer distribution depends strongly upon Reynolds number and also Prandtl number. The average heat transfer coefficient exhibits a great variation with Prandtl number but only a slight dependency upon the width–height ratio of the step.

INTRODUCTION

Prediction of heat transfer and flow in the separated, reattached and redeveloped regions of incompressible or compressible fluid is very important in relation to many types of heat exchangers and to the improvement of their heat transfer performance, and there have been numerous experimental works [1][2]. Recent great advance of computer makes a numerical analysis on such a complicated heat transfer and flow problem possible and many papers have previously been published [3][4]. It has been noticed in these works that the flow structure in the separated and reattached regions is extremely complex, and the details of heat transfer mechanism are not clarified yet.

Recent manufacturing technology has capabilities to produce complicated heat transfer surfaces such as various extended surfaces in order to improve the heat transfer performance. On the other hand, the recent rapid development of electronic equipments such as computers brings about an increase of heat flux. Therefore advances of cooling technology in these electronic equipments are expected. A surface mounted step or a surface mounted rib in a parallel walled channel is a basic model simulating such flow situations. There have been several papers concerning its heat transfer problem. Hsieh et al. [5]–[7] reported numerical results on the flow and heat transfer around a surface step in an infinite flow. Experimental and numerical investigations have been reported on a single step or a number of steps [8]–[12]. In these previous studies, details of the heat transfer and flow around a surface step in a parallel walled channel are not clear yet, especially in case of narrow space.

The purpose of the present study is to predict numerically details of a two-dimensional laminar heat transfer and flow around a surface step located in a channel with two parallel walls. The finite difference method was employed to solve Navier-Stokes, energy and continuity equations.

FUNDAMENTAL EQUATIONS AND NUMERICAL PROCEDURES

Fundamental equations analyzed in the following are the continuity, momentum and energy equations for two-dimensional unsteady laminar flow of incompressible fluid with constant properties.

$$\nabla \cdot u = 0 \tag{1}$$

$$u_t + u \cdot \nabla u = -\nabla p + \Delta u / Re \tag{2}$$

$$\theta_t + u \cdot \nabla\theta = \Delta\theta/(Re \cdot Pr) \tag{3}$$

where fluid velocity u, pressure p, time t and space coordinates are nondimensionalized by u_{im}, ρu_{im}^2, h/u_{im} and h, respectively, and the non-dimensional temperature θ is defined as

$$\theta = (T - T_\infty)/(T_w - T_\infty) \qquad \text{for uniform wall temperature} \tag{4}$$

$$\theta = \lambda(T - T_\infty)/(q_w h) \qquad \text{for uniform heat flux} \tag{5}$$

Hereafter the symbol UWT or UHF is used to denote the uniform wall temperature condition or the uniform heat flux one for brevity respectively. Treated in the present study is a surface step located in a channel of two parallel walls as illustrated in figure 1.

The governing equations described above are expressed in curvilinear coordinates and their finite difference forms are obtained by the second order central difference for the space differential, the first order forward difference for the time differential, and the third order upwind difference for the convection terms. Resulting finite difference equations were solved using the time splitting method [13][14] based on MAC method. The steady state solution was determined as the converging one of the unsteady calculations. The divergence of the momentum equation (2) with the continuity one (1) leads to the Poisson equation for the pressure, and it was solved with the Tschebyscheff SOR method using the staggered grid.

The boundary conditions are as follows. The parabolic velocity profile at the inlet of channel and the no-slip condition at walls are assumed. The mean velocity at the inlet is u_{im}. As for the temperature, it is assumed to be uniform at the inlet T_∞, and only three side walls of the surface step to be heated at T_w in case of UWT or at q_w in case of UHF. Other parts of the channel walls except the step are adiabatic. At the outlet of the channel, the second order gradients are assumed to be zero for the velocity and also temperature. The pressure is fixed at the outlet and its first order gradient at the inlet is zero.

The solutions to the momentum and pressure equations are considered to converge when the magnitude of maximum relative change at any grid point is less than 10^{-5}, and that to the energy equation when it is less than 5×10^{-6} respectively. As to the thermal boundary condition on the adiabatic wall or the heat flux, the second order difference scheme on one side is employed.

NUMERICAL RESULTS AND DISCUSSION

Numerical calculations were conducted for a surface step located in a channel of two parallel walls as represented in figure 1. The channel height H was varied from $2h$ to $4h$, where h denotes the step height, and the side length of step w from h to $4h$. The Reynolds number based on u_{im} and h, $Re = u_{im}h/\nu$, is 100, 200, 300, and 400, and the Prandtl number 0.01, 0.1, 0.7, 2, 7, and 10.

Computational grids were generated by solving the Poisson equation and were altered from 121×31 to 136×31 depending on the channel height. The computational region extends from $15h$ upstream of the step to $30h \sim 70h$ downstream. In the preliminary numerical computations, several cases of computational region were examined and the above regions were confirmed to be satisfactory for simulating the heat transfer and flow around the surface step.

Present numerical results on the velocity profile for $Re = 96$, $w/h = 4$, and $H/h = 2$ are compared with measured ones by Tropea and Gackstatter [15] in figure 2. It is clear that the present numerical scheme simulates well the real flow characteristics. The reattachment length X_r/h is 11.9 in the present result and 11.6 in the experiment. In the present paper, the reattachment of flow is defined as a point of zero skin friction on the wall.

Shown in figure 3 is the effect of channel height upon the flow pattern around the surface step for $Re = 100$ and $w/h = 2$. The decrease of channel height brings about a large variation of flow ahead of the step and an increase of recirculating fluid mass. General flow characteristics downstream of the step are similar to those for the downstep and a variation of X_r/h with H/h is quite small, as shown in figure 4. It shows X_r increases almost linearly with H but its increasing rate is small.

The local Nusselt number distribution around the step is illustrated for $w/h = 2$, $Re = 100$, and $Pr = 0.7$ in figure 5, which shows the results for UWT condition along the step surface. In general, the decrease of H/h results in an increase of Nu since it brings about an acceleration of the flow around the step. Especially upstream of the front face, the flow approaches nearer to it and the thermal boundary layer there becomes thin resulting in an increase of Nu. On the other hand, there exists a separation bubble on the upper surface of the step for $H/h = 2$ as described previously. However its effect upon the local heat transfer is not clear in the figure. A high value of Nu on the rear side of the step for $H/h = 2$ may be originated from a large amount of the recirculating fluid. The local Nusselt number distribution for UHF condition is calculated, but the general characteristics are almost the same as those for UWT condition as shown in figure 5.

Effects of the streamwise length of the step w upon the heat transfer and flow behaviors are shown in the following. It is found that non-dimensional step length w/h results in no essential change of the flow pattern. But the length of recirculation region decreases with an increase of w/h, approaching a value for the downstep, as shown in figure 6. The local Nusselt number distribution for UWT condition is clarified, but w/h brings about no visual change of Nu except on the rear face, which is in contact with the separation bubble and its length varies with w/h. It results in a small effect upon Nu there. Results for UHF condition show almost the same as those for UWT condition.

Effects of the Reynolds number are summarized in the following. The streamlines are illustrated in figure 7. A small separation bubble on the upstream corner diminishes with an increase of Re and disappears at $Re = 300$. In a paper by Davalath and Bayazitoglu [12] on the forced convection across three rectangular blocks, they reported there is no separation bubble on the bottom corner of the upstream block in a range of Re from 25 to 375. On the other hand, results by Achenbach [16] on an upward facing step show that the separation bubble ahead of the step increases its size as Re increases. It seems there are several problems to be considered about the grid generated near the corner. On the upper surface of the step, there is no separation bubble at $Re = 100$. However it appears at Re greater than 200 and the separated shear layer reattaches on to the upper surface at $Re = 200$ and 300. At $Re = 400$, a weak unsteady flow feature appears and then its time averaged results are included in figure 7. It is clear that the reattachment point on the channel wall shifts downstream with an increase of Re. In case of $Re = 400$, the shear layer separated near the upstream corner reattaches onto the downstream channel wall as forming only one separation bubble. The reattachment point moves with time and its streamwise range is about $0.3h$. The undulation of the flow pattern is slow and no vortex shedding is found. It is interesting to notice that the separation point on the upper surface of the step locates somewhat downstream of the upstream corner. A variation of the reattachment point on the downstream channel wall with Re is represented in figure 8. An increase of Re results in a large downstream shift of X_r.

The local Nusselt number distribution for $Pr = 0.7$, $w/h = 2$ and $H/h = 3$ is illustrated at various Reynolds numbers in figure 9. The flow pattern changes largely as shown in figure 7. It results in a complicated variation of Nu, especially on the upper surface where Nu decreases steeply with the streamwise distance from the corner and reaches a minimum near the separation point. At $Re = 200$ and 300, the separated shear layer reattaches to the upper surface of the step, as described previously. However there is no obvious peak of Nu there. On the other hand, Nu on the rear face of the step increases with Re, since the recirculation region grows up with Re. Results for UHF condition show qualitatively the same characteristics as those for UWT condition shown in figure 9.

Effects of the Prandtl number upon the heat transfer behaviors are represented in figure 10. An increase of Pr brings about a large increase of the local Nusselt number all over three sides. Figure 10 shows the results for UWT condition. Results for UHF condition are almost the same as those in figure 10. Nu attains a minimum for $Pr = 7$ and 10 on the upper surface where the surface skin friction becomes minimum. An increase of Pr over 2.0 results in an occurrence of maximum of Nu on the rear face. It may be originated from the convection in the recirculation region.

Mean Nusselt number along three surfaces of the step is represented in figure 11 through 13. Figure 11 shows a variation of Nu_{av} with w/h for $Re = 100$ and $Pr = 0.7$. In case of $H/h = 2$, Nu_{av} decreases linearly with w/h in a range of w/h from 1 to 4. However for $H/h = 3$ and 4, Nu_{av} decreases with w/h but its decreasing rate becomes small at large values of w/h. Illustrated in figure 12 is Nu_{av} as a function of Re. Nu_{av} increases with an increase of Re and it is clear that Nu_{av} for UWT is a little higher than that for UHF at $Re = 400$. Figure 13 demonstrates a variation of

Nu_{av} with Pr. Nu_{av} increases largely with Pr and at Pr greater than about 1.0, Nu_{av} for UHF is higher than that for UWT.

CONCLUDING REMARKS

Laminar heat transfer and flow around a surface step in a parallel walled channel were predicted numerically through solving the finite difference forms of Navier-Stokes and energy equations. Results obtained in the present analyses are as follows.

A decrease of the channel height brings about a large variation of the flow pattern around the step as resulting in an increase of the local Nusselt number.

Effects of the streamwise step length upon the flow and local Nusselt number around the step are generally small.

An increase of Reynolds number produces the separation of flow on the upper surface of the step, and the reattachment of the flow on it appears at Re smaller than 300. However at $Re = 400$, the shear layer separated from the upper surface reattaches onto the downstream channel wall as forming one separation bubble. These large changes of the flow pattern around the step with Re result in a large variation of the local Nusselt number distribution there.

The local Nusselt number increases with an increase of Prandtl number all over three sides of the step resulting in a steep increase of the mean Nusselt number.

Difference of the thermal boundary condition on the step surface brings no essential change of the local and mean heat transfer characteristics, though some differences are found at large Reynolds number or high Prandtl number.

The present authors express their sincere thanks to Prof. Daiguji, H. and Dr. Yamamoto, S. of the Tohoku University for their guidance on the numerical schemes in the calculations.

NOMENCLATURE

C_f : surface skin friction coefficient $= \tau_w/(\rho u_{im}^2/2)$

C_p : pressure coefficient $= P/(\rho u_{im}^2/2)$

h, H : step height, channel height

Nu : local Nusselt number $= \alpha h/\lambda$

Nu_{av} : mean Nusselt number

Pr : Prandtl number

q_w : heat flux

Re : Reynolds number $= u_{im}h/\nu$

P, T : pressure and temperature

T_∞ : uniform temperature at channel inlet

T_w : wall temperature

u : velocity vector

u_{im} : mean velocity at channel inlet

w : streamwise length of step

X_r : reattachment length

α : local heat transfer coefficient
$$= -\lambda(\partial T/\partial y)_w/(T_w - T_\infty) \quad \text{for UWT}$$
$$= q_w/(T_w - T_\infty) \quad \text{for UHF}$$

θ : non-dimensional temperature
$$= (T - T_\infty)/(T_w - T_\infty) \quad \text{for UWT}$$
$$= \lambda(T - T_\infty)/(q_w h) \quad \text{for UHF}$$

λ, ν, ρ : thermal conductivity, kinematic viscosity and density of fluid

REFERENCES

1. Aung, W., Separated forced convection, 1983, Proc. 1983, ASME-JSME Thermal Eng. Joint Conf., 2, 499-515.
2. Ota, T. and Nishiyama, H., A correlation of maximum turbulent heat transfer coefficient in reattachment flow region, 1987, Int. J. Heat Mass Transfer, 30, 1193-1200.
3. Anderson, D.A., Tannehill, J.C. and Pletcher, R.H., 1984, Computational Fluid Mechanics and Heat Transfer, Hemisphere.
4. Kondoh, T. and Nagano, Y., Computational study of laminar heat transfer behind backward-facing steps, 1988, Trans. JSME, 54, 1760-1767.
5. Hsieh, S.-S. and Huang, D.-Y., Numerical computation of laminar separated forced convection on surface-mounted ribs, 1987, Numerical Heat Transfer, 12, 335-348.
6. Hsieh, S.-S. and Huang, D.-Y., Flow characteristics of laminar separation on surface-mounted ribs, 1987, AIAA J., 25, 819-823.
7. Hsieh, S.-S., Shin, H.-J. and Hong, Y.-J., Laminar forced convection from surface-mounted ribs, 1990, Int. J. Heat Mass Transfer, 33, 1987-1999.
8. Yanagida, T., Nakayama, W. and Nemoto, T., Heat transfer from a longitudinal row of heat dissipating rectangular bodies standing on the wall of a cooling duct, 1984, Trans. JSME, 50, 1294-1301.
9. Igarashi, T. and Yamasaki, H., Fluid flow and heat transfer of a rectangular cylinder in the turbulent boundary layer on a plate, 1989, Trans. JSME, 55, 3157-3165.
10. Igarashi, T. and Takasaki, H., Fluid flow around three rectangular cylinders in a flat plate laminar boundary layer, 1989, Trans. JSME, 55, 3341-3348.
11. Aiba, S., Heat transfer around small square ribs mounted on an adiabatic plane channel, 1990, Wärme-und Stoffübertragung, 25, 85-91.
12. Davalath, J. and Bayazitoglu, Y., Forced convection cooling across rectangular blocks, 1987, Trans. ASME, J. Heat Transfer, 109, 321-328.
13. Daiguji, H. Ed., 1988, Fundamentals of Computational Fluid Dynamics, Corona Pub.
14. Daiguji, H. and Yamamoto, S., An efficient time-marching scheme for solving compressible Euler equations, 1986, Trans. JSME, 52, 248-254.
15. Tropea, C.D. and Gackstatter, R., The flow over two-dimensional surfacemounted obstacles at low Reynolds numbers, 1985, Trans. ASME, J. Fluids Eng., 107, 489-494.
16. Achenbach, E., Mass transfer downstream a backward or a forward-facing steps, 1990, Heat Transfer 1990, 5, 305-310.

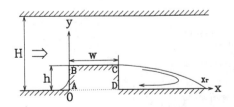

FIGURE 1. Surface step and coordinate system

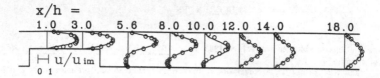

FIGURE 2. Comparison of velocity profile. $Re = 96$, $w/h = 4$, $H/h = 2$, — ; present, o ; Tropea et al.

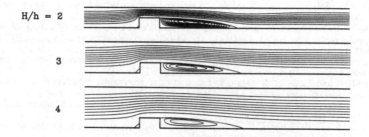

FIGURE 3. Effect of H/h on streamline. $Re = 100$, $w/h = 2$

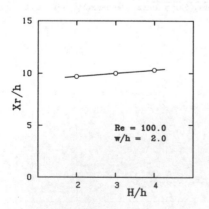

FIGURE 4. Variation of reattachment length with H/h

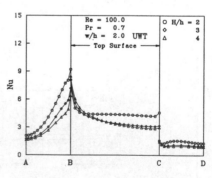

FIGURE 5. Local Nusselt number distribution for UWT

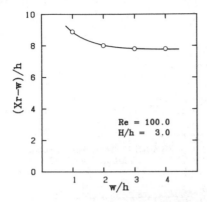

FIGURE 6. Variation of reattachment
length with w/h

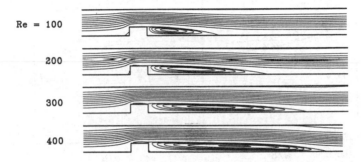

FIGURE 7. Effect of Re on streamline. $w/h = 2$, $H/h = 3$

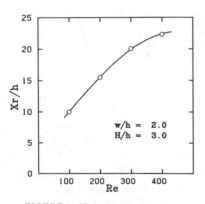

FIGURE 8. Variation of reattachment
length with Re

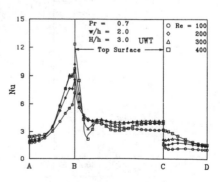

FIGURE 9. Local Nusselt number
distribution for UWT

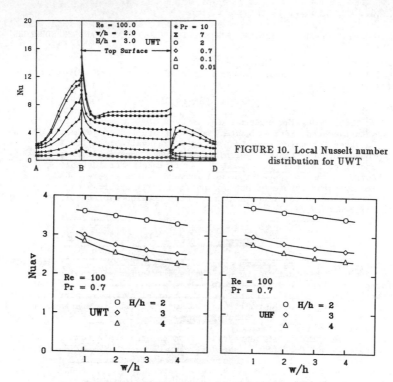

FIGURE 10. Local Nusselt number
distribution for UWT

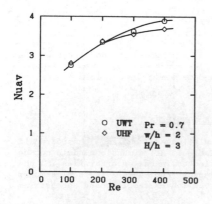
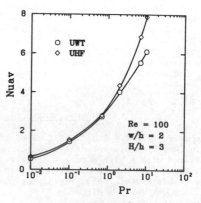

FIGURE 11. Variation of mean Nusselt number with w/h

FIGURE 12. Variation of mean
Nusselt number with Re

FIGURE 13. Variation of mean
Nusselt number with Pr

CORRELATIONS OF j - AND f - FACTORS FOR MULTILOUVERED HEAT TRANSFER SURFACES

Bengt Sunden and Jerker Svantesson
Chalmers University of Technology, Department of Thermo and Fluid Dynamics
41296 Göteborg, Sweden

Available correlations of the j - and f - factors for multilouvered heat transfer surfaces are tested for some surfaces for which experimental data are available. It is found that these correlations may overpredict or underpredict the j - and f - factors considerably. Correlations adjusted to each louvered surface were found to give good agreement but are not generally applicable. A new set of correlations have also been developed predicting satisfactory results.

INTRODUCTION

Multilouvered surfaces are used in compact heat exchangers, e.g. automotive heat exchangers, in order to enhance the heat transfer. The objective is to get a large and effective area per unit volume of the heat exchanger. At the design stage and for the evaluation of the thermal hydraulic performance of compact heat exchangers using multilouvered surfaces, it is important to have correlations of the j - (Colburn) and f - factors of satisfactory reliability. In the open literature two sets of correlations have been presented. These have been suggested (based on measured data) by Davenport (1) and Achaichia and Cowell (2), respectively. However, these correlations are not generally applicable although sometimes give acceptable prediction. In other cases unsatisfactory results are achieved. There is thus a demand for more general correlations. Manufacturing and developing companies may have their own correlations but are usually not released publicly.

In the present paper, the available correlations have been tested for some multilouvered surfaces for which experimental heat transfer and pressure drop data have been reported previously, see Svantesson and Sunden (3), Sunden and Svantesson (4) and Sunden and Svantesson (5). Simple correlations adjusted to each louvered surface have been developed with good agreement. A new successful set of correlations for the j - and f - factors was developed. The constants in these correlations were determined by multiple regression analysis.

LOUVERED SURFACES UNDER CONSIDERATION

The basic geometry and dimensions of the multilouvered surfaces considered in this work are presented in Figure 1 and Table 1. All dimensions could not be measured with a high accuracy. This is reflected in the uncertainties in the louver angles which are given within a range. From Table 1 it is clear that the dimensions of the surfaces, the number of louver sections and number of louvers per section differ.

CORRELATIONS BY DAVENPORT

The correlations for the j - and f - factors suggested in (1) are:
300 < Re < 4000

$$j = \frac{\alpha}{\rho c_p u_m} \cdot Pr^{2/3} = 0.249 \cdot Re^{-0.42} \cdot l_h^{0.33} \cdot (l_1/H)^{1.1} \cdot H^{0.26} \qquad (1)$$

70 < Re < 900

$$f = 5.47 \cdot Re^{-0.72} \cdot l_h^{0.37} \cdot (l_1/H)^{0.89} \cdot l_p^{0.2} \cdot H^{0.23} \qquad (2)$$

where H is the so-called fin height, l_h the louver height, l_1 the louver length and l_p the louver pitch. Re is the Reynolds number based on the louver pitch and the mean velocity through the minimum free flow area.

The pressure drop Δp is related to the friction factor by

$$\Delta p = f \cdot \frac{A_t}{A_f} \cdot \frac{\rho u_m^2}{2}$$

where A_t is the total air-side area, A_f the minimum free flow area, ρ the fluid density and u_m the mean velocity through the minimum free flow area.

CORRELATIONS BY ACHAICHIA AND COWELL

Achaichia and Cowell suggested correlations for the Stanton number (St) and the friction factor based on experiments, see (2). These correlations are (nomenclature as in Fig. 1):

150 < Re < 3000

$$St = j \cdot Pr^{-2/3} = 1.54 \cdot Re^{-0.57} \cdot (f_p/l_p)^{-0.19} \cdot (H/l_p)^{-0.11} \cdot (2l_h/l_p)^{0.15} \qquad (3)$$

150 < Re < 3000

$$f = 0.895 \cdot f_A^{1.07} \cdot f_p^{-0.22} \cdot l_p^{0.25} \cdot H^{0.26} \cdot (2l_h)^{0.33} \qquad (4)$$

$$f_A = 596 \cdot Re^{(0.318 \cdot \log Re - 2.25)} \qquad (5)$$

where f_p is the fin pitch.

All dimensions in equations (1) to (4) are in mm (millimeters).

CORRELATIONS ADJUSTED TO EACH LOUVERED SURFACE

In this work, correlations adjusted to the experimental data of each tested multilouvered surface were considered. These correlations are in the form:

$$j = a \cdot Re^b \tag{6}$$

$$f = c \cdot Re^d \tag{7}$$

where a, b, c and d are different for the various louvered surfaces and are determined by matching equations (6) and (7) to the experimental data. The values of a, b, c and d are given in (3) but are omitted here due to space limitation.

NEW CORRELATIONS

The available correlations, equations (1) - (4), do not take the depth of the heat exchanger core (L in Fig. 1) into account. Also the influence of the variables do not appear as ratios but have to be inserted in certain dimensions. The following correlations are suggested in this work

$$St = j \cdot Pr^{-2/3} = c_1 \cdot Re^{c_2} \cdot (f_p/l_p)^{c_3} \cdot (H/l_p)^{c_4} \cdot (l_h/l_p)^{c_5} \cdot (L/l_p)^{c_6} \tag{8}$$

$$f = b_1 \cdot Re^{b_2} \cdot (f_p/l_p)^{b_3} \cdot (H/l_p)^{b_4} \cdot (l_h/l_p)^{b_5} \cdot (L/l_p)^{b_6} \tag{9}$$

The constants c_1 to c_6 and b_1 to b_6 were determined by multiple regression analysis, see Draper and Smith (6), and are given in Table 2, see also Sunden (7).

TABLE 2: Constants c_1 to c_6 and b_1 to b_6 (95 % confidence)

$c_1 = 3.67$	$2.64 \le c_1 \le 5.10$	$b_1 = 9.2$	$6.9 \le b_1 \le 12.3$
$c_2 = -0.591$	$-0.62 \le c_2 \le -0.57$	$b_2 = -0.540$	$-0.561 \le b_2 \le -0.518$
$c_3 = 0.0206$	$-0.07 \le c_3 \le 0.11$	$b_3 = -0.022$	$-0.10 \le b_3 \le 0.05$
$c_4 = -0.285$	$-0.42 \le c_4 \le -0.15$	$b_4 = -1.085$	$-1.19 \le b_4 \le -0.98$
$c_5 = 0.0671$	$0.003 \le c_5 \le 0.13$	$b_5 = 0.067$	$0.01 \le b_5 \le 0.12$
$c_6 = -0.243$	$-0.32 \le c_6 \le -0.17$	$b_6 = 0.310$	$0.25 \le b_6 \le 0.37$

RESULTS

Figures 2 (j - factors) and 3 (friction factors) show our experimental data for six multilouvered surfaces compared with the correlations of Davenport and Achaichia and Cowell. It is obvious that these correlations do not fit our data very well.

In Figures 2 and 3 are also the results of the new correlations (8) and (9) provided. Excellent agreement has been achieved and it is likely that these new correlations are of more general applicability. To extend their usage, it would be preferable to determine the constants c_1 to c_6 and b_1 to b_6 by matching to a much larger data base.

CONCLUSION

New correlations, including the heat exchanger depth, were suggested and proved to be satisfactory. The constants in these new correlations were obtained by multiple regression analysis.

ACKNOWLEDGEMENT

The present project was financially supported by AB Volvo and the former National Swedish Board for Technical Development (STU).

REFERENCES

1. DAVENPORT, C.J., 1983 Correlations for Heat Transfer and Flow Friction Characteristics of Louvred Fin, AIChE Symp. No. 225, Vol. 79, pp. 19-27.
2. ACHAICHIA, A. and COWELL, T., 1988 Heat Transfer and Pressure Drop Characteristics of Flat Tube and Louvered Plate Fin Surface, Experimental Thermal and Fluid Science, Vol. 1, pp. 147-157.
3. SVANTESSON, J. and SUNDEN, B., 1990 Heat Transfer and Pressure Drop from Six Standard Multilouvered Surfaces, Publ. No. 90/7, Chalmers University of Technology, Department of Thermo and Fluid Dynamics, Göteborg.
4. SUNDEN, B. and SVANTESSON, J., 1992 Thermal Hydraulic Performance of Some Standard Multilouvered Surfaces, Int. J. of Heat and Technology, accepted for publication.
5. SUNDEN, B. and SVANTESSON, J., 1991 Heat Transfer and Pressure Drop from Louvered Surfaces in Automotive Heat Exchangers, Exp. Heat Transfer, Vol. 4, pp. 111-125, 1991.
6. DRAPER, M.R. and SMITH, H., 1966 Applied Regression Analysis, J. Wiley & Sons.
7. SUNDEN, B., 1991 Correlations of the Stanton number and friction factor for the air-side of automotive compact heat exchangers, Chalmers University of Technology, Department of Thermo and Fluid Dynamics, Göteborg (in Swedish).

NOMENCLATURE

a	constant
A_f	minimum free flow area
A_t	total air-side area
b	constant
b	louver length in cross-plane
b_1 to b_6	constants
c	constant
d	constant
c_1 to c_6	constants
f	friction factor
f	function
f_A	fin pitch, see Fig. 1
H_p	fin height, see Fig. 1
j	Colburn factor = $St \cdot Pr^{2/3} = \dfrac{\alpha}{\rho c_p u_m} \cdot Pr^{2/3}$
L	core length, heat exchanger depth, see Fig. 1
l_h	louver height, see Fig. 1
l_p	louver pitch, see Fig. 1
NL	number of louvers

NS   number of louver sets
Pr   Prandtl number
Re   Reynolds number $= u_m l_p/\nu$
St   Stanton number $= j \cdot Pr^{2/3}$
t   fin thickness, see Fig. 1
u_m mean velocity through the minimum free flow area

Greek symbols

α heat transfer coefficent
Δp pressure drop
θ louver angle
ν kinematic viscosity
ρ density

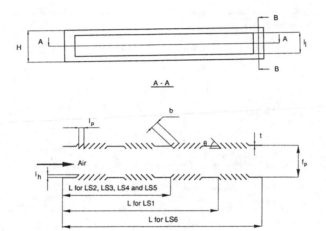

Figure 1. Basic geometry of the considered louvered surfaces.

Table 1. Dimensions of the louvered surfaces.

	L	H	l_h	l_1	l_p	b	t	θ	f_p	f_p/l_p	NL	NS
	mm	mm	mm	mm	mm	mm	mm	o	mm			
LS1	57.4	12.5	0.16	10.2	1.4	1.5	0.06	18-26	1.5	1.1	10	3
LS2	57.4	12.4	0.38	10.3	1.4	1.5	0.06	14-23	2.0	1.4	17	2
LS3	37.0	12.4	0.26	10.0	1.3	1.3	0.06	23-26	2.0	1.5	11	2
LS4	37.0	8.6	0.19	6.8	1.2	1.1	0.04	21-27	1.8	1.5	13	2
LS5	50.0	9.6	0.25	6.8	1.1	1.1	0.06	21-30	1.8	1.6	19	2
LS6	47.8	8.0	0.18	5.0	0.5	0.8	0.04	23-34	1.9	3.8	12	4

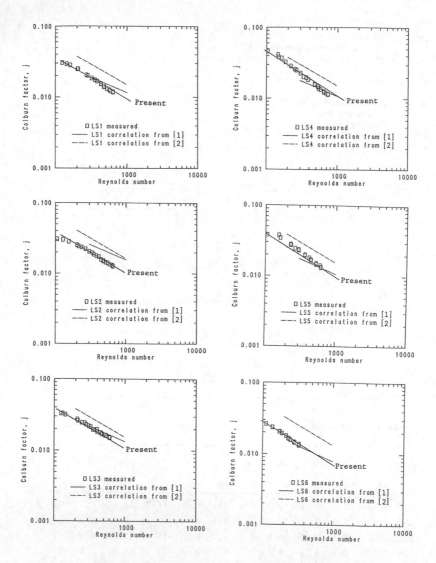

Figure 2. Colburn factor j vs Re for six different louvered surfaces.
Experimental data and correlations.

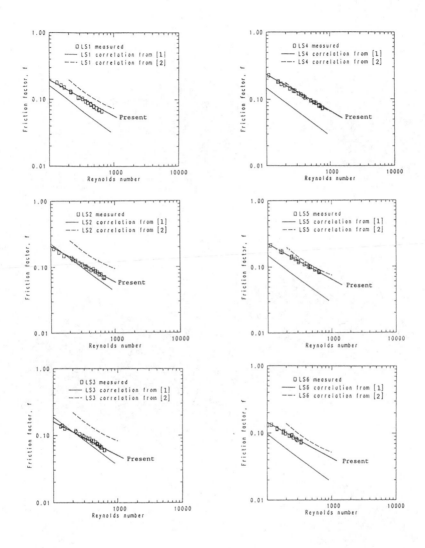

Figure 3. Friction factor vs Re for six different louvered surfaces.
Experimental data and correlations.

CALIBRATION OF A PHOTO–EVAPORATIVE MASS TRANSFER MEASUREMENT TECHNIQUE AND ITS APPLICATION

T. Pekdemir*, T.W. Davies*, and P. Flohr[†]
* School of Engineering, University of Exeter, North Park Rd., Exeter EX4 4QF
[†] Faculty of Aerospace Engineering, University of Stuttgart, Germany

A convective mass transfer measurement technique based on the infra–red reflective properties of drying surfaces has been validated. The convection surface of interest is covered with chromatographic paper and wetted with 1–butanol. The drying rate was determined from measurements of changes in reflected infra–red light intensity using an infra–red emittor/detector. The technique was tested by calibration in four different convection systems. The technique was then used to determine the circumferential distribution of mass transfer rate for circular cylinders in crossflow.

INTRODUCTION

When direct heat transfer measurements are difficult, mass transfer analogues can often provide useful information e.g. the electrochemical technique (1) and the naphthalene sublimation technique (2). Recently a photo–evaporative technique was described by Utton and Sheppard (3), and Davies et al. (4, 5) applied this technique in a study of the distribution of local mass transfer rates in obstructed channels and in the impingement zone of a free air jet. These studies used visible light for illumination and a photodiode for measuring the intensity of light transmitted through the paper covered convection surface. Difficulties in controlling ambient light conditions and variability in calibration technique produced an unsatisfactory range of paper calibration constants for essentially similar systems. In the present study the reflection of infra–red light was used and two basically different kinds of calibration experiments were used to establish confidence in the value of the crucial paper calibration constant.

THE PHOTO–EVAPORATIVE TECHNIQUE

The principal of the technique lies in the change in light reflection or transmission of the absorbent paper with its liquid content. Thus a completely wetted paper is more transparent than a completely dry one and, given the right paper, the change in transmitted or reflected light intensity with degree of wetness is linear. Certain blotting and chromatographic papers have been found to possess such a characteristic. The ideal, parallel, changes in reflected light intensity and in liquid content of the absorbent paper are shown in Figure 1. A typical variation in measured reflection intensity is shown in Figure 2. Referring to Figure 1 (a) the mass flux N $(kgm^{-2}s^{-1})$ during the constant–rate drying period may be combined with the definition of a mass transfer coefficient giving:-

$$N = k_m(\rho_s - \rho_\infty) = \frac{dm}{dt}/A = \left[\frac{(m_4 - m_3)}{(t_2 - t_1)}\right]/A \tag{1}$$

where A is the area of the evaporating surface, k_m mass transfer coefficient, ρ_s vapour density of the solvent at the surface condition and ρ_∞ ($\simeq 0$, because the concentration of the solvent in the free stream equals 0) that in the free stream.

If the output of the transducer is linear during the constant drying rate period then:-

$$\frac{dI}{dt} = \frac{(I_4 - I_3)}{(t_2 - t_1)} \tag{2}$$

In order to eliminate effects caused by possible changes in the intensity of infra–red light emitted by the transducer between single measurements or different experimental setups, the intensity difference $(I_4 - I_3)$ is normalised with respect to the overall intensity change $I_o = I_1 - I_2$.

Since, during the constant drying rate period, the mass flux is also a linear function of time we may combine eqns. (1) and (2) to give:-

$$N = \frac{dm}{dt}/A = c\left[\frac{dI}{dt}/I_o\right] \tag{3}$$

where c (kgm^{-2}) is the calibration constant of the paper and represents the liquid holding capacity of the absorbent paper.

Once the paper has been calibrated, local or point measurements of evaporation rate at a paper covered convection surface may be made by simply measuring I_o, $(I_4 - I_3)$ and $(t_2 - t_1)$.

CALIBRATION OF THE PAPER

The technique was calibrated directly and indirectly using four different convection systems, namely; stagnation flow on a cylinder, parallel flow over a flat plate, free convection at a horizontal flate plate and free convetion at a vertical plate.

Stagnation flow on a cylinder (indirect method)

The laminar stagnation flow on a circular cylinder in crossflow has been widely studied and numerous mass and/or heat transfer correlations have been proposed in the literature. Therefore the technique may be calibrated using one of these well established correlations to give k_m as a function of flow velocity and relating this value to the temporal change in I at the forward stagnation point.

For the heat transfer at the stagnation point Van Driest (6) recommends the correlation:-

$$Nu = 1.14Re^{0.5}Pr^{0.367} \tag{4}$$

Using the analogy between heat and mass transfer, the Sherwood number $(\frac{k_m D}{D_{AB}})$ may be correlated by:-

$$Sh = 1.14Re^{0.5}Sc^{0.367} \tag{5}$$

Experimental studies such as that by Kestin and Wood (7) have confirmed this equation. Inserting physical properties and the definition of k_m from eqn (1) into eqn (5) gives:-

$$\frac{ND}{\rho_s D_{AB}} = 1.14(\frac{u_\infty D}{\nu})^{0.5}(\frac{\nu}{D_{AB}})^{0.367} \tag{6}$$

where D is the diameter of the cylinder, u_∞ and ν velocity and kinematic viscosity of the free stream, and D_{AB} the mass diffusion coefficient of the solvent in air, given by:-

$$D_{AB} = D_o\left\{\frac{T}{T_o}\right\}^2\left\{\frac{P_o}{P}\right\} \tag{7}$$

where D_o is the diffusion coefficient at standard conditions ($T_o = 273.15 \,^\circ K$, $P_o = 1$ atm.) and equals $7.03\times10^{-6}m^2s^{-1}$ for 1-butanol.

Combining eqn. (3) with eqn (6) gives:-

$$c = 1.14\frac{\rho_s D_{AB}}{D}(\frac{u_\infty D}{\nu})^{0.5}(\frac{\nu}{D_{AB}})^{0.367}(I_o\frac{\Delta t}{\Delta I}) \tag{8}$$

Three factors were expected to have an effect on this equation, namely; surface roughness of the cylinder, free stream turbulence and blockage effects in the calibration duct. Achenbach (8) experimentally showed that surface roughness had negligible or no effect on the heat transfer from the stagnation point at a circular cylinder in crossflow.

The experimental study of Kestin et al. (9) showed that the stagnation point heat transfer is highly sensitive to free stream turbulence. However, Lowery and Vachan (10) showed that free stream turbulence up to a level of 0.75% had no significant effect. The turbulence in the wind tunnel used in the present study was about 0.6 %. Therefore the effect of the free stream turbulence was neglected.

The blockage is characterised by the ratio of the diameter of the cylinder D to the height of the duct H (i.e D/H). The blockage ratios in the present investigation were 0.15 and 0.30 for small cylinder and large cylinder respectively. Zhukauskas (11) showed that stagnation point heat transfer is affected by the blockage ratio (D/H) and using his correlation the lead constant in eqn (8) was set at 1.17 and 1.19 for the small cylinder and the large cylinder respectively.

Flat plate in parallel flow (indirect method)

For a flat plate in parallel flow the well–known Sherwood number correlation is:-

$$Sh = 0.0296 Re_x^{4/5} Sc^{1/3} \tag{9}$$

So the equation for the calculation of the calibration constant takes the same form as that used for the stagnation flow case:-

$$c = 0.0296 \frac{\rho_s D_{AB}}{x} \left(\frac{u_\infty x}{\nu}\right)^{4/5} \left(\frac{\nu}{D_{AB}}\right)^{1/3} (I_o \frac{\Delta t}{\Delta I}) \tag{10}$$

where now x is the distance from the leading edge of the flat plate on which measurements were made.

Free convection from a horizontal plate (direct method)

For natural convection mass flux measurements involving vapour heavier than air the downward facing surface of a horizontal flat plate is used as the convection surface in order to approach a homogeneous distribution of the local mass flux over the entire surface. Using eqn (3) the calibration constant may be calculated by directly measuring the total mass transfer rate with an electronic balance.

Free convection from a vertical plate (indirect method)

For the free convection from a vertical plate the Sherwood number is correlated in the form (12):-

$$Sh = a(Gr_{m,x} Sc)^b \tag{11}$$

Thus the calibration constant may be calculated using the following equation.

$$c = a\left(\frac{\rho_s D_{AB}}{x}\right) \left[\left(\frac{y \rho_s x^3}{\rho \nu^2}\right)\left(\frac{\nu}{D_{AB}}\right)\right]^b (I_o \frac{\Delta t}{\Delta I}) \tag{12}$$

where a and b are constants having the values of 0.387 and 0.25 respectively.

EXPERIMENTAL EQUIPMENT

In the case of forced convection, tests were carried out in a wind tunnel having a test section of width 0.305 m, height 0.305 m, and length 0.61 m. Eqn (8) was used to interpret the results obtained from measurements at the forward stagnation point of two cylinders at different flow velocities. The cylinders used were 0.05 m and 0.10 m in diameter and made of transparent Perspex tubes with a wall thickness of 4 mm. The transducer was fixed midway on the inner side of the tube. A thermocouple was placed near the transducer and underneath the paper through a small hole in the cylinder wall.

Eqn (10) was used to interpret the results of measurements made along a flat plate fixed across the midplane of the working section and 25 cm long. The position of the thermocouple and the transducer (underneath the plate) could be changed in the direction of the flow between 0.05 m and 0.2 m from the leading edge of the plate. A schematic experimental setup is illustrated by Figure 3. The wind tunnel fittings for the circular cylinders in crossflow and the flat plate in parallel flow calibration systems are shown in Figure 4.

The free convection experimental equipment consisted of a simple Perspex plate (vertical or horizontal) resting on an electronic balance having 10 mg accuracy.

The cylinders used for the calibration experiments were also used to investigate the circumferential mass transfer rate distribution around the circular cylinders in crossflow.

The paper used was chromatographic paper of type Whatman filter paper No 1 of 0.18 mm thickness. This paper has some advantages, such as more homogeneous and lower surface roughness, over the blotting paper which was used in previous studies (4, 5). The paper was 'cured' by repeatedly wetting and drying so that it became dimensionally stable. Thereafter any influence of dimensional change during the drying process on the reflecting was included in the calibration constant. The glue Copydex, a transparent paper-plastic glue, was selected for glueing the absorbent paper onto the surface of interest. The solvent used was 1–butanol. It is less volatile and less toxic than propanol which was used by previous investigators (4, 5). The 1–butanol was applied to the surface of interest by a hand held spray gun. The sensor used was a combined infra-red emmitor and detector of type RS 307–913 operating at the wavelength of approximately 0.94μm.

Since the calibration constant is a function of temperature any differences between the calibration temperature and the temperature at which subsequent experiments are carried out taken into account via eqn (7).

RESULTS AND DISCUSSION

The results obtained from the calibration experiments are summarised in Table 1.

Calibration system	Eqn No	$10^2 \cdot c$ (kgm^{-2})	No. of reptns.	Error %
Stagnation flow, small cylinder	8	5.1330	70	12
Stagnation flow, large cylinder	8	5.1946	55	17
Flat plate in parallel flow	10	4.2011	49	14
Horizontal plate in free convection	3	6.2384	8	43
Vertical plate in free convection	12	5.8379	10	17

Table 1: Experimentally obtained calibration constants

In the case of parallel flow over the flat plate, the leading edge of the mass transfer boundary layer moves downstream during the experiments as the upstream section dries out. This unsteadiness in the upstream boundary conditions invalidates eqn. (10).

In the case of the free convection from a horizontal flate plate, error may be due to difficulties in achieving uniform mass flux over the entire surface of the plate and relating this to a single point measurement of reflectivety change.

It is very difficult to replicate in an experiment the conditions under which eqn. (12) describes the average mass transfer rate from an ideal vertical plate system.

The stagnation flow on a circular cylinder, as discussed above, is relatively insensitive to disturbances. From the number of experiments carried out in this case and the agreement between the results obtained from two different cylinders, and relatively small deviations between individual experiments, it is suggested that this flow system is the preferred way of calibrating the paper. Averaging the experimental results obtained from the two cylinders gives a calibration constant of $5.16 \times 10^{-2} kgm^{-2}$.

Local mass transfer rate distributions around circular cylinders in crossflow

Some measurements of local mass transfer rate around circular cylinders in crossflow were made in order to investigate the applicability of the experimental technique.

The results obtained are shown in Figure 5. Giedt (13) reported that for $Re_D < 1.0 \times 10^5$, which may be called subcritical flow, the laminar boundary layer separates at about $\theta = 80°$. The local Nusselt number Nu_θ decreases with incresing θ as a result of boundary layer development. Beyond the separation point ($\theta \simeq 80°$) the local Nusselt number increases with increasing θ because of mixing in the wake region. For $Re_D > 1.0 \times 10^5$, which may be called critical flow, a transition from laminar to turbulent flow in the boundary layer occurs at about $\theta = 80°$. The position of the transition point moves towards the forward stagnation point with increasing Reynolds number (8). The transition causes a sudden increase in the local Nusselt number. Then the separation point occurs at about $\theta = 130°$.

The results obtained from the small cylinder in the present study show, in general, good agreement with Giedt's data. The results obtained from the larger cylinder show that transition has moved further up towards forward stagnation point and occurs at about $\theta = 60°$. This disagreement between the present data and that of Giedt (13) is attributed to the effect of the surface roughness (8). The different mass transfer behaviour of the small cylinder and the large cylinder within about the same Reynolds number range suggests that the effect of the surface roughness is more noticeable for larger cylinders.

REFERENCES

1. Patrick, M.A., Pembery, J.G.A. and Wragg, A.A., 1984, Proc. 1st. UK Nat. Conf. on Heat Transfer, Vol. 2, pp 879–892
2. Walker, V. and Wilkie, D., 1967, Proc. I.M.E. Sypm. on High Pressure Gas as Heat Transfer Medium, pp 190–197
3. Utton, D.B. and Sheppard, M.A., 1985, Proc. Int. Centre for Heat and Mass Transfer, No 18, pp 155–166, Eds. Soloukhin R.I. and Afgan N.H., Hemisphere Publishing Corp.
4. Davies, T.W., Patrick, M.A. and Tanyildizi, V., 1990, 18th Australasian Chem. Eng. Symp., Auckland, New Zealand, Vol 1 (13), pp 408–414
5. Davies, T.W., Patrick, M.A., Balint, T.S. and Tanyildizi, V., 1991, Proc. Symp. Heat Transfer, IChemE, Birmingham

6. Van Driest, 1959, Turbulent Flows and Heat Transfer, High Speed Aerodynamics and Jet Propulsion, Vol. 5, Ed. by Lin, C.C., Princeton University Press, New York
7. Kestin, J. and Wood, R.T., 1971, J. Heat Transfer, pp 321–327
8. Achenbach, E., 1977, Int. J. Heat Mass Transfer, Vol. 20, pp 359–369
9. Kestin, J., Maeder, P.F. and Sogin, H.H., Z. Angew., 1961, Math. Phys., Vol. 12, pp 115–132
10. Lowery, G.W. and Vachan, R.I., 1975, Int . J. Heat Mass Transfer, Vol. 18, pp 1229–1242
11. Zhukauskas, A., 1972, Advances in Heat Transfer, Vol. 8, pp 93–157 Eds. Harnett, J.P. and Irvine T.F., Academic Press, New York
12. Martin, B.W., 1984, Int. J. Heat Mass Transfer, Vol. 27, No. 9, pp 1583–1586
13. Giedt, W.H., 1949, Transaction of A.S.M.E., Vol. 71

NOTATION

A	area of the evaporating surface, m^2
a	constant in eqn (12)
b	constant in eqn (12)
c	calibration constant, $kg m^{-2}$
D	diameter of the cylinder, m
D_{AB}	mass diffusion coefficient of the solvent in air, $m^2 s^{-1}$
g	gravitational acceleration, $9.81 m s^{-2}$
$Gr_{m,x}$	mass transfer Grashof number
H	height of the duct, m
k_m	convection mass transfer coefficient, $m s^{-1}$
I	intensity of the reflected infra–red light, V
I_0	overall light intensity change, V
m	absorbed mass of the solvent in the paper, kg
N	mass flux of species A, $kg s^{-1} m^{-2}$
Nu	Nusselt number
P	pressure, hPa or Pa
Pr	Prandtl number
Re	Reynolds number
Sc	Schmidt number
Sh	Sherwood number
T	temperature, $°C$
t	time, s
u	fluid velocity, $m s^{-1}$
x	distance from the leading edge of the flat plate, m

Greek Letters

ν	kinematic viscosity of the free stream fluid, $m^2 s^{-1}$
ρ	density of the free stream fluid, $kg m^{-3}$

Subscripts

s	surface conditions
o	standard conditions
∞	free stream conditions, bulk

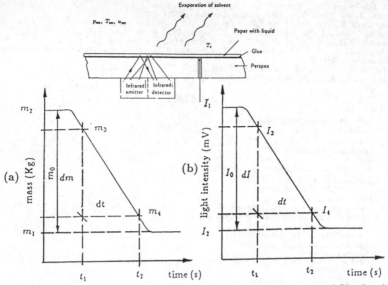

Figure 1: Operation of the photo-evaporative technique. Changes in (a) liquid content and (b) reflected light intensity during the paper drying process.

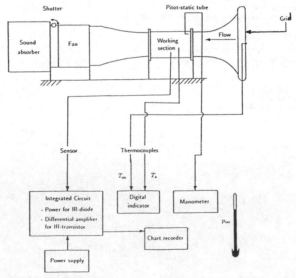

Figure 3: The general experimental setup

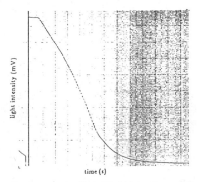

Figure 2: A typical chart recorder output; reflected light intensity vs drying time.

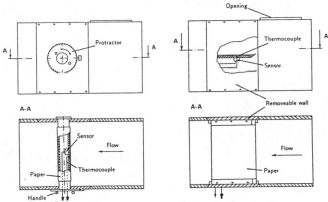

Figure 4: Wind tunnel fittings; (a) cylinder in crossflow (b) flat plate boundary layer

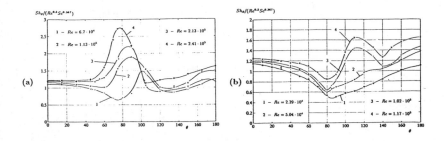

Figure 5: Local Sherwood number distributions around the circumference of circular cylindrical tubes in crossflow having diameters of (a) 0.10 m and (b) 0.05 m

EXPERIMENTAL TURBULENT MIXED CONVECTION CREATED BY CONFINED BUOYANT WALL JETS

D.Blay, S.Mergui, J.L. Tuhault, F.Penot
Université de Poitiers, Laboratoire d'Etudes Thermiques, 40 Avenue du Recteur Pineau, 86022 Poitiers, France

This paper deals with an experimental study of the mixed turbulent air convective flow created by plane jets within a square cavity. Two flow configurations were studied. In the first one, the flow was created by one horizontal jet and a horizontal heated surface (floor); in the second one, the flow was generated by two jets (one vertical + one horizontal) at different temperatures. We first classify the different types of flow according to the values of the characteristic parameters — mainly the Reynolds and Froude numbers of jets. This was achieved with flow visualizations using laser tomography for several values of Reynolds and Froude numbers. These visualizations showed how sensitive the flow structure was to the jet Froude number, and they pointed out the existence of a critical value from which important changes in the flow appear. Values of this critical Froude number are given for several flow configurations.

The second objective was to measure flow velocity, turbulence kinetic energy and temperature. These measurements were carried out for several parameter configurations, 300 points in the cavity being scanned in each case.

INTRODUCTION

Up to now, in the field of mixed convective internal flows, most of the works which were carried out on air flows in ventilated rooms were mainly numerical works [1], [2], [3], [4]. These numerical simulations were performed by integration of the Navier-Stokes equations completed by a k-ε turbulence model. But very little pertinent experimental work is still available to test the accuracy of these numerical works. This is one of the reasons why we decided to carry out experiments with the following goals in mind :
- a better understanding of the physics of internal mixed convective flows,
- the setting up of an experimental database (velocities, turbulence fluctuations, temperatures) to enable comparisons with numerical calculations.
Two main flow configurations were studied:
- firstly the flow was created by one horizontal buoyant jet discharging into the cavity at the upper left corner with a discharge temperature equal to T_h (fig.1). Both vertical walls as well as the top wall were maintened at the same uniform temperature which was equal to the inlet jet temperature ($T_w =T_h$). The flow was heated by the bottom wall (floor) which was kept at a greater temperature ($T_{fl}> T_h$).

- secondly, the flow was generated by two buoyant jets (fig.2). One inlet was located at the upper corner of the left vertical wall and the second inlet was located at the left corner of the bottom wall. The vertical jet was warmer than the horizontal jet. Each wall was at the same temperature than the horizontal jet (Tw=Th).

TEST APPARATUS

Model

The test apparatus was designed so as to generate a two-dimensional flow, in order to ensure that experiments and corresponding calculations were not too cumbersome.
Experiments were performed on a laboratory model composed of a 1.04mx1.04mx 0.7m cavity equipped with two inlet slots (18mm wide) and one outlet slot (24mm wide). This cavity was divided into three identical smaller cavities, the central working cavity (0.30m wide) where measurements were performed and two side cavities (0.20m wide each) where the same flow as in the working cavity was reproduced. This design made it possible to obtain a

fairly good two-dimensional flow in the central cavity by eliminating end effects. Radiative heat transfer between wall surfaces and conductive heat transfer within the walls were eliminated by using imposed temperature conditions on the four active walls as boundary conditions. The two other walls separating the central cavity from each guard cavity were considered to be adiabatic.

Air inlet conditions

There were two independent air circuits. They respectively fed the horizontal (cold) jet and, should the occasion arise, the vertical (warm) jet.

The air flow passed through a porous medium and a honeycombed structure in order to ensure uniform velocity all along the slot. The air was cooled down or warmed up using a water-air heat exchanger. Water temperature was controlled by a cryothermostat with a precision of 0.25°C. Preliminary velocity and temperature measurements were made to qualify the air flow at the slot exit. Fig.3a shows transversal jet velocity profiles measured at the slot exit, at several locations. Fig.3b shows one transversal jet temperature profile measured at mid-length. In order to check the 2D behaviour of the inlet conditions, longitudinal velocity and temperature profiles were measured all along the slot centerline.

It was found that the standard deviation of the longitudinal discharge velocity was less than 5% (fig.4). The longitudinal mean air discharge temperature profile was found to be uniform with a precision better than 1%.

This device allowed to impose the following air inlet temperature and velocity ranges:
- 10°C to 40°C for the horizontal cold jet and 20°C to 70°C for the vertical warm jet.
- 0.1 m/s up to 0.6 m/s for both jets.

Wall conditions

Each active wall was made of flat aluminium heat exchangers maintained at a constant and uniform temperature by the use of temperature controlled water. The water flowrate was oversized to ensure that no significant temperature gradient could appear anywhere on the wall surface. Under these conditions, the imposed wall surface temperature could be chosen between 10°C and 50°C with a precision of 0.25°C.

SCALING PARAMETERS

Mixed convective flows are physically interesting for the balance between inertial and buoyancy forces. Depending on the relative importance of these forces, the flow structure can vary significantly.

In the one jet configuration, the characteristic temperature difference was chosen as $(T_{fl} - T_h)$. Then, the governing independent parameters are:

- horizontal jet Reynolds number: $Re_h = \dfrac{U_h \, e}{\nu}$

- horizontal jet Froude number: $Fr_h = \dfrac{U_h}{\sqrt{g\beta e \, (T_{fl} - T_h)}}$

- non dimensional wall temperature: $\Theta_w = \dfrac{(T_w - T_h)}{(T_{fl} - T_h)}$.

In the two-jet configuration, the horizontal jet is taken as the reference jet while the vertical jet is seen as a perturbation of the previous one. The characteristic temperature difference was chosen as $(T_v - T_h)$.

The governing independent parameters are:

- horizontal jet Reynolds number: $Re_h = \dfrac{U_h \, e}{\nu}$

- horizontal jet Froude number: $Fr_h = \dfrac{U_h}{\sqrt{g\beta e \, (T_v - T_h)}}$

- jet discharge velocity ratios: $\dfrac{U_v}{U_h}$

- non dimensional wall temperature: $\Theta_w = \dfrac{(T_w - T_h)}{(T_v - T_h)}$

FLOW VISUALIZATION

The first objective of this experimental work was to classify the different types of flow according to the values of the characteristic parameters. Flow visualizations by laser tomography were conducted using cracked oil smoke. These visualizations showed the great sensitivity of the flow structure to the Froude number and they pointed out the existence of a critical value around which important changes in the flow occur, see fig.5.

The visualizations revealed the existence of a flow bifurcation characterized by a critical parameter set around which it could be seen that the overall stream circulation could vary from a clockwise direction to an anti-clockwise direction. Due to the number of parameters driving the flow, it was impossible to carry out a systematic parametrical study to determine the critical parameter domain. However, several test cases were explored with a view to characterizing the phenomenon. Several conditions seemed necessary for a change in flow direction :

- the horizontal jet had to be colder than the cavity core,
- the jet(s) had to have a low momentum level,
- a sufficient amount of energy had to be transferred from the walls to the cavity core.

Under these conditions, the horizontal jet had a tendency to drop below warmer core layers and the overall flow direction could be reversed.

One-jet + heated floor configuration

The whole set of test cases presented here was characterized by a zero value of Θ_w. The flow was only governed by two parameters, the Reynolds and Froude numbers, which varied in the following range:

$$200 < Re_h < 800 \qquad 1.08 < Fr_h < 4.09$$

Results tended to show that the critical Froude number was independent of the Reynolds number. This critical value was found to be approximately 3.2 . However, these values had to be corrected due to the higher density of the oil smoke used to visualize the flow.

To determine the value of the critical Froude number more accurately, local laser Doppler velocity measurements were performed for two different temperature conditions:

case 1 : $T_h = 15°C$ $T_{fl} = 54.3°C$

case 2 : $T_h = 15°C$ $T_{fl} = 35°C$

Several monitoring points were chosen in the vicinity of the ceiling and measurements were carried out for different values of the Froude number which was progressively reduced or increased by either decreasing or increasing the inlet jet velocity. For each case, care was taken to ensure that the thermal steady state was reached. For a particular value of the Froude number, Fr_c, called the critical value, a sudden change in the flow direction was noticed.

Similar results were found for both cases. The critical Froude number values which were found are given below.

- case 1   Decreasing Froude number: $1.93 < Fr_c < 2.06$

          Increasing Froude number: $3.12 < Fr_c < 3.31$

- case 2   Decreasing Froude number: $1.92 < Fr_c < 2.45$

          Increasing Froude number: $2.89 < Fr_c < 3.41$

These results reveal a significant hysteresis effect characteristic of subcritical flow bifurcations.

Two-jet configuration

It was impossible to obtain an inversion of the flow direction as long as the wall temperature was equal to the inlet horizontal jet temperature ($\Theta_w > 0$), whatever the Reynolds and Froude numbers. This was only achieved when the wall temperature was increased ($T_w > T_h$), which had the effect of increasing the cavity core temperature with little subsequent increase of the buoyant vertical jet momentum.

For instance, this inversion was obtained for the following parameter set:

$$Re_h = 240 \qquad Fr_h = 1.65 \qquad \frac{U_v}{U_h} = 0.55 \qquad \Theta_w = 0.51$$

VELOCITY AND TEMPERATURE MEASUREMENTS

The second objective of this work was to provide pertinent experimental data useful for the validation of numerical methods and turbulence models. We measured the velocity and temperature fields for several parameter sets and for both configurations.

At steady state, both mean velocity and turbulence fluctuation components were measured with a two-colour laser Doppler velocimetry device. Temperature measurements were performed with a 20μm diameter Cr-Al thermocouple. Velocity, temperature and turbulence intensity measurements were carried out in more than 300 locations within the working cavity. Some results are given for the 'one-jet + heated floor' configuration. They correspond to horizontal and vertical profiles at mid-height (y=0.52m) and mid-length of the cavity (x=0.52m). Temperature boundary conditions were chosen such as to obtain a core temperature as close as possible to the surroundings temperature, in order to minimize thermal losses.

The configuration is defined as follows:
- the discharge temperature of the horizontal jet is T_h = 15°C, its mean discharge velocity is U_h=0.57m/s.
- the bottom wall temperature is equal to 35.5°C
- the three other wall temperatures are equal to 15°C.
This case corresponds to the following values of the parameters:
$$Re_h = 684 \qquad Fr_h = 5.31 \qquad \Theta_w = 0.$$

The 2D behaviour of the flow structure was firstly checked. It appeared that the flow was highly 2D (the difference between the local velocity measured at z=0. and the mean velocity (averaged in the z direction) was found to be less than 4% for both profiles.

Figures 6 to 8 show the different profiles respectively obtained at y=0.520m and at x=0.520m. It can be noted that the recirculation flow was important and approximately equal to six times the jet discharge flowrate. Mass balances resulting from the integration of theses profiles were achieved with a discrepancy of 1% for the horizontal v-velocity profile and of 6% for the vertical u-velocity profile.

Similar results were found by Costa & alii [8] for the two-jet configuration.

CONCLUSION

This experimental study of internal mixed convective flows created by confined buoyant wall jet(s) pointed out the existence of a flow bifurcation. In the configuration with one horizontal wall jet and a heated bottom wall, it was shown that this bifucartion was completly determined by a critical value of the Froude number which was found to be comprised between 1.92 and 3.41.

Velocity and temperature measurements were carried out in order to set-up a base of pertinent experimental data. The recirculation flowrate was found to be important and about six times the discharge flowrate of the horizontal jet.

ACKNOWLEDGEMENTS

The work reported here was supported by the Department of Electricity Applications of Electricité de France.

REFERENCES

1. Kapoor K., Jaluria Y., 1989 Heat transfer from a negatively buoyant wall jet. Int. Journal of Heat and Mass Transfer, vol.32, n°4, pp. 697-709 .

2. Mukarami S., Kato S., 1989 Numerical and experimental study on Room Airflow - 3D predictions using the k-ε turbulence model. Buiding and Environment, vol.24, n°1, pp. 85-97.

3. Awbi H.B. , 1989 Application of computational fluid dynamics in room ventilation. Buiding and Environment, vol.24, n°1, pp. 73-84.

4. Gosman A.D., Nielsen P.V., Restivo A., Whitelaw J.H., 1980 The flow properties of rooms with small ventilation openings. J.of Fluid Engineering (ASME), vol.102, p. 316-323.

5. Jones D.D. , Launder B.E., 1982 The prediction of laminarization with a two-equation model of turbulence. Int. J.of Heat & Mass Transfer, vol.15, pp.301-314.

6. Ciofalo M., Collins M.W., 1989 Prediction of Heat Transfer in turbulent recirculating flows using an improved wall treatment. Numerical Heat Transfer, Part.B, vol. 15, p. 21-47.

7. Blay D., Brun C., 1989 Numerical study of a non isothermal jet within a cavity. 4th Int. Heat Transfer Conf., Algiers.

8. Costa J.J., S. Mergui, Tuhault J.L. , Penot F., Blay D., Oliveira L.A., 1992 Test of tubulence models for the numerical simulation of internal mixed convection flows. ROOMVENT'92, Aalborg, Denmark.

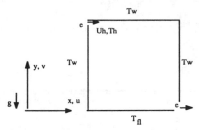

Figure 1 One jet + Heated floor. $T_h = T_w$; $T_{fl} > T_w$ Figure 2 Two jets. $T_h = T_w$; $T_v > T_w$

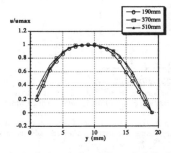

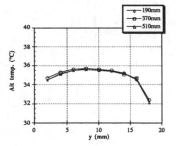

Fig. 3a Vertical u-velocity profiles at the inlet slot exit   Fig. 3b Vertical temperature profiles at the inlet slot exit

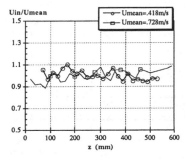

Fig.4 Longitudinal discharge velocity profiles along the inlet slot exit

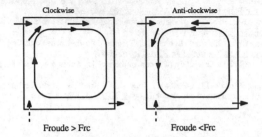

Figure 5 Flow structure vs Froude number

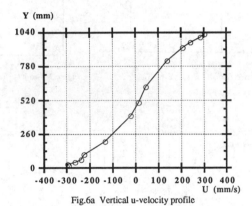

Fig.6a Vertical u-velocity profile

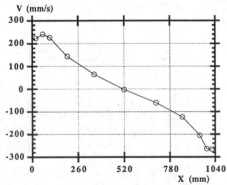

Fig.6b Horizontal v-velocity profile

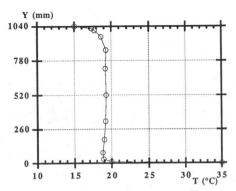

Fig.7a Vertical temperature profile

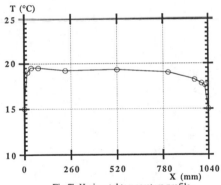

Fig.7b Horizontal temperature profile

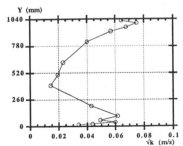

Fig.8a Vertical turbulent kinetic energy profile

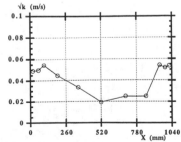

Fig.8b Horizontal turbulent kinetic energy profile

FORCED CONVECTION CONDENSATION ON HORIZONTAL LOW INTEGRAL FIN TUBES

A.Cavallini, B.Bella, G.A.Longo and L.Rossetto
Istituto di Fisica Tecnica dell'Università, Padova, Italy.

The experimental apparatus, used to measure heat transfer coefficients for condensation of refrigerants on the outside of a single horizontal finned tube, is described. Experimental data, obtained during condensation of refrigerant 11 with vapour pressure ranging from 109 to 198 kPa, average vapour to wall temperature difference varying from 4.4 to 11.7 °C and maximum vapour velocity ranging from 1.9 to 26.2 m/s, is presented. The comparison between present experimental data and existing correlations shows that the model of Honda et al. (1,2) reproduces data at stationary conditions very well. The heat transfer coefficients increase with vapour flow rate reaching a 50% enhancement at the highest vapour velocity.

INTRODUCTION

The refrigeration, air-conditioning and process industries widely utilise condensers with low integral fin tubes, because of their high heat transfer performance.
Many research-works have investigated, both experimentally and theoretically, the effects of fin geometry and condensing fluid properties on the heat transfer coefficient realized when a stationary vapour condenses on a single low integral fin tube.
Only few experimental studies present data taken with vapour condensing under high velocity, (3,4), and no theoretical model available in the open literature considers the effect of the vapour shear stress at the interface.
The present work describes an experimental apparatus to measure heat transfer coefficients during condensation of refrigerant 11 on a single low integral fin tube with vertical downflow vapour. The effect of vapour velocity was experimentally investigated and the results obtained are reported.

EXPERIMENTAL APPARATUS

The experimental set-up used to measure heat transfer coefficients is shown schematically in Figure 1. Saturated vapour of refrigerant 11 flows downward through a test section consisting of ten dummy tubes and one cooled tube, 150 mm long (see Figure 2).
The horizontal cooled tube, copper made and integral finned, (geometry as reported in Table 1) is instrumented to measure the surface temperature. Eight copper-constantan thermocouples, 0.128 mm in diameter and teflon coated, are embedded and soldered in four equidistant axial holes, 1 mm in diameter and 27 mm deep, drilled on both sides (inlet and outlet) of the tube

wall. A twisted tape insert is placed inside the cooled tube to mix the flowing water and promote turbulence.

The dummy tubes, not cooled, are smooth tubes partially housed in the test section wall with the same length and outside diameter of the instrumented tube.

The walls of the test section are made of PVC, 40 mm thick and insulated to prevent heat losses.

The inlet cooling water has been kept around +5°C and its flowrate has been derived from the weight of the water collected in a fixed time. The water temperature gain across the instrumented tube has been measured with a differential four junction copper-constantan thermopile, while the perfect mixing of the water has been obtained in two mixing chambers placed in series at the outlet of the pipe.

Refrigerant 11 is evaporated in an electrical boiler with 40 kW maximum power, supplied by controlled voltage, and then the vapour passes through a wire-mesh demister and a centrifugal liquid separator to eliminate completely the entrained liquid.

Dry-saturated vapour enters the test section through a perforated plate distributor and then flows through a 300 mm long settling chamber before reaching the test tube.

Along the external wall of the test section, above and below the instrumented tube, two manometric connections to a mercury U tube manometer are placed to determine the pressure profile of the condensing vapour. The manometer is heated by hot air to avoid vapour condensation.

The refrigerant 11 at the outlet of the test section is completely condensed in an auxiliary condenser and then it returns to the boiler passing through one of the three microturbine flowmeters.

HEAT TRANSFER COEFFICIENTS AND EXPERIMENTAL PROCEDURE

Some preliminary tests have been performed to evaluate the heat losses through the test section and to check the thermocouples readings.

The test section was fed by refrigerant 11 saturated vapour at constant pressure while the instrumented tube was not cooled. In steady state conditions the difference between the vapour saturation temperature, derived from the pressure reading, and the tube wall temperature, measured by the thermocouples, during these tests ranged from 0.03 to 0.09 K.

Then the experimental apparatus, described above, has been used to determine heat transfer coefficients during refrigerant 11 pure vapour condensation on the single instrumented cooled integral fin tube.

TABLE 1 - Geometrical characteristics of the test tube

Tube Material	Copper
Outside diameter at the fin tip (mm)	16.40
Outside diameter at the fin root (mm)	15.00
Inside diameter (mm)	10.00
Fin pitch (mm)	0.75
Fin height (mm)	0.70
Average fin thickness (mm)	0.22
Fin half tip angle (rad)	0.174

TABLE 2 - Operative conditions of the experimental runs

Fluid	R-11
Runs	34
Vapour inlet pressure (kPa)	109-198
Mean temperature difference (°C)	4.4-11.7
Maximum vapour velocity (m s^{-1})	1.9-26.2
Heat flux density (kW m^{-2})	53.8-117.2
Maximum mass flux (kg m^{-2}s^{-1})	20-203
$Re_L \cdot 10^{-5}$	1.24-16.0

The heat flow rate Q exchanged in the tube is calculated from the heat balance over the cooling water:

$$Q = c_{pc}\dot{m}_c\Delta T_c \tag{1}$$

where ΔT_c is the water temperature gain measured by the thermopile, $\dot{m}_c$ is the cooling water flow rate and c_{pc} is the specific heat capacity of the cooling water.
The average condensation heat transfer coefficient has been referred to the fin envelope surface area S (nominal surface area of a smooth tube having the diameter of the finned tube at the fin tip) and to the logarithmic average temperature difference ΔT_{ln} between vapour and tube wall:

$$\alpha = Q / (S \Delta T_{ln}) \tag{2}$$

with $S = \pi d_0 L$ and

$$\Delta T_{ln} = (T_{WO}-T_{WI})/ln[(T_S-T_{WI})/(T_S-T_{WO})] \tag{3}$$

where L and d_0 are respectively the tube length and the tube outside diameter at the fin tip, T_S is the average vapour saturation temperature derived from the pressure profile, T_{WI} and T_{WO} the average surface temperatures evaluated from the thermocouples placed in two distinct sections, 27 mm from the inlet and from the outlet of the tube, respectively.
The operating conditions relative to the data obtained with refrigerant 11 are reported in Table 2.

CORRELATIONS AND EXPERIMENTAL RESULTS

The first theoretical model for condensation on a single horizontal finned tube was proposed by Beatty and Katz in 1948 (5); it applies a Nusselt type analysis, considering only the effect of gravity in the drainage of condensate. The results of this model are inadequate to predict condensation heat transfer with high surface tension fluids and high fin density tubes (6). The surface tension reduces the condensate film thickness on the fin flanks but it produces also a retention of condensate in the lower part of the tube.
Among the several correlations developed in the last years for condensation on finned tubes, two models show the best performance in predicting heat transfer coefficients (6): the Honda et Al. 1987 model (1,2) and the Adamek and Webb 1990 model (7).

Both these models refer to condensation on a single horizontal finned tube and consider the effect of surface tension, fin efficiency, conduction in the tube wall, but not the effect of vapour velocity.

Figure 3 reports present experimental data plotted on the coordinates of Nu_{exp}/Nu_{cal} vs. Re_L, where Nu_{exp} is the experimental Nusselt number, Nu_{cal} is the Nusselt number evaluated by the above theoretical models and Re_L is the Reynolds number referred to maximum vapour velocity u_{MAX} and to the properties of the liquid phase. Table 3 gives the mean percentage deviation E1 and the mean percentage absolute deviation E2 defined as follows:

$$E1=(100/Np) \sum_{1}^{Np} i \ (1-\alpha_{cal}/\alpha_{exp}) \qquad E2=(100/Np) \sum_{1}^{Np} i \ |(1-\alpha_{cal}/\alpha_{exp})| \qquad (4)$$

This comparison shows that Honda et Al. 1987 model (1,2) reproduces very well the present data when the vapour velocity (and therefore Re_L) is low.

Figure 4 reports the present experimental data plotted on the coordinates Nu_{exp}/Nu_{st} vs. Re_L, where Nu_{st} is the Nusselt number at negligible vapour velocity, $u_{MAX} = 2$ m/s. For comparison the trend for condensation on a smooth tube, where the Nusselt number under vapour velocity is computed by the 1986 model of Honda et Al. (8) is also reported, while the Nusselt number at stationary vapour Nu_{st} is derived from the classical Nusselt equation (9). This plot shows that the effect of vapour velocity for condensation on a finned tube is less pronounced than for a smooth tube but it is still remarkable for $Re_L > 400000$. The available experimental data for condensation of halogenated refrigerants on a horizontal finned tube under not negligible vapour velocity (present data for R11 and Honda et Al. (3,4) data for R113) indicate a dependence of $Nu/Pr_L^{0.4}$ on $Re_L^{0.33}$.

CONCLUSIONS

The heat transfer enhancement during outside condensation on horizontal low integral fin tubes due to vapour shear is evident, even if less pronounced than with smooth tubes.

For the finned tube investigated, the Honda et Al. 1987 model (1,2) for stationary vapour keeps valid up to $Re_L = 400000$. At higher values of Re_L the experimental data shows the dependence of $Nu/Pr_L^{0.4}$ on $Re_L^{0.33}$.

Further experimental work is in progress to test other fluids and different geometries.

TABLE 3 - Comparison between theoretical models and present work
experimental data

Model	E1	E2
Beatty and Katz 1948	31.94	31.94
Honda et Al. 1987	4.08	11.74
Adamek and Webb 1990	-13.38	16.38

ACKNOWLEDGEMENTS

Research carried out with partial financial aid of C.N.R.

SYMBOLS USED

c_p = Specific heat capacity $(J \ kg^{-1}K^{-1})$
d = Tube diameter (m)
E1   = Mean percentage deviation
E2   = Mean percentage absolute deviation
h   = Fin height (m)
L   = Tube length (m)
$\dot{m}$ = Mass flow rate $(kg \ s^{-1})$
Np   = Number of experimental runs
Nu   = $\alpha d_0/\lambda_L$ = Nusselt number
p   = Fin pitch (m)
Pr_L = $c_{pL}\mu_L/\lambda_L$ = Prandtl number
Q   = Heat flux (W)
Re_L = $\rho_L u_{MAX} d_0/\mu_L$ = Reynolds number
S   = Heat transfer surface (m^2)
t   = Fin thickness (m)
T   = Temperature (K)
u_{MAX} = Maximum vapour velocity, referred to the minimum cross-sectional area $(m \ s^{-1})$
α = Film heat transfer coefficient $(W \ m^{-2}K^{-1})$
Δ = Difference
ρ = Density $(kg \ m^{-3})$
λ = Thermal conductivity $(W \ m^{-1}K^{-1})$
μ = Dynamic viscosity $(kg \ m^{-1}s^{-1})$

Subscripts

C   = coolant side
cal   = calculated
exp   = experimental
ln   = logarithmic
L   = condensate
O   = outside of finned tube at the fin tip
S   = saturated vapour
st   = stationary vapour
W   = tube wall

REFERENCES

1. HONDA H., NOZU S., UCHIMA B., 1987, ASME J. of Heat Transfer 109, 218-225.
2. HONDA H., NOZU S., UCHIMA B., 1987, Proc. 2nd ASME-JSME Joint Conference, Vol.4, pp.385-392.
3. HONDA H., UCHIMA B., NOZU S., NAKATA H., TORIGOE E., 1989, ASME HTD 108, 117-125.
4. HONDA H., UCHIMA B., NOZU S., TORIGOE E., IMAI S., 1991, Proc. ASME-JSME Joint Conference, Reno (Nevada).
5. BEATTY K.O., KATZ D.L., 1948, Chemical Eng. Progress 44, 55-70.
6. CAVALLINI A., LONGO G.A., ROSSETTO L., 1991, Proc. 18th Int. Cong. of Refrigeration, Montreal.
7. ADAMEK T., WEBB R., 1990, Int. J. of Heat and Mass Transfer 33, 1721-1735.
8. HONDA H., NOZU S., UCHIMA B., FUJII T., 1986, Int. J. of Heat and Mass Transfer 29, 429-438.
9. NUSSELT W., 1916, Z. Ver. dt. Ing. 60, 541-546, 569-575.

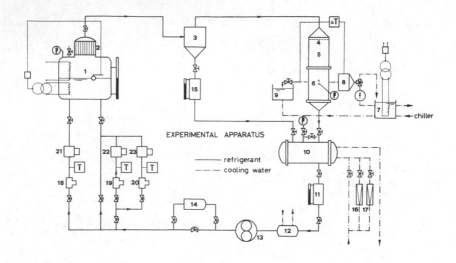

1 Boiler, 2 Wire-mesh, 3 Centrifugal separator, 4 Perforated plate, 5 Settling chamber, 6 Test section, 7 Cooling water, 8,9 Cooling water distributor and receiver, 10 Auxiliary condenser, 11,15 Liquid receiver, 12 Cooler, 13 Rotary pump, 14 Drier, 16,17 Water rotameter, 18,19,20 Filter, 21,22,23 Microturbine flowmeter.

Figure 1 Diagram of the experimental apparatus.

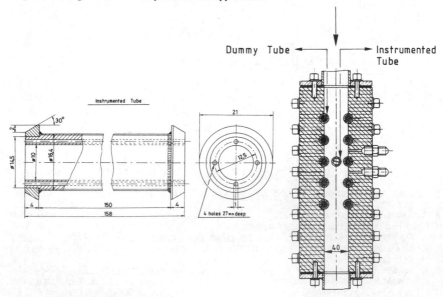

Figure 2 Diagram of the test section (measures in millimeters).

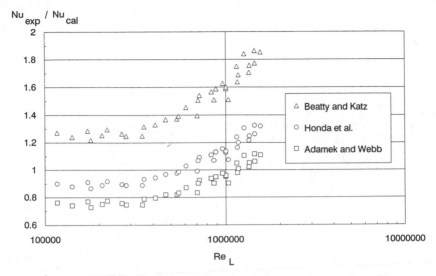

Figure 3 Present experimental data plotted on the coordinates Nu_{exp}/Nu_{cal} vs. Re_L: Nu_{exp} is the experimental Nusselt number while Nu_{cal} is the Nusselt number evaluated by each of the three theoretical models.

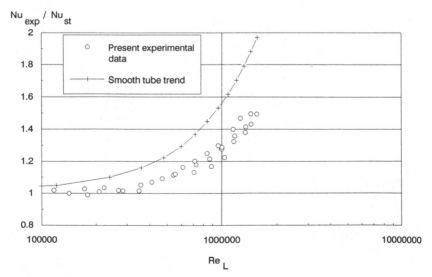

Figure 4 Present experimental data plotted on the coordinates Nu_{exp}/Nu_{st} vs. Re_L and compared with smooth tube trend: Nu_{st} is the Nusselt number for quiescent vapour.

BUOYANCY EFFECTS ON THE LAMINAR FORCED CONVECTION IN THE THERMAL ENTRANCE REGION OF INCLINED DUCTS WITH VARIOUS HEATING CONDITIONS

Dr. R. Smyth
Department of Mechanical and Process Engineering, University of Sheffield, England.

Buoyancy effects on the laminar mixed convection of assisting flows in the thermal entrance region of inclined rectangular ducts with an aspect ratio AR (width-to-depth) of 5:1 have been numerically studied to examine the effect on local and average Nusselt numbers. Three heated situations have been studied, (a) one pair of parallel sides of the duct heated only, (b) one upper side heated only, and (c) one lower side heated only. The fluid was air with a Rayleigh number Ra of 1×10^7 and Reynolds number Re from 800 to 1800. The results have been compared with experimental data from a parallel study. It was found that due to the combined effects of buoyancy and the particular heated situation, the heat transfer was significantly removed from that for pure forced convection.

INTRODUCTION

Mixed or combined convection heat transfer inside a duct is either **assisting** or **aiding** when the forced convection flow direction is upward and assisted by the fluid buoyancy or **opposing** when the forced convection flow direction is downward and is opposed by the fluid buoyancy. From experimental observation by other workers it has been found that in vertical tubes the laminar forced convection heat transfer coefficient in the aiding situation can be as much as 2.5 times that for pure forced convection.

Examples of situations where mixed convection is important are in the cooling channels (flat plate bundle) in a nuclear power reactor which could be prone to mixed convection flow in certain situations, industrial heat exchangers in chemical and food processing industries where problems may be encountered involving combined convection in channels, cooling of electrical equipment and the design of certain types of solar energy collectors (flat plate air heaters).

The effects of buoyancy on the flows in these situations can be important. For example where the flow rates vary widely giving rise to mixed convection. Again this phenomena may exist where processes are fully automatic and precise information is required regarding the minimum flow rate necessary to fulfil safety conditions in the event of component failure.

A number of theoretical studies by other workers have been made recently of convection in rectangular cross-sectioned channels. Cheng and Hwang [1] have analyzed combined natural and forced convection heat transfer for steady fully developed laminar flow in a horizontal rectangular channel with constant wall heat flux and at two Prandtl numbers Pr.

Abou-Ellail and Morcos [2] have investigated numerically the buoyancy effect on laminar forced convection heat transfer in the entrance region of uniformly heated horizontal rectangular ducts having AR of 1 and 4 for different values of Pr. Results for water have revealed that the buoyancy effects are negligible up to a certain axial distance from entrance depending on the magnitude of Ra. Rahman and Colbourne [3] have analytically studied the secondary flow effect of incompressible laminar flow in a viscous Newtonian fluid in a square horizontal duct. The secondary flow has revealed the existence of two vortices in the duct cross-section.

An analysis of fully-developed combined forced and free convection in vertical rectangular channels has been reported by Agrawal [4]. Combined free and forced laminar convection has been investigated, in non-circular vertical ducts by Igbal et al [5], comprising rectangular, eliptical and rhombic cross-sections with uniform axial heat input and either uniform peripheral heat flux or uniform peripheral temperature distribution. Combined free and forced convection results show that for AR from 1 to 4, Ra varying from 0 to 3000 and uniform circumferential heat flux, Nu increases with AR for a slight amount of buoyancy. Under higher buoyancy rate, this increase becomes appreciable. Chow et al [6] have studied numerically the effects of free convection and axial conduction on forced convection heat transfer to fully developed flow in a vertical channel at low Peclet numbers Pe.

Specific objections of the present study were as follows:-
1. To observe the buoyancy effect on the velocity and temperature field in the duct under different angles of inclinations and different thermal boundary conditions.
2. To determine the effect of buoyancy induced secondary motion caused by one side being heated or when both sides are heated.
3. To determine the effect of the primary flow direction on the heat transfer in the duct for various inclined duct situations.

THEORETICAL ANALYSIS

The numerical analysis scheme of Patankar [7] which has been found to be applicable to a wide range of practical engineering problems and has been encapsulated in commercially available fluid dynamic simulation computer codes was used to solve the governing differential equations of continuity, momentum and energy for steady flow in three dimensions.

The boundary conditions for the solution of these equations were (1) uniform axial velocity and temperature at the duct inlet (2) prescribed constant heat flux at one or both of the wider sides of the duct and (3) insulated narrow sides of the duct.

RESULTS AND DISCUSSION

The numerical study was carried out for assisting flows with horizontal, inclined and vertical rectangular ducts having an AR of 5, giving a hydraulic diameter d_H = 0.067 m, and having the wider sides heated and the narrower sides unheated. The heated sides had three situations; upper side heated only, lower side heated only and both sides heated with equal heat flux. The side, opposite to the heated side in the one side heated situation, was assumed adiabatic. The temperature of air at the duct entrance was taken as 19°C with a Pr of 0.71. The surface heat flux was fixed for all cases at q = 200 W/m² whilst two different inlet air velocities were used which gave an Re of 800 and 1800. This gave a Richardson number Ri of 23 and 4.3. The above range of parameters have been covered by 16 case studies. Sample flow fields are presented below and have been calculated to exhibit the effect of duct inclination on both wall temperature and flow field and to compare with experimental results.

The velocity vectors in the transverse plane at an axial distance of $9d_H$ are given in Fig. 1 and Fig. 2. For both sides heated, Fig. 2 shows a clear existence of 4 vortices in the section and reveals that, compared to Fig. 1, the effects of this secondary flow is delayed due to increasing Re. In Fig. 1 as Re is reduced, the vortices dominate the flow field for all the angles of inclination ψ. For $\psi = 0°$, the vector scale reveals the strong acceleration of the flow, driven by the secondary flow. The secondary flow pattern, represented by the number of vortices in the section, seems to vary along the duct and appears to have 2 major and 4 minor vortices in the whole section. For higher values of duct inclination angle ($\psi = 45°$ and $60°$), it was found that the strong secondary flow at upper plate is gradually reduced and the lower plate secondary flow seems to be able to penetrate the upper plate boundary layer.

The contours of isotherms in the x-y plane, for the same cases shown in Fig. 1 and Fig. 2 and for the same axial position are displayed in Fig. 3 and Fig. 4. In these figures is shown the average temperature of the horizontal surfaces, the field isotherms in °C and the bulk air temperature T_B. The both sides heated case with a relatively high Re is shown in Fig. 4. For $\psi = 0°$, there is a thin thermal boundary layer. As was indicated in Fig. 2, the appearance of the vortices accelerate the development of temperature distribution. The isotherms reveal the irregularity of the air temperature in the transverse direction. For $\psi = 30°$, the results show a great improvement in the upper plate heat transfer as the average plate temperature reduces by approximately 9°C. The lower plate boundary layer becomes more regular especially at the plate centre while at the plate edges there is still a transverse temperature irregularity due to upward secondary flow depicted in Fig. 1.

Fig. 3 shows the contour of isotherms in the same x-y plane as for Fig. 4 but for a lower Re. This figure corresponds to the velocity vector shown in Fig. 2. The air temperature variation seems to be larger as Re reduces. The upward flows created by two adjacent vortices restrict the main flow and divert it to the vortex core so producing longitudinal rolls having a cold core separated by hot boundaries. Therefore, the upper plate average temperature increases rapidly for low Re and changes from 82.8°C for Re = 1803 to 97.6°C for Re = 801.

The lower plate behaviour is completely opposed to that for the upper plate, i.e. reducing Re enhances the secondary flow vortices and improves the heat transfer process. Therefore the lower plate average temperature shows a small reduction for a low Re. The temperature difference between the upper and lower plates increases rapidly as Re decreases and the heat flux increases and the 9°C shown for Re = 1803 increases to 29°C for Re = 801. For $\psi = 30°$, the upper plate average temperature reduces by 14°C while the lower plate average temperature increases by 3°C. This behaviour can attribute to the reduction of the secondary flow velocity on the lower plate and the reduction of the upper plate isolation due to the improvement in the axial flow velocity in the upper plate boundary layer.

COMPARISON OF PREDICTION RESULTS WITH EXPERIMENTAL WORK

Theoretical predictions at Re = 900 have been compared with an approximately corresponding sets of experiments, Re = 870 to 940. The comparison comprised the axial velocity profiles from laser Doppler anemometer LDA tests for the assisting flow cases with the three heating situations. The comparison of the predicted axial velocity with the LDA test, for ψ = 0°, 60° and 90°, are shown in Fig. 5. In Fig. 5(b) at ψ = 0° for upper plate heated only, the LDA result shows good agreement with the predicted result while in Fig. 5 (a), ψ=0° for lower plate heated only, the LDA result is skewed a little toward the lower plate as the result of a lower test Reynolds number than that used for the prediction case. At these lower Reynolds numbers the secondary flow is more significant and accounts for this profile difference at the lower plate. Figure 5 (c), shows that the mixed convection prediction profile deviates from the pure force convection profile and is skewed towards the lower plate while the experimental results show more skewing than the prediction in the lower plate boundary layer region.

For ψ=60° for all three heating conditions, the comparison shows good agreement with the small discrepancy between the LDA results and the predictions probably due to the small differences in the inlet and boundary conditions and measurement error. For the vertical duct in the both sides heated case, the LDA results shows the reduction in the flow velocity across the duct and near to both plates. The theoretical predictions are in fair agreement with the LDA results for the duct centre position. The trends revealed by the LDA test is predicted in the theoretical solution and the slight differences can be attributed to the heat loss by conduction to the side wall.

CONCLUSIONS

For the horizontal duct the flow adjacent to the upper plate both for the upper plate heated only and both sides heated conditions, is mainly a laminar, forced convection boundary layer but at a high heat flux and low Re the probability of having recirculation is high. The flow adjacent to the lower plate, for the lower plate heated only and both sides heated, is greatly affected by secondary flow which creates longitudinal vortices. The number of vortices and their configuration varies across and along the duct while the position of the onset of the vortices and their ability to penetrate into upper plate boundary layer is determined greatly by the plate heat flux and Re.

For the inclined duct the biasing of the velocity profile towards the heated surface increases as ψ moves towards the vertical, increasing the flow at the duct centre and reducing it at the duct sides and corners.

For the vertical duct a completely one-dimensional flow with a very clear skewing of the velocity profile toward the heated surface is obtained with a very high degree of symmetry for the both sides heated case. Increasing the plate heat flux and reducing Re may lead to a reversed flow at the duct core as the above two factors accelerate the flow at the heated surfaces.

REFERENCES

1. K C Cheng and G J Hwang, 1969 <u>Trans. ASME, J of Heat Transfer</u> 53.
2. M M M Abou-Ellail and S M Morcos, 1980 <u>ASME paper</u> 80-HT-139.
3. M Rahman and P Colbourne, 1984 <u>Int J of Heat and Mass Transfer</u> <u>5</u> 167.
4. H C Agrawal, 1962 <u>Int J of Heat and Mass Transfer</u> <u>5</u> 439.
5. M Igbal, A K Khatry and B D Aggarwal, 1972 <u>App. Sci. Res.</u> <u>26</u>, 183.
6. L C Chow, S R Husain and A Campo, 1984 <u>Trans. ASME, J of Heat Transfer</u> <u>106</u> 297.
7. S V Patenkar, 1980 <u>Numerical Heat Transfer and Fluid Flow</u> McGraw-Hill Book Co.

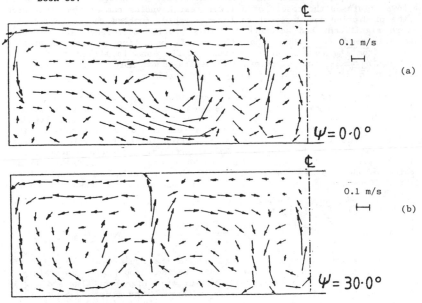

Fig. 1 (a) Transverse velocity vectors ψ = 0° Re =801
(b) Transverse velocity vectors ψ = 30° Re = 801

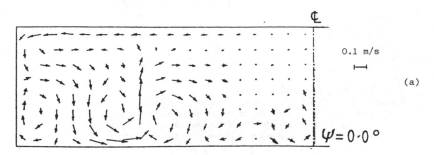

Fig. 2 (a) Transverse velocity vectors ψ = 0° Re = 1803

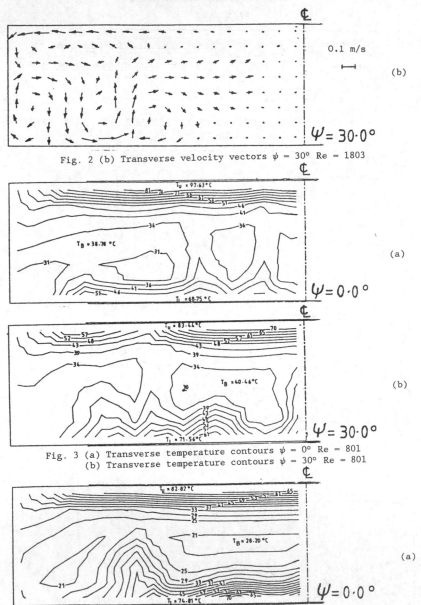

Fig. 2 (b) Transverse velocity vectors ψ = 30° Re = 1803

Fig. 3 (a) Transverse temperature contours ψ = 0° Re = 801
(b) Transverse temperature contours ψ = 30° Re = 801

Fig. 4 (a) Transverse temperature contours ψ = 0° Re = 1803

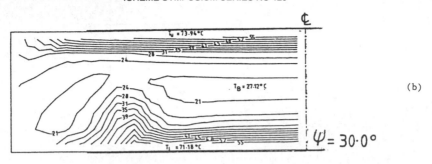

Fig. 4 (b) Transverse temperature contours ψ = 30° Re =1803

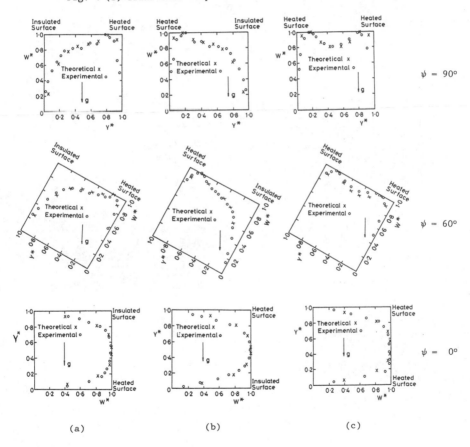

Fig. 5   Duct centre plane dimensionless axial velocity W*
(a) bottom side heated only (b) top side heated only (c) both sides heated

TRANSIENT ANALYSIS OF A DIESEL ENGINE IN THE WARM-UP PHASE: EFFECTS OF THE INTERCONNECTED FLUID LOOPS

H. ZITOUNI[*], A. ALEXANDRE [**]

The purpose of the present work is to study the thermal behavior of the fluid loops associated to a 2 litres diesel engine running in the warm-up phase. Emphasis is placed on the use of an original supercharging device (named V.T.): a compressor driven by an automatic speed variator. Using the nodal method, models of the various components (engine, heat exchangers, compressor, ...) have been developed based upon the coupling of the heat transfer and thermodynamics processes involved in each component. To support the analytical work, steady state and transient tests were carried out. The comparison of predicted results to experimental data shows good agreement. A diesel engine supercharged with a V.T. system coupled with an air bypass appears to be a solution to heat-up the engine as quickly as possible especially in extreme cold conditions.

INTRODUCTION

The thermal analysis of a diesel engine, which is running in the warm-up phase, provides that low efficiency, high fuel consumption and pollutant emissions are directly associated with the engine's low temperature. This temperature occurs an incomplete combustion and excessive friction losses. These explain our present interest in the transient thermal behavior of a supercharged 2 litres diesel engine after a cold start. We have studied a quick warm up solution for a supercharged engine associated to a stable compressor mechanically driven by the engine. This device named "Vario-Turbo" or V.T. [5] has a rotation speed among 50000-80000 rpm on all the range of the engine's speed. Besides its high pressure ratio at low engine's speed, the compressor can be run to achieve low nominal efficiency in view of increasing the heat transferred from the supercharging loop to the cooling loop by mean of a heat exchanger (after cooler), as shown in figure 1. The paper is arranged as follow. First, the modelling of the various components (engine, compressor, heat exchangers) is descibed. then the association of the sub-models in a complete cooling system. Finally, we present the experimental data and the validation of the model.

MODEL OF THE ENGINE

The modeling of all the thermal processes involved within a diesel engine requires the solution of coupled heat transfer equation [1, 2]. To avoid such a complex study, a simplified one-node model of the engine has been developed where the heat rejected to the coolant is approached by an empirical expression.Steady state experimentation was carried out to establish a relationship between the heat transferred from the engine's structure to the coolant and the operating conditions. The most influent parameters are: the engine's speed, load, coolant temperature, oil temperature and ambient temperature. The mathematical expression of the heat release to the coolant is then formulated as follow:

$$\frac{\overset{\circ}{Q}_e}{\overset{\circ}{Q}_{e,max}} = f\,(T_{ext},\, T_h,\, T_e,)\,[\,\kappa + (\frac{N}{N_{max}})^{0.8}\,]\,[\,\phi + \frac{q_c}{q_{c,max}}\,] \qquad (1)$$

[*] Direction de la Recherche - RENAULT - 92500 RUEIL MALMAISON.
[**] Laboratoire d'Etudes Thermiques URA CNRS 1403- ENSMA - F86034 POITIERS.

With:
$$f = \alpha + \beta\, T_{ext} + \gamma T_e + \gamma T_e^2 + \theta T_h + \theta' T_h^2$$

The coefficient α, β, γ,... of the equation (1) are determined using the steady state experimental data. The identification of the overall heat inertial of the engine is achieved using the transient tests. Figure (2) represents the engine's model where only one node placed at the engine's coolant outlet.

MODEL OF THE COMPRESSOR

The most important modelling aspect relative to the compressor model is the coupling of the thermodynamics equations and the heat balance. The operating conditions of the compressor are calculated from the performance map that gives the relationship between the air mass flow rate variables, pressure ratio and the compressor speed variable as shown in figure (4). The map variables of mass flow rate and compressor speed are corrected by factors relating the inlet conditions to the standard conditions. In our modelling the compressor map is introduced in a tabular form. We associate during the transient simulations the thermal equations based on the compressor inertia, mass flow rate and the temperature evolution with the compressor thermodynamical evolution. The actual compressor work is related to the operating conditions by:

$$\overset{\circ}{W}_c = \frac{\overset{\circ}{m}_c}{\eta_{is}} \frac{\gamma}{\gamma-1} r\, T_1 \left[\left(\frac{P_2}{P_1}\right)^{\gamma-1/\gamma} - 1 \right] \tag{2}$$

In diesel engine's, the air of admission depends on the engine's speed and air inlet temperature (see figure 3):

$$\overset{\circ}{m}_{2'} = \rho_{2'}\, D\, \frac{N}{120}\, \tau_{2'} \tag{3}$$

The association of an air by-pass with the supercharging loop implies one more equation that gives the mass flow rate of the by-pass:

$$\overset{\circ}{m}_b = \rho_2 \left(\frac{1}{\tau_c}\right)^{1/\gamma} S_d \sqrt{2\, C_p\, T_2 \left[1 - \left(\frac{1}{\tau_c}\right)^{\gamma-1/\gamma} \right]} \tag{4}$$

Therefore the air flow rate of the compressor is:

$$m_c = m_a + m_b \tag{5}$$

To increase the density of the air, heat exchanger is placed between the compressor outlet 2 and the intake manifold 2'. The air pressure drop in this heat exchanger is not negligible and it is given experimentally in function of the air mass flow rate. We take account of this pressure drop in the calculation of the by-pass mass flow rate. The air density at the intake manifold is then given by:

$$\rho_{2'} = \rho_0 \frac{P_{2'}}{P_0} \frac{T_0}{T_{2'}} \tag{6}$$

This equation governs directly the compressor and the engine mass flow rates given by the equations 3 to 5.

MODEL OF THE HEAT EXCHANGERS

The heat exchangers include the radiator, the cab heater and aftercooler. They play the role of interface between the different fluid loops. The energy balance for each heat exchanger taking into account the overall heat transfer between the two fluids, the effectiveness and the heat inertia yields to the following equations:

$$C_h \frac{dT_{h2}}{dt} = (\overset{\circ}{m}c_p)_h (T_{h1} - T_{h2}) - \varepsilon (\overset{\circ}{m}c_p)_{min} (T_{h1} - T_{c1}) \tag{7}$$

$$C_c \frac{dT_{c2}}{dt} = (\overset{\circ}{m}c_p)_c (T_{c1} - T_{c2}) + \varepsilon (\overset{\circ}{m}c_p)_{min} (T_{h1} - T_{c1}) \tag{8}$$

With : $(\overset{\circ}{m}c_p)_{min} = Min [(\overset{\circ}{m}c_p)_h , (\overset{\circ}{m}c_p)_c]$

$\varepsilon = \varepsilon (\overset{\circ}{m}_c, \overset{\circ}{m}_h)$

Subscripts "h" and "c" indicate respectively the hot and cold fluids. The effectiveness of the heat exchangers is given experimentally and introduced in our model in function of the mass flow rates of the two fluids.

THE COMPLETE SYSTEM

Experiments :
Four configurations was tested in climatic chamber using the same vehicle:
Test 1: A diesel engine equiped with a classical turbocharged
Test 2: A diesel engine equipped with the Vario-Turbo device ($P_2 = 1.6$ bar)
Test 3: A diesel engine equipped with the Vario-Turbo device ($P_2 = 1.9$ bar)
Test 4: A diesel engine equiped with the Vario-Turbo device and an air by-pass
      after the aftercooler ($P_2 = 1.4$ bar)

Figure 1 is one of the configuration tested. These experiments was carried out at $T_{ext} = -18$ °C and N=2250 rpm.The results of these experiments are presented in figure 4 and 5. Figure 4 shows us that the use of the Vario-Turbo device associated with an air by pass provides the quickest and the highest compressor outlet air temperature. The amount of heat transfered to the coolant loop is then increased. The coolant engine outlet temperature reaches 80°C after 16 minutes for the classical turbocharged system, see figure 5. This warm-up phase is substantially improved by the V.T. system and it is about 7 minutes for the configuration with an air by-pass. This causes an increasing of about 8% only of the fuel consumption.

Model validation :
The components sub-models described previously are assembled into the tested vehicle cooling system. In the first step the complete model was run using the same inputs as during the experimentations. The results of these calculations was compared to the experimental data. Figure 6 is a comparison between the predicted and the experimented thermal response of the cooling loop in the test 1. As can be seen the simulation program provides a good prediction of the heat-up of the cooling fluid (figure 6-a). Nevertheless a little inadequacy of the heat rejection to the coolant is observed in the transient phase after a cold start (figure 6-b). This is certainly due to the size of the engine model (only one node) which is insufficient to simulate all the engines's dynamic.
After the validation of the model, the second step is to use it with the same inputs to allow a real comparison between the various configurations. Figures 7 and 8 shows the predicted heat-up performances of the supercharging and the cooling loops for the following configurations:

configuration (i) : A diesel engine equiped with a classical turbocharged
configuration (ii) : A diesel engine equipped with the Vario-Turbo device
configuration (iii): A diesel engine equipped with the Vario-Turbo device an air by-pass
      after the aftercooler
configuration (iv): A diesel engine equiped with the Vario-Turbo device and an air by-pass
      befor the aftercooler, as shown 3.

As can be seen the configuration (iv), using a Vario-Turbo device and an air by-pass just after the compressor, allows the highest and the quickest heat-up of the supercharging loop as well as the cooling loop.This is due to the the degradation of the operating conditions of the compressor ($\eta_{is} = 0.30$). The amount of heat transfered of the coolant by means of the aftercooler is about 5kW.

CONCLUSION

A complete modelling of the fluid loops (air and coolant) related to a supercharged 2 litres engine was developed.the model was validated by comparison of the predicted results with the experimental data. From the experimental data an the results of the model, the use of the Vario-turbo device associated with an air by-pass appears to be the solution to warm-up quickly the engine's coolant. In fact, the additional heat transfered to the cooling loop (5kW) is of the same order of the heat rejected to the coolant after a cold start.

ACKNOWLEDGMENT

This work has been supported financially and technically by "VALEO-Thermique Moteur" within the context of an M.R.T. contract. "Le Moteur Moderne" company has provided technical assistance for the development of the engine's model. In particular the authors want to express our appreciation to P. Palier (VALEO) and B.Besson (Le Moteur Moderne) for their assistance in the form of test data and suggestions for the engine's model.

REFERENCES

1. D.N. Assanics and J.B. Heywod, SAE paper n° 860329, 1986.
3. G. Sitkei , G.V. Ramanaiah, SAE paper n°720027,1972.
2. V.J. Ursini, E.C. Chiang and J.H. Johnson, SAE paper n° 821049, 1982.
3. H. Zitouni et A. Alexandre, Eurotherm 15, Toulouse (France).4-6 Déc. 1991.
4. H. Zitouni, Thèse de doctorat d'université - LET - ENSMA - Poitiers(France), 1991.

NOMENCLATURE

c_p	Constant pressure specific heat	J/kg/K
C	Heat capacitance	J/K
$\overset{o}{m}$	Masse flow rate	kg/s
P	Pressure	N/m^2
N	Engine speed	rpm
$\overset{o}{Q_e}$	Heat release to the coolant	W
q_c	Fuel consumption	g/s
T	Temperature	K
t	Time	s
W_c	Compressor actual work	KW
ρ	Density	Kg/m^3
ε	Effectiveness of the heat exchanger	
γ	Ratio of specific heats	
η_{is}	Isentropic efficiency	
τ_c	Pressure ratio	

Subscripts:

1	compressor inlet
2	compressor outlet
2'	engine's admission
c	compressor, cold, fuel consumption
e	Coolant
ext	external
h	oil, hot

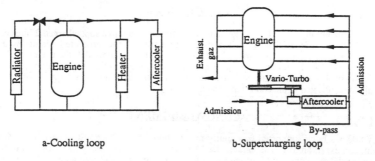

a-Cooling loop   b-Supercharging loop

Fig.1- Supercharged diesel engine fluid loops

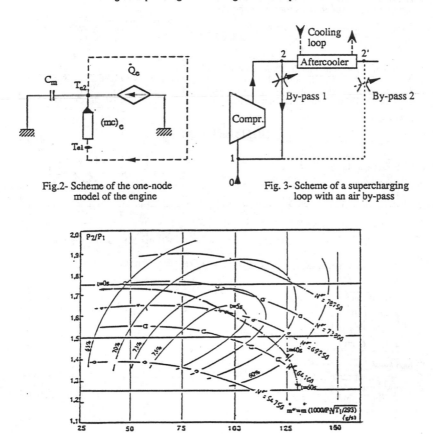

Fig.2- Scheme of the one-node
model of the engine

Fig. 3- Scheme of a supercharging
loop with an air by-pass

Fig.4 -Transient evolution of the of the compressor operating conditions
on the performance map (————)

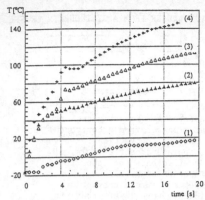

Fig.4- Experimental air temperature at the compressor outlet for the four tested configurations

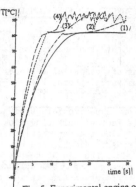

Fig. 5- Experimental engine outlet coolant temperature for the four tested configurations

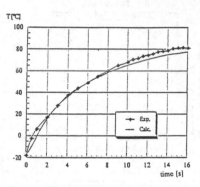

a - Engine outlet coolant temperature

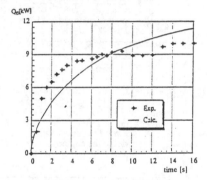

b- Heat flux rejected to the coolant

Fig.6' - Compáraison of measered and calculated results in a warm-up phase

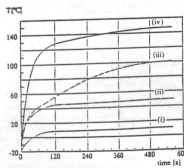

Fig.7 - Predicted air temperature at the compressor outlet for four configurations

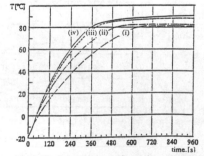

Fig.8 - Predicted engine outlet coolant temperature for four configurations

COMBINED FORCED AND FREE-CONVECTIVE HEAT TRANSFER FROM A CYLINDER IN CROSSFLOW

Y. Kikuchi, N. Itoh and Y. Morikawa

Department of Mechanical Engineering, Hiroshima University
4-1, Kagamiyama 1-Chome, Higashi-Hiroshima 724, Japan

An experimental study was conducted to investigate the
effect of pulsating flow on heat transfer from a horizontal
cylinder under combined forced and free convection for both
upward (aiding) and downward (opposing) crossflow. In calm
flow the local heat transfer coefficients on the front
portion of the cylinder agreed fairly well with the
correlation of Schmidt and Wenner for forced convection.
The mean heat transfer coefficients were lower than the
calculated values by a linear combination of conventional
forced and free convection correlations. This discrepancy
was more accentuated for opposing flow. The pulsating flow
produced a great enhancement in heat transfer for both
aiding and opposing flows.

INTRODUCTION

In recent years there has been an increased interest in predicting the
temperature and flow fields in many engineering applications under low flow
conditions, where buoyancy effects become important and combined forced and
free convection phenomena are encountered. These applications include heat
exchangers for low grade energy recovery and solar and geothermal energy
systems, and steam generators in nuclear and fossil power plants.
Many investigators have conducted both experimental and analytical
studies of combined convective heat transfer, but most of their studies have
been limited to heat transfer of longitudinal flow along tubes. There are a
few studies(1-4) which treat heat transfer from tubes in crossflow.
This led to the present authors to carry out an experimental study of
combined forced and free convetive heat transfer from a horizontal cylinder in
crossflow. In the present experiment, local temperature distributions are
measured on the cylinder surface in both upflow and downflow. This paper
gives the experimental results of local and mean heat transfer coefficients in
various ranges of Reynolds number and Rayleigh number, with particular
emphasis on the effect of pulsating flow on heat transfer.

EXPERIMENTAL APPARATUS AND PROCEDURES

A schematic diagram of the experimental apparatus is shown in Fig.1. It
is an upflow (aiding flow) system.† The main circulation and the overflow
lines are made vinyl plastic piping of 40 mm nominal diameter. The working
fluid (water) is delivered from a reservoir by a pump and stored in a head

---

† In the case of downflow (opposing flow) system the flow lines are
rearramged to be upset.

tank to reduce its flow and pressure fluctuations. Water first flows downward through the main line and then turns upward at the inlet of a contracted section. Water experiences the flow conditioning at the contracted section before it enters the test section which is a rectangular channel with a cross section of 75 or 150 mm in width L and 100 mm in depth. The water flows upward through the test section and returns to the reservoir for recircula- tion. The overflow line is used to maintain the water level of the head tank constant. A heat exchanger of the reservoir is used to cool the working fluid. An auxiliary pump is equipped in the inlet of the contracted section to add pulsations into the main flow.

A horizontal cylinder of 15 mm in diameter D is set at the 215 mm downstream of the entrance of the test section. In order to heat the cylinder, direct current is supplied to a stainless-steel foil (10 μm thick) which is tightly in contact with the thermally insulated wall of a vinyl plastic tube. Dissipated power is measured with a voltmeter-ammeter combination system with a 1% rated accuracy. The heat flux q is, therefore, kept uniform.

Local surface temperatures are measured by two Chromel-Alumel thermocouples (50 μm in diameter) which are welded to the back of the stainless-steel foil at the angle interval of π. The circumferential temperature distribution can be obtained by rotating the cylinder every π/18 in angle.

The temperature of the inlet of the test section is also measured by a thermocouple. Flowrate is obtained with accumulating the working fluid in a specially measuring vessel during a given time.

Each run is performed in the following manner. The flowrate of the test section is first set at a fixed value. Power is supplied to the test cylinder. In order to maintain the working fluid at a constant temperature, coolant is supplied to the heat exchanger in the reservoir. The output signals from thermocouples, which measure the temperatures of the surface of the test cylinder and the inlet of the test section, are recorded on a strip chart. Sufficient time is allowed for the convection development to reach a steady

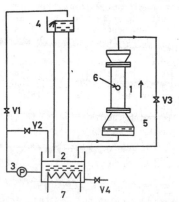

1. Test section; 2. Reservoir; 3. Pump;
4. Head tank; 5. Contracted section;
6. Test cylinder; 7. Heat exchanger

V1~V4: Valves

Fig. 1 Schematic diagram of experimental apparatus

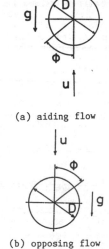

(a) aiding flow

(b) opposing flow

Fig. 2 Coordinate system

state conditon. Upon reaching a steady state all the temperature readings are taken by a data acquistion system which is controlled by a personal computer.
The working fluid is held nearly at room temperature during the experiment to minimize heat losses. Other conditons are as follows:
Reynolds number Re: 0-2000
Modified Rayleigh number Ra^*: $4x10^5-1x10^7$
where the Re and Ra^* are defined as $Re=uD/\nu$ and $Ra^*=\beta gqD^4/\lambda\nu\kappa$, respectively.
Coordinate system is shown in Fig.2 for (a) aiding flow and (b) opposing flow.

RESULTS AND DISCUSSION

Temperature fluctuation

Figure 3 shows typical records of temperature fluctuation which are measured at several circumferencial positions of the cylinder surface in aiding flow for different flow cases: (a) calm flow (L=150 mm) and (b) pulsating flow (L=75 mm). In each case the Re and Ra are 300 and $7.5x10^6$, respectively.
For the calm flow the temperature fluctuations are fairly calm on the front portion ($\phi<\pi/2$) of the cylinder. On the rear portion ($\phi>\pi/2$), however, violent oscillations of temperature are dominant, which are accompanied with intermittent release of hot thermals from the boundary layer on the rear portion of the cylinder.
For the pulsating flow small fluctuations are observed on the front portion of the cylinder. On the rear portion, however, the fluctuation signals indicate a long-period swell supperposed with many ripples.
The above-mentioned fact is indicated more clearly in Fig.4, which carried plots of the present data as probability density distributions. A sharp symmetrical distibution appears on the front stagnation point of $\phi=0$. This sharpness is more accentuated for the calm flow in which the fluctuation is extremely small. On the rear portion of the cylinder, however, the

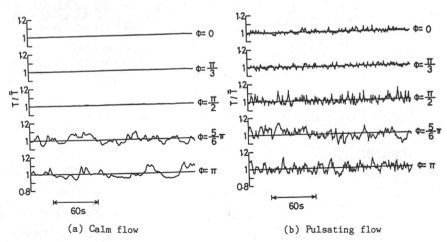

(a) Calm flow   (b) Pulsating flow

Fig. 3 Typical records of temperature fluctuation

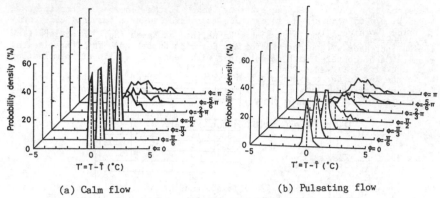

(a) Calm flow   (b) Pulsating flow

Fig. 4 Probability density of temperature fluctuation

distribution is distorted to become wide since the thermals are intermittently released from the boundary layer on the rear portion of the cylinder. In order to verify these phenomena a simultaneous measurment of both temperature and velocity distributions of the working fluid around the heated cylinder are needed.

Similar fluctuations are also observed in opposing flow.

Heat transfer characteristics

Figure 5 shows the effect of Reynolds number on the distributions of local heat transfer around a cylinder for different flow systems: (a) aiding flow and (b) opposing flow. In each case the Ra is 7.5×10^{6}. The abscissa is the circumferencial position ϕ of a cylinder, and the ordinate the dimensionless heat transfer coefficient expressed by $Nu_{\phi}/Pr^{0.4}$. The Nu_{ϕ} is the local Nusselt number defined as $Nu_{\phi}=\alpha D/\lambda$. It is seen that with any Re, the heat transfer coefficient first tends to decrease downstream (with increasing ϕ) on the front portion of the cylinder, and after attaining a minimum at the separation point increases somewhat to remain more or less constant thereafter. This particularity is more accentuated as the Re becomes higher.

In this figure are also shown the experimental results of Schmidt & Wenner(5) for forced convective heat transfer on the front portion of a cylinder in crossflow. They derived the following relation from their experimental data:

$$Nu_{\phi}=1.14Re^{0.5}Pr^{0.4}[1-(2\phi/\pi)^{3}] \tag{1}$$

In calm flow the present data agree fairly well with the calculations by Eq.(1) within the scattering of data. In pulsating flow, however, heat transfer is much enhanced on the front portion of the cylinder. This enhancement is more accentuated for aiding flow.

In order to more clarify the effect of buoyancy effect on heat transfer, the experimental results of mean heat transfer coefficient for aiding flow and opposing flow are brought together in Fig.6. The ordinate is the mean Nusselt number which is determined by integrating the local Nusselt number relation

$$Nu = \int_0^\pi Nu_\phi \, d\phi / \pi \qquad\qquad (2)$$

The abscissa is the Rayleigh number Ra defined as $Ra=Ra^*/Nu$. In aiding flow the Nu increases with higher Ra for any Reynolds number Re. The experimental results are compared with calculations with means of a linear combination

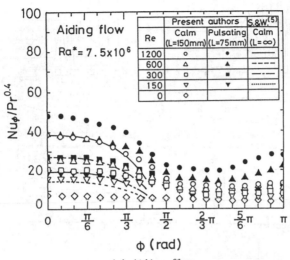

(a) Aiding flow

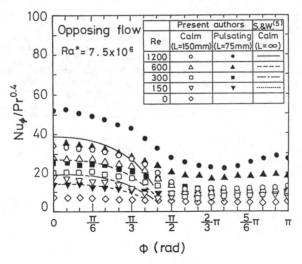

(b) Opposing flow

Fig. 5 Effect of Reynolds number on local heat transfer coefficient

method as the following expression

$$Nu = Nu_f + Nu_n \qquad (3)$$

where Nu_f and Nu_n are the Nusselt numbers related to forced convection and free convection, respectively, which are given by Zukauskas(6) and Churchill & Chu(7) as

$$Nu_f = 0.52 Re^{0.5} Pr^{0.37} \qquad (4)$$

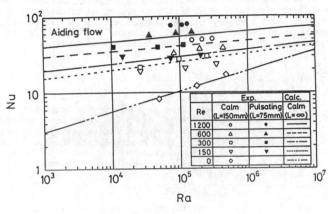

(a) Aiding flow

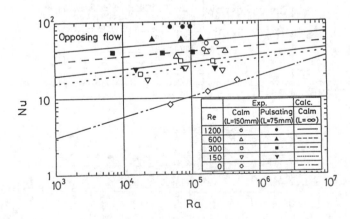

(b) Opposing flow

Fig. 6 Effect of Rayleigh number on mean heat transfer coefficient

$$Nu_n = \left[0.6+0.387 \left(\frac{Ra}{[1+(0.559/Pr)^{9/16}]^{16/9}} \right)^{1/6} \right]^2 \tag{5}$$

The calculated results are higher than the present data for calm flow, except with Re=0 (free convection). Equation (3), therefore, tends to overestimate the combined forced and free convective heat transfer. The pulsating flow produces a great enhancement in heat transfer for any Re and Ra.

In opposing flow, however, a complicated tendency is observed in the effect of Ra on Nu since local buoyancy forces act against the main flow. For low Re, the Nu first increases with higher Ra and after attaining a maximum, decreases slightly. For high Re, the Nu increses with higher Ra in the range of the present experimental conditions. In order to more clarify these complicated phenomena, higher Ra experiments should be conducted.

CONCLUSIONS

The effect of pulsating flow on heat transfer from a horizonal cylinder in combined forced and free convection crossflow has been investigated experimentally. The working fluid flows upward (aiding flow) and downward (opposing flow). Comparison of the experimental results with some calculations has yielded the following conclusion:

(1) In calm flow the local heat transfer coefficient on the front portion of the cylinder agrees fairly well with the empirical correlation of Schmidt & Wenner. In pulsating flow, however, the heat transfer coefficients are higher than their correlation.

(2) The mean heat transfer coefficient is lower than calculations by Eq.(3) which is a linear combination of forced and free convection in calm flow. This discrepancy is more accentuated for opposing flow. The pulsating flow produces a great enhancement in heat transfer and the heat transfer coefficients are higher than calculated values of Eq.(3).

ACKNOWLEDGEMENTS

The authors wish to acknowledge the technical contributions of Messrs. K. Tokui and K. Matsubayashi at all the stages of the experiments.

REFERENCES

1. Oosthuizen, P.H. and Madan, S., 1970, Trans. ASME, J. Heat Transfer, 92, p.194.
2. Fand, R.M. and Keswani, K.K., 1973, Int. J. Heat Mass Transfer, 16, p.1175.
3. Kitamura, K. et al., 1990, Proc. National Heat Transfer Symp. of Japan, pp.136-138 (in Japanese).
4. Noto, K. et al., 1990, ibid, pp.454-456 (in Japanese).
5. Shmidt, E. and Wenner, K, 1941, Forsh. Geb. Ingenierwes., 12, p.65.
6. Zukauskas, A., 1987, "Advances in Heat Transfer", Academic Press, Vol.18, pp.87-159.
7. Churchill, S.W. and Chu, H.H.S., 1975, Int. J. Heat Mass Transfer, 18, p.1049.

COMPUTATIONAL MODELLING ASPECTS OF AGITATED VESSELS

E.A. Foumeny[*], P.J. Heggs[**] and P.J. Bentley[***]

[*]Department of Chemical Engineering, The University of Leeds.
[**]Department of Chemical Engineering, University of Bradford.
[***]I.C.I., C & P, Wilton.

Computational modelling into the performance evaluation of agitated vessels can be of great benefit for accurate design. Unfortunately, existing methodologies fail to provide accurate design. Empirical approach based on dependable measurements and representative mathematical description can facilitate the desired quality of design data. The significance of these issues are being discussed.

INTRODUCTION

The thermal characteristics of mixing vessels have been the subject of extensive research in the process engineering disciplines. There are two main types of heat transfer designs for agitated vessels. These are vessels with external jackets or vessels with internal coils. Typical examples of the jacket type are shown in Figure 1. The objective of most design exercises is to calculate a value for the overall heat transfer coefficient for a particular vessel design and evaluate the possible heat transfer rate and response time that can be expected. The thermal resistances in a mixing system can be added together using the summation of resistances rule.

$$R_{total} = R_i + R_v + R_j + R_{foul} \tag{1}$$

The values of the convective and fouling resistances in the system are difficult to quantify. The convective film resistances, R_i and R_j, are calculated from empirical design correlations which express the heat transfer characteristics of the convective film in terms of various dimensionless collections of geometrical and physical properties of the system. The fouling resistances are more difficult to quantify as they dynamically vary with operating conditions of a piece of equipment. The use of resistance data obtained from literature sources is advised but caution should be employed in the treatment of fouling. Neglecting it can lead to the under-designing of equipment and thereby failure to achieve the required duty. Over-design on the other hand is expensive and in some instances can produce a design which is difficult to control within the required process limits.

Computational modelling approach into the thermal characteristics of agitated vessels can be of great benefit for accurate design. The heuristics that are prevalent in current design methodologies are prone to inaccuracies. Designers often follow a strict design procedure to estimate the expected performance of a piece of heat transfer equipment and finally adjust their findings by introducing a safety factor. This mistrust of existing methodologies is founded on the unreliability of the available design information.

Descriptive models based on realistic assumptions, together with supportive experimental data can provide dependable data. The obvious advantage of good computer simulation is that several different types of process equipment design can be considered for the same task without the necessity of costly and exhaustive experimentation.

One of the implications of computer simulation on research is that the development of design correlations can be geared towards the flexibility of computer implementation. It is possible to consider the variation of many different parameters with respect to time and temperature if required. This implies that correlations are not required to be generalised for large variations in working conditions, but that several more specific correlations can be used within restricted ranges.

The purpose of this paper is to describe the role of descriptive mathematical models in agitated vessels and the way in which they could be used not only with respect to performance prediction and parameter estimation but also in seeking alternative design for enhanced performance. Theoretically predicted and experimentally observed information will be presented and a comparative study between a commonly used plain vessel configuration and a dimpled type will be given.

COMPUTATIONAL MODELLING

Computational modelling is the prediction of performance of systems from their mathematical description using computational resources. Models are normally solved by means of suitable numerical techniques. In the case of, for example, heating of materials in a stirred system, the simplest way of predicting the temperature of the vessel contents is by considering the heating medium and process fluid to be in intimate contact.

The equation that can be derived from such an idealization is given by:

$$T = T_s - (T_s - T_i) \, e^{-UA\Delta\theta / m_f c_f} \qquad (2)$$

which can be used either to predict the thermal response of the agitated fluid or to evaluate the overall coefficient, U. The profile produced by this consideration overestimates the temperatures of the contents when compared with the experimental data and the predictions, found from more sophisticated models. Its one advantage is the relative ease with which a predicted temperature profile can be produced and, in addition, Equation (1) can be used as a useful heuristic in the absence of experimental data to validate more complex models.

A descriptive model should consider the existence of the vessel wall by incorporating the appropriate conduction/convection terms in the model equations. When a temperature gradient is considered in the wall then the solution of the problem becomes much more complex. The accumulation of heat in the wall is difficult to quantify as the temperature varies throughout the wall.

In such a case typical equations would be of the forms:

Vessel Wall:
$$\frac{\partial^2 T_v}{\partial r^2} + \frac{1}{r} \frac{\partial T_v}{\partial r} = \frac{\rho_v C_v}{k_v} \frac{\partial T_v}{\partial \theta} \qquad (3)$$

Vessel Contents:
$$\rho_f C_f V_f \frac{\partial T_f}{\partial \theta} = h_i A_i \, (T_v - T_f) \qquad (4)$$

Boundary conditions:

$$\text{B.C.1} \quad r = r_o \qquad h_j \, (T_s - T_v \Big|_{r_o}) = -k \left.\frac{\partial T_v}{\partial r}\right|_{r_o} \tag{5}$$

$$\text{B.C.2} \quad r = r_i \qquad h_i (T_v \Big|_{r_i} - T_c) = -k_v \left.\frac{\partial T_v}{\partial r}\right|_{r_i} \tag{6}$$

The solution of the model equations, (3) - (6), is best achieved through the use of finite difference techniques. The Crank-Nicolson scheme is used to numerically solve the partial differential equations complete with the associated boundary conditions. The first order differential term is fully implicit but the second order derivative is calculated as the numerical average of the fully implicit expression and the fully explicit method. The main advantage of this technique is that the next time row can be predicted at each stage in a similar way as the explicit technique but the scheme is valid for all values of r. The compromise position of the Crank-Nicolson scheme produces a stable and accurate numerical solution of the model equations which require relatively short computational time. The final form of the finite difference representation of the model equations (3) - (6) are outlined below:

$$(1+\alpha+\tfrac{1}{2}\alpha\beta)T_v(1,n+1)-\alpha T_v(2,n+1)=(1-\alpha-\tfrac{1}{2}\alpha\beta)T_v(1,n)+\alpha T_v(2,n)+\alpha\beta T_s \tag{7}$$

$$-\tfrac{1}{2}\alpha T_v(i-1,n+1)+(1+\alpha)T_v(i,n+1)-\tfrac{1}{2}\alpha T_v(i+1,n+1) =$$
$$(1-\alpha)T_v(i,n)+\tfrac{1}{2}\alpha(T_v(i+1,n)+T_v(i-1,n)) \tag{8}$$

$$-\alpha T_v(I-1,n+1)+(1+\alpha+\tfrac{1}{2}\alpha\gamma)T_v(I,n+1)-\tfrac{1}{2}\alpha\gamma T_c(n+1) =$$
$$\alpha T_v(I-1,n)+(1-\alpha-\tfrac{1}{2}\alpha\gamma)T_v(I,n)+\tfrac{1}{2}\alpha\gamma T_c(n) \tag{9}$$

$$(2+w)T_c(n+1)-wT_v(I,n+1) = (2-w)T_c(n)+wT_v(I,n) \tag{10}$$

where $\quad \alpha = \dfrac{k_v \, \Delta \theta}{\rho_v C_v (\Delta x)^2} \qquad \beta = \dfrac{2 \, \Delta x \, h_s}{k_v}$

$\qquad\qquad \gamma = \dfrac{2 \, \Delta x \, h_i}{k_v} \quad \text{and} \quad \omega = \dfrac{h_i A_i \Delta \theta}{\rho_f C_f V_f}$

Equations (7)-(10), which have to be solved simultaneously, provide numerical information on vessel wall temperatures as well as contents.

COMPUTATIONAL RESULTS

The numerical form of the model equations solved and a range of physical and operating conditions have been studied. The objective here is to establish the sensitivity of system performance to various factors. As far as the case under consideration, i.e. heating of liquids, water and steam are chosen as the media. The physical properties of water vary greatly between the temperatures of 25°C and 125°C. Therefore, it is reasonable to assume that the convective film resistances would also change. The problem in varying boundary conditions within a finite difference scheme is that the matrix of temperature coefficients will alter with the change in boundary conditions. This implies that the equations have to be re-defined within the computer model and a new solution obtained every time the boundary conditions are altered. The computational time required for such a model is much higher than for the simpler type with constant boundary conditions. The computational program of the numerical equations of the system contains its own database of the physical properties of water through the temperature range encountered in the system and uses linear interpolation to determine the properties at the process temperatures.

This model is the most realistic consideration to date and the accuracy of the prediction is dependent only on the quality of the heat transfer correlations used in the calculation of the convective film resistances. Figure 2 shows quite clearly that whilst the internal film heat transfer coefficient, h_i, is relatively stable, the jacket side coefficient, h_j, increases greatly with time. A comparison between predicted temperature profiles of fixed and changing boundary conditions are shown in Figure 3. The influence of this change on the performance appears to be noticeable. This finding clearly indicates that the adjustment of boundary conditions to account for the variation of the physical properties is desirable as the model becomes more realistic. It is, however, worth noting that the accuracy of the temperature prediction is dependent on the reliability of the employed design correlations. The next step in model development is to vary the value of the jacket side coefficient without re-evaluation of this coefficient with temperature. The jacket side heat transfer coefficient is the most difficult to determine and as a result the least reliable. In order to establish the sensitivity of the prediction to this coefficient the performance of the system is simulated using different values of h_j. The results showed similar trends to those of Figure 3. This, again, signifies the importance of accurate data.

An object that is not at thermal equilibrium with it's surroundings will undergo heat transfer with it's environment. In the case of a vessel at a temperature above ambient conditions the heat transfer with the surroundings is referred to as heat loss. If the vessel is being heated, then the situation is more difficult to assess. Only the heat loss from the process fluid itself need be considered. Heat loss through the jacket to the environment will increase the steam requirements of the system but will not affect the heating time of the vessel contents. The heat loss term, $h_1 A_{f_s} (T_f - T_a)$, appears in the fluid phase heat balance equation and for practical values of h_1, < 100 W/m^2 K, its effect on the predicted performance is marginal.

Inspection of the simulated results, Figures 2 and 3, reveals noticeable differences between predicted and observed performances. This is despite the refined nature of the mathematical description of the process. This clearly proves the weakness of the employed heat transfer coefficient correlations. In a further attempt, a series of experimental data have been used to establish the best estimates of the coefficients through the matching of the model prediction to the observed profiles. Extensive analysis revealed that the existing correlations overestimate the convective coefficients. For example, the vessel side coefficient, h_i, is overpredicted by at least 50%. This is partly attributed to the fact that fouling cannot be quantified so easily. Typical results are being given in Figure 4. However, as indicated earlier, heat loss is found to have very little effect on the predicted temperature profiles (Bentley, 1988).

IMPROVED DESIGNS

The models presented so far do not make any assumptions about the outer jacket wall geometry. The inner jacket wall, separating the steam from the process fluid, is defined as a plain vertical cylinder of uniform surface and thickness. In this way the inner jacket can be assumed to have uniform heat transfer properties and so no allowance is made for angular or longitudinal heat transfer abnormalities. Dimpled jackets have a dimpled surface on the outer jacket wall. These dimples intrude into the jacket cavity and enhance heat transfer by promoting greater degrees of turbulence in the heating or cooling medium in the jacket.

Dimpled jackets have been shown to have higher performance for both heating and cooling when compared with conventional jacket designs.

The heat transfer coefficient for the jacket is estimated in the models using the Nusselt theory for film condensation. The correlation that results from this classical approach is only valid for specific geometries, in this case a vertical plane surface. This is still valid in a dimpled jacket when considering the inner jacket wall, as this is a vertical plane surface.

The temperature history profiles produced by the two phase model, described earlier, have been compared with experimental data obtained from a dimpled jacketed vessel. This vessel has the same configuration and size as the plain jacketed vessel investigated earlier but with a dimpled exterior jacket surface. It is found that dimpled vessels provide higher heat transfer coefficients than the plain one by at least 20%. The exact magnitude of this difference is influenced by factors such as dimple size and spacing as well as the operating and other conditions of the system. However, the sensitivity of performance to an increase in the value of the thermal coefficient has already been illustrated, Figure 3.

CONCLUSIONS

A mathematical model of heat transfer in agitated vessels has been formulated and solved numerically for the prediction of their thermal behaviour. The model is used to examine the sensitivity of performance to system parameters. Amongst all the issues studied here, the reliability of the heat transfer coefficient data can be said to be the most important single factor influencing the accuracy of the design. As far as improved performance is concerned certain refinement of the configuration can provide the desired enhancement. A practical approach would be, for example, to insert flow guides in the jacket side of dimpled systems.

NOMENCLATURE

A	surface area (m^2)	Subscripts	
C	heat capacity (J/kg K)		
h	heat transfer coefficient (W/m^{2}K)	a	ambient
k	thermal conductivity (W/m K)	c	content
m	mass of liquid (kg)	f	fluid
R	resistance (K/W)	fs	fluid surface
r	radius (m)	i	inside
T	temperature (K)	j,o	jacket side or outside
T_i	initial temperature (K)	l	loss
U	overall heat transfer coefficient (W/m^{2}K)	s	steam
V	volume (m^3)	v	vessel wall
θ	time (s)		
ρ	density (kg/m^3)		

REFERENCES

Ayazi Shamlou, P. and Edwards, M.F., (1986). Chem. Eng. Sci. 41, 8, 1957.

Bentley, P.J., (1988). M.Eng. Internal Report, The University of Leeds.

Chilton, T.H., Drew, T.B. and Jebens, R.H., (1944). Ind. Eng. Chem., 36, 510.

Fogg, R.M. and Uhl, V.W., (1973). Chem. Eng. Prog., 69, 7,76.

Foumeny, E.A., (1979). M.Sc. Dissertation, The University of Leeds.

Markovitz, R.E., (1971). Chem. Engng., 15, 156.

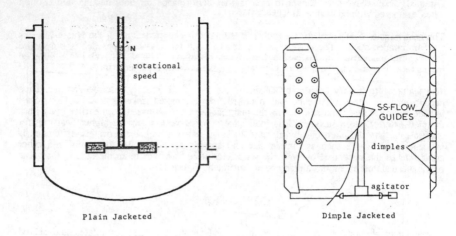

Figure 1: Schematic diagram of agitated vessels

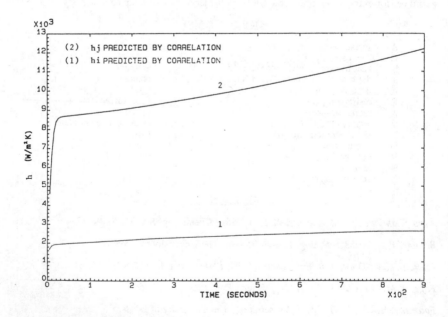

Figure 2: Effect of temperature on convective coefficients

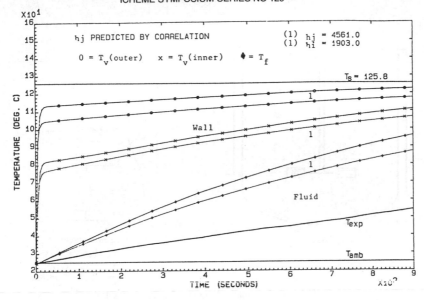

Figure 3: Effect of changing boundary conditions on the system performance

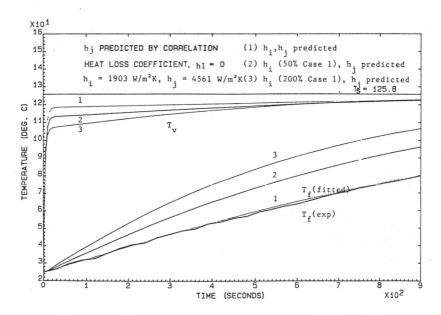

Figure 4: Predicted temperature profiles using different inside coefficient

HEAT TRANSFER IN MECHANICALLY AGITATED BIOREACTION VESSEL

P. Mohan* , A.N.Emery* and T.Al-Hassan+

Heat flux probes, modified to avoid boundary layer discontinuity effects, have been mounted in an 800 litre bioreactor of diameter 0.79 m, equipped also with a range of temperature sensors to enable determination of local temperature gradients as well as both local and global heat transfer coefficients. The experimental programme uses; 1) non-aerated Carboxy Methyl Cellulose (CMC), 2) the same fluid aerated, and 3) real multi-phase fungal fermentation broths. Parallel experiments aim to characterise both the reactor hydrodynamics and the morphology of the solid microbial phases, and these can be related to the heat transfer performance.

Axial local variations of heat flux and temperature at the wall have been shown using a Rushton Turbine in a standard vessel configuration working in the above fluids, and interpreted in terms of observed hydrodynamic behaviour.

INTRODUCTION

Intensive bioreaction processes are increasingly found to be rate-limited by inadequate heat transfer capability, a situation arising equally from the poor applicability of conventional heat transfer prediction methods in bioreaction environments and from the lack of a body of data covering the appropriate operating regimes. In particular the modification of heat transfer by the presence of the microbial solid phase itself, both in its effects on bulk flow properties and in boundary layer modification have received little attention. The modification is described of a pilot-scale bioreactor to allow heat transfer studies to be conducted in real and simulated bioreaction environments as well as the correlation of the data with relevant operating parameters.

EQUIPMENT

This study has used a well instrumented LSL-Biolafitte Bioreactor of 600 l working volume which has been modified as follows for heat transfer studies:

In-situ measurement of Heat Flux

The micro-foil heat flux sensors (Rhopoint Inc) used are of the differential thermocouple type which utilise a thin foil thermopile bonded to both sides of a known thermal barrier as shown in fig. 1. This also shows that the "skirt" of the sensor has been extended by 21mm in each direction so that the boundary layer is fully developed over the exposed sensor when it is mounted on the inner vessel wall. This modification ensures that the probe is hydrodynamically non-invasive – otherwise disruption of the boundary layer will adversely affect heat transfer results [1]. Initially two heat flux probes were used mounted at different heights on the vessel wall; after validation a third heat flux probe was placed in between

*School of Chemical Engineering, University of Birmingham, Birmingham B15 2TT
+Health and Safety Executive, Ferguson House, 15 Marylebone Road, London NW1 5JD

the above two probes. A fourth heat flux probe is stuck to the outside of the jacket at the same height as heat flux probe 2 to monitor the heat lost to the environment.

Temperature profiling

A range of thermocouples and resistance temperature detectors has been added to measure process fluid and wall temperature at various positions in the vessel as shown in Figure 2. In addition two thermocouples and a turbine flow meter in the heating/cooling water circuit allow completion of the heat balance. Simultaneous measurement of the power transmitted through the agitator can be effected using the strain gauges installed. All process data is passed through a TCS model T-100, type 6438 interface to the SETCON process management system mounted on a MicroVax 2 .

Methodology

The heat flux probe measures heat flow per unit area, i.e. Q/A; in millivolts, this data is then computer processed to obtain the corresponding value in W/m^2. The heat flux probe also has a thermocouple point in it. This thermocouple measures the temperature of the wall. One of the mobile thermocouples on the baffles is fixed at the same plane as that of the heat flux probe, and thus measures the temperature of the bulk at that plane. The heat transfer coefficient then at that plane is :

Heat tr. coeff.=(Heat flux)/(Temp. difference between the wall and the bulk)

The heat transfer experiment in the pilot plant is conducted under unsteady state conditions. This unsteady state is approximated in terms of 1 min. steady state steps. Shorter step lengths (down to 3 secs) would in some cases be desirable, but the length of fermentation processes, and the need to follow an extensive part of that time course means that the data handling capacity becomes limiting if such step lengths are used. It should be noted here that, since the heat flux probe has an adhesive pad for attachment to the stainless steel wall surface, an allowance for this has been made in the heat transfer calculations. The experiment was initially setup for a standard vessel configuration (SVC) with a single Rushton turbine.

EXPERIMENTAL PROGRAMME

The experimental programme aims to develop a tool for studying heat transfer in actual fermentations. The intention is to integrate the study of heat transfer with that of rheology, power and morphology. To achieve this the experimental programme has been conceived in three stages using; 1) non-aerated non-Newtonian fluids, Carboxy Methyl Cellulose (CMC), 2) the same fluid aerated and 3) real multi-phase fermentations.

As a first step in heat transfer measurement a limited number of experiments was conducted with water in the SVC with a single Rushton turbine. The heat transfer coefficients calculated from these water experiments showed that, with water in the process side the heat transfer limiting factor was the jacket side heat transfer rate due to the limited cooling water flow rate. The heat transfer resistance offered by the stainless steel vessel wall, which is 5 mm in thickness, was also substantial. When the heat transfer is jacket side controlled then this leads to three problems :

       a) The process side heat transfer can no longer be assumed to be at steady state even in small intervals of time.
       b) The process side driving force, i.e. the temperature difference between the wall and the bulk, is very small.
       c) The heat flux probe no longer measures changes in heat transfer performance resulting from changes in process-side determinants.

This condition is however avoided when high viscosity process fluids are used such as CMC and fungal fermentation broths due to the much lower process side film heat transfer coefficients which are then obtained.

CMC Experiments

CMC is widely used as a model (non-Newtonian) viscous fluid. The whole experimental setup was first tested with 0.1% (w/w) CMC at an impeller speed of 200 rpm (Re=9000). The fluid was heated from 30 ^{0}C to 45^{0}C and then cooled back to 30^{0}C in a batch. The heating and cooling was done gradually to prevent sudden changes of temperature.

The raw data show fluctuations resulting from the effects of the following factors:

a) The local heat transfer coefficient is influenced by flow patterns and local hydrodynamics; this has been recently confirmed, using a similar probe, by Fasano et al [2]. The fluctuation is mainly caused by the fluctuating velocity of the fluid near the boundary layer. However in our case due to the long sampling time these fluctuations are randomly picked up by the probes and hence they do not show a uniform pattern.

b) There are temporal fluctuations in heat transfer, specially noticeable near the jacket inlet, because the inlet temperature is changing with time.

c) The signal from the heat flux probe is small and the circuit is susceptible to pick up of electrical noise from other equipment.

However the raw data can be successfully analysed if the data is smoothed using a second order regression technique after it is split into heating and cooling data for independent processing. The axial variation of heat transfer coefficient as measured by the two probes is shown in figure 3. Figure 3 also shows a variation between the heat transfer coefficient values measured by the two probes of approximately 50% for a difference of height of only 0.31m, the higher value being recorded, as one might expect, at the plane closest to the impeller; this emphasises the fact that heat transfer is a local phenomenon and a local study will give a better picture than an overall one.

For 0.28% CMC a range of impeller speeds was used, i.e 100, 150, 200 and 250 rpm. The processing of the process side signals was similar to that described for 0.1% CMC but with the additional heat flux probe. Figure 4 shows the local heat transfer profile. This local heat transfer should ideally be linked to a suitable hydrodynamic model, as also suggested by Brain and Man [5] and this is the intention in the next phase of the work (in collaboration with the Mixing Group of Prof. A.W. Nienow) . Such a model would enable the prediction of flow velocities adjacent to the vessel wall as determined by vessel and impeller geometry, power input, gas flow rate and the rheology of the fluid. Such flow velocities are clearly the major determinants of process side film heat transfer coefficients.

The results obtained from the above experiments were validated by applying a suitable correlation from the literature to predict heat transfer coefficients for comparison with those measured. To calculate the global heat transfer the average value of the heat flux probes was taken. A recent correlation of Kai et al [6], (equation 1), was used this having been based on a wide range of vessel and impeller geometries.

$$h_j D_T/K_T = 0.456(\epsilon D_T^4/\nu_a^3)^{2/9}(c_p \mu_a/K_T)^{1/3}(D/D_T)^{0.58}(\Sigma w_i \sin\theta/H_L)^{0.71}(D_T/H_L)^{-1.63} \qquad (1)$$

The comparison (see the following Table) shows that the experimental results agree with the predictions based on this correlation. A heat balance (including the heat dissipated by the impeller) between the process side and jacket side heat transfer rates shows that the unaccounted losses amount to about 5% of the total heat flux.

Conc. of CMC (w/w)	Impeller speed (rpm)	Heat tr. coeff., W/m^2K	
		Kai et al	Present Study
0.1%	200 (Re=9000)	410	432
0.28%	100 (Re=980)	122	160
0.28%	150 (Re=1750)	195	250
0.28%	200 (Re=2620)	271	300
0.28%	250 (Re=3560)	350	390

The opportunity next arose to move straight to validation of the techniques in a real fermentation, prior to a return to study of aerated CMC solutions at a later date.

Fermentation Experiment

Heat evolution during growth and metabolism is a universal characteristic of living organisms, and it has been suggested [7] that its measurement can be used as an on-line indicator of cell biomass. However, such determinations may be compromised by variation in local heat transfer performance over the course of the fermentation which may arise from changes both in cell concentration and in morphology. In order to investigate the scale of such changes, a fungal fermentation, known to give a wide range of rheological behaviour was used in this experimentation. A strain of Penicillium chrysogenum known as PC 8 was obtained from SmithKline Beecham Pharmaceuticals Ltd. . The major constituents of the medium were 25 g/l of glucose, 10 g/l of $(NH_4)_2SO_4$ and 7.5 g/l of KH_2PO_4. The air flow rate was initially 200 l/m but this was raised, first to 250 l/m at 25.5 hours and then to 300 l/m at 29 hours in order to meet oxygen demand. Similarly the agitator speed was increased over the same period over the range 250 - 350 rpm. Penicillium grows as a mycelium, and the morphology varies from discrete filaments of hyphae to pellets of highly entangled hyphal mass.

The temperature setpoint for the fermentation was 26 ^0C (for optimum growth and Penicillin production) and this was maintained by the jacket heating / cooling loop. For temperature control there is a secondary cooling loop which is activated by a control (PID) valve which injects mains water into the recirculation loop hence increasing the jacket flow rate. This change in jacket flow rate induces some fluctuations in the heat flux signal. The data processing technique was however exactly the same as that used in the unsteady state experiments with 0.1% CMC. The results of the calculation of local heat transfer coefficients are presented in figure 5, which covers the period during which the cell growth passes from the exponential to the stationary phase. During the early part of the growth phase the system was, as for water, jacket - side controlled. Our experience with unaerated CMC confirmed that with an increase in apparent viscosity the heat transfer coefficient decreased rapidly. However in this fermentation, the heat transfer coefficient actually declined only slightly during the exponential growth-phase, even though both cell mass and apparent viscosity were increasing at this time (data not shown). It was apparent also that even quite sharp changes in power input and gas flow rate were not immediately reflected in the changes in heat transfer performance that would be predicted.

The factors which enhance heat transfer are not only aeration rate (at least below the flooding rate), impeller speed and bulk flow but also the morphology of the solid phase (i.e. the fungal mycelium). Blakeborough et al [3,4] showed for an air lift fermenter that at higher biomass concentrations the solid phase tends to scrape the boundary layer thus enhancing the heat transfer, but that the effect was dependent upon the (qualitatively assessed) morphology. They found that heat transfer enhancement by the presence of the fungal mycelium of *A. niger* was dependent on the morphology of the mycelium (as for example determined by changes in medium osmolarity). Fermentation conditions therefore add an additional level of complexity to understanding and prediction of heat transfer in non-Newtonian fluids – and for the biochemical engineer it is an important objective to relate the effects on heat transfer to quantifiable morphological determinants exhibited by the fungal mycelium. The work of Thomas et al [8,9] now enables the latter to be quantified using methods based on image analysis and, in a further stage of the work, the development of appropriate prediction methods will be completed.

Figure 6 shows the profile of heat flux generation by the biomass which shows, as might be expected, a linear increase. However the slopes based on the data of the two probes (HF1 and HF2) differ, the respective values being 27 and 16 $(W/m^2)/(g/l)$. The higher value, obtained at the point closer to the impeller may result both from local changes in metabolic heat production (due to spatial variations in oxygen transfer and uptake) and to axial variations in the effects of impeller-induced flow at the heat transfer surface. Clearly these factors critically affect the choice of position in any potential process identification and control based on heat flux measurements.

For these initial experiments oxygen uptake rate could be determined using off-gas analysis by mass spectrometer, and it could be demonstrated, using an average value of heat-flux measured for the two probes, an assumption partially justified by the near linear axial variation in heat flux between regimes of high and low intensity shown for this speed range in figure 4, that the heat production by this bioreaction closely matches that predicted by Cooney et al [10]. According to them, irrespective of the growth rate, substrate, and type of organism, the heat production rate is related to the oxygen uptake by the factor 0.124 + 0.003 kcal/mmole of oxygen. The experimentally determined factor for this study was found to be 0.1134 kcal/mmole of oxygen (see figure 7), allowance having been made for heat input by the impeller and air flow and heat loss by the evaporation of water, though the correlation is not very convincing.

CONCLUSIONS

Initial experiments using heat flux probes in real fermentation systems have proved both the feasibility and value of the approach. Care to avoid disturbance of the boundary layer by the sensor itself, and to apply appropriate signal processing results in reliable and apparently accurate on-line determination of process-side heat transfer coefficients during actual fermentations.

Early data already shows that such coefficients are not determined solely by the interaction of power input, geometry and flow properties but also by the structure of the fungal mycelium itself. Further work using simultaneous determination of both power uptake and morphological characteristics as well as heat transfer coefficient is intended to classify and characterise these interactions and provide a proper guide for heat transfer design in bioreaction environments.

REFERENCES

1. Aske, H., Beek, W.J, Van Barrel and Graauw, (1967), Chem. Eng. Sci., 22, 135-146.

2. Fasano, J.B., Brodkey, R.S. and Haam, S.J. , (1991), 7th European Conference on Mixing, Ed. Bruxelmane, Belgium, vol. 2, pp. 497-505, 18-20 Sep. .

3. Blakebrough, N., McManamey, W.J. and Tart, K.R., (1978), Trans IChemE, 56, 127-135.

4. Blakebrough, N., McManamey, W.J. and Walker, G., (1983), Chem Eng Res Des, 61, 264-266.

5. Brain, T.J.S. and Man, K.L., (1989), Chem Eng Prog, July 1989, 76-80.

6. Kai, W. and Shengyao, Y., (1989), Chem. Eng. Sci., 44, (1), 33-40.

7. Wang, H., Wang, D.I.C. and Cooney C.L., (1978), European J. Appl. Microbiol. Biotechnol. 5, (5), pp. 207-214.

8. Packer, H.L., Keshavarz-Moore, E., Lilly, M.D. and Thomas, C.R., accepted for publication in Biotechnol Bioeng.

9. Packer, H.L. and Thomas, C.R., (1990), Biotechnol Bioeng., 35, 870-881.

10. Cooney, C.L., Wang, D.I.C. and Mateles, R.I., (1968), Biotechnol. Bioeng., 11, 269-281.

NOMENCLATURE

h_J : Jacket heat transfer coefficient (process side) , W/m^2K

c_p : Specific heat of the fluid , J/kg K

D_T : Diameter of the vessel , m

K_T : Thermal conductivity of the process fluid , W/m K

D : Diameter of the impeller , m

H_L : Height of the liquid level inside the vessel , m

w : Width of the impeller blade , m

ε : Power consumed per unit mass = $P / [(\pi/4) D_T^2 H_L \rho]$, W/kg

ρ : Density , kg/m^3

ν_a : Apparent kinematic viscosity = μ_a/ρ , m^2/s

μ_a : Apparent viscosity , Pa. s

θ : Angle of the impeller blade , rad

ACKNOWLEDGEMENT

The authors wishes to acknowledge the financial support from SERC Biotechnology Directorate, the British-German Academic Research Collaboration (ARC) Programme, the ORS Awards Scheme and Hufton Postgraduate Fund.
The fermentation was carried out in collaboration with members of the Image Analysis Group (Dr. C.R.Thomas).

FIG. 1: MODIFIED HEAT FLUX PROBE

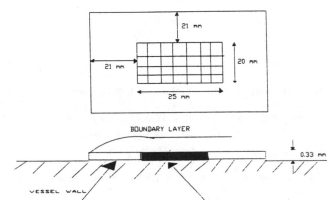

THE EXTRA AREA FOR THE BOUNDARY LAYER DEVELOPMENT

FIG. 2: 800 LITRE PILOT PLANT

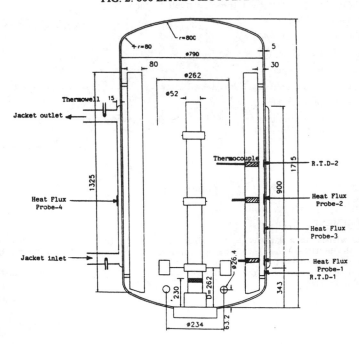

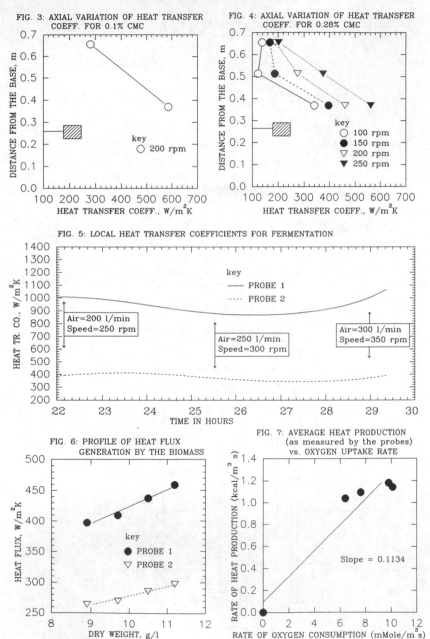

FIG. 3: AXIAL VARIATION OF HEAT TRANSFER COEFF. FOR 0.1% CMC

FIG. 4: AXIAL VARIATION OF HEAT TRANSFER COEFF. FOR 0.28% CMC

FIG. 5: LOCAL HEAT TRANSFER COEFFICIENTS FOR FERMENTATION

FIG. 6: PROFILE OF HEAT FLUX GENERATION BY THE BIOMASS

FIG. 7: AVERAGE HEAT PRODUCTION (as measured by the probes) vs. OXYGEN UPTAKE RATE

CHARACTERIZING OF FLOW STRUCTURE INSIDE A RECTANGULAR CHANNEL BY USE OF ELECTROCHEMICAL METHOD. CASE OF NON NEWTONIAN FLUIDS.

D.BEREIZIAT, R.DEVIENNE, M.LEBOUCHÉ

SYNOPSIS

Experiments based on an electrochemical method were carried out to determine the local wall shear rate and local flow structure in corrugated channels, for Newtonian and shear-thinning fluids. The range of the experiments covered following regimes : laminar, transitional and low REYNOLDS number turbulent. Experiments have shown the limits of the different flow regimes and the related fluid flow phenomena. These included appearance of secondary flows and associated spanwise   wall shear rate variations and destruction of these secondary flows by the onset of turbulence.

1) INTRODUCTION

The channel with corrugated walls is one of the devices employed for enhancing heat or mass transfer efficiency of processes at high PECLET numbers ; this is the case with plate heat exchangers used for food processing.

Several studies of heat and momentum transfer were developed to obtain flow patterns, pressure drop and heat transfer correlations in such channels,for Newtonian fluids at different steady flow regimes [1], [2], [3].

The application of these complex geometries in heat transfer equipment calls for an optimum in heat transfer , with an acceptable pressure drop. For an efficient use of the heating surface, the distribution of local heat transfer coefficient should be homogeneous. According to the above consideration, in order to find out adequate configurations of corrugated wall channels, FOCKE et al [4], MOLKI et al [5], GAISER et al [6] and SPARROW [7] examined experimentally the overall fluid flow states and heat or mass transfer characteristics, for various geometrical parameters of the corrugation.

A class of channel configurations frequently studied is characterized by wavy walls. NISHIMURA et al [8] obtained experimental and theoretical relations between the SHERWOOD numbers and the REYNOLDS numbers for a wavy channel, in the laminar and turbulent flow regimes for a Newtonian fluid. NISHIMURA et al [9] also calculated and measured the variation of pressure drop and wall shear stress with the REYNOLDS numbers for various wavy channels, at different flow regimes.

LEMTA - CNRS URA 875 - UNIVERSITE DE NANCY I- 2, Av.de la Forêt de Haye - BP 160 54504 NANCY. FRANCE

A search of the literature did not reveal results about the case of non-Newtonian fluids, in spite of the fitness of corrugated channels to treat very viscous fluids. To our knowledge, only EDWARDS et al [10] and LEULIET [3] obtained some results about the corrugated channel thermohydrodynamic behaviour with shear-thinning fluids.

Also, the original feature of present investigation resides in the attempt to give more accurate informations about the local structure of the flow, observed from the wall, for Newtonian and shear-thinning fluids at different flow regimes.

Also, we try to determine clearly the separation between the laminar, transitional and turbulent regimes by using the electrochemical "limiting current" method, giving the wall velocity gradient and particularly, its mean and fluctuating values.

2) EXPERIMENTAL, APPARATUS AND PROCEDURE.

2.1. Electrochemical Technique.

The principle of this technique [11], [12] is based on the diffusion-controlled reaction of an ion at the electrode-electrolyte interface. The solution conductivity is increased by adding a large amount of an inert electrolyte; consequently, migration of the reacting species becomes negligible ; limiting current condition are indicated by a plateau (see fig.1) on the current-potential scan. Under these conditions, the surface concentration of the reacting species may be assumed to be zero and the relation between the limiting (diffusion controlled) current I and the wall shear rate S may the evaluated by :

$$I = 0,807 \; z_0 f.C_0.A \left(\frac{|S|^2}{D} \right)^{1/3} \tag{1}$$

Where D is the molecular diffusion coefficient. The only hypothesis is a constant shear rate or velocity gradient through out the concentration field, near the wall.

Validity of the technique for the case of shear thinning fluids.

A preliminary work has consisted to verify the validity of a relation like (1) with a shear thinning fluid and so, to measure the molecular diffusivity of these solutions. These experiments were carrried out with a Couette flow.

For a shear -thinning fluid obeying to the generalised power law model :

$$|\tau| = \eta \; |S|^n \tag{2}$$

the diffusion current over a circular micro-electrode imbedded into the wall of the fixed cylinder of the Couette apparatus, calculated from (1), becomes :

$$I = 0,678.z_0.f.C_0 \; D^{2/3} \; d^{5/3} \left[\frac{2}{n} \; \frac{R_2^{2/n}}{R_2^{2/n} - R_1^{2/n}} - 1 \right]^{1/3} \; \Omega^{1/3} \tag{3}$$

2.2 Procedure.

The equinolar ferrocyanide/ferricyanide redocx couple was choosen as the electrochemical system, with potassium chloride as inert electrolyte.

Voltage scans were made before each experiments to determine the extent of limiting current plateau (Fig.1). A potential in the midrange of the current plateau, around - 600 mV ,was applied for all experiments.

All these experiments were made at constant temperature, aqual to $20°C \pm 0,1°C$.

Test liquids.

Both water and six shear-thinning aqueous solutions of sodium carboxymethylcellulose (CMC) at high molecular weight, at concentrations ranging from 0.25 to 3% by weight, where used to represent the purely viscous fluids.

A plate-cone constant shear stress viscometer (CARRI-MED) was used to charecterize the rheological properties of the test fluids. All shear-thinning liquids closely obeyed to the power law relationship over a wide range of shear rates, from $10 \ s^{-1}$ up to $1200 \ s^{-1}$.

2.3 Couette experiments. Results.

The mass transfer results, for two different test fluids, are presented in Fig. 2, where the limiting current I is plotted against the angular velocity , in fact : $\Omega^{1/3}$. At generalised Taylor numbers less than 24,6 (CMC 0.75%) and 28,7 (CMC 1%) the experimental data fall on a straight line, indicating the validity of the typical laminar mass transfer relationship (3). On this basis, we can define a molecular diffusion coefficient for either shear- thinning fluid.

2.4 Experimental device ; flow loop.

Description.

The description of the flow loop is facilitated by reference to Fig.3, where different components are shown schematically : test section (5) upstream (3) and downstream (7) tanks, flowmeter (6), exchanger (2) and volumetric pump (1). The most important part is the test section into which one smooth and three periodically corrugated channels were fitted. Pertinent dimensions are listed in Table 1 for the smooth and the Test 1 channel.

Test 1 channel is a scale model of an industrial prototype plate heat exchangers.

Measurements.

The diffusion limiting current was obtained with aid of embedded platinum electrodes. Circular ones (8 mm diameter : fig. 4-a) were used to obtain absolute value of the wall velocity gradient at different locations along the surface (as shown in Fig.5 for the Test 1 channel) The use of differential double-probes (Fig.4-b) sensitive to the direction of the flow was essential to determine the recirculating zones and their stability. Test 1 channel contains two fixed differential probes (Fig.5).

Signal processing.

The limiting current collected on the probes, which is function of the wall velocity gradient, is stored and processed by a specific numerical analyser ; the calculations concern mean and fluctuating values.

The ratio between the standard deviation (RMS) of the wall shear rate and it's mean value, called the fluctuation rate T, is a good way to show clearly the separation between the different flow regimes. According to relation (1) giving $|S| \alpha I^3$, we can write, for weak frequencies and fluctuation rate lower than 40% [13] :

$$\frac{i}{I} = \frac{1}{3}\frac{s}{|S|} \quad \text{and} \quad \frac{\sqrt{i^2}}{I} = \frac{1}{3}\frac{\sqrt{s^2}}{|S|} = \frac{T}{3} \tag{4}$$

3) EXPERIMENTAL RESULTS AND DISCUSSION.

The results presented in this paper concern the smooth and the Test 1 channel.

Experiments cover the generalised REYNOLDS number range from 10 to 17 000.

3.1 Case of the smooth channel.

Local mean wall shear rate distributions.

The flow characterization inside the smooth channel has boadly contributed to validate the experimental procedure. The relationship between the limiting current (proportional to the wall shear rate to the power 1/3) and the REYNOLDS number defined by LEULIET [3] is shown in Fig. 6 for laminar flow and different CMC concentrations. It was found that the influence of the CMC concentration is notable for solutions with concentrations lower than 1.25% ; beyond, there is no effect. Indeed, the wall velocity gradient is closely linked to the structural index of the power law fluid, which is varying for low concentrations of CMC. It is possible to show, from these results, that S is effectively proportionnal to R_{eg}.

Local fluctuating wall shear rate distributions.

The analyse of fluctuating wall velocity gradient presented in Fig. 7 shows clearly the transition between laminar and turbulent regimes , which happens at REYNOLDS number equal to 2.000. The maximum fluctuation rate of 28% is reached very soon and coincides with that obtained by SOURLIER [14] , by LDV.

The continuous and slow decreasing of fluctuation rate when REYNOLDS number increases indicates that the flow organizes itself up to the fully turbulent regime.

Statistical and spectral analysis of the motion is not presented in this paper ; the results were commented in a previous study [15].

3.2 Case of the Test 1 channel.

Local wall shear rate distribution.

For exemple, with a CMC solution at 0.75% by weight, Fig. 8 shows clearly, for a frontal probe (N°9 on fig.5), the end of laminar regime for R_{eg1} = 400, the transition extending up to R_{eg2} = 600. In the second case (Fig.9), for a probe embedded on the top of the corrugation (N°11 on Fig.5) , the observed critical values for REYNOLDS number varies slightly : R_{eg1} = 500 and R_{eg2} = 700 and the transition is less marked than in the first case.

Laminar flow :

In this case, we observe on fig.8 that I increases like $R_{eg}^{0,35}$, i-e that the wall velocity gradient or shear-rate is quite proportionnal to the generalised REYNOLDS number. For the top probe, on fig.9, this shear rate (and consequently the wall shear stress) increases more rapidly with REYNOLDS number.

Transitional and turbulent flow.

The corrugation induces turbulence at low values of REYNOLDS number, compared with smooth channel. This is significant for non-Newtonian fluids, because their high consistency means that they will generally be processed under low REYNOLDS number conditions.

Of major importance is the observation that the transitional flow shows various form of secondary flows. These recirculation vortices exhibit an intermittent reversed flow and a strong non uniformity of the flow, in the spanwise and streamwise directions. This flow intermittency is closely related to heat and mass transfer, because it induces a steep fall of the wall velocity gradient in the frontal zone of the corrugation (probe n°9). Moreover, this corresponds to a great increase of the fluctuation rate (Fig.10), up to 85% for a CMC solution at 0.5% by weight. Then, the high peak which is clearly pointed out for REYNOLDS number near 450 is followed by the decay of the fluctuation rate (30%) when the REYNOLDS number increases.

So, the flow intermittency is closely related to the wall shear rate. When the turbulent regime is established (recirculating vortices have deasappeared), wall shear rate increases very rapidly with REYNOLDS number on the frontal zone, less rapidly on the top of the corrugation.

3.3 Study of recirculating zones.

From differential double probes, sensitive to the direction of the flow, we observe very well the direct effect of the corrugation on the local flow structure. These probes are embedded in the hole of the corrugation. They are sensible, respectively to the velocity component parallel (probe A(1-2)Fig.5) and perpendicular (probe B(3-4)Fig.5) to the bottom line of the corrugation. Fig.11 gives results for a CMC solution at 0.75% by weight. For REYNOLDS number equal to 536 (curve a), one observe a velocity component parallel to the bottom line of the corrugation, oriented from 1 toward 2 (probe A). Besides, this velocity component conserves the same direction for all REYNOLDS numbers, demonstrating clearly the directive effect of the corrugation on the flow. The probe B shows that, in the direction perpendicular to the corrugation, the flow near the wall is opposite to the main flow, which proves the existence of a stable recirculating vortex.

As soon as REYNOLDS number reaches 578, the velocity component perpendicular to the corrugation may change his direction during time. When this REYNOLDS number increases, the unstable character increases also, and disappears for REYNOLDS number equal to 954 ; one observe only direct flow (no vortex) ; it is the end of the transitional regime.

4. CONCLUSION.

The results obtained here correspond to a specific geometry, which contains small zones of recirculation.We have found qualitative and quantitative informations about the local wall shear rate characteristics and about the local structure of the flow. These informations are usefull to understand the mechanisms which lead to increasing of heat and mass transfer. In the future, the effect of geometric parameters on the wall shear rate, in particular the role of the corrugation inclination angle, will be considered with other test channels.

NOMENCLATURE

Latin symbols.

A	Electrode area	m^2
b	Electrode length	m

C_o	Bulk concentration of electrochemical reactive	$mole.m^{-3}$
d	Micro-electrode diameter	m
D	Molecular diffusion coefficient	$m^2.s^{-1}$
Dh	Hydraulic diameter	m
E_c-E_a	Potential difference between the cathode and the anode	V
f	Faraday number	$C.mole^{-1}$
$I = \overline{I} + i$	Limiting current	A
K	Mass transfer coefficient	$m.s^{-1}$
l	Electrode width	m
n	Structural index	
R_1, R_2	Respectively, inner cylinder radius and external one (Couette apparatus)	m
$S = \overline{S} + s$	Instantaneous wall shear rate	s^{-1}
S_x, S_z	Wall shear rate components	s^{-1}
T	Fluctuation rate	
V	Mean velocity	$m.s^{-1}$
Z_o	Electron number in the electro-chemical reaction.	

Greek symbols.

η	Fluid consistency	$Pa.s^n$
Ω	Angular velocity of the inner cylinder	$rad.s^{-1}$
ρ	Volumic mass of the fluid	$kg.m^{-3}$
τ	Shear stress	$kg.m^{-1}s^{-2}$

LITERATURE CITED.

[1]   AMBLARD A.
        Comportement hydraulique et thermique d'un canal plan corrugué. Application aux
        échangeurs de chaleur à plaques.
        Thesis. Institut National Polytechnique de GRENOBLE. 1984.

[2]   HUGONNOT P.
        Etude locale de l'écoulement et performances thermohydrauliques à faible nombre de
        REYNOLDS d'un canal plan corrugué. Applications aux échangeurs de chaleur à
        plaques.
        Thesis. Université de NANCY I. 1989.

[3]   LEULIET J.C.
        Comportement hydraulique et thermique des échangeurs à plaques traitant des produits
        non-newtoniens.
        Thesis. Université de NANCY I. 1988.

[4]   FOCKE W.N.
        Turbulent convective transfer in plate heat exchangers.
        Int.Comm.Heat Mass Transfer, Vol.10, pp. 201-210. 1983.

[5]   MOLKI M. and YUEN C.M.
        Effect of interwall spacing on heat transfer and pressure drop in a corrugated-wall
        duct.
        Int.J.Heat Mass Transfer, Vol.29, n°7, pp.987-997. 1986.

[6] GAISER G. and KOTTKE V.
Effects of corrugation parameters on local and integral heat transfer in plate heat exchangers and regenerators.
IXth Int.Heat Mass Transfer Conference. JERUSALEM. 19-24 August. pp.85-90. 1990.

[7] SPARROW E.M. and COMB J.W.
Effect of interwall spacing and fluid flow inlet conditions on a corrugated-wall exchanger.
Int.J.Heat Mass Transfer. Vol.26, n°7, pp. 993-1005. 1983.

[8] NISHIMURA T., OHORI Y., KAJIMOTO Y and KAWAMURA V.
Mass Transfer characteristics in a channel with symmetric wavy wall for steady flow.
J.Chem.Eng. Japan, 18, n°6, pp. 550-555. 1985.

[9] NISHIMURA T., OHORI Y. and KAWAMURA Y.
Flow characteristics in a channel with symmetric wavy wall for steady flow.
J.Chem.Eng.Japan, 17, n°5, pp. 466-471. 1984.

[10] EDWARDS M.F., CHANGAL A.A. and PARROTT D.L.
Heat transfer and pressure drop characteristics of a plate heat exchanger using newtonian and non-newtonian liquids.
The Chemical Engineer, May, pp.286-293. 1974.

[11] LEBOUCHÉ M.
Contribution à l'étude des mouvements turbulents des liquides par la méthode polarographique.
Thèse d'Etat. Université de NANCY. 1968

[12] COGNET G.
Contribution à l'étude de l'écoulement de COUETTE par la méthode polarographique.
Thèse d'Etat. Université de NANCY. 1968.

[13] FORTUNA G. and HANRATTY T.J.
Frequency response of the boundary layer on wall transfer probes.
Int.J.Heat Mass Transfer. Vol.14, pp. 1499-1507. 1971.

[14] SOURLIER P.
Contribution à l'étude des propriétés convectives de fluides thermodépendants. Cas de l'écoulement en canal de section rectangulaire.
Thèse. Université de NANCY I. 1988.

[15] BEREIZIAT D., LEBOUCHE M. and DEVIENNE R.
Structure de l'écoulement turbulent entre plaques ondulées. Mise en oeuvre de techniques électrochimiques.
Traitement Industriel des Fluides Alimentaires non-Newtoniens. (TIFAN) Tome III, Actes du 3ème colloque, NANCY, 19-21 Septembre. 1990.

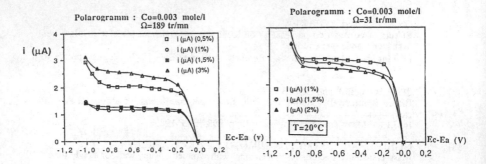

Figure 1

Scans current- potential.
Micro-electrode polarogramms.

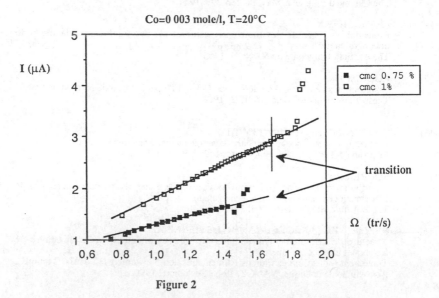

Figure 2

Evolution of the limiting current in the Couette apparatus.

Details of measure zone 9

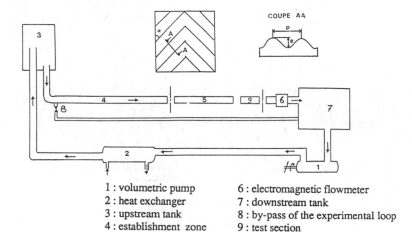

1 : volumetric pump
2 : heat exchanger
3 : upstream tank
4 : establishment zone
5 : corrugated zone

6 : electromagnetic flowmeter
7 : downstream tank
8 : by-pass of the experimental loop
9 : test section

Figure 3

EXPERIMENTAL APPARATUS

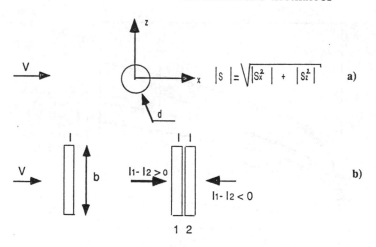

$$|S| = \sqrt{|S_x^2| + |S_z^2|}$$ a)

$I_1 - I_2 > 0$

$I_1 - I_2 < 0$

b)

Figure 4

ELECTROCHEMICAL MEASUREMENTS

a) **circular micro-electrode.**
b) **differential rectangular probe.**

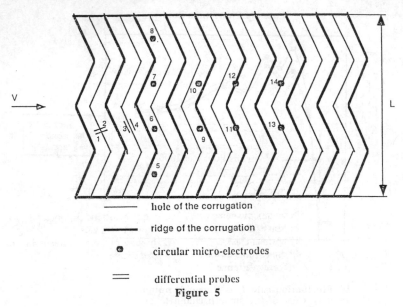

—————— hole of the corrugation

—————— ridge of the corrugation

● circular micro-electrodes

= differential probes

Figure 5

Details of test section and positions of electrodes

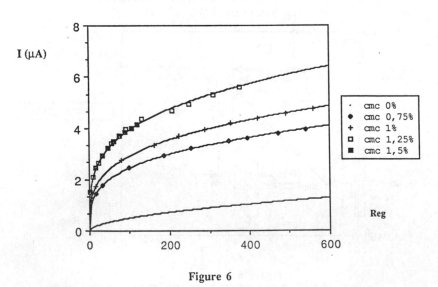

Figure 6

Mean wall shear rate $S^{1/3} \propto I$ vs generalised Reynolds number. Case of the smooth channel.

Case of a Newtonian fluid

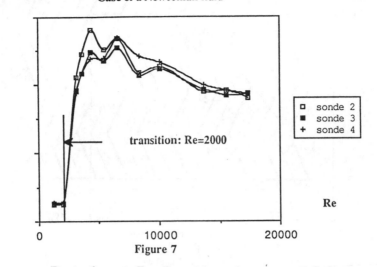

Figure 7

**Fluctuation rate T vs Reynolds number.
Case of the smooth channel.**

Gap 3 mm ; cmc 0.75%

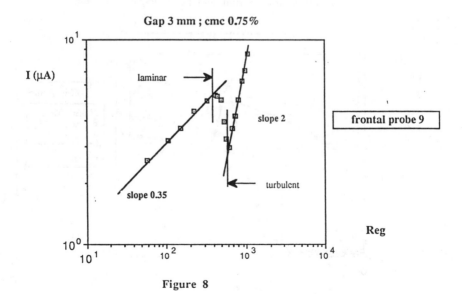

Figure 8

**Mean wall shear rate $S^{1/3} \propto I$ vs generalised Reynolds number.
Case of a frontal probe N°9.**

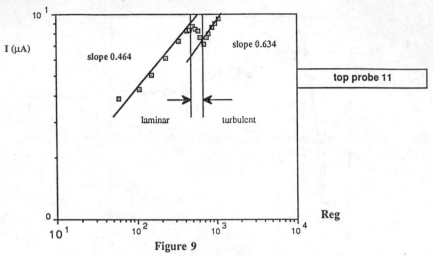

Figure 9

Mean wall shear rate $S^{1/3} \alpha$ I vs generalised Reynolds number.
Case of a top probe N° 11.

Case of a frontal probe

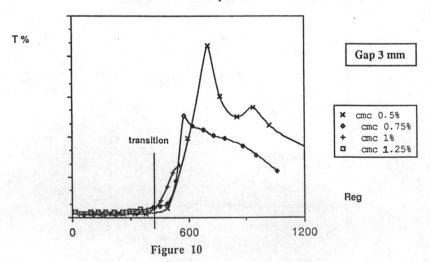

Figure 10

Fluctuation rate vs generalised Reynolds number.
Case of the Test1 corrugated channel.

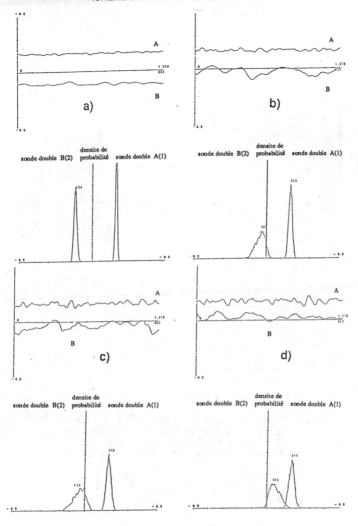

Figure 11

Study of recirculating zones from differential probes
for a CMC solution at 0.75 % by weight.
Signal and probality density for
different generalised Reynolds numbers.

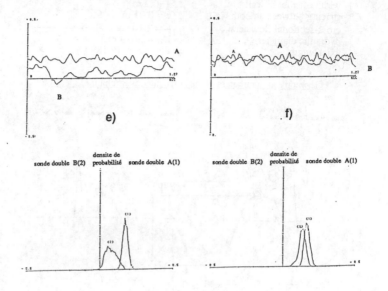

Figure 11

Study of recirculating zones from differential probes
for a CMC solution at 0.75 % by weight.
Signal and probality density for
different generalised Reynolds numbers.

a) Reg = 536   d) Reg = 679
b) Reg = 578   e) Reg = 734
c) Reg = 622   f) Reg = 954

probe A: $(I_1 - I_2) / (I_1 + I_2)$.
probe B: $(I_3 - I_4) / (I_3 + I_4)$

Geometrical characteristics of the smooth channel

- width of smooth wall : $l = 0,2$ m
- spacing between smooth walls : $e = 0.01$ m
- cross-sectional flow aera : $S = 2.10^{-3} m^2$
- hydraulic diameter : $D_h = 0.019$ m
- axial length : $L_t = 3.6$ m or $189 \, D_h$

Geometrical parameters of the corrugated wall channel : test 1

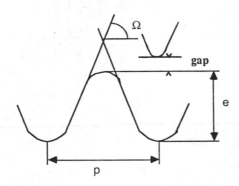

- material : PMMA
- plate length : $L = 0.78$ m
- plate width : $l = 0.2$ m
- depth of corrugations : $e = 3.5 . 10^{-3} m$
- cross-sectional area with
  gap equal to zero : $S = 7.15 . 10^{-4} m^2$
- corrugation inclination angle to
  main flow direction : $\alpha = 60°$
- corrugation angle : $\Omega = 45°$
- corrugation pitch : $p = 11$ mm
- hydraulic diameter with gap
  equal to zero : $D_h = 2e/\varepsilon = 7.10^{-3} m$
- ratio of developed to projected
  aera : $\varepsilon = 1.21$

Table 1

GEOMETRICAL CHARACTERISTICS OF THE TEST CHANNELS

HEAT TRANSFER AND PRESSURE DROP IN TUBE BANKS AT HIGH TURBULENCE INTENSITY

M. Beziel and K. Stephan

Institut für Technische Thermodynamik und Thermische Verfahrenstechnik, Universität Stuttgart, Pfaffenwaldring 9, D - W 7000 Stuttgart 80 (Germany)

Results of experimental investigations on the influence of turbulence intensity, transversal and longitudinal pitch-to-diameter ratio on heat transfer and pressure drop in tube bundles in inline arrangement are presented. The vertically arranged tubes with pitch-to-diameter ratios between 1.26 and 3.07 were heated by saturated steam and cooled outside by air in cross flow. The turbulence intensity in the entrance cross-section was enhanced by means of biplanar grids. The mean streamwise turbulence intensity behind the grids varied between 0.008 and 0.455. The measurements show an increase of the Nusselt number for a given turbulence intensity. The drag coefficient is only a weak function of the inlet turbulence intensity. The use of turbulence grids thus leads to higher efficiencies of tube bundle heat exchangers.

INTRODUCTION

Compact heat exchangers consisting of tube bundles with a few rows of plain tubes in cross flow are often used in industry. Heat transfer coefficients in the first rows of these heat exchangers are much lower than further downstream. Hence, the heat transfer rate of tube bundles with a few rows is mainly determined by the heat transfer of the first and seond row. In order to increase the heat transfer some investigators installed turbulence generators in the oncoming fluid. Bressler (1) reported that the heat transfer coefficient thus could be increased from the first to the fifth row of tube bundles with ten rows. Schellerich (2) correlated the measured Nusselt numbers for a single row and for the first row of a tube bundle consisting of four rows with the turbulence intensity. Stephan (3) determined the effect of a high turbulence on heat transfer and flow resistance experimentally. He found that the effect of turbulence on heat transfer is very pronounced for the first row and becomes smaller further downstream. Traub (4) examined four different tube bundles with three rows in inline arrangement at turbulence intensities up to 0.25. He found an increase of the heat transfer at constant Reynolds number with increasing turbulence intensity whereas the enhancement of heat transfer falls off from the first to the the the third row.

This paper presents results of systematical experimental investigations on heat transfer and pressure drop in tube bundles consisting of three rows of plain tubes in inline arrangement with equal transversal and longitudinal pitch-to-diameterer ratios between a=b=1.26 and a=b=3.07 at turbulence intensities up to 0.455.

EXPERIMENTAL PROCEDURE

Apparatus

The tubes were tested in an open wind tunnel. The wind tunnel was described in detail in (5). The experimental set-up is shown schematically in Figure 1. The cross-section of the channel was 258 mm x 258 mm. These dimensions allowed tube arrangements presented in Table 1.

TABLE 1 - Tube arrangements.

transversal and longitudinal pitch-to-diameter ratio $a=s_q/d_a$, $b=s_l/d_a$ [-]	tube diameter d_a [mm]	tranversal and longitudinal distance between tubes s_q, s_l [mm]	number of tubes in each row [-]
1.26	34	43.0	6
1.29	20	25.8	10
1.43	18	25.8	10
1.54	28	43.0	6
1.61	16	25.8	10
1.72	15	25.8	10
2.15	20	43.0	6
2.39	18	43.0	6
2.53	34	43.0	3
2.69	16	43.0	6
2.87	15	43.0	6
3.07	28	86.0	3

The tubes were heated by saturated steam, condensing inside, and cooled by air in cross flow. The turbulence intensity of the air in the entrance cross-section could be varied by means of different grids, placed at two different distances before the row.

Instrumentation

The mass flow rate of the air was measured by a nozzle at the inlet of the wind tunnel. All temperatures were measured by calibrated resistance thermometers. The air-side pressure drop across the tube bundle was obtained from the difference of the average static pressures before and behind the turbulence grid and the tube bundle (see p_1 and p_2 in Fig. 1). For heat transfer studies the mass flow rate of the condensate was determined. Five different biplanar grids were used, with the dimensions given in Table 2. Grid G 0 means no grid and thus the turbulence intensity of Tu=0.008 is the minimum streamwise turbulence intensity of the wind tunnel. New measurements of the mean streamwise turbulence intensity as a function of the distance x behind the grids by means of a X-wire probe and a constant temperature hot wire anemometer result in a higher turbulence intensity than reported in (5, 6). This is due to the fact that high velocity fluctuations perpendicular to the mean flow velocity exist quite close behind the grids and therefore the assumption of isotropic turbulence holds no longer even in the distance of $x_1=100$ mm. Thus turbulence intensities from 0.067 to 0.455 were obtained, Table 2.

TABLE 2 - Dimensions of turbulence generating grids.

grid	M	D	M/D	Tu ($x_1=100$mm)
	[mm]	[mm]	[-]	[-]
G 2/20	40	20	2	0.455
G 4/10	40	10	4	0.252
G 5/6	30	6	5	0.166
G 5/4	20	4	5	0.107
G 4/10	40	4	10	0.067
G 0	-	-	-	0.008

RESULTS

Heat transfer

In Figure 2 the Nußelt number Nu is shown as a function of the Reynolds number Re_e for each row and the tube bundle with a=b=2.87. The data for constant turbulence intensity can be approximated by straight lines. In Figure 2 only the curves for the minimum and the maximum turbulence intensity are shown. The data for the tube bundle are compared with an equation from Gnielinski (7) represented by the full line. This equation represents the data for a turbulence intensity of nearly 0.08 in the whole range of Reynolds-numbers. Furthermore it can be seen that the influence of the turbulence intensity decreases with the number of rows. In (5, 6) it was shown that the turbulence intensity increases the heat transfer of single rows up to 50%, whereas the present experiments show that the heat transfer for the second row increases much less rapidly with increasing turbulence intensity. For the third row of this inline arrangement no increase of the heat transfer with the turbulence intensity was observed. This is due to turbulent wake behind the first an the second row. The first two rows act as turbulence generator that equalizes the inlet turbulence intensity. Therefore no improvement of the heat transfer can be achieved.

Pressure drop

Figure 3 shows the drag coefficient ζ_{RB} as a function of the Reynolds number for the tube bundle with a=b=1.43. In this case the drag coefficient is independent of the Reynolds number and turbulence intensity, except for the highest turbulence intensity of Tu=0.455. The data are compared with equations from Jakob (8), Zukauskas (9) and Gaddis (10) represented by the full lines, respectively. It can be seen that these equations agree well for low Reynolds numbers for turbulence intensity Tu=0.008 and for high Reynolds numbers for turbulence intensities between 0.008 and 0.107.

DISCUSSION

It could be demonstrated that a turbulence grid can considerably intensify the heat transfer of tube bundles with rows of plain tubes in inline arrangement. Simultaneously the drag coefficient of the turbulence grid and the tube bundle increases and it is still an open question whether the enhanced heat transfer goes at the expense of the higher pressure drop across the system. As Stephan and Mitrovic (11) reported the dimensionless number St^3/ζ_{tot} is a criterion for the enhancement of heat transfer of a given heat exchanger with different turbulence intensities but same ventilation power. In Figure 4 this number is plotted as a function of the Reynolds number for the tube bundle with a=b=1.29. The curve with the open circles represents the data for the minimum turbulence intensity. All data above this reference curve represent heat exchangers with grids. For all grids an enhancement of the efficiency could be achieved. Due to the additional high pressure drop of the grid which generates a turbulence intensity of Tu=0.455 the efficiency of this exchanger is nearly the same as for the exchanger with Tu=0.008 in the whole range of Reynolds numbers.

CONCLUSIONS

1. The free stream turbulence intensity influences the heat transfer of tube bundles with plain tubes in inline arrangement. A considerable enhancement of the heat transfer could be achieved for the first and second row. For the third row the free stream turbulence intensity has no influence on the heat transfer.

2. The drag coefficient of the tube bundle turned out to be weak function of the inlet turbulence intensity.

3. As a criterion for the enhancement of the efficiency of heat exchangers with different turbulence grids the dimensionless number St^3/ζ_{tot} is used. It is shown that for tube bundles with an inline arrangement and small transversal and longitudinal pitch-to-diameter ratios the use of turbulence grids leads to higher efficiencies.

ACKNOWLEDGEMENT

We gratefully acknowledge the financial support by the Deutsche Forschungsgemeinschaft (DFG).

REFERENCES

(1) Bressler, R., 1958, Forsch.-Ing.-Wes. 24, 90-103.
(2) Schellerich, W., 1983, Forsch.ber. Dtsch. Kältetechn. Verein No. 8.
(3) Stephan, K., 1968, Verfahrenstechnik 4, 158-160.
(4) Traub, D., 1991, Chem. Eng. Process. 28, 173-181.
(5) Beziel, M., Stephan, K., 1990, Proc. 9th Int. Heat Transfer Conf., Jerusalem, Israel, vol.
     5, 67-71, G. Hetsroni, Hemisphere, Washington D. C.
(6) Beziel, M., Stephan, K., 1991, Chem.-Ing.-Tech. 5, 508-509.
(7) Gnielinski, V., 1991, VDI-Wärmeatlas, Gf1-Gf3.
(8) Jakob, M., 1938, Trans. ASME 60, 384-386.
(9) Zukauskas, A. A., Ulinskas, R., 1983, Heat Exchanger Design Handbook, vol.2 sec.
     2.2.4, E. U. Schlünder, Hemisphere, Washington D.C.
(10) Gaddis, E. S., 1991, VDI-Wärmeatlas, Ld1-Ld7.
(11) Stephan, K., Mitrovic, J., 1984, Chem.-Ing.-Tech. 6, 427-431.

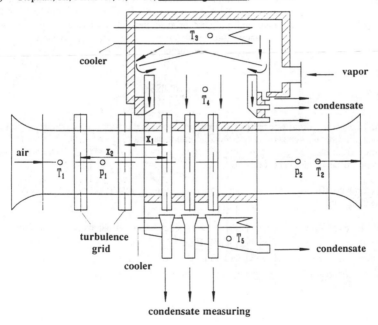

Figure 1. Experimental set-up

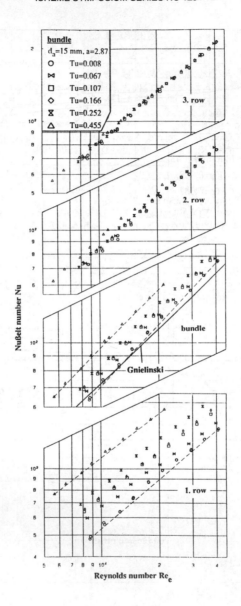

Figure 2. Heat transfer for the tube bundle with a=b=2.87.

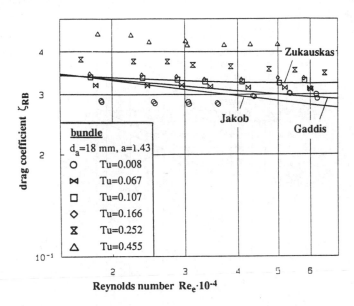

Figure 3. Pressure drop for a tube bundle with a=b=1.43.

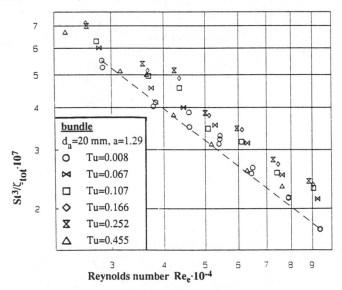

Figure 4. Criterion for the tube bundle with a=b=1.29.

FREE CONVECTION ALONG A VERTICAL FLAT PLATE WITH UNIFORM SUCTION OR INJECTION AND CONSTANT SURFACE HEAT FLUX

I. Pop
University of Cluj, Faculty of Mathematics, R-3400 Cluj, CP 253, Romania

T. Watanabe
Department of Mechanical Engineering, Faculty of Engineering, Iwate University, Morioka 020, Japan

Steady laminar free convection with uniform suction or injection along a semi-infinite vertical flat plate is studied for the case of constant surface heat flux rate. The partial differential equations are reduced to a system of nonlinear ordinary differential equations by means of the difference-differential method. The solution of these ordinary differential equations is expressed in a form of integral equations, which are then solved by iterative numerical quadratures. The results for the flow and heat transfer characteristics are reported for a wide range of suction or injection parameter and for different values of Prandtl numbers.

INTRODUCTION

The problem of laminar free convection flow and heat transfer from a vertical flat plate has been studied extensively by many authors. They mainly investigated the case of impermeable surface with constant temperature, but provided little information about the case of permeable surface subjected to constant heat flux, see Sparrow and Gregg[1], Roy[2], Wilks[3], Berezovsky and Sokovishin[4], Mahajan and Gebhart[5], Martynenko et al.[6], Carey and Mollendorf[7], Merkin[8,9], Yang and Lin[10], Merkin and Mahmood[11]. It is well known that suction or injection of fluid through the surface, as in mass transfer cooling, can significantly modify the flow field and affect the rate of heat tarnsfer in free convection, see Hartnett[12]. On the basis of previous work, it is realized that there is no complete theoretical solution to the problem of free convection from a permeable vertical flat plate. It is thus the primary objective of the present paper to invesitgate the effects of uniform suction or injection on the laminar free convection from a vertical flat plate. The plate is subjected to a uniform heat flux condition, which is often approximated in practical applications, and is easier to measure in a laboratory. Following the procedure similar to the one successfully employed by Watanabe and Kawakami[13] and Watanabe[14] to this class of problem, the nonsimilar equations are solved by using the difference-differential method. Sample results of the skin friction and heat transfer coefficients are presented in tabulation forms for a wide range of suction or injection parameter s (from -0.7 to 0.7) and for different values of Prandtl number (from 0.3 to 15). Such tabulations serve as a reference, against which other approximate solutions or experimental data can be compared in the future. Comparisons have been made with the available results for the case of zero suction or injection parameter. Our results are in reasonable agreement with those of other authors for $Pr = 0.733$.

ANALYSIS

Consider the steady-state laminar free convection flow from a permeable semi-infinite vertical flat plate maintained at a constant heat flux q_w, while the ambient fluid consisits of one at a uniform temperature T_∞. The fluid properties are assumed constant except for the density variations contributing to the buoyancy force. The governing boundary layer equations for the problem are

$$\frac{\partial u}{\partial x} + \frac{\partial v}{\partial y} = 0 \tag{1}$$

$$u \frac{\partial u}{\partial x} + v \frac{\partial u}{\partial y} = \nu \frac{\partial^2 u}{\partial y^2} + g\beta\,(T - T_\infty) \tag{2}$$

$$u \frac{\partial T}{\partial x} + v \frac{\partial T}{\partial y} = \frac{\nu}{Pr} \frac{\partial^2 T}{\partial y^2}\,. \tag{3}$$

We propose to solve these equations with the boundary conditions

$$y = 0 \;:\; u = 0,\; v = v_0, \qquad k \frac{\partial T}{\partial y} = -q_w \tag{4a}$$

$$y \to \infty \;:\; u = 0,\; T = T_\infty \tag{4b}$$

where v_0 is the velocity of suction or injection, when either $v_0 < 0$ or $v_0 > 0$, respectively.

As a solution of eqs. (1) to (4) it is now to seek one of the form

$$\eta = \frac{y}{x}\,(Gr_x/5)^{1/5}\,, \qquad f(x,\eta) = \frac{\psi}{5\nu}(Gr_x/5)^{-1/5} \tag{5a}$$

$$\theta(x,\eta) = \frac{T - T_\infty}{x\,q_w/k}\,(Gr_x/5)^{1/5} \tag{5b}$$

where ψ is the stream function defined indirectly as $u = \partial\psi/\partial y$ and $v = -\partial\psi/\partial x$. Thus, the velocity components u and v are written as

$$u = Ue\,\frac{\partial f}{\partial \eta} \tag{6a}$$

$$v = -\frac{\nu}{x}\,(Gr_x/5)^{1/5}\,(\,4f + 5x\,\frac{\partial f}{\partial x} - \eta\,\frac{\partial f}{\partial \eta}\,) \tag{6b}$$

where $Ue = (5\nu/x)\,(Gr_x/5)^{2/5}$.

Substituting (6) into eqs. (2) and (3), we get

$$\frac{\partial^3 f}{\partial \eta^3} + 4 f \frac{\partial^2 f}{\partial \eta^2} - 3 \left(\frac{\partial f}{\partial \eta} \right)^2 + \theta = 5 x \left(\frac{\partial f}{\partial \eta} \frac{\partial^2 f}{\partial x \partial \eta} - \frac{\partial f}{\partial x} \frac{\partial^2 f}{\partial \eta^2} \right) \qquad (7)$$

$$\frac{1}{Pr} \frac{\partial^2 \theta}{\partial \eta^2} + 4 f \frac{\partial \theta}{\partial \eta} - \frac{\partial f}{\partial \eta} \theta = 5 x \left(\frac{\partial \theta}{\partial x} \frac{\partial f}{\partial \eta} - \frac{\partial f}{\partial x} \frac{\partial \theta}{\partial \eta} \right) \qquad (8)$$

and the boundary conditions (4) become

$$\eta = 0 : \quad \frac{\partial f}{\partial \eta} = 0, \quad 4 f + 5 x \frac{\partial f}{\partial x} = 5 s x^{1/5} , \quad \frac{\partial \theta}{\partial \eta} = - 1 \qquad (9a)$$

$$\eta \to \infty : \quad \frac{\partial f}{\partial \eta} = 0, \quad \theta = 0 \qquad (9b)$$

where s is the parameter of suction ($s > 0$) or injection ($s < 0$). If we further put $x^* = C x^{1/5}$ ($\equiv s$), eqs. (7) and (8) can then be written as

$$\frac{\partial^3 f}{\partial \eta^3} + \left(4 f + x^* \frac{\partial f}{\partial x^*} \right) \frac{\partial^2 f}{\partial \eta^2} - x^* \frac{\partial^2 f}{\partial x^* \partial \eta} \frac{\partial f}{\partial \eta} - 3 \left(\frac{\partial f}{\partial \eta} \right)^2 + \theta - 0 \qquad (10)$$

$$\frac{1}{Pr} \frac{\partial^2 \theta}{\partial \eta^2} + \left(4 f + x^* \frac{\partial f}{\partial x^*} \right) \frac{\partial \theta}{\partial \eta} - x^* \frac{\partial f}{\partial \eta} \frac{\partial \theta}{\partial x^*} - \frac{\partial f}{\partial \eta} \theta = 0 \qquad (11)$$

while the boundary conditions (9) read

$$\eta = 0 : \quad \frac{\partial f}{\partial \eta} = 0, \quad 4 f + x^* \frac{\partial f}{\partial x^*} = 5 x^*, \quad \frac{\partial \theta}{\partial \eta} = - 1 \qquad (12a)$$

$$\eta \to \infty : \quad \frac{\partial f}{\partial \eta} = 0, \quad \theta = 0. \qquad (12b)$$

In technological applications, it is often the skin friction and heat transfer coefficients which are of the greatest interest. With the definition of these coefficients as

$$\frac{1}{2} C_f = \frac{\tau_w}{\rho U e^2} , \quad Nu = \frac{x \, q_w}{k (T - T_\infty)} \qquad (13)$$

we obtain

$$C_f \, (Gr_x/5)^{1/5} = \frac{2}{5} f''(s,0) \qquad (14a)$$

$$Nu \, (Gr_x/5)^{-1/5} = [- \theta(s,0)]^{-1} \qquad (14b)$$

where primes denote differentiation with respect to η.

Equations (10) and (11), subject to the boundary conditions (12), are solved by employing the difference-differential method proposed by Hartree and Womersley[15]. To do this, we shall approximate these equations by a system of ordinary differential equations by replacing the partial derivatives with respect to x^* at $x^* = x_i^* = ih$ (i = 0, 1, ⋯⋯) by finite-differences using, for example, a four-point formula of Gregory-Newton backward difference. A full description of this method is presented in rfs.[13,14] and it is unnecessary to repeat the details here.

RESULTS AND DISCUSSION

The numerical results were obtained for the suction or injection parameter s ranging from -0.7 to 0.7 and for different values of the Prandtl number P_r. Tables 1 and 2 list results for $f''(s,0)$ and $\theta(s,0)$ which are representative for the skin friction and heat transfer coefficients given by eq.(14) for P_r = 0.733(air) and P_r = 6.7 (water). We also present in Table 3 results for these coefficients in the case of an impermeable plate(s = 0) and for different values of P_r (from 0.3 to 15). In order to examine the accuracy acquired in the present method, the results are compared against the exact similarity solutions of Mahajan and Gebhart[5] for s = 0. Thus, for P_r = 0.733 Mahajan and Gebhart have obtained $f''(0)$ = 0.8089 and $\theta(0)$ = 1.4798, which are in excellent agreement with our results shown in Table 3.

Velocity and temperature profiles for P_r = 0.733 and for various values of s are plotted in Figs.1 and 2. As is expected, these profiles decrease with the increase of s. However, the accompanied figures are obtained for P_r = 0.733, when even if P_r and s are changed, similar profiles can be generated.

Table 1.　Values of $f''(s,0)$ and $\theta(s,0)$
for P_r = 0.733

s	$f''(s,0)$	$\theta(s,0)$
0.7	0.15181	0.38973
0.6	0.20677	0.45473
0.5	0.29337	0.54389
0.4	0.40855	0.66189
0.3	0.52809	0.81045
0.2	0.63659	0.99346
0.1	0.72994	1.21526
0	0.80893	1.47981
-0.1	0.87607	1.79068
-0.2	0.93426	2.15093
-0.3	0.98619	2.56313
-0.4	1.03403	3.02919
-0.5	1.07948	3.55042
-0.6	1.12368	4.12747
-0.7	1.16738	4.76045

Table 2. Values of $f''(s,0)$ and $\theta(s,0)$ for $Pr = 6.7$

s	$f''(s,0)$	$\theta(s,0)$
0.4	0.00557	0.07463
0.3	0.00990	0.09950
0.2	0.02228	0.14925
0.1	0.08699	0.29731
0	0.35633	0.84170
-0.1	0.75853	1.89262
-0.2	1.19247	3.31563
-0.3	1.60887	4.98283
-0.4	1.98909	6.82017

Table 3. Values of $f''(0)$ and $\theta(0)$

Pr	$f''(0)$	$\theta(0)$
0.3	1.11753	1.92699
0.5	0.92976	1.65217
0.733	0.80893	1.47981
1	0.72197	1.35470
2	0.55923	1.13001
3	0.48117	1.02050
5	0.39769	0.90172
6.7	0.35633	0.84170
7	0.35051	0.83316
10	0.30636	0.76755
15	0.26263	0.70055

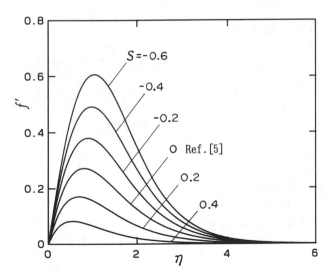

Fig. 1. Velocity profiles for $Pr = 0.733$

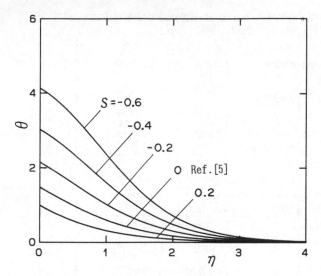

Fig. 2. Temperature profiles for $Pr = 0.733$

REFERENCES

1. SPARROW, E.M. and GREGG, J.L., Trans. ASME 78, pp. 435-440, 1956.

2. ROY, S., J. Heat Transfer 95, pp. 124-126, 1973.

3. WILKS, G., Int. J. Heat Mass Transfer 17, pp. 743-753, 1974.

4. BEREZOVSKY, A.A. and SOKOVISHIN, Yu.A., Izv. Akad. Nauk SSR, Mekh. Zhid. Gaza No 2, pp. 129-136, 1977.

5. MAHAJAN, R.L. and GEBHART, B., Int. J. Heat Mass Transfer 21, pp. 549-556, 1978.

6. MARTYNENKO, O.G., BEREZOVSKY, A.A. and SOKOVISHIN, Yu.A., The Asymptotic Method in the Free Convection Heat Transfer Theroy (in Russian). Izd. Nauki i Tekhnika, Minsk, 1979.

7. CAREY, V.P. and MOLLENDORF, J.C., Int. J. Heat Mass Transfer 23, pp. 95-109, 1980.

8. MERKIN, J.H., J. Engng. Math. 19, pp. 189-201, 1985.

9. MERKIN, J.H., J. Engng. Math. 23, pp. 273-282, 1989.

10. YANG, J.-Y. and LIN, C.-N., Wärme- und Stoffübertr. 23, pp. 213-217, 1988.

11. MERKIN, J.H. and MAHMOOD, T., J. Engng. Math. 24, pp. 95-107, 1990.

12. HARTNETT, J.P., Mass Transfer Cooling. In: Handbook of Heat Transfer Applications, 2nd ed., edited by W.M.Rohsenow, J.P. Hartnett and E.H. Ganic. Hemisphere, Washington, D.C., pp. 1-111, 1985.

13. WATANABE, T. and KAWAKAMI, H., Trans. Japan Soc. Mech. Engrs. (in Japanese) 55, pp. 3365-3369, 1989.

14. WATANABE, T., Acta Mechanica 87, pp. 1-9, 1991.

15. HARTREE, D.R. and WOMERSLEY, J.R., Proc. Roy. Soc. A161, pp. 353-366, 1973.

NOMENCLATURE

C	constant, $-(v_0/5\nu)\ (g\beta q_w/\ 5k\nu^2)^{-1/5}$
C_f	skin friction coefficient
f	reduced stream function
g	acceleration due to gravity, m/s^2, ft/s^2
Gr_x	local Grashof number, $g\beta q_w\ x^4/\ k\nu^2$
h	uniform step size of difference, m, ft
k	thermal conductivity, W/(m·K), Btu/(hr·ft·°F)
Nu	Nusselt number
Pr	Prandtl number, $\nu/\ \alpha$
q_w	uniform heat flux at the plate, W/m^2, Btu/(hr·ft^2)
s	parameter of suction or injection, $(v_0/5\nu)\ (g\beta q_w/\ 5k\nu^2)^{-1/5}$
T	temperature, °C, K, °F, °R
u	velocity component in x direction, m/s, ft/s
Ue	velocity scale, m/s, ft/s
v	velocity component in y direction, m/s, ft/s
v_0	velocity of suction or injection, m/s, ft/s
x	coordinate along the plate, m, ft
x^*	transformed coordinate, $C\ x^{1/5}$
y	coordinate normal to the plate, m, ft

Greek Symbols

α	thermal diffusivity, m^2/s, ft^2/s
β	coefficient of thermal expansion, K^{-1}, °R^{-1}
η	non-similarity variable
θ	non-dimensional temperature
μ	dynamic viscosity, Pa·s, lb/(s·ft)
ν	kinematic viscosity, m^2/s, ft^2/s
ρ	density, kg/m^3, lb$_m$/ft^3
τ_w	shear stress, Pa, lb$_f$/ft^2
ψ	stream function, m^2/s, ft^2/s

CALCULATION OF HEAT TRANSFER ON A FLAT PLATE FOR HIGH PRANDTL AND REYNOLDS NUMBERS

P. Hrycak

New Jersey Institute of Technology, Newark, N.J., 07102, USA

Expressions have been derived for local and average flow-resistance coefficients, for fully-developed turbulent flow over a smooth, flat plate. These results have been extended to calculations of turbulent thermal boundary-layer thickness, with an unheated starting length, and to calculation of heat transfer for high Prandtl and Reynolds numbers. The present results are, by necessity, all based on a semi-empirical theory and extrapolation of experimental results, in the absence of a comprehensive theory on the subject matter. Certain aspects of the discussion are based on solution of generally accepted, differential equations, however, and may be of more fundamental significance.

INTRODUCTION

From the historical perspective, one has to admit that methods of dealing successfully with turbulent flow over extended, flat surfaces have been developed mainly with the help of experimental results originally obtained for flow in round, smooth pipes. Such experimental data were the foundations on which, through the refined analysis, empirical theories were based. Eventually, extensive measurements have been carried out also on flow on smooth, flat plates, in line with the developments in the boundary-layer theory (cf. Schlichting, (1)). The boundary-layer equation, integrated over the boundary-layer thickness, δ, yields the expression

$$\rho \frac{d}{dx} \int_0^\delta (u - u_\infty)u \, dy = -K(p,n) \, \rho \, u_\infty^2 (\nu/u_\infty \delta)^p \qquad (1)$$

Here, the form of the right-hand side is taken over from the pipe flow, as τ_w, the wall shear stress, represented as a function of two parameters p and n, that are themselves related to the flow Reynolds number. As suggested by numerous experimental observations (1), the velocity profile in the boundary layer is considered to be of the $(1/n)$-th power type:

$$u/u_\infty = (y/\delta)^{1/n} \text{ or } u/u^* = C(n)(u^* y/\nu)^{1/n} \qquad (2)$$

and the local shear stress at the wall can be described by the formula

$$\tau_w = K(p,n) \rho \, u_\infty^2 (\nu/u_\infty \delta)^p, \quad p = 2/(1 + n), \text{ and} \qquad (3)$$

$$K(p,n) = 1.013 \, C(n)^{-pn}, \quad C(n) = 0.92 \, n + \ln 10, \quad n > 5 \qquad (4)$$

The parameter n is linked to Re_d (the pipe Reynolds number as defined in (1)) by a simple formula

$$n = 2 \log Re_d - 2, \quad Re_d > 10^3 \qquad (5)$$

that is based on experimental results of Nikuradse [1] and Hrycak [2]. We also let $u^* = (\tau_w/\rho)^{\frac{1}{2}}$, the friction velocity and note $u/u^* = u^+$, and $yu^*/\nu = y^+$ that are linked together in the form of the "law of the wall" [1]

$$u^+ = (1/k) \ln y^+ + 5.5 \tag{6}$$

that represents the "universal velocity distribution" in the boundary layer, and is considered as generally applicable to flat-plate flow. Here, k = 0.4, is known as von Karman constant. It is of interest to note that the u^+ curve from eq. (6) represents also very nearly an envelope to the family of curves of the $(1/n)$-th power type as per eq. (2). Equation (2), for n = 7, has been originally discovered by Prandtl [1]; also, $C(7) = 8.74$, and $K(\frac{1}{4},7)/1.013 = 0.0225$, is obtained from the well-known Blasius formula for flow-resistance coefficient for flow in round pipes, supported by a large body of experimental results. Lately, $K(\frac{1}{4},7) = 0.0228$ has been used in the literature, as a more representative result for flow over a flat plate (cf. Eckert and Drake, (3), p. 324). The formula for the generalized form of $C(n)$, eq. (4), has been obtained from the condition that the resulting $(1/n)$-th power curves be tangent to the generalized velocity distribution curve above (eq. (6) here). Thus, the parameter n represents a link between the flow Reynolds number and the local wall shear stress, as suggested by eqs. (2), (4), and (5). As a consequence of this, a very accurate representation of wall shear stress becomes possible by simple power-law expressions, for restricted intervals of the applicable flow-Reynolds numbers. Only integral n's are used here, however. Examples are seen in Fig.2(2).

On substitution of eq. (2) in (1), there results the expression

$$\delta^{2/(1+n)}\frac{d\delta}{dx} = \left[(n + 1)(n + 2)/n\right] K(p,n) \, (\nu/u_\infty)^{2/(1+n)} = A, \text{ and} \tag{7}$$

$$\delta^{(n+3)/(n+1)} = (n + 3)/(n + 1) \, Ax + C, \text{ that yields} \tag{8}$$

$$\delta = \left[(n + 2)(n + 3)K(p,n)/n\right]^{(n+1)/(n+3)}(\nu/u_\infty x)^{2/(n+3)} \, x \tag{9}$$

on assuming, at first, that for $\delta = 0$, the constant of integration $C = 0$. The local flow-resistance coefficient, C_f, at a distance x from the plate's origin can be defined then as

$$C_f = 2\tau_w/(\rho u_\infty^2) \tag{10}$$

On substituting eqs. (3) and (9) in eq. (10), there results the expression

$$C_f = 2\left[K(p,n)\right]^{(n+1)/(n+3)}\left[(n + 2)(n + 3)/n\right]^{-2/(n+3)}(u_\infty x/\nu)^{-2/(n+3)} \tag{11}$$

The average flow resistance coefficient, C_D, is then obtained by integration, for a plate of lengt L (assuming here an early transition, cf. (1)), as

$$C_D = (1/L) \int_0^L C_f \, dx = (n + 3)/(n + 1) \, C_f(L) \tag{12}$$

The link existing between the pipe Reynolds number $Re_d = \bar{u}d/\nu$ and the plate Reynolds number $Re_L = u_\infty L/\nu$ is based on the relation $d/2 = \delta$, here by eq. (9), and on the ratio of the average to the maximum pipe velocity, $\bar{u}/u_\infty = \alpha$,

$$u_\infty L/\nu = \left[(\bar{u}d)/(2\alpha\nu)\right]^{(n+3)/(n+1)} n/\left[(n + 2)(n + 3) K(p,n)\right] \tag{13}$$

such that $u_\infty L/\nu$ is proportional to $\bar{u}d/\nu$ raised to $(n + 3)/(n + 1)$ power; also,

$$\alpha = 2n^2/\left[(n + 1)(2n + 1)\right] \tag{14}$$

Equations (14) and (4) are useful in finding the proper value of n to be used in calculations of flow resistance coefficients by power-law methods. The accuracy of flow-resistance coefficients calculated here depends on the constants used in eq. (6), which are based on the results of Nikuradse (1), that represent the average values, good for the entire cross-section of the pipe. In Fig. 1, the results of calculations by eq. (12) for n = 10 and 11 are compared with experimental findings (cf. (1), p. 639), that are well represented by the "Schlichting formula" for the total drag,

$$C_D = 0.455/(\log Re_L)^{2.84} \tag{15}$$

For $Re_L > 10^8$, C_D by eq. (12) differs less than 2 % from that by the Schlichting formula. The respective formulas here,

$$C_D = 0.03654 \, Re_L^{-2/13}, \text{ and } C_f = 0.03092 \, Re_x^{-2/13} \text{ (for n = 10)} \tag{16}$$

$$C_D = 0.03014 \, Re_L^{-1/7}, \text{ and } C_f = 0.02583 \, Re_x^{-1/7} \text{ (for n = 11)} \tag{17}$$

are deemed useful for $10^7 < Re < 10^9$, in particular in the situations where they appear in various integral expressions or differential equations related to fluid-flow and heat-transfer problems because of their mathematical simplicity. C_f by eq. (16) is compared with experimental results by several investigators in Fig. 2 (cf. Monin and Yaglom (4), p. 324, for example, or (1), p. 143). The agreement is again quite satisfactory. In general, as n is slowly varying function of the Reynolds number, average n-values may be used in computations. Thus, for the cases at hand, n → 10.8 from 10, and n → 10.9 from 11.

HEAT TRANSFER APPLICATIONS

Because of similarity of the basic differential equations, it appears that the previous discussion can also be extended to some heat transfer problems. The integrated energy equation for turbulent flow regime, with the $Pr^{-2/3}$ factor suggested by Chilton and Colburn (cf. (3), p. 324), and δ_T = thermal boundary-layer thickness, may be stated as a relation complementary to eq. (1) above:

$$\varsigma c_p \frac{d}{dx} \int_0^{\delta_T} \theta u \, dy = \varsigma c_p u_\infty \theta_w \, K(p,n)(u_\infty \delta/\nu)^{-p}(Pr)^{-2/3} \tag{18}$$

with the velocity by eq. (2), and the temperature distribution given by

$$\theta = \theta_w [1 - (y/\delta_T)^{1/n}], \quad \theta_w = (T_w - T_\infty) \tag{19}$$

The form of eq. (19) is suggested by the equation for the universal temperature distribution, parallel to the "law of the wall" above, and stated here as

$$u_\infty^+ [(T_w - T_\infty)/q_w \varsigma c_p u_\infty] = (Pr_t/k) \ln (\delta_T u^*/\nu) + 13 (Pr)^{2/3} - 7 \tag{20}$$

(cf. White (5), p. 485). It is also well verified experimentally (Kader (6), for example). On letting $\delta_T < \delta$ and $\xi = \delta_T/\delta$, and with A by eq. (7), and δ by eq. (9), with the additional substitution $y = \xi^{(n+1)/n}$, eq. (19) yields

$$\delta^{2/(n+1)} \frac{d}{dx} (\delta \, \xi^{(n+1)/n}) = A \, (Pr)^{-2/3}, \text{ and} \tag{21}$$

$$y + (n + 3)/(n + 1) \, x \, dy/dx = (Pr)^{-2/3} \tag{22}$$

that has the solution, for the heating process starting at x = x_0,

$$y = C \, x^{-\frac{(n+1)}{(n+3)}} + (Pr)^{-2/3}, \quad C = -(Pr)^{-2/3} x_0^{(n+1)/(n+3)}, \text{ and} \tag{23}$$

$$\delta_T/\delta = \{(Pr)^{-2/3} [1 - (x_0/x)^{(n+1)/(n+3)}]\}^{n/(n+1)} \tag{24}$$

Of general interest is the case where the heating process starts still in the laminar region: then, $\xi_o = (\delta_T/\delta)_o \neq 0$ at the origin, during integration of eq. (22). We now let $x \to x - x_t + x_{o.t}$, where $x_{o,t}$ is the virtual origin for the start of the turbulent boundary layer, equal to the laminar boundary-layer thickness that started at the origin the plate, while x_t = coordinate where actually is the true starting point for the turbulent regime. In this case,

$$y = (Pr)^{-2/3}\{1 + (\xi_o^{(n+1)/n}(Pr)^{2/3} - 1)[x_{o,t}/(x-x_t+x_{o,t})]^{(n+1)/(n+3)}\} \tag{25}$$

Thus, for $x = x_t$, $y = \xi_o^{(n+1)/n}$, and $(\delta_T/\delta)_o$ is to be calculated as is shown in (3), p. 175. Equations (9) and (24) show dependence of the thermal boundary-layer thickness on various parameters of the problem, in particular on the Prandtl number and on the start of the heating process.

For fully-turbulent heat and momentum transfer here, we let

$$\dot{q}_w = -(\lambda + \lambda_t)\ dT/dy, \text{ and } \tau_w = (\eta + \eta_t)\ du/dy \tag{26}$$

which expressions, when integrated side-by-side, generate the relations

$$\dot{q}_w \approx <\lambda + \lambda_t> (T_w - T_\infty)/\delta_T, \text{ and } \tau_w \approx <\eta + \eta_t> u_\infty/\delta \tag{27}$$

In the expression $\dot{q}_w/\tau_w$, the ratio $<\lambda + \lambda_t>/<\eta+\eta_t>$ is a complicated function of Prandtl number, of the turbulent diffusivities of heat and momentum, and of the ratio of the boundary-layer thicknesses, ξ. It is approximated as

$$(Pr)^{\pm \beta}(u_\infty/u^*)^{\pm \sigma}(\lambda/\eta), \text{ lim } (\beta,\sigma) \to 0 \text{ as } n \to 1 \tag{28}$$

On combining eqs. (27) and (28), and noting $u_\infty/u^* = (C_f/2)^{-\frac{1}{2}}$, there results

$$\dot{q}_w/\tau_w = (Pr)^{-\beta}(C_f/2)^{\sigma/2}(\lambda/\eta)(T_w - T_\infty)/u_\infty \tag{29}$$

When eq. (24) is introduced, eq. (29) transforms into the Nusselt number expression for the problem at hand

$$Nu_x = \frac{(Pr)^{2n/3(n+1) - \beta}(Re_x)\ (C_f/2)^{1-\sigma}}{[1 - (x_o/x)^{(n+1)/(n+3)}]^{1/(n+1)}} \tag{30}$$

As $u^+ = y^+$ for $y^+ < 5$, (cf. (1)), we may consider that $n \to 1$ for small Re_x. In this case eq. (30) reduces to a simple formula

$$Nu_x = \frac{(Pr)^{1/3}\ Re_x\ C_f/2}{[1 - (x_o/x)^{\frac{1}{2}}]^{\frac{1}{2}}} \tag{31}$$

that is in a very good qualitative agreement with the well-known formula for laminar heat transfer for the present geometry (cf. (3), p. 176). On the other hand, as both Pr, Re $\to \infty$, Stanton number = $Nu/Re\ Pr = C_h$ becomes in the limit our eq. (32) below. $C_{h,T}$ is defined as Stanton number where T_w changes at x_o.

$$C_h = (Const)\ (Pr)^{-2/3}(C_f/2)^{1/2} \text{ (asymptotic)} \tag{32}$$

that indicates: as $n \to \infty$, $\beta \to 1/3$, and $\sigma \to 1/2$. It appears that the effect of changes in the experimental parameters β and σ on the Nusselt number, as a consequence of substantial changes in Pr and Re_x, could be expressed as $(Pr)^a$, with the parameter "a" to be evaluated experimentally or numerically, based on the semi-empirical theory. The results of calculations are seen in Fig.3.

$$Nu_x = \frac{(Pr)^a\ Re_x\ C_f/2}{[1 - (x_o/x)^{(n+1)/(n+3)}]^{1/(n+1)}}, \text{ and} \tag{33}$$

$$C_{h,T} = (Pr)^{a-1}C_f/2[1 - (x_o/x)^{(n+1)/(n+3)}]^{1/(n+1)} \tag{34}$$

All calculations of "a" have been carried out with $x_0 = 0$, without loss of generality, as only the Prandtl number was directly affected. By combining the universal velocity and temperature distribution (eqs. (6) and (20) here), with $Pr_t = 1$, and using our formula for δ_T/δ, as per eq. (24), there results

$$C_h = \frac{C_f/2}{1 + [13((Pr)^{2/3} - 1) + 0.5 - (1/k) \ln(\delta/\delta_T)](C_f/2)^{1/2}} \tag{35}$$

as originally suggested by ref. (5), p. 485, but observing $\delta_T \neq \delta$, as characteristic for large Prandtl numbers. A similar relation for the computation of C_h has been obtained by Kader and Yaglom (7),

$$C_h = \frac{(C_f/2)^{1/2}}{2.12 \ln \left[(C_f)(Re_x)(Pr)\right] +]2.5(Pr)^{2/3} - 7.2} \tag{36}$$

Equations (35) and (36) yield very similar results, that were used in calculations of "a". Relatively few experimental results are available for heat transfer on a flat plate for $Pr > 1$ (cf. those by Slanciauskas et al., listed in (7)). Experimental data for turbulent heat transfer in pipes are quite numerous. Relations similar to eq. (36) have been developed, for example, in refs. (6) and (7), that are well supported experimentally, and show trends very similar to our eqs. (33) and (34), as summarized in Table 1. The effect of x_0/x can be estimated from the comparison with the so-called "Seban's formula," cf. (3), p. 217, that is based on $(1/7)$-th velocity and temperature profiles. When compared to eq. (33), $Pr = 1$, for $x_0/x < 0.9$, the relative error was less than 4.6 %, with $n = 7$. For other, more representative n-values, agreement should be even closer. Seban's formula agrees with experimental results, while experimental error is estimated here as about 5 %. Table 1 shows the calculated a-values.

Table 1 The exponent "a" of the Prandtl number in eqs. (33) and (34)

Re_x Pr	10^4	10^5	10^6	10^7	10^8	10^9
10	0.36	0.44	0.51	0.56	0.60	0.66
10^2	0.346	0.41	0.445	0.49	0.52	0.55
10^3	0.343	0.39	0.42	0.445	0.47	0.495
10^4	0.34	0.377	0.40	0.42	0.44	0.455

CONCLUSIONS

Starting with the experimental results, originally obtained with round-pipe flow for wall shear stress, the universal velocity distribution, and the Chilton-Colburn analogy for the turbulent heat transfer, a formula for the thermal boundary-layer thickness has been derived that included the effect of the unheated starting length. Also, the local and the average flow-resistance coefficients have been calculated, based on the concept of $(1/n)$-th power velocity distribution in the boundary layer. An improved expression for the Stanton number has been proposed, and on the functional dependence of the Nusselt number on the Prandtl number. The numerical results were calculated through the analysis of the available experimental data and on the extension of a semi-empirical theory. An additional degree of confidence in the present results was found in our use of methods that were similar to those found successful in the analysis of very numerous data on heat transfer in round, smooth pipes, and in channels.

It is believed, therefore, that the present results show correctly the trends to be expected in the analysis of heat transfer in fully-turbulent boundary layers for zero pressure gradient, for high Prandtl numbers. The power-law approach made it possible, in principle, to treat more difficult initial conditions while solving the governing diff. equations.

REFERENCES

1. Schlichting, H., 1979, "Boundary-Layer Theory," 7th Ed., pp. 635-643, McGraw-Hill, New York, USA
2. Hrycak, P., 1991, "Proceedings," 13th Canadian Congress of Applied Mechanics, Vol. 2, pp. 432-433, Winnipeg, Man., Canada
3. Eckert, E.R.G., and Drake, R.M., Jr., 1959, "Heat and Mass Transfer," McGraw-Hill, New York, USA
4. Monin, S.A., and Yaglom, A.M., 1975, "Statistical Fluid Mechanics," Vol. 1, MIT Press, Cambridge, Mass., USA
5. White, F.M., 1991, "Viscous Fluid Flow," 2nd Ed., McGraw-Hill, New York, USA
6. Kader, B.A., 1981, <u>International J. of Heat and Mass Transfer</u>, Vol. 24, pp. 1541-1544
7. Kader, B.A., and Yaglom, A.M., 1972, <u>International J. of Heat and Mass Transfer</u>, Vol. 15, pp. 2329-2351

NOMENCLATURE

a	exponent of Prandtl number in eqs. (33) and (34)
C	dimensionless coefficient, a constant of integration
d, L	diameter of pipe, total length of plate, m
k	$= 0.4$, von Karman costant
n, p	exponents in power-law expressions; n defined in eq. (5)
$\dot{q}$	heat flux, W/m^2
T	temperature, K
T^+	dimensionless temperature, $(T_w - T_\infty)(\rho c_p u^*)/q_w$
u	velocity parallel to plate, m/s
u^*	friction velocity, $(\tau_w/\rho)^{\frac{1}{2}}$, m/s
u^+	dimensionless velocity, u/u^*
u_∞^+	$u_\infty/u^* = (2/C_f)^{\frac{1}{2}}$, dimensionless velocity
y, x	coordinates, perpendicular and parallel to plate, m
y^+	dimensionless coordinate, yu^*/ν
Nu_x	Nusselt number, hx/λ; h = film heat transfer coefficient, W/m^2K
Pr	Prandtl number, $c_p\eta/\lambda$
Re_x	Reynolds number, ux/ν
St	Stanton number, $Nu/(Pr\, Re)$

Greek Symbols

α	velocity ratio, $\bar{u}/u_\infty$, in pipe flow
β, σ	experimental parameters occurring in eqs. (28) to (30)
δ, δ_T	boundary-layer thicknesses, hydrodynamic and thermal, m
η	dynamic viscosity, kg/(m s)
θ	reduced temperature, $T - T_\infty$, K
ν	kinematic viscosity, m^2/s
ρ	density, kg/m^3
τ	shearing stress, N/m^2
ξ	boundary-layer thickness ratio, δ_T/δ

Subscripts

f, D, h	referring to local friction coefficient, drag, Stanton number
o	location where heating starts or T_w change occurs
t	turbulent-regime related
w	location at surface of plate
∞	location outside of boundary layer; $<\ >$ signifies an average

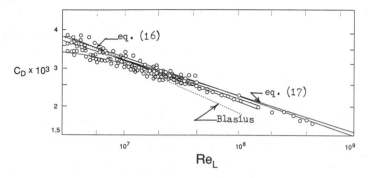

Fig. 1 Comparison of Calculations of Total-Drag Coefficient by Eqs. (16) and (17) with Experimental Data by Several Investigators

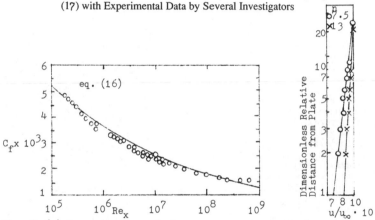

Fig. 2 (a) Comparison of Calculation of Local Flow-Resistance Coefficient with Experimental Data by Dhawan, Kempf, and Schultz-Grunow (1)
(b) Determination of Power-Law Coefficient n from Velocity Profile

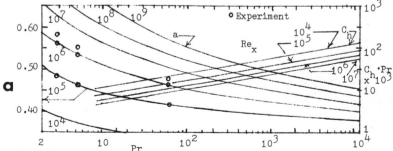

Fig. 3 Comparison of Experimental Results by Slanciauskas et al. with Calculations of Exponent "a" in Eqs. (33) and (34). C_h from Eq. (34) x Prandtl Number ($x_o = 0$)

STUDY ON RAPID COOLING OF A HOT STEEL SURFACE WITH MULTI WATER NOZZLES

Michiharu Hariki and Akira Onishi

Sumitomo Metal Industries, Hasaki-machi, Kashima-gun, Ibaraki, Japan

In a Thermo Mechanical Control Process various cooling devices for controlling the temperature of a hot steel surface have been used. However, the useful devices for rapid cooling are few. Moreover, the heat transfer behaviour of these cooling devices has hot been clar -ified. An experimental study of heat transfer with multi water nozzles using high water flux on a high temperature surface was made. The results showed that the new water-only nozzles could achieve high cooling rates equivalent to those obtained with exist -ing mist jet cooling systems.

INTRODUCTION .

To improve the mechanical properties of hot rolled plates and strips, the Thermo Mechanical Control Process(TMCP) has been developed in steel mills. In this process various cooling devices for controlling the temperature of a hot steel surface have been used. However, the useful devices for rapid cooling are few except slit jet cooling and multi mist jet cooling (1). Moreover, the heat transfer behaviour of these cooling devices has not been clarified sufficiently. On tne other hand, by arranging small water nozzles densely between each roller as in multi mist jet cooling, high water flux and large impingement area will be realized. Therefore, cooling capacity and cooling uniformity will be improved. The relation between water flux and heat transfer coefficient of the area where the water strikes the surface was studied. by Mitsutsuka(2). However, in the case of a high water flux (above 0.033 m^3/m^2 s), the characteristics of heat transfer was not clarified sufficiently. As a first step, the characteristics of heat transfer of multi water fullcone nozzles cooling was investigated on the impingement surface for the case of a high water flux and a high surface temperature. And then comparison of the heat transfer coefficient of the area where the water strikes the surface and where it flows over the surface was made between the multi water fullcone nozzles cooling and the other rapid cooling techniques, namely slit jet cooling and multi mist jet cooling. From these experimental results, the characteristics of heat transfer with multi water fullcone nozzles are discussed. The cooling capacity of multi water fullcone nozzles cooling was not superior to multi mist jet cooling, however it was equivalent to slit jet cooling. As a second step, the characteristics of heat transfer of the newly developed water atomizing nozzle in place of fullcone spray nozzle were investigated. The features of the newly developed nozzle are to realize high impact pressure and nearly conical spray pattern by mixing a straight water stream and a diagonal one inside the nozzle. Therefore, by increasing water flux (in place of air flux), high cooling capacity equivalent to multi mist jet cooling was obtained. In this case the heat transfer coefficient of the newly developed . nozzles installed in the real plant agreed with the experimental results obtained using those nozzles.

CHARACTERISTICS OF HEAT TRANSFER
WITH MULTI WATER NOZZLES

Experimental procedure

The schematic diagram of the system for measuring the specimen temperature in the model cooling devices is shown in Fig. 1. After a uniformly heated specimen(265x265x3mm, stainless steel)had been taken out of an electric furnace, it was put into the model cooling device in which the nozzles were spraying under the set conditions and was cooled on one side only. The temperature was measured from the other side by spot welded thermocouples. Then, the relaton between surface temperature and heat transfer coefficient was calculated by simulation of the cooling curves. The specifications of the model cooling devices is shown in Table 1. The dimensions of the devices was about 350x1000mm. The design water flux was 0.067 m^3/m^2 s. In order to estimate the characteristics of heat transfer at high water flux and high temperature with the multi fullcone nozzles, the first experiment was conducted under the conditions of impinging water flux, W, from 0.017 to 0.12 m^3/m^2s, spray distance, D, from 50 to 150mm, and waterlayer height on the impingement area, h, from 0 to 50mm. And then, the second experiment was conducted under the conditions of impinging water flux, W, from 0.03 to 0.12 m^3/m^2 s, spray distance, from 50 to 150 mm, and water temperature of 23 ±4.5°C. The point where the water struck the hot surface was selected as the water striking zone and 200-325 mm area from the center of water striking zone was selected as the flowing water zone.

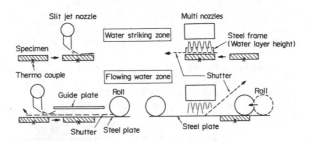

Fig. I Measuring system for the specimen temperature
in the model cooling devices.

Table 1   Specifications of the model cooling devices.

Cooling method	Slit jet	Multi mist jet	Multi spray
Design pressure	0.3 MPa		0.12
Amount of water	0.05m³/s·m	1.3×10⁻⁴ m³/s·piece	0.8 X 10⁻⁴
Nozzle pitch	——	50 X 40mm	40 X 30
Number of nozzles	1 piece	114	184
*Water flux	0.067 m³/m²·s		

✳ Cooling length of slit jet = 700mm

RESULTS AND DISCUSSION

Water flux and heat transfer

Relation between water flux and heat transfer on the impinging surface is shown in Fig. 2. In the case of high water flux, W, from 0.05 to 0.12m³/m²s, the heat transfer coefficient difference between the upper surface cooling and the lower surface cooling was small. It was about 15%. The relation is expressed by the following equation:

$$(\text{heat transfer coefficient } \alpha) \propto (\text{Water flux, } W)^{0.51} \tag{1}$$

Therefore high water flux cooling enables more rapid cooling. However the cooling capacity doesn't increase so much, in the case of high water flux, as for low water flux (W<0.033 m³/m²S) (2).

Water layer height and heat transfer

Some water collected on the surface in the case of high water flux cooling. So the water layer height was changed from 0 to 50 mm under condition of spray distance 50 and 90mm respectively. In the case of high water flux, the heat transfer coefficient hardly changed when the water layer height was changed from 0 to 50 mm. (Fig. 3)

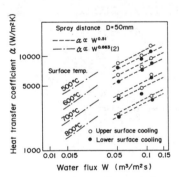

Fig. 2   Effect of water flux on the heat transfer coefficient.

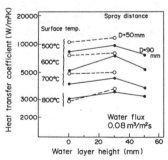

Fig. 3   Effect of water layer height on the heat transfer coefficient.

Heat transfer coefficient of water striking zone and flowing water zone

The ranking of heat transfer coefficient for the water striking zone and flowing water zone for these rapid cooling methods was slit jet, multi mist jet, and multi water spray cooling. (Fig. 4) Heat transfer coefficient ratio of (flowing water zone/water striking zone) was 0.2 in the high temperature range and it was 0.5 in the low temperature range. Heat transfer characteristics of the flowing water zone were equivalent to immersed cooling with rapidly stirred water (3).

Mean heat transfer coefficient between each roller

Slit jet cooling had the largest heat transfer coefficient for the water striking zone among these rapid cooling methods, but the effective length of water striking zone was about 100mm. Therefore, it has the disadvantage that it does not enlarge the ratio of(water striking zone/flowing water zone) berween each roller as compared with muti muti nozzles type. From the results above, comparing an equal length of water striking zone and flowing water zone, the cooling capacity of mult water spray was not superior to multi mist jet cooling. However, it was equivalent to slit jet cooling. (Fig. 5)

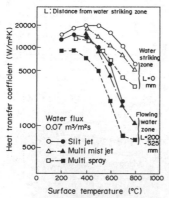

Fig. 4 Influence of the distance from water striking zone on the heat transfer coefficient.

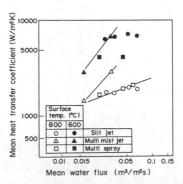

Fig. 5 Relation between mean water flux and mean heat transfer coefficient.

INVESTIGATION OF THE NEWLY DEVELOPED NOZZLEZ

Features of the newly developed nozzles

The schematic diagram of newly developed nozzles is shown in Fig. 6 . In order to realize high impact pressure and nearly conical spray pattern, a straight water stream is mixed with a diagonal one inside of the both nozzles.   The assembly type nozzle is available to change the spray pattern, and the integral type is useful for real operation in a TMCP line . Water flow rate of both types of nozzles depends on the outlet diameter and the diagonal gap .   In comparison with the straight type nozzle, the ratio of water flow rate is about 50-70%. At a nozzle pressure of 0.3MPa, changes in water flow rate from 0.0003 to 0.0014 m^3/s is possible by changing the outlet diameter from 4.5 to 9.75 mm.   The relation between impinging water flux and impact pressute is shown in Fig. 7 .   The high impact pressure equivalent to multi mist jet cooling was obtained by increasing water flux in place of air flux.

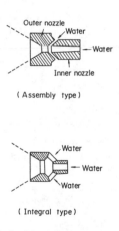

(Assembly type)

(Integral type)

Fig. 6   Schematic diagram of
newly developed nozzles.

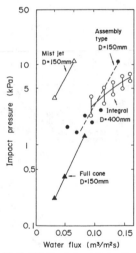

Fig. 7   Relation between water flux
and impact pressure.

Characteristics of heat transfer coefficient

The heat transfer coefficient of the newly developed nozzles over the area where the water strikes the surface was examined by means of the previously described experimental procedure.   The experimental conditions are shown in Table2. In Fig. 8, the relation between impinging water flux and heat transfer coefficient is plotted on the line described by extrapolating the well known regression formula of spray and jet cooling (3)   toward high water flux at high temperature.   The relation between representative velocity which was defined by Sugiyama et al. (4) and heat transfer coefficient is shown in Fig. 9. It is possible to estimate heat transfer coefficient of various cooling methods by using the representative velocity in the case of high water flux. Fig.10 shows the regression formula of a previous relationship with multi water nozzles on the area where the water strikes the surface. Finally the cooling device incorporating the newly developed nozzle was investigated in a hot strip mill at Wakayama. The test headers were installed at the outlet of the finisher mill stand and under the runout table as shown in Fig. 11. The measured data were in agreement with the previous data . The variation of a coiling temperature was about 30℃ by using these cooling devices.

Table 2 Experimental conditions.

Nozzle type	Assembly		Integral
Case No.	A	B	C
Outlet diameter	∅ 3.3 mm	∅ 6.5	∅ 9.75
Nozzle pitch	30 X 40mm	60 X 80	65 X 130
Spray distance	150 - 400 mm		.
Water flux	0.05 - 0.25 m³/m²s		

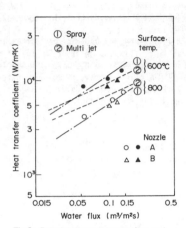

Fig.8 Relation between water flux and heat transfer coefficient.

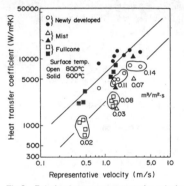

Fig.9 Relation between representative velocity and heat transfer coefficient.

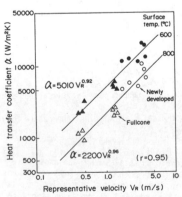

Fig.10 Regression formula of heat transfer coefficient.

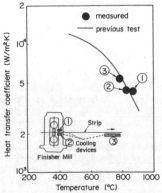

Fig.11 Comparison of heat transfer coefficient between measured and previous test data.

CONCLUSIONS

Investigations on the heat transfer with multi water nozzles in a situation with a high water flux and a high surface temperatute were carried out and the following results were obtained.

1. In case of high water flux (above $0.05m^3/m^2s$), heat transfer coefficients hardly changed when water layer height was changed from 0 to 50 mm.
2. Heat transfer coefficient ratio of (flowing water zone/water striking zone)was 0.2 in the high temperature range and 0.5 in the low temperature range.
3. The cooling capacity of multi fullcone nozzles was not superior to multi mist jet cooling, however it was equivalent to slit jet cooling.
4. The equivalent high cooling capacity to multi mist jet cooling was obtained by increasing water flux of the newly developed nozzles.
5. Heat transfer performance of the rapid cooling devices installed in real plant agreed with the experimental results obtained in the laboratory tests.

NOMENCLATURE

α =heat transfer coefficient , W/m^2K
W =water flux, m^3/m^2s
D =spray distance , mm
h =water layer height , mm
P =impact pressure , KPa
V_R =representative velocity , m/s

REFERENCES

1) S.Wilmotte,P.Simon, " Le refroidissement par brouillard d'eau et ses   applications dans le domaine des laminoirs " . Revue de metallurgie-CIT. Avril, 1981. pp. 319-327
2) M.Mitsutsuka, " Study on the Water Spray Cooling of Steel Plate at High Temperature " , Tetsu to Hagane. Vol. 54, No. 14, 1968, pp. 33-47.
3) The Iron and Steel Institute of Japan. " Kozai no Kyousei Reikyaku " , 1978.
4) Shunichi Sugiyama, Takao Noguchi, Hiroshi Kamio and Kazuo Kunioka. " Analytical Arrangement and Application of Cooling Ability in Various Water Jet Cooling Methods" , Nihon Kokan Giho, No. 88, 1981. pp. 39-48

EXPERIMENTAL STUDY ON MASS TRANSFER IN THE ENTRANCE REGION OF RECTANGULAR DUCT

M. Hirota, H. Fujita, H. Yokosawa, and T. Nonogawa

Dept. of Mechanical Eng., Nagoya University, Chikusa-ku, Nagoya 464-01, Japan

In the entrance region of a rectangular duct, the mechanism of heat (mass) transfer is complicated due to interference between boundary layers developing along each wall and transition of each boundary layer. When fluid enters the duct uniformly through a bellmouth nozzle, for large local Reynolds numbers, the transition of the boundary layer first occurs at the duct corner, then the turbulent boundary layer develops toward the duct center as the flow proceeds downstream. Downstream of this transition, the mass transfer rate increases rapidly owing to turbulence, and reaches a constant value. Under the sharp-edged inlet condition, in contrast, the boundary layer is turbulent over the entire region of the duct. The local mass transfer rate is maximum at the entrance of the dcut, then decreases monotonously downstream.

INTRODUCTION

Ducts used in a compact heat exchanger are often so short that the characteristics of their velocity and the temperature field are those of an entrance region. In the entrance region of a duct with a rectangular cross-section, the mechanism of heat transfer is much more complicated than that of a circular pipe because the laminar boundary layers developing along each wall interact with one another and the transition to the turbulent boundary layers occurs downstream. Moreover, the configuration of the duct entrance exerts a great influence on the development and the transition of the boundary layers [Mills (1)].

The purpose of this study is to make clear experimentally the fundamental characteristics of heat transfer in the entrance region of the rectangular duct. The naphthalene sublimation method [Mendes (2)] was used to simulate the heat transfer based on the heat and mass transfer analogy. Detailed distributions of the local mass transfer coefficients are presented for two kinds of entrance configurations: one is the bellmouth inlet and the other the sharp-edged inlet.

EXPERIMENTS

Figure 1 shows a schematic diagram of an experimental apparatus. Flow circuit is operated in the suction mode. Air flows through a settling chamber designed to set up the flow inlet conditions (① in Fig. 1), then through a test duct ②, a venturi nozzle flow meter ③, and is then exhausted outside by a turbo fan ④. The test duct has a rectangular cross-section of 50 mm × 25 mm (aspect ratio = 2 : 1) and a total length of 600 mm (=18 d: d = hydraulic diameter of the duct = 33.3 mm). The test section shown in Fig. 1, in which the mass transfer rate and velocity distributions have been measured, is 300 mm (=9 d) in length and accounts for about the upper-stream half of the test duct. The origin of the test section is located 33.3 mm (i.e., 1 d) downstream of the test duct entrance. The flow field in the test section, therefore, presents entrance region conditions.

The naphthalene sublimation method has been used to measure the mass transfer rate. In the test section, the surfaces of both long side walls of the test duct (50 mm × 300 mm) are made of naphthalene, which simulate the constant wall temperature heating in heat transfer experiments[2]. The short side walls (25 mm × 300 mm) consist of smooth acrylic resin plates, which correspond to the adiabatic walls. All but the test section of the duct is also made of smooth acrylic resin plates.

As shown in Fig. 2, two kinds of entrance configurations were employed to set up two types of flow inlet condiions: namely, the "bellmouth inlet" condition and the "sharp-edged inlet" condition. The settling chamber ① in Fig. 1 was exchanged at every flow inlet condition tested. In the bellmouth inlet condition shown in Fig. 2(a), air enters the test duct through a bellmouth nozzle with a uniform velocity distribution. It was ascertained that the uniformity of the primary flow velocity distribution at the duct entrance was extremely high and the boundary layer thickness of the inlet flow was almost zero. The turbulence intensity of the inlet flow was less than 0.2% over the whole of the cross-section. On the other hand, in the sharp-edged inlet condition shown in Fig. 2(b), a baffle plate is fixed at the upstream end of the test duct [Sparrow (3)]. The inlet flow separates at the duct entrance, thereby producing a very high turbulence at the entrance. Under both flow inlet conditions, the starting point of the mass boundary layer almost coincides with that of the hydraulic boundary layer.

The experiments were conducted at Reynolds number $Re_d = U \cdot d / \nu$ of $(2.0 \sim 8.0) \times 10^4$, where U and ν denote the bulk flow velocity and the kinematic viscosity of air, respectively. The naphthalene surface profiles before and after the sublimation were measured by a digital linear gauge with an accuracy of 1μ m. It was traversed over the naphthalene surface by means of a two-dimensional positioner. A computer-controlled data acquisition system was used in order achieve precise positioning and quick data processing. The distributions of the primary flow velocity and turbulence intensity were measured by the normal-wire hot-wire probe. Figure 3 shows the coordinate system defined in the test section. The X_1-axis coincides with the central axis of the test duct, and its origin is located at the entrance of the test section, i.e., 1 d downstream of the duct entrance. The naphthalene surfaces are perpendicular to the X_3-axis ($X_3/B = \pm 0.5$).

RESULTS AND DISCUSSION

Velocity distributions

Figures 4(a) and 4(b) show the velocity distributions obtained under the bellmouth inlet condition at the streamwise locations of $X_1/d=4.0$ and $X_1/d=8.0$, respectively. The value of Re_d is 3.5×10^4 The left half of each figure shows the distribution of the primary flow velocity U_1, and the right half shows that of the turbulent velocity in the X_1-direction $\sqrt{u_1'^2}$. The primary and turbulent velocities are normalized by the bulk flow velocity U.

It is clear from the distributions of U_1 that the boundary layer thickness is increased as the short side wall ($X_2/B = 1.0$) is approached and/or the measuring station is located more downstream. The distributions of $\sqrt{u_1'^2}$ shown in Figs. 4(a) and 4(b) closely correspond to those of U_1. Near the duct center in which the distributions of U_1 are almost uniform, the values of $\sqrt{u_1'^2}$ are as low as those at the test duct entrance ($\sqrt{u_1'^2}/U = 0.2\%$). As the adjacent wall is approached, however, the values of $\sqrt{u_1'^2}$ are increased, and at $X_2/B=0.8$ the flow is turbulent over the entire boundary layer.

From the results presented above, it is considered that, in the bellmouth inlet condition, the laminar boundary layer is first formed along the corner near the duct entrance, and then turbulent transition of the boundary layer originates from the duct corner. The "turbulent region" (not a fully developed turbulent flow but a transitional flow to the fully developed turbulence) develops from the corner toward the duct center as the flow proceeds downstream.

Figure 5 shows the velocity distributions at $X_1/d=8.0$ under the sharp-edged inlet condition. Since the flow separation at the duct entrance produces turbulence of a very high level, the value of $\sqrt{\overline{u_1'^2}}$ in this inlet condition is much larger than that in the bellmouth inlet condition. The entire area of the duct cross-section is covered by the turbulent boundary layer.

Distributions of mean Sherwood number

Figure 6 shows the distribution of mean Sherwood number $\overline{Sh} = \overline{h_m} \cdot d/D$ obtained under each inlet condition, where $\overline{h_m}$ and D denote the mean mass transfer coefficient averaged over the naphthalene surface and the molecular diffusivity of naphthalene vapor into air, respectively. The values of $\overline{Sh}$ for the sharp-edged inlet condition are $1.4 \sim 1.8$ times as large as those of the bellmouth inlet condition. However, the difference between the values of $\overline{Sh}$ obtained under the sharp-edged and the bellmouth inlet conditions are decreased at larger values of the Reynolds number Re_d. The present $\overline{Sh}$ results may be formulated as follows.

$$\overline{Sh} = 0.0034(Re_d)^{0.98} \qquad \text{(bellmouth inlet condition)} \qquad (1)$$
$$\overline{Sh} = 0.0380(Re_d)^{0.80} \qquad \text{(sharp-edged inlet condition)} \qquad (2)$$

In the sharp-edged condition, the boundary layer is turbulent over the whole region of the test section, and therefore, the values of $\overline{Sh}$ in this inlet condition are proportional to $Re_d^{0.80}$ similar to those in a fully developed turbulent pipe flow [Kays and Crawford (4)].

In Fig. 6, the results of $\overline{Sh}$ are also shown which have been obtained in the test duct with the naphthalene surface on one long side wall only. The results are essentially the same as those obtained for the duct with naphthalene surfaces on both long side walls. This suggests that the mass boundary layer developing along each long side wall does not merge in the present length of the test section yet.

Distributions of local Sherwood number

Figures 7 show the longitudinal distributions of the local Sherwood number $Sh_d = h_m \cdot d/D$ obtained under the bellmouth inlet condition, where h_m denotes the local mass transfer coefficient. The abscissa shows the longitudinal distance from the entrance. The duct center line (the bisector of the long side wall) is located at $X_2/B=0$ (lowest distribution indicated in Fig. 7), and the adjacent wall is approached as the values of X_2/B are increased (see Fig. 3).

The results obtained at $Re_d = 3.5 \times 10^4$ are shown in Fig. 7(a). The distribution of Sh_d shows the local maximum at the entrance of the test section ($X_1/d=0$), and then it decreases in the flow direction. After this relatively slow decrease of Sh_d, it begins to increase rapidly and approaches a constant value asymptotically downstream. As the values of X_2/B are increased, that is, as the adjacent wall is approached, this rapid increase of Sh_d occurs at a more upstream location. Therefore, it follows that, under the bellmouth inlet condition, the mass transfer exhibits extremely three-dimensional characteristics near the entrance of the test section. On the other hand, near the exit of the test section ($X_1/d=9.0$), the asymptotic values of Sh_d are almost constant irrespective of the values of X_2/B, i.e., the spanwise locations. This suggests that, in contrast to Sh_d distributions near the entrance, the characteristics of the local mass transfer near the exit of the test section can be treated like those in a two-dimensional channel.

From the velocity distributions shown in Fig. 4, it is thought that the rapid increase of Sh_d shown above is caused by the transition of the laminar boundary layer to the turbulent one. The present distribution of Sh_d clearly reveals the three-dimensional nature of the boundary layer transition in the test duct: that is, the turbulent transition of the boundary layer occurs at more upstream locations as the duct corner is approached, and the turbulent region spreads from the corner toward the duct center as the flow proceeds downstream.

Figure 7(b) shows the distributions of Sh_d obtained with the same flow inlet condition as indicated Fig. 7(a) but at a larger value of $Re_d = 6.5 \times 10^4$. Although the characteristics of the distributions are qualitatively similar to those in Fig. 7(a), the starting point of the rapid increase of Sh_d caused by the boundary layer transition shifts more upstream than in Fig. 7(a).

The distributions of Sh_d under the sharp-edged inlet condition at $Re_d=3.5 \times 10^4$ are shown in Fig. 8. The value of Sh_d is maximum at the entrance, then it decreases monotonously in the flow direction. The longitudinal distributions of Sh_d obtained at different spanwise locations are similar to one another. This shows that the adjacent wall does not affect the distributions of Sh_d in the sharp-edged inlet condition. The characteristics of the mass transfer in the present case can be regarded as similar to those in a two-dimensional channel.

Laminar mass transfer region under the bellmouth inlet condition

A detailed comparison of Fig. 7(a) and Fig. 8 reveals that the values of the maximum Sh_d observed at the entrance of the sharp-edged inlet condition are about twice as large as those of the bellmouth inlet condition. However, the asymptotic values of Sh_d observed near the exit of the test section of Fig. 7(a) are almost the same as those in Fig. 8. This suggests that the longitudinal distribution of the local Sherwood number Sh_d in the turbulent region of the test section is independent of the entrance configurations. From these results, it appears that the relatively low mean Sherwood number Sh for the bellmouth inlet condition shown in Fig. 6 may be due to the laminar mass transfer region which appears near the entrance of the test section. The distribution of this laminar region is examined in detail.

In convective heat transfer in a laminar boundary layer over a flat plate, local Sherwood number $Sh_X = h_m \cdot X/D$ is proportional to the square root of the Reynolds number $Re_X = U \cdot X/\nu$, where X denotes the distance from a leading edge of the plate [4]. Hence, the local mass transfer coefficients h_m obtained under the bellmouth inlet condition have been rearranged with $Sh_X = h_m \cdot X_l/D$ and $\sqrt{Re_X} = \sqrt{U \cdot X_l/\nu}$. Then, the starting point of the transition to turbulent mass transfer has been defined as the point at which the linear distribution of Sh_X and $\sqrt{Re_X}$ has broken up. Figure 9 shows examples of the distributions of Sh_X against $\sqrt{Re_X}$ obtained at various Reynolds numbers Re_d. At any Re_d, Sh_X near the entrance of the test section (low $\sqrt{Re_X}$ values) is proportional to $\sqrt{Re_X}$. At $\sqrt{Re_X} = 400 \sim 500$, however, Sh_X suddenly deviates from the linear distribution. The values of $\sqrt{Re_X}$ at which this deviation is observed have been examined for each Re_d and X_2/B, and these values of $\sqrt{Re_X}$ have been converted into those of the longitudinal coordinate X_l/d.

Figure 10 is a distribution map of the starting points of the transition plotted over the naphthalene surface for various Re_d. The shadowed areas correspond to the laminar mass transfer region, and a line connecting circular symbols indicates the location at which the transition to the turbulent mass transfer has started. As described in the preceding sections, the transition originates from the duct corner, and then the turbulent region spreads toward the duct center downstream. Thus, in Fig. 10, the laminar mass transfer region forms a triangular area near the entrance, in which the local Sherwood number Sh_d is relatively low. As the values of Re_d are increased, this triangular area of low mass transfer rate becomes smaller. This result closely corresponds to the result of the mean Sherwood number $\overline{Sh}$ shown in Fig. 6; that is, the values of $\overline{Sh}$ under the bellmouth inlet condition approach those of the sharp-edged one as the turbulent mass transfer region expands to the larger part of the test section at larger values of Re_d.

CONCLUSIONS

(1) Under the bellmouth inlet condition, the laminar boundary layer is first formed along the corner near the duct entrance, and then the transition to the turbulent boundary layer originates from the duct corner. On the other hand, under the sharp-edged inlet condition, an entire area of the duct cross-section is occupied by

the turbulent boundary layer.

(2) Local Sherwood number Sh_d on the bellmouth inlet condition is maximum at the entrance of the test section, and it then decreases in the flow direction. Thereafter, Sh_d begins to increase due to the turbulent transition of the boundary layer. This increase starts at a more upstream location as the adjacent wall of the duct is approached. The triangular region of laminar mass transfer formed near the entrance of the test section becomes smaller at larger Reynolds number Re_d.

(3) The value of Sh_d on the sharp-edged inlet condition peaks at the entrance, and decreases monotounously in the flow direction. The longitudinal distributions of Sh_d obtained at different spanwise locations of the test duct are similar to one another.

REFERENCES

[1] Mills, A. F., 1962, J. Mechanical Engineering Science, 4-1 , 63.

[2] P. R. S. Mendes, 1991, Int. J. Experimental Thermal Fluid Science, 4 , 510.

[3] Sparrow, E. M., and Cur, N., 1982, ASME J. Heat Transfer, 104, 82.

[4] Kays, W. M., and Crawford, M. E., 1980, Convective heat and mass transfer, McGraw-Hill Book Co., New York.

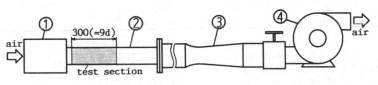

① settling chamber
② test duct
③ venturi nozzle flow meter
④ turbo fan

Figure 1 Schematic diagram of an experimental apparatus

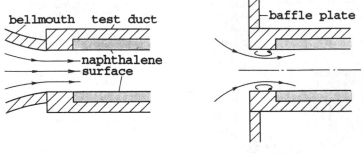

(a) bellmouth inlet condition   (b) sharp-edged inlet condition

Figure 2 Flow inlet conditions

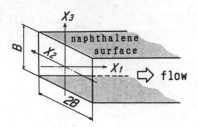

Figure 3 Coordinate system

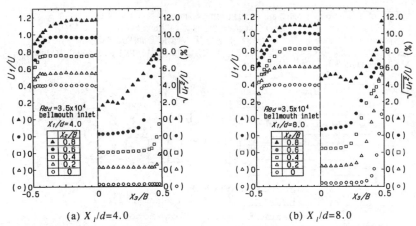

(a) $X_1/d=4.0$ (b) $X_1/d=8.0$

Figure 4 Velocity distributions under the bellmouth inlet condition

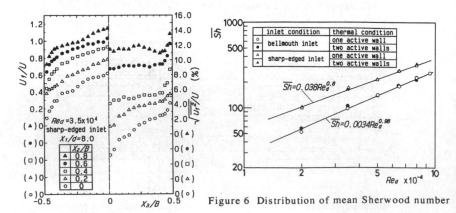

Figure 6 Distribution of mean Sherwood number

Figure 5 Velocity distributions under
the sharp-edged inlet condition ($X_1/d=8.0$)

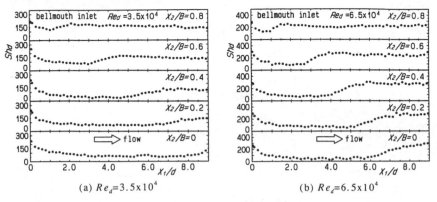

(a) $Re_d = 3.5 \times 10^4$ (b) $Re_d = 6.5 \times 10^4$

Figure 7 Distributions of local Sherwood number under the bellmouth inlet condition

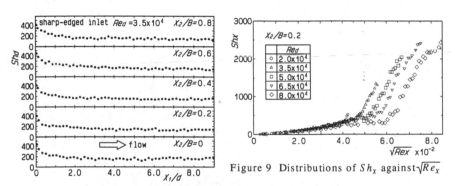

Figure 9 Distributions of Sh_x against $\sqrt{Re_x}$

Figure 8 Distributons of local Sherwood number
under the sharp-edged inlet condition at $Re_d = 3.5 \times 10^4$

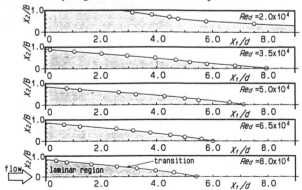

Figure 10 Distribution map of the starting point of the transition

A NUMERICAL ANALYSIS OF LAMINAR FLOW HEAT TRANSFER IN ECCENTRIC ANNULI

J.S.Szmyd[*], K.Suzuki[#], Z.Sz.Kolenda[*] and Y.Hagiwara[#]

[*]Staszic Univ., 30059 KRAKOW, Poland, [#]Kyoto Univ., KYOTO 606, Japan

A numerical analysis has been made of steady-state convective heat transfer for hydrodynamically and thermally fully developed laminar flow in eccentric annuli. Computations were made for two types of boundary conditions of the second kind. Peripheral distribution of temperature and local Nusselt number are discussed. It is shown that classical definition of local Nusselt number is not always convenient. In such a case, it is shown that new definition of local Nusselt number is more useful.

INTRODUCTION

Convective heat transfer in a thermally and hydrodynamically fully developed laminar flow in an eccentric annulus has been treated by several researches. However, attention was paid mostly to the value of average Nusselt number, average over the whole heat transfer surface, so that the discussion on the peripheral nonuniformity of heat transfer characteristics is rather scare. Therefore the local heat transfer characteristics will be discussed in detail in the present paper.

ANALYSIS

Steady-state heat transfer is calculated for a hydrodynamically and thermally fully developed laminar flow in eccentric annuli. The fluid is assumed to be incompressible, Newtonian, with constant properties. However, the calculated results will be applied to fluids of any Prandtl number. Body forces such as gravity and electromagnetic force are ignored, and it is assumed that any type of secondary flow does not appear. Also assumed is the absence of free convection, mass diffusion, chemical reaction and change of phase.

The governing equations to be solved are the continuity equation, the momentum equation and the energy equation.

The geometry of the passage is described in Fig.1(a) and the bipolar coordinate system (ξ, η) to be used in the present calculation is shown in Fig.1(b). The flow and temperature fields are assumed to be symmetric with respect to the $\eta = 0$ and π axis.

Transformation of the governing equations into the forms appropriate for bipolar coordinate system gives:

$$\frac{\partial U}{\partial z} = 0 \tag{1}$$

$$\frac{1}{\rho}\frac{\partial p}{\partial z} = \frac{1}{B^2}\left[C\frac{\partial}{\partial \xi}(Cv\ \frac{\partial U}{\partial \xi}) + C\frac{\partial}{\partial \xi}(Sv\ \frac{\partial U}{\partial \eta}) + S\frac{\partial}{\partial \eta}(Cv\ \frac{\partial U}{\partial \xi}) + S\frac{\partial}{\partial \eta}(Sv\ \frac{\partial U}{\partial \eta})\right.$$

$$\left. + S\frac{\partial}{\partial \xi}(Sv\ \frac{\partial U}{\partial \xi}) - S\frac{\partial}{\partial \xi}(Cv\ \frac{\partial U}{\partial \eta}) - C\frac{\partial}{\partial \eta}(Sv\ \frac{\partial U}{\partial \xi}) + C\frac{\partial}{\partial \eta}(Cv\ \frac{\partial U}{\partial \eta})\right] \tag{2}$$

$$U\frac{\partial T}{\partial z} = \frac{1}{B^2}\left[C\frac{\partial}{\partial \xi}(Ca\ \frac{\partial T}{\partial \xi}) + C\frac{\partial}{\partial \xi}(Sa\ \frac{\partial T}{\partial \eta}) + S\frac{\partial}{\partial \eta}(Ca\ \frac{\partial T}{\partial \xi}) + S\frac{\partial}{\partial \eta}(Sa\ \frac{\partial T}{\partial \eta})\right.$$

$$\left. + S\frac{\partial}{\partial \xi}(Sa\ \frac{\partial T}{\partial \xi}) - S\frac{\partial}{\partial \xi}(Ca\ \frac{\partial T}{\partial \eta}) - C\frac{\partial}{\partial \eta}(Sa\ \frac{\partial T}{\partial \xi}) + C\frac{\partial}{\partial \eta}(Ca\ \frac{\partial T}{\partial \eta})\right] \tag{3}$$

where z is the axial coordinate and U is the streamwise velocity and

$$C = \cosh \xi\ \cos \eta\ -\ 1, \qquad S = \sinh \xi\ \sin \eta$$

$$B = \frac{1}{2d}\ \sqrt{(R_o^2 + R_i^2 - d^2)^2 - 4\ R_o^2\ R_i^2}$$

The finite-difference equivalents of Eqs.(1), (2) and (3) were obtained by applying the scheme described in [2] and were solved numerically. At the start of computation, a specific value was assigned to the pressure gradient and the corresponding peripheral distributions of wall shear stress, therefore the spatial distribution of streamwise velocity, were calculated with an iterative procedure. Convergence of the computation was checked by using the following overall momentum equation:

$$(2\pi R_i \bar{\tau}_i + 2\pi R_o \bar{\tau}_o) = - \pi(R_o^2 - R_i^2)(dp/dz)$$

The value of Reynolds number Re was calculated from the mass flow rate which was evaluated based on the computed velocity profile ($Re \approx 1000$). 50 x 50 grid points in total were allocated in ξ - η space and the iterative calculation was made using the line-by-line method [2].

Non-slip boundary conditions are adopted at the inner and outer walls for flow analysis. An annulus contains two surfaces on which thermal conditions may generally be specified independently from each other. Therefore there are a large number of heat transfer problems depending on the types of thermal boundary conditions on both surfaces. In this paper two types of the second kind of boundary conditions [1] are treated; the case (A): the outer cylinder surface being kept thermally adiabatic (q_o = 0.0) and the heat flux on the inner cylinder surface being uniform both in streamwise and peripheral directions (q_i = constant), and the case (B): the inner cylinder surface being kept adiabatic (q_i = 0.0) and the outer cylinder surface being uniform both in streamwise and peripheral directions (q_o = constant). The axial temperature gradient appearing in Eq.(3) can be calculated from the following equation:

$$(dT/dz) = 2\ R_k q_k\ /\ \rho c_p \bar{U}\ (R_o^2 - R_i^2)$$

where either of the subscripts i (for inner cylinder) or o (for outer cylinder) should be assigned for k in the above equation.

PRESENTATION AND DISCUSSION OF RESULTS

First, attention will be paid to the average Nusselt number, Nu, averaged over all the heat-transfer surfaces. The average Nusselt number, $\overline{Nu}_k$ which appear in the following discussion is defined as follow:

$$\overline{Nu}_k = \frac{q_k \, D_h}{\lambda \, (\overline{T}_k - T_b)}$$

where either of the subscript i (for inner cylinder) or o (for outer cylinder) should be assigned for k in the above equation, q_k is the magnitude of the wall heat flux, $\overline{T}_k$ is the peripheral mean wall temperature. Fig.2. shows the average Nusselt number, calculated for the two boundary conditions A and B [4],[5]. The value of Nu is higher for smaller values of α and e. That is, better heat transfer occurs with a flow passage of wider flow space and closer to the axisymmetric case. The present results of Nu agree with those previously reported by Trombetta [3].

Now the cross-sectional and peripheral distributions of local values will be discussed. The peripheral distribution of the local Nusselt number calculated for case A, $\alpha=0.3$ and $\alpha=0.7$ are shown in Fig.3.. The local Nusselt number $Nu_k(\theta_k)$ is defined as [1],[6]

$$Nu_k(\theta_k) = \frac{h_k(\theta_k) \, D_h}{\lambda} \qquad h_k(\theta_k) = \frac{q_k}{T_k(\theta_k) - T_b} \qquad (4)$$

where the subscript i (for inner cylinder) or o (for outer cylinder) should be assigned for k in the above equation, q_k is the magnitude of the wall heat flux, $T_k(\theta)$ is the local wall temperature, h_k is the local heat transfer coefficient, θ_i and θ_o are the angular coordinates respectively around the centers of the inner and outer cylinders. Heat transfer coefficient in Eq.4 is defined with reference to the fluid bulk mean temperature; this is quite arbitrary, but normally very convenient [6]. When $\alpha=0.7$ (Fig.3.), the local Nusselt number has a negative value at the positions around $\eta=0$. A negative value of the local Nusselt number does not suggest either that heat transfer is low or that the direction of heat transfer is opposite to that at the position corresponding to a positive local Nusselt number. The wall temperature is lower where the local Nusselt number is negative so that the heat transfer is actually better there. The normal gradient of fluid temperature has the same sign as that corresponding to the direction of heat transfer at other positions. However, the local wall temperature can become lower than the fluid bulk temperature, T_b, at some peripheral positions because the nonuniformity of the wall temperature becomes quite large in some cases for large values of α and e. Therefore the local Nusselt number defined as in Eq.(4) can have negative values there. In such situations, the conductance concept of the heat transfer coefficient loses its significance [1] and the local Nusselt number defined as in Eq.(4) is not very convenient.

For convective heat transfer in eccentric annuli of the second kind of boundary conditions, local Nusselt number of different definition must be proposed. Fig.4. shows heat-flow lines for convective heat transfer in eccentric annulus for boundary condition A. Point $T_o(0)$ is the lowest temperature of adiabatic wall and all heat-flow lines lie almost tangentially very close to the point having the temperature $T_o(0)$. If we now define a local heat-transfer coefficient $h_k^+(\theta_k)$ with reference to the temperature $T_z(0)$ equation (4) becomes

$$Nu_k^+(\theta_k) = \frac{h_k^+(\theta_k) \, D_h}{\lambda} \qquad h_k^+(\theta_k) = \frac{q_k}{T_k(\theta) - T_z(0)} \qquad (5)$$

where if $k = i$, $z = o$, if $k = o$, $z = i$.

The peripheral distributions of the newly defined local Nusselt number calculated with Eq.(5) for cases A and B are illustrated in Figs. 5 and 6. It is seen that the peripheral nonuniformity of the local Nusselt number becomes more noticeable with an increase of radius ratio α and eccentricity e, while the nonuniformity is small at $\alpha = 0.1$. It shows everywhere positive values. With this form of local Nusselt number, direct illustration [5] of local wall temperature is not necessary. Thus, it is more convenient for practical use and of more physical meaning.

CONCLUSIONS

In this paper, laminar convective heat transfer has been studied numerically for thermally and hydrodynamically fully developed flow in eccentric annuli. Computations were made for two types of boundary conditions of the second kind. Peripheral distributions of wall temperature and the local Nusselt number are discussed. It is demonstrated that the non-uniformity of the wall temperature is very conspicuous in the case of large radius ratio and eccentricity. It is also shown that classical definition of local Nusselt number is not always convenient. In such a case, it is shown that new definition of local Nusselt number is more useful.

Acknowledgments
This research was supported in part by Stefan Batory Foundation and by State Committee for Scientific Research (Grant No.KBN 3 0814 91 01). It was carried out at Kyoto University when the first author (J.Szmyd) stayed there under the support from TORAY Science Foundation (Grant No.88-2906). The authors sincerely acknowledge the support received from the foundations mentioned in the above.

References
[1] Shah R.K., London A.L.; (1978), _Laminar Flow Forced Convection in Ducts_, Academic Press, New York.
[2] Suzuki K.; (1984), _Advances in heat transfer_ (in Japanese), ed Y.Kato et al. Yokendo, Tokyo, 269-289.
[3] Trombetta M.L.; (1971), _Int.J.Heat Mass Transfer_, **14** (8), 1161-1173.
[4] Suzuki K., Szmyd J.S., Ohtsuka H.; (1990), _Trans.JSME_, **56** (531), 3445-3450, (in Japanese).
[5] Suzuki K., Szmyd J.S., Ohtsuka H.; (1991), _Heat Transfer, Japanese Research_, **20** (2), 169-183.
[6] Kays W.M, Crawford M.E.; (1980), _Convective Heat and Mass Transssfer_, McGraw-Hill, New York.

Nomenclature

A : cross-sectional area of annuli [m^2], a : thermal diffusivity [m^2/s], c_p: specific heat at constant pressure [kJ/(kg K)], d : displacement of inner cylinder [m], D_h: equivalent hydraulic diameter = $2(R_o-R_i)$ [m], e : eccentricity = $d/(R_o-R_i)$, $\overline{Nu}$, $Nu(\theta)$: average, local Nusselt number, q : heat flux [W/m^2], p : pressure [Pa], R : radius [m], Re: Reynolds number = $U_m D_h/\nu$, T : temperature [K], U : velocity [m/s], z : axial distance [m], α : radius ratio = R_i/R_o, θ : angle (see Fig.1) rad, ν : kinematic viscosity of fluid [m^2/s], λ : thermal conductivity of fluid [W/(m K)], ρ : density of fluid [kg/m^3], τ : shear stress [Pa].

Subscripts and Superscripts

b: bulk mean, i: outer surface of inner cylinder, m: cross-sectional average, o: inner surface of outer cylinder, $-$: average over heat-transfer surface.

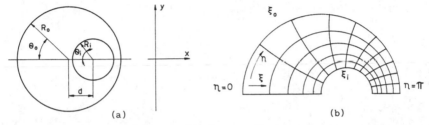

(a)   (b)

Fig.1. Eccentric annulus and bipolar coordinate system.

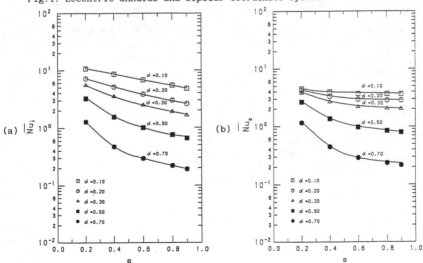

Fig.2. Average Nusselt number: (a) boundary conditions A (b) boundary conditions B, (——— computation, symbols [3]).

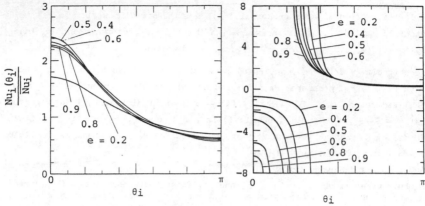

Fig.3. Peripheral distributions of local Nusselt number

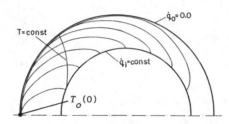

Fig.4. Heat-flow lines for convective heat transfer en eccentric annulus, thermal boundary condition A.

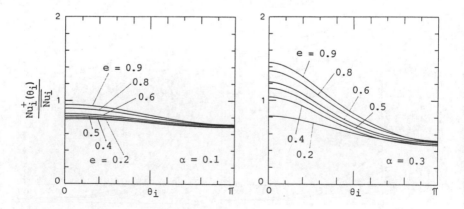

Fig.5.1. Peripheral distributions of local Nusselt number for case A.

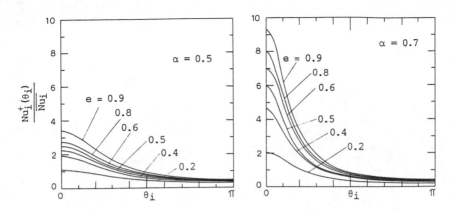

Fig.5.2. Peripheral distributions of local Nusselt number for case A.

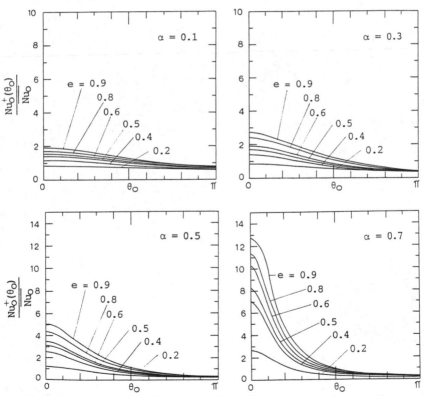

Fig.6. Peripheral distributions of local Nusselt number for case B.

A TWO DIMENSIONAL MODEL FOR LAMINAR SUPERCOOLED LIQUID FLOW

K.A.R. Ismail* and A. Padilha**
* UNICAMP-FEM-DETF, C.P. 6122, 13081 Campinas - SP, Brazil
** UNESP - Eng. Mecânica, C.P. 31, 15378 Ilha Solteira - SP, Brazil

In order to investigate the solidification process involving phase change of laminar flow submitted to superfreezing, a theoretical and experimental study is developed. A numerical unsteady model with temperature dependent physical properties for the solid region is proposed, based upon the energy balance equation in the solid and liquid region, as well as across the solidification front, and using interface dependent coordinate grid (the Landau transformation). A rig has been built using liquid nitrogen as refrigerant. The numerical results compare well with the experimental ones.

INTRODUCTION

The process of supercooling of stationary or flowing liquids involving movement of the solidification front, creates important technical problems in the industrial field such as glass, steel and ice industries and more recently in thermal storage systems and food conservation.

The theoretical and numerical pionner study due to Zerkle and Sunderland Ref. [1] presents pressure and temperature distributions, interface positions for the flow of liquid in steady laminar conditions inside a tube of circular section. Cho and Sunderland Ref.[2] presented a conduction based analytical solution for the problem of phase change in a semi-infinite region considering the thermal conductivity as temperature dependent. The same problem was treated by Olivier and Sunderland [3], and also considered the specific heat as varying linearly with temperature. The numerical solution for the one-dimensional stationary phase change problem with constant properties and using fixed grid developed in Rao and Sastri Ref.[4] has shown good agreement with other numerical models.

The objective of this paper is to present a numerical solution for a mathematical model able to simulate the two dimensional phase change process in a finite region involving a moving fluid in laminar regime inside a duct.

THEORETICAL TREATMENT

The problem considered is shown in Figure 1 where the liquid (water) flows in laminar regime inside a duct of rectangular cross section and side walls subjected to super-cooling. The liquid flows initially in an entry region free of solidification $\bar{x} < 0$ where the wall temperature is maintained slightly higher than the phase change temperature and along which the flow becomes fully developed. For $\bar{x} > 0$, the wall temperature is suddenly reduced to a value $T_w < T_f$ constant along the duct. In this region the water is supercooled forming an ice layer along the duct with thickness increasing in the direction of flow.

The formulation of the problem is based upon the following simplifying assumptions:

1. The interface solid-liquid temperature is constant and equal to the phase change temperature;

2. Heat is transferred from the liquid region to the interface by forced convection and the local Nusselt number correlations of Saah Ref.[5] are used;

$$Nu_{x,T} = \begin{cases} 1,233 \ (x*)^{-1/3} + 0,4 & \text{for} \quad x* \leq 0,001 \\ 7,541 + 6,874 \ (10^3 \ x*)^{-0,488} \ e^{-245} \ x* & \text{for} \quad x* > 0,001 \end{cases}$$

3. The physical properties in the liquid (water) region are considered constants and evaluated at the phase change temperature whereas those for the solid (ice) regions are considered to be temperature dependent and calculated as follows,

specific volume (V), [m^3/kg], Perry Ref.[6],

$$V(\overline{T}) = 1.0907 \times 10^{-3} + 1.4635 \times 10^{-7} \ \overline{T} \ ; \tag{1}$$

specific heat (C_p), [J/kg^{o}C], Weast and Astle Ref.[7],

$$C_p(\overline{T}) = 2116.5607 + 7.2845 \ \overline{T} \ ; \tag{2}$$

thermal conductivity (K), [W/m^{o}C], Powell Ref.[8],

$$K(\overline{T}) = 2.09 - 0.003553 \ \overline{T} \tag{3}$$

4. Axial heat conduction (8500 < Pe < 28000), viscous energy dissipation and free convection effects are ignored;

5. The liquid is pure, newtonian and incompressible;

6. The solid layer is smooth, homogeneous and isotropic.

The mathematical formulation of the problem for the solid phase region is

$$C_p(\overline{T}) \cdot \frac{\partial \overline{T}}{\partial \overline{t}} = V(\overline{T}) \ \frac{\partial}{\partial} \ [K(\overline{T}) \cdot \frac{\partial \overline{T}}{\partial \overline{y}}] \ , \tag{4}$$

in $0 < \overline{x} \leq \overline{L}$, $\overline{s} < \overline{y} < H$ and $\overline{t} > 0$;

and for the liquid phase region as

$$\frac{\partial \overline{T}}{\partial \overline{t}} + \overline{u} \ \frac{\partial \overline{T}}{\partial \overline{x}} + \overline{v} \ \frac{\partial \overline{T}}{\partial \overline{y}} = \alpha_1 \ \frac{\partial^2 \overline{T}}{\partial \overline{y}^2} \ , \tag{5}$$

in $0 < \overline{x} \leq L$, $0 \leq \overline{y} < \overline{s}$ and $\overline{t} > 0$,

with the following initial and boundary conditions:

$$\overline{T} = T_i \ , \quad 0 \leq \overline{x} \leq L \ , \quad 0 \leq \overline{y} \leq H \ , \quad \overline{t} = 0 \ , \tag{6}$$

$$\overline{T} = T_w \ , \quad 0 < \overline{x} \leq L \ , \quad \overline{y} = H \quad , \quad \overline{t} > 0 \ , \tag{7}$$

$$\frac{\partial \overline{T}}{\partial y} = 0 \ , \quad 0 < \overline{x} \leq L \ , \quad \overline{y} = 0 \quad , \quad \overline{t} > 0 \ , \tag{8}$$

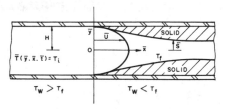

FIG. (1) SCHEME OF THE PROBLEM

$$\overline{T} = T_i \quad , \quad \overline{x} = 0 \quad , \quad 0 \leq \overline{y} \leq H \quad , \quad \overline{t} > 0 . \tag{9}$$

For the solid-liquid interface involving convection, Ozisik Ref.[9], we have

$$\overline{T}_1 = \overline{T}_s = T_f \quad , \quad 0 < \overline{x} \leq L \quad , \quad \overline{y} = \overline{s} \quad , \quad \overline{t} > 0 , \tag{10}$$

$$[1 + (\frac{\partial \overline{s}}{\partial \overline{x}})^2] \cdot [K_s \frac{\partial \overline{T}_s}{\partial \overline{y}} - h_m (\overline{T}_m - T_f)] = -\rho_1 \cdot C_L \cdot \frac{\partial \overline{s}}{\partial \overline{t}}$$

$$0 < \overline{x} \leq L \quad , \quad \overline{y} = \overline{S} \quad , \quad \overline{t} > 0 . \tag{11}$$

Following Zerkle and Sunderland Ref.[1], one may assume that the velocity profile in the test section remains parabolic and can be admitted as

$$\overline{u} (\overline{x}, \overline{y}, \overline{t}) = (3/2) \cdot (H/\overline{s}) U_m [1 - (\frac{\overline{y}}{\overline{s}})^2] ,$$

$$\overline{v} (\overline{x}, \overline{y}, \overline{t}) = (\frac{3}{2}) (\frac{H\overline{y}}{\overline{s}^2}) U_m [1 - (\frac{\overline{y}}{\overline{S}})^2] \frac{\partial \overline{S}}{\partial \overline{x}} \tag{12}$$

$$0 \leq \overline{x} \leq L \quad , \quad 0 \leq \overline{y} \leq S \quad , \quad \overline{t} \geq 0 .$$

where

$U_m = Q/(2.a.H)$ (average velocity at the entry cross section),

$$\overline{T}_m - T_f = \frac{4}{5} \cdot (\overline{T}_o - T_f) \quad , \quad 0 < \overline{x} \leq L \quad , \quad \overline{y} = 0 \quad , \quad \overline{t} > 0$$

where $\overline{T}_o$ is the temperature along the channel center line.

In order to obtain a dimensionless formulation the following variables are defined:

$$x = \overline{x}/H \quad , \quad x^* = \overline{x}^*/H \quad , \quad y = \overline{y}/H \quad , \quad s = \overline{s}/H$$

$$T = \frac{\overline{T} - T_f}{T_f - T_w} \quad , \quad t = \alpha_1 \overline{t}/H^2 \quad , \quad u = H.\overline{u}/\alpha_1 \quad \text{and} \quad v = H.\overline{v}/\alpha_1$$

To immobilize the moving interface a coordinate transformation (Landau transforms), is used resulting in

$$\xi = \overline{y}/\overline{s} \quad \text{and} \quad \eta = (\overline{y} - \overline{s})/(H - \overline{s}) ,$$

and hence the regions are caracterized by

solid region: $T(x, \eta, t) \quad , \quad 0 \leq x \leq L/H \quad , \quad 0 \leq \eta \leq 1 \quad , \quad t \geq 0$

liquid region: $T(x, \xi, t) \quad , \quad 0 \leq x \leq L/H \quad , \quad 0 \leq \xi \leq 1 \quad , \quad t \geq 0$

Landau transforms are used in order to facilitate the numerical procedure and the solution of the problem. After transformation, the resulting equations can be written in the form.

For the solid region:

$$C_p(T)\left[\frac{\partial T}{\partial t} - \left(\frac{1-\eta}{1-s}\right)\left(\frac{\partial s}{\partial t}\right)\left(\frac{\partial T}{\partial \eta}\right)\right] = \frac{V(T).K(T)}{\alpha_1}\left(\frac{1}{1-s}\right)^2 \frac{\partial^2 T}{\partial \eta^2} - \frac{\beta(T).(T_f - T_w)}{\alpha_1} \cdot \left(\frac{1}{1-s}\right)^2 \left(\frac{\partial T}{\partial \eta}\right)^2$$

$$0 < x \leq L/H \quad , \quad 0 < \eta < 1 \quad , \quad t > 0 \; ;$$

(13)

For the liquid region:

$$\frac{\partial T}{\partial t} - \left(\frac{\xi}{s}\right)\left(\frac{\partial s}{\partial t}\right)\left(\frac{\partial T}{\partial \xi}\right) + u\left[\frac{\partial T}{\partial x} - \left(\frac{\xi}{s}\right)\left(\frac{\partial s}{\partial x}\right)\left(\frac{\partial T}{\partial \xi}\right)\right] + \frac{v}{S}\frac{\partial T}{\partial \xi} = \left(\frac{1}{s}\right)^2 \frac{\partial^2 T}{\partial \xi^2}$$

$$0 < x \leq L/H \quad , \quad 0 \leq \xi < 1 \quad , \quad t > 0 \quad ,$$

(14)

subject to the following initial and boundary conditions:

$$T(x, \xi, t) = T_i \quad , \quad 0 < x \leq L/H \quad , \quad 0 \leq \xi \leq 1 \quad , \quad t = 0 \quad ,$$

(15)

$$\frac{\partial T}{\partial \xi}(x, \xi, t) = 0 \quad , \quad 0 < x \leq L/H \quad , \quad \xi = 0 \quad , \quad t > 0 \quad ,$$

(16)

$$T(x, \eta, t) = -1 \quad , \quad 0 < x \leq L/H \quad , \quad \eta = 1 \quad , \quad t > 0 \quad .$$

(17)

As for the interface we have

$$T(x, \eta, t) = T(\xi, x, t) = 0 \quad , \quad 0 < x \leq L/H \quad ,$$

(18)

$$\eta = 0 \quad , \quad \xi = 1 \quad , \quad t > 0 \quad .$$

and

$$\left[\left(\frac{1}{1-s}\right)\frac{\partial T}{\partial \eta} - Nu_m\left(\frac{H}{D_h}\right)\left(\frac{k_1}{k_s}\right)T_m\right]\left[1 + \left(\frac{\partial s}{\partial x}\right)^2\right] = -\frac{\alpha_1}{\alpha_s} \cdot \left(\frac{1}{Ste}\right)\frac{\partial s}{\partial t}$$

$$0 < x \leq L/H \quad , \quad \eta = 0 \quad , \quad \xi = 1 \quad , \quad t > 0 \quad .$$

(19)

Following the same assumption as Zerkle and Sunderland Ref.[1], the velocity profile and bulk temperature are given respectivelly as

$$u(x, \xi, t) = \frac{3.Q}{4a\alpha_1 s}(1 - \xi^2) \quad \text{and} \quad v(x, \xi, t) = \frac{3.Q\,\xi}{4a\alpha_1 s}(1 - \xi^2)\frac{\partial S}{\partial x}$$

(20)

$$0 \leq x \leq L/H \quad , \quad 0 \leq \xi \leq 1 \quad , \quad t \geq 0$$

and

$$T_m = \frac{4}{5} T(x, \xi, t) \quad , \quad 0 < x \leq L/H \quad , \quad \xi = 0 \quad , \quad T > 0 \quad ,$$

(21)

where

$$\beta(T) = 0,003553.V(T) \quad .$$

The term $(\partial s/\partial t)$ in Eqs. (13) and (14) can be interpreted as velocity and the terms in which it appears are called psudoconvective ones. This pseudoconvection is created by the imobilization of the interface. Clearly Eqs. (13) and (14) are more complex than the original ones, reflecting the penalty for the immobilization of the interface.

METHOD OF SOLUTION

The system of equations and the associated initial and boundary conditions, Eqs. (13-19), were approximated by finite diferences using rectangular grids in the transformed domain (x, ξ, t) and (x, η, t), with a uniform step in the x-direction and variable time step. Equation (19) of the interface was solved using Crank-Nicolson explicit method and hence eliminated the necessity for iteration between the temperature field and the interface. Sparrow and Chuck Ref. [10] used this scheme and found no limitation on the time step. In case of the system of Eqs. (13-19), the term $(\partial T/\partial t)$ in Eq. (13), was calculated explicity by Crank-Nicolson method to ensure convergence. The next step was to use implicit Crank-Nicolson scheme in Eq. (13) to evaluate the temperature field with the derivatives $(\partial T/\partial t)$, and the interface position s(t). The implicit Crank-Nicolson method was also applied to Eq. (14) to evaluate the temperature field in the liquid region using the previously calculated $(\partial s/\partial t)$ and s(t).

After several numerical tests involving numerical stability, precision and computational time, a grid of 50x5 nodes was used for each region of the phase change problem.

EXPERIMENTAL ANALYSIS

The experiments were realized on a closed circuit rig with a transparent test section measuring 1000 mm in length, 100 mm in width and 34 mm in height. The water flows to the test section after an entry length of 5000 mm to ensure fully developed laminar flow. The top part of the test section had scale markings along the channel for photografic measurements while the bottom part had 25 Copper-Constantan thermocouples of 0.3 mm diameter distributed over the test section. These thermocouples had an axial spacing of 200 mm and transversal distance of 25 mm. The test section was cooled by liquid nitrogen flowing through independent circuit.

RESULTS AND DISCUSSIONS

To enable experimental comparisons, the numerical predictions, were realized for the same regime and wall temperature. The calculations were started with a thin initial layer s = 10^{-6}m admitting a linear temperature variation within this region.

This procedure was found necessary to avoid difficulties to start the numerical calculations when t = 0 with abrupt temperature variation adjacent to the wall. In the start the solidification velocity reaches its maximum value due to the high temperature gradient, needing initial time steps of order of 10^{-6} seconds. This problem was solved using a variable relaxation factor which increases the time step with the evolution of the solidified thickness.

Figure(2) shows the effects of neglecting the ice-layer thickness $(\overline{s})$ the term $(\partial s/\partial t)$, and/or the term $(\partial T/\partial \eta)^2$ in Eqs.(13) and (14). Ignoring the term $(\partial T/\partial \eta)^2$ in the energy equation for the solid region can lead to maximum error of 8%. On the other hand if the term $(\partial s/\partial t)$ is ignored, as it was done in Ref. [11] the error goes up to about 13%. When both terms were maintained in the equations the results are found to be close to the experiments as shown in Fig. (3). The position of the solidification front predicted by the model and the corresponding experimental results agree well within maximum error of about 4%.

Figure (4) shows the relative error ε as function of the wall temperature. As it can be seen the relative error increases with decreasing T_w. Hence the simplification used in Ref.[11] is actually a function of wall temperature.

Figure (5) shows a comparison between predicted and measured temperature fields in the solid region. The maximum error is 5%.

CONCLUSIONS

The semi-analytical model proposed here and the use of Landau transforms lead to very good predictions of the temperature field and position of the solidification front. Also the model is rather general and other models can be

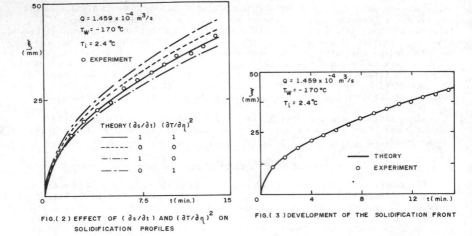

FIG.(2) EFFECT OF $(\partial s/\partial t)$ AND $(\partial T/\partial \eta)^2$ ON SOLIDIFICATION PROFILES

FIG.(3) DEVELOPMENT OF THE SOLIDIFICATION FRONT

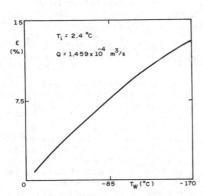

FIG.(4) RELATIVE ERROR WHEN IGNORING $(\partial s/\partial t)$

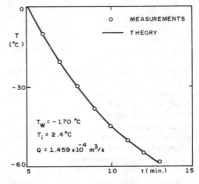

FIG.(5) TEMPERATURE PROFILES FOR THE SOLID REGION

derived from it.

REFERENCES

[1] Zerkle, R.D. & Sunderland, J.E., Journal of Heat Transfer 90: 183-190, 1968.

[2] Cho, S.H. & Sunderland, J.E., Journal of Heat Transfer 96: 214-217, 1974.

[3] Oliver, D.L.R. & Sunderland, J.E., International Journal of Heat Mass Transfer 30: 2657-2661, 1987.

[4] Rao, P.R. & Sastri, V.M.K., Journal of Heat Mass Transfer 27: 2077-2084, 1984.

[5] Saah, R.K. & London, A.L., Laminar Flow Convection in Ducts. Academic Press. New York, 1978.

[6] Perry, J.H., Chemical Engineers Handbook, McGraw-Hill. Kogakuska, 1973.

[7] Weast, R.C. & Astle, M.J., Handbook of Chemistry and Physics. CRS Press Inc., Florida.

[8] Powell, R.W.,Advances in Physics 25: 276-297, 1948.

[9] Ozisik, M.N., Heat Conduction. John Willey. 1980. 687 p.

[10] Sparrow, E.M. & Chuck, Numerical Heat Transfer 7: 1-15, 1984.

[11] Ramachandran, N.; Gupta, J.P.; Jaluria, Y., Numerical Heat Transfer, 1989.

NOMENCLATURE

a   height of duct, [m]
C_L latent heat of solidification, [J/kg]
C_p specific heat, [J/kg °C]
D_H hydraulic diameter for a duct of rectangular cross-section, [m]
h_m average heat transfer coefficient, [W/m² °C]
H   distance measured from centerline to solid-mold interface, [m]
K   thermal conductivity, [W/m °C]
L   reference length, [m]
Nu_m average Nusselt number
Pe = Re.Pr: Peclet number
Q   flow rate, [m³/s]
Pr   Prandtl number
Re   Reynolds number
$\bar{s}$ half-width of melt region
$\bar{t}$ time, [s]
$\bar{T}$ temperature, [°C]
T_m bulk fluid temperature
u   velocity in x direction, [m/s]
U   average velocity over duct cross section, [m/s]
ρ   density, [kg/m³]
η   transformed coordinate for solid region, dimensionless
ξ   transformed coordinate for liquid region, dimensionless
α   thermal difusivity, [m²/s]
ν   kinematic viscosity, [m²/s]

subscripts

e   entrance of duct
f   interface
I   initial
l   liquid
o   centerline of the duct
s   solid
w   wall

AN EXTENSION OF MULTIGRID TECHNIQUE TO THE ANALYSIS OF CONJUGATE HEAT TRANSFER

H.Y. WANG - J.B. SAULNIER

The present paper discusses a numerical study by a multigrid technique, of conjugate conduction/convection problems. The performance of this recently developed calculation procedure is evaluated in two laminar two-dimensional elliptic flows, such as an air cooling of electronic equipment and a particular heat exchanger. The effect of the presence of obstacles on the thermal and dynamic fields is clearly demonstrated by examining the velocity and temperature distribution. The multigrid procedure is observed to converge rapidly to an acceptable accuracy.

I - INTRODUCTION

The problem of conjugate heat transfer related to the cooling of electronic equipments and heat exchanger is of considerable technological interest. It is well known that the thermal design in these fields is clearly highly dependent on the knowledge of the heat transfer phenomena.

The heat flow through a partially blocked channel (Fig.1) is both a classical flow problem and a problem of practical importance because it is widely used in air cooling of electronic equipment. However, the trend for miniaturization of the Integrated Circuits (ICs) in the electronic industry is setting challenging problems for the heat transfer analyst. The life time, the mean time between failures, and a lot of electronic failures are effectively controlled by the thermal design of this system. Currently, the study of the cooling of electronic equipments has been an active area of experimental or numerical /1-3/ research in order to understand the influence of the presence of the dissipating obstacles (ICs) on the flow and thermal fields. The second configuration considered is a concentric-pipe heat exchanger with discs and doughnuts (Fig.2). Recent numerical studies /4-5/ have reported the importance of the conjugate effects between the solid and the fluid in this type of heat exchanger, however, there are only few computational studies related to this complex heat exchanger. A detailed analysis of the thermal and dynamic fields is yet a critical step in improving current design techniques. A quantitative knowledge of local velocity and temperature will lead to a better assessment of equipment reliability and performance.

The flow situations mentioned above are quite complex with recirculations and reattachments even in a very simplified 2D representation, and they constitute good test cases for a numerical algorithm. The geometries are complicated involving many obstacles like for instance ICs located on

LABORATOIRE D'ETUDES THERMIQUES URA 1403 C.N.R.S.
E.N.S.M.A. POITIERS FRANCE

channel wall (Printed Circuit Boards: PCBs), and discs and doughnuts appearing in the shell heat exchanger, and these act as blockages in the flow. It is well known that the task of solving such problems is a large and complex one when using a classical single grid iterative method. Typically, a big number of nodes must be required in each direction in order to retrieve precisely the flow field in the recirculating zones and the temperature field in the region where high gradients are present. In addition, the governing equations of the fluid are nonlinear and strongly coupled between the momentum and continuity equations through the common pressure field. As a consequence, the numerical simulation may consume a lot of time, and it is truly necessary to find a method which solves the equations accurately and cheaply. The multigrid technique pioneered by Brandt /6/ has been found to be a powerful tool for the iterative resolution of elliptic and quasi-elliptic partial differential equations discretized for instance by a control volume aproach. Finally, thanks to an extension of this multigrid technique to conjugated problems, our attention can focus on some characteristic results in the fields of an electronic cooling equipment, and of a discs and doughnuts roughened concentric-pipe heat exchanger.

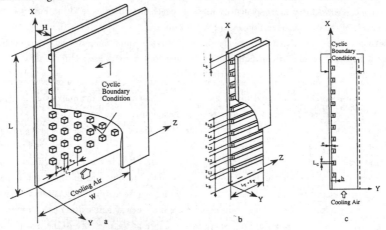

Fig. 1 - The electronic board (1a) and its 2D scheme (1b and 1c)

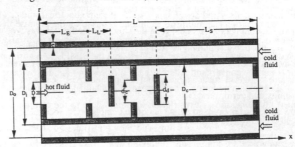

Fig.2 - Longitudinal cross section of the discs and doughnuts roughened concentric-pipe heat exchanger

II - METHODOLOGY

The flow situations described in the section I can be mathematically represented by the steady laminar Navier-Stokes equations in a fully elliptic form. The general form of the steady axisymmetric ($n = 1$ and $y = r$) or in cartesian coordinates ($n = 0$) laminar flow can be written as

$$\frac{1}{r^n}\left[\frac{\partial}{\partial x}(\rho u r^n \phi) + \frac{\partial}{\partial y}(\rho v r^n \phi) - \frac{\partial}{\partial x}(r^n \Gamma_\phi \frac{\partial \phi}{\partial x}) - \frac{\partial}{\partial y}(r^n \Gamma_\phi \frac{\partial \phi}{\partial y})\right] = S_\phi$$

where ϕ stands for any dependent variable as u, v, p and T, Γ_ϕ is the appropriate transport coefficient for ϕ, and S_ϕ is the source term of the transport equation for ϕ. As for the natural convection in laminar regime, all fluid properties are assumed constant, and the fluid is assumed to satisfy the Boussinesq approximation of Navier-Stokes equations, which relates the temperature to the density through the following equation :

$$\rho = \rho_0[1 - \beta(T - T_0)]$$

The differential equations are discretized by using the method of control volumes developed by PATANKAR /7/. A staggered mesh system is used to locate the variables, and the Power-Law scheme is used to discretize the convection and diffusion terms.

These problems are clearly conjugate ones, involving simultaneous heat transfer by conduction in the structure, and by convection in the laminar coolant fluid. For a laminar flow, such conjugate problems can be analyzed at the interface between fluid and solid by using the harmonic averaging of conductivities. Sometimes, the ratio of the thermal conductivities of the materials (solid and fluid) may be as large as 1×10^3, which gives rise to very strong anisotropies in the discretized coefficient. This numerical difficulty has been efficiently solved using the Block Adjustment Procedure /8/. For forced convection, since the solution of the energy equation does not affect the flow field, the set of discretized energy equations is solved for T only, after the velocity distribution has been obtained. However, for natural convection, the thermal and flow equations should be solved simultaneously because the temperature and velocity fields are coupled.

The basic concepts of multigrid have been clearly described in detail by HACKBUSCH in /9/, and we recall the main ideas. For a classical iterative method, the first iterations are most efficient for reducing the high-frequency components of the error, but the following iterations are inefficient in annihilating low frequency components. However, the low frequencies on fine grid can become high frequencies on a coarser grid. The multigrid technique is based on the premise that each frequency range of the error must be smoothed on the grid where it is most suitable to do so. Consequently, it is the combination of the smoothing iterations (we used here the SIMPLEC algorithm) and coarse-grid corrections that is rapidly convergent, whereas each of these two actions by themselves converges slowly or not at all. In concrete terms, a set of several embedded grids, which become coarser and coarser is defined. Prolongation and restriction operators help to the communication between the coarse and fine grids in order to accelerate the convergence. This technique has shown the advantage of consuming less time (in our case by a factor between 15 and 30) as compared to that of a classical single grid method. More details concerning the multigrid technique and its application to conjugated problem can be found in reference /10/.

III - CALCULATIONS AND DISCUSSION OF RESULTS

Calculations of each configuration are now discussed. All our analysis was performed using a computer program on a DEC3100 (15 Mips), running under UNIX. Table 1 summarizes the geometrical characteristics and the physical properties of the two tested cases :

Case 1 : air cooling of electronic equipment

Case 2 : discs and doughnuts roughened concentric-pipe heat exchanger

parameter	case 1	case 2	parameter	case 1	case 2
b(mm)	2.6	/	L_L(mm)	5.08	55
D(mm)	/	40	L_E(mm)	22.86	105
D_i(mm)	/	64	Ls(mm)	24	125
D_o(mm)	/	84	L_T+S_T(mm)	20.32	/
D_c(mm)	/	60	L_C(mm)	1	/
d_c(mm)	/	40	Q(W)	0.45	/
d_d(mm)	/	44	S_{L1-5}(mm)	14,13,15,	/
e(mm)	1.6	2		14,25	/
H(mm)	20.32	/	T_0(°C)	23	300,20
L(mm)	233.68	400	u_0(m/s)	/	/
W(mm)	147.32	/	λ (W/m.K)	0.65	40

Table 1 - The geometrical and thermophysical parameters

III - 1- Air *cooling of electronic Integrated Circuits (natural convection)*

Even though we attempt here to simulate one single electronic board (Fig. 1a), the problem considered has still a three-dimensional characteristic. So we shall first comment the way we have schematized the real three-dimensional (3D) problem by a simplified two-dimensional(2D) representation (Fig.1c). The main problem, when shifting from 3D to 2D, concerns essentially the estimation of the heat dissipation. We transform the actual power dissipated on the chip, Q(W), into a surface density q (W/m^2), given by : $q = \dfrac{Q}{(L_T + S_T) L_C}$.

This heat is then supposed to be uniformly dissipated at a constant rate (W/m^2) inside the junction region represented by the black layer in Fig.1b. We finally retain the 2D model presented by Fig.1c, which is representative of the previous 3D configuration (Fig.1b) where all the characteristics were supposed uniform in the z direction. The boundary conditions can be expressed as :

for x=0 $\quad \dfrac{\partial u}{\partial x} = \dfrac{\partial v}{\partial x} = 0 \quad$ and $\quad T = T_0$

for x=L $\quad \dfrac{\partial u}{\partial x} = \dfrac{\partial v}{\partial x} = \dfrac{\partial T}{\partial x} = 0$

At the vertical walls

for y=0=H $\quad u = v = 0 \quad$ and $\quad T(x,-e)=T(x,H)$

The locations at y=-e and y=H represent cyclic thermal boundary conditions which have been dealt with, by using the Cyclic Tridiagonal Matrix Algorithm (CTDMA) /11/.

Figure 3 shows the convergence plot for four types of grid, for which both the single grid (SG) and the multigrid (MG) have been applied at the finest level. The convergence of the iteration procedure is monitored by observing an averaged residual, which was based on the Euclidean norm /9/ for the flow and thermal equations. Let us first observe that the rate of convergence of multigrid is nearly the same for all the four grids, and thus, the number of fine grid iterations is practically independent to the increase of mesh points. From these plots, it is clearly seen also that the multigrid technique converges more rapidly, typically in 12 fine grid iterations than the single grid method, which needs here 420 iterations to reach the same level of accuracy ($|R| = 1.5 \times 10^{-6}$).

The maximum chip temperature obtained by the two numerical methods (MG and SG) is shown in Fig.4. We find that the single grid method leads to no significant change of the maximum temperatures that were predicted by the multigrid technique, but it consumes 27 times more than the multigrid technique. When comparing with experimental results, it is clear that the four grids lead to a thermal field that is rather near the experimental one. The major difference occurs near the first two ICs, where the numerical results are a bit lower than the experimental ones (about 10 °C).

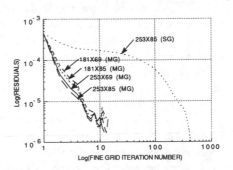

Fig.3 - Evolution of residual obtained by single grid and multigrid techniques

The velocity profile across the section of the channel exit is illustrated in Fig.5. The profiles obtained by the four types of grid are a little different from the profiles obtained by the finest grid. Moreover, the buoyancy effect induces a higher velocity near the two heated plates as compared to that of centre part of the channel.

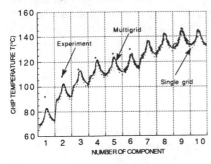

Fig.4 - Temperature distribution along the PCB, at the chip level

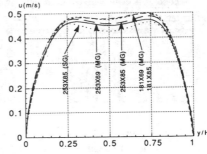

Fig.5 - Velocity distribution in the exit section

III - 2 - *Application to the discs and doughnuts roughened heat exchanger (forced convection)*

The physical system to be modelled was a concentric-pipe heat exchanger with discs and doughnuts in the inner pipe as shown in Fig.2. Calculations for this geometry are made for four Reynolds numbers ($Re_i = \rho u_0 D/\mu$ in the inner pipe, and $Re_o = \rho u_0 (D_o - D_i)/\mu$ in the annular pipe) of 500, 1000, 1500 and 2000. The inlet temperature of the two fluids is imposed to 20°C for the cold fluid and 300°C for the hot fluid, and all the outermost surfaces of heat exchanger are treated here as insulated. A grid optimization test was carried out, and it was found that the grid size of 117X145, and the one of 153 X 167 give a maximum local difference less than 2% between the two solutions. The gain when using the multigrid was found of the order of 15.

A flow visualization study (Fig.6) was performed to provide qualitative information on the effects of the Reynolds number in the discs and doughnuts roughened heat exchanger. For clarity, only a few velocity vectors for Re=2000 are plotted. We observe that the direction of the flow vectors varies strongly, and the flow field presents a recirculation zone behind each blockage (discs and doughnuts). Moreover, the open cavity flow in the inlet zone is dominated by a weakly driven eddy. There is a small mass transport between core and cavity as evidenced by relatively small velocity in the recirculating region.

Figure 7 shows the Nusselt number distribution along the inner wall of the annular pipe as a fonction of the Reynolds number in order to provide some insight into the heat exchange behaviour. It is clear that the increase of the Reynolds number efficiently enhances the heat transfer as evidenced by a higher value of the Nusselt number. However, in all cases, the Nusselt number distributions have similar shapes. The Nusselt number decreases from a high value to reach a relatively smaller value near the recirculating regions. Away from these regions, the Nusselt number varies alternately due to the presence of the discs and dough nuts: in general, the local flow accelerations in the disc zones and the presence of the fin result in a high local value, and the recirculating zones lead to a low value.

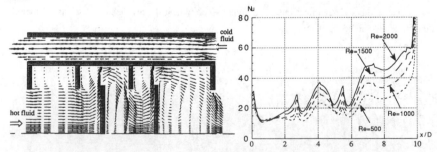

Fig.6 - Flow visualization
for $Re_i = Re_o = 2000$

Fig.7 - Evaluation of the Nusselt number
along the inner wall of the annular pipe
($Re_i = Re_o = Re$)

IV - CONCLUSION

We have tried to demonstrate the performance of the multigrid technique in two examples of practical importance, such as the cooling of electronic equipment and the concentric-pipe heat exchanger. We observed that the multigrid procedure converges rapidly, with a gain of computing time in relation to single grid method, between 15 and 30, depending on the configuration.

Besides these exercices have brought a rather good insight to complex conjugate heat transfers, for which both the role of the conduction in the solid parts and the flow structures in the fluid are rather difficult to imagine or to check. Furthermore, we plan to extend our multigrid procedure to a three-dimensional model in order to describe 3D effects, which can not be predicted by the current two-dimensional model, particularly for the cooling of electronic board problem.

Nomenclature

b :   height of IC
C_p :   specific heat
D :   inlet diameter of the hot fluid
D_i :   inner diameter of the annual pipe
D_o :   outer diameter of the annual pipe
D_c :   inner diameter of the pipe with discs and doughnuts
d_c :   diameter of the doughnut
d_d :   diameter of the disc
e :   wall thickness of the PCB or the pipe
H :   channel height
L :   channel or pipe length
L_L :   streamwise dimension of an IC or between the disc and doughnut
L_E :   dimension of the inlet zone
L_s :   dimension of the outlet zone
L_T :   transversal dimension of an IC
L_C :   streamwise chip length
NC :   number of component
p :   pressure
Q :   power dissipated within an IC (W)
q :   flux surface density (W/m^2)
S_L :   streamwise spacing between the ICs
S_T :   transversal spacing between the ICs
T :   temperature
T_0 :   inlet temperature of the fluid
u :   streamwise fluid velocity (m/s)
u_0 :   inlet velocity of the streamwise fluid
v :   transversal fluid velocity (m/s)
W :   channel width
x :   streamwise coordinate
y :   transversal coordinate
λ :   solid thermal conductivity (W/m.K)
ρ :   volumetric mass (kg/m^3)

REFERENCES

1/ A. ORTEGA and R. J. MOFFAT, 1986, 'Buoyancy induced convection in a non-uniformly heated array of cubical elements on a vertical channel wall', Heat Transfer In Electronic Equipment, HTD-Vol. 57, pp. 123-134 .

2/ D. TOROK and R. GRONSETH, 1988, 'Developing Flows in Narrow Channels, Containing Heated Obstacles', Int. J. for Numerical Methods in Fluids, Vol. 8, pp. 1543-1561.

3/ A. WIETRZAK and D. POULIKAKOS, 1990, 'Turbulent forced convective cooling of microelectronic devices', Int. J. Heat and Fluid Flow, Vol. 11, No.2, pp.105-113.

4/ S. MORI et al, 1974, 'Heat Transfer to Laminar Flow in a Circular Tube with Conduction in the Tube Wall', Heat Transfer: Jap. Res., Vol.3, No.2, pp. 37-46.

5/ T. M. NGUYEN et al, 1989, 'Laminar Flow and Conjugate Heat Transfer in Rib Roughened Tubes', Numerical Heat Transfer, Part A, Vol.15, pp. 165-179.

6/ A. BRANDT, 1977, 'Multi-Level Adaptive Solution to Boundary - Value Problems', Mathematics of computation, Vol. 31, Num. 138, pp. 333-390.

7/ S. V. PATANKAR, 1980, Numerical Heat Transfer and Fluid Flow, Mc Graw Hill Book Company.

8/ A. D. Gosman et al, 1985, Computer-aided engineering - heat transfer and fluid flow, pp. 27-29.

9/ W. HACKBUSCH, 1985, 'Multi-Grid Methods and Applications', Printed in Germany.

10/ H. Y. WANG, 1991, 'Simulation Numérique par Multigrilles des Transferts Conjugués dans les Cartes de Composants Electroniques', Thèse N° d'ordre : 405, Université de Poitiers, pp.188-190.

11/ S. V. PATANKAR et al, 1977, 'Full Developed Flow and Heat Transfer in Ducts Having Streamwise - Periodic Variations of Cross-Sectional Area', Trans. ASME, Ser. C,J. of Heat Transfer, Vol. 99, pp. 180-186.

TURBULENT CONVECTIVE HEAT TRANSFER PREDICTIONS WITH TWO-EQUATION SCALAR TURBULENCE MODEL

M. EL HAYEK and J. HENRIETTE
Faculté Polytechnique de Mons, B-7000 Mons, Belgium.

A high-Reynolds two-equation scalar turbulence model, based on the variance of temperature fluctuation and its rate of dissipation, is developed to predict the turbulent convective heat transfer mechanism without the help of the restrictive analogy concept. A scalar time scale is used to constitute an eddy diffusivity for heat. Thus, no assumption about the hypothetical universal turbulent Prandtl number, used in the classical two-equation turbulence models to link the eddy diffusivity to the turbulent viscosity, is required. A new set of model constants is given and an extended two-layer wall function is developed. Comparison with experimental data is performed for some applications of practical interest and shows good agreement for the considered flows.

INTRODUCTION

In engineering practice, most equipment involving heat or mass transfer associated with fluid flow work in turbulent regime. Solving related problems by use of the basic governing equations implies therefore the solving of the turbulence closure problem. This is practically done by using a turbulence model in more or less elaborate form.

Many turbulence models have been developed, ranging from the most simple - the mixing length model - to the most complete - the Reynolds stress model (RSM). Despite its shortcomings, the two-equation $k - \epsilon$ model is the most widely used for engineering calculations, owing to its simplicity together with its ability to correctly predict some common practical situations. All these models may be identified as "mechanical" models because they involve only mechanical quantities such as turbulent kinetic energy, rate of dissipation, etc...

For turbulent convective heat transfer calculation, the most current procedure, used in conjunction with the $k - \epsilon$ model, consists of invoking the analogy between momentum and heat transfer mechanisms. Accordingly, an eddy diffusivity is introduced, related to the eddy viscosity by the so-called turbulent Prandtl number. Furthermore, the latter is generally considered as a universal constant. Experiments show, however, that the analogy fails in most practical situations. An alternative is to use a separate model for heat transfer. Such an alternative is natural in the algebraic models which solve the problem by using simple algebraic relations deduced directly from experiments and extensively reviewed by Arpaci and Larsen (1). The same situation prevails for the second order models which use specific equations to determine turbulent heat flux. At the level of first order two-equation models, it is, also, possible to remove the analogy concept by using specific equations for the variance of temperature fluctuations and its rate of dissipation. These parameters are used to constitute a thermal time scale from which an expression for the eddy diffusivity is deduced by dimensional analysis.

Despite the rapid expansion of turbulence modelling in the literature during the last 20 years, reliable scalar turbulence models based on scalar time scale have not been reported extensively and when they have been , it was, at best, for very simple flows (homogeneous turbulence, mixing layers, etc...) and at second order modelling level. The main motivation was the prediction of buoyancy affected flows where the variance of temperature fluctuations is of prime importance : Newman *et al* (2), Elghobashi and Launder (3), Chung and Sung (4), Jones and Musonge (5) and

Malin and Younis (6). To the authors knowledge, the only works about two-equation scalar turbulence model are those of Nagano and Kim (7) and Yoshizawa (8). The former proposes a low Reynolds numbers version with isotropic eddy diffusivity based on a mixed time scale and shows impressive results for the situations considered (wall affected flows). The latter develops an anisotropic model using the statistical "two-scale direct interaction approximation" theory.

The present paper describes the development of a scalar turbulence model and its numerical implementation and testing against situations of engineering interest.

MATHEMATICAL FORMULATION

The steady-state turbulent convective heat transfer problem can be described by the following equations governing mean-fields and expressing the mass, momentum and energy conservation, respectively :

$$\frac{\partial \rho U_i}{\partial x_i} = 0 \tag{1}$$

$$\frac{\partial \rho U_j U_i}{\partial x_j} = \frac{\partial}{\partial x_j}\left[\mu\left(\frac{\partial U_i}{\partial x_j} + \frac{\partial U_j}{\partial x_i}\right) - \rho \overline{u_i u_j}\right] - \frac{\partial P}{\partial x_i} + S_{Ui} \tag{2}$$

$$\frac{\partial \rho U_j T}{\partial x_j} = \frac{\partial}{\partial x_j}\left(\alpha\frac{\partial T}{\partial x_j} - \rho\overline{u_j t}\right) + S_T \tag{3}$$

where the upper-case letters stand for mean quantities and lower-case letters for fluctuations. The Reynolds stress $\overline{u_i u_j}$ and turbulent heat flux $\overline{u_j t}$ must be modelled in order to close the set of equations. S_{Ui} and S_T are source terms used to take into account the effect of other phenomena not included in the equations (buoyancy effect, radiation effect, etc...). α is a molecular diffusivity defined as the ratio of molecular conductivity and specific heat.

Reynolds Stresses Modelling

As was noted in the introduction, the present study uses the standard high Reynolds number $k - \epsilon$ model of Launder and Spalding (9). This model uses the generalised Boussinesq eddy viscosity hypothesis to link the Reynolds stresses to mean strains :

$$-\rho\overline{u_i u_j} = \mu_t\left(\frac{\partial U_i}{\partial x_j} + \frac{\partial U_j}{\partial x_i}\right) - \frac{2}{3}\rho k \delta_{ij} \tag{4}$$

where μ_t is the eddy viscosity related, by dimensional analysis, to turbulent kinetic energy k and its rate of dissipation ϵ by :

$$\mu_t = \rho C_\mu k^2 / \epsilon \tag{5}$$

k and ϵ are governed by the following two transport equations, respectively :

$$\frac{\partial \rho U_j k}{\partial x_j} = \frac{\partial}{\partial x_j}\left[\left(\mu + \frac{\mu_t}{\sigma_k}\right)\frac{\partial k}{\partial x_j}\right] - \rho\overline{u_i u_j}\frac{\partial U_i}{\partial x_j} - \rho\epsilon \tag{6}$$

$$\frac{\partial \rho U_j \epsilon}{\partial x_j} = \frac{\partial}{\partial x_j}\left[\left(\mu + \frac{\mu_t}{\sigma_\epsilon}\right)\frac{\partial \epsilon}{\partial x_j}\right] - C_{\epsilon 1}\frac{\epsilon}{k}\rho\overline{u_i u_j}\frac{\partial U_i}{\partial x_j} - C_{\epsilon 2}\rho\frac{\epsilon^2}{k} \tag{7}$$

The five constants involved in the model are fixed at their following standard values $(C_\mu = 0.09; C_{\epsilon 1} = 1.44; C_{\epsilon 2} = 1.92; \sigma_k = 1.0; \sigma_\epsilon = 1.3)$.

Turbulent Fluxes Modelling

As for the $k - \epsilon$ model, the generalized Boussinesq hypothesis links the turbulent fluxes to mean temperature gradients via an eddy diffusivity α_t :

$$-\rho\overline{u_j t} = \alpha_t\frac{\partial T}{\partial x_j} \tag{8}$$

By invoking the analogy between momentum and heat transfer mechanisms, most modellers have used a linear relation between μ_t and α_t. The proportionality factor is the so-called turbulent Prandtl number which is considered of the order of 0.7 to 0.9 in most engineering problems. Experiments show, however, a large data scattering and a problem dependence for this parameter and none of the empirical correlations devoted to its prediction and reviewed by Reynolds (10) seem to work universally.

The alternative, is to model α_t separately with a specially designed model. As for the turbulent viscosity, the turbulent diffusivity may be expressed as the product of a velocity scale and a scalar length scale. The velocity scale can be related to turbulent kinetic energy in the usual way. But the length scale requires choosing between two main possibilities : the thermal length scale and the mixed length scale. These scales are both used in the literature. In the present work, the thermal scale is adopted and related to the variance of temperature fluctuations $g \, (= \overline{t^2}/2\,)$ and its rate of dissipation $\chi \, (= \alpha \overline{(\partial t/\partial x_k)^2}\,)$ via a thermal time scale. The turbulent diffusivity becomes then :

$$\alpha_t = \rho \, C_\alpha k g / \chi \tag{9}$$

The determination of g and χ can be conducted as for k and ϵ by using transport-like equations, obtained from energy conservation equation, which express the variation of g and χ levels resulting from the imbalance of the generation and dissipation processes and the diffusive and convective transport, Launder (11).

The modelled forms adopted for these equations can be written as :

$$\frac{\partial \rho U_j g}{\partial x_j} = \frac{\partial}{\partial x_j}\left[\left(\alpha + \frac{\alpha_t}{\sigma_g}\right)\frac{\partial g}{\partial x_j}\right] - \rho \overline{u_j t}\frac{\partial T}{\partial x_j} - \rho \chi \tag{10}$$

$$\frac{\partial \rho U_j \chi}{\partial x_j} = \frac{\partial}{\partial x_j}\left[\left(\alpha + \frac{\alpha_t}{\sigma_\chi}\right)\frac{\partial \chi}{\partial x_j}\right] - C_{\chi1}\frac{\chi}{g}\rho \overline{u_j t}\frac{\partial T}{\partial x_j} - C_{\chi2}\frac{\chi}{k}\rho \overline{u_i u_j}\frac{\partial U_i}{\partial x_j} - C_{\chi3}\rho\frac{\chi^2}{g} - C_{\chi4}\rho\frac{\epsilon\chi}{k} \tag{11}$$

The seven constants ($C_\alpha, \sigma_g, \sigma_\chi, C_{\chi1}, C_{\chi2}, C_{\chi3}$ and $C_{\chi4}$) have to be determined using experimental results for some simple flows which allow analytical solution of different turbulence equations. Comparison with experimental data for these flows yield direct information about the constants in the form of simple relations relating each constant to other or to experimentally known quantities.

Dissipation Constants : The turbulence generated by a grid decays in the downstream direction, diminishing in intensity and increasing in scale. This results from the absence of generation, the main cause to maintain the turbulence state. In such situation, the equations 6,7,10 and 11 can be simplified and solved analytically in conjunction with the hypothesis of a constant time scale ratio $R \, (= (g/\chi)(\epsilon/k))$ to obtain :

$$C_{\chi3} = R(C_{\epsilon2} - 1) \quad ; \quad C_{\chi4} = 1/R \tag{12a,b}$$

The experimental value of R is of the order of 0.8. However, it was found (2) that a value of 1.25 is more suitable for numerical predictions of homogeneous scalar turbulence. This latter, adopted here, gives 1.1 and 0.8 for $C_{\chi3}$ and $C_{\chi4}$, respectively.

Production Constants : Opposite to decaying turbulence, where there is no generation, the rapid distortion turbulence is generated by flows where the production is the main process governing the flow. Such situation occurs, for example, in very thin shear layers. In this case, also, the governing equations can be simplified and solved with the same hypothesis of constant time scale ratio to give :

$$C_{\chi1} = 1.0 \quad ; \quad C_{\chi2} = C_{\epsilon1} - 1 \tag{13a,b}$$

These values compare very well with those collected from the literature and determined in different ways. The value of $C_{\chi1}$ is identical to that given in (2) and subsequently adopted by others (3,7). The constant $C_{\chi2}$ is slightly different from its value used by Nagano and Kim (7), who numerically optimized it at 0.72. Yoshizawa (8) has found, by statistical considerations, that a value of the order of 0.5, which is very close to that given by 13b, is more suitable.

<u>Diffusivity Constant :</u> From the definitions of the time scale ratio and the turbulent Prandtl number, the following expression for C_α can be deduced :

$$C_\alpha = C_\mu / (Pr_t R) \qquad (14)$$

Near a wall, the turbulent Prandtl number is close to unity and R is of the order of 0.5. The resulting C_α value is then of the order of 0.2.

<u>Diffusion Constants :</u> By invoking the local equilibrium hypothesis, it is easy to build a relation for σ_χ, function of known parameters ($Pr_t, R, ...$). In the present study, however, those constants are fixed at unity, a value consistent, as stated by Nagano and Kim (7), with the generalized gradient diffusion model adopted by Elghobashi and Launder (3) to model the turbulent diffusion term of turbulent heat flux equations.

The final set of constants adopted in the present work, after numerical adjustments, is the following :

$$C_\alpha = 0.25; \quad C_{\chi 1} = 1.0; \quad C_{\chi 2} = 0.44; \quad C_{\chi 3} = 1.12; \quad C_{\chi 4} = 0.82; \quad \sigma_g = 1.0; \quad \sigma_\chi = 1.0$$

BOUNDARY CONDITIONS AND NUMERICAL METHOD

A special treatment of the near-wall zone is required because the high-Reynolds turbulence model developed above and applicable to bulk flow are no longer valid in the wall region. Adapting the model by means of damping coefficients is probably the best approach, but the numerical solution requires then a very refined mesh, which is hardly acceptable for engineering calculations. The alternative is to use algebraic expressions (wall functions) to describe the behaviour in this near-wall region. But the derivation of these wall functions has to be done very carefully in order to correctly capture all the features of the transfer mechanism, particularly in the case of heat transfer which is generally concentrated in the near wall zone.

It is well known that the standard wall function, with the hypothesis of local equilibrium to determine boundary conditions for mean field and turbulent quantities, fails to reproduce the correct heat transfer behaviour near walls. A more reliable technique is the two-layer wall function developed by Chieng and Launder (12). This latter uses the following modified logarithmic law for velocity in which $k^{1/2}$ is used as velocity scale instead of the friction velocity :

$$\frac{U k^{1/2}}{\tau_w / \rho} = \frac{1}{\kappa^*} \ln(E^* y^*) \qquad (15)$$

where $\kappa^* = C_\mu^{1/4} \kappa$, $\quad E^* = C_\mu^{1/4} E$, $\quad y^* = \rho k^{1/2} y / \mu$, $\quad \kappa = 0.41$ and $E = 9.8$.

For the turbulent quantities, assumptions about their variations in each layer are made (Fig.1) and used to integrate the k and ϵ equations in the wall-adjacent control volume. The details of the procedure can be found in (12). We give here the development for the g and χ equations.

The logarithmic law for mean temperature adopted in the present work is that of Kader (13), which may be written, after velocity scaling transformation, as :

$$\frac{(T_w - T) k^{1/2}}{q_w / \rho c_p} = \frac{1}{\kappa_t^*} \ln(Pr y^*) + \beta^*(Pr) \qquad (16)$$

where $\kappa_t^* = C_\mu^{1/4} \kappa_t = C_\mu^{1/4} \kappa / Pr_t$ and $\beta^* = C_\mu^{-1/4}(3.85 Pr^{1/3} - 1.3)^2$.

The following hypotheses concerning the variation of g, χ and the turbulent heat flux are considered (Fig.1) :

$$y < y_v \Rightarrow \left\{ \begin{array}{l} g = g_v \left(\dfrac{y}{y_v}\right)^2 \\[2mm] \chi = 2\alpha \left(\dfrac{\partial g^{1/2}}{\partial y}\right)^2 \\[2mm] q = 0 \end{array} \right\} \quad y \geq y_v \Rightarrow \left\{ \begin{array}{l} g = \dfrac{g_n - g_v}{y_n - y_v} y + \left(g_P - \dfrac{g_P - g_N}{y_P - y_N}\right) = b_g y + a_g \\[2mm] \chi = \dfrac{1}{C_t} \dfrac{k^{1/2}}{y} g \\[2mm] q = q_w + (q_n - q_w) y / y_n \end{array} \right\} \quad (17)$$

where C_t is a constant (~ 1.0) and y_v is the nondimensionalized viscous sublayer thickness defined by a constant turbulent Reynolds number :

$$Re_v = \frac{\rho k_v^{1/2} y_v}{\mu} = 20 \tag{18}$$

The integration of the production and dissipation terms in the g-equation over the wall control volume gives :

$$\overline{P_g} = \frac{q_w}{c_p} \frac{T_n - T_v}{y_n} + \frac{q_w}{\rho c_p} \left[\frac{q_n - q_w}{c_p \kappa^* k_v^{1/2} y_n} \right] \left(1 - \frac{y_v}{y_n} \right) \tag{19}$$

$$\overline{\chi} = \frac{2 a g_v}{y_n y_v} + \frac{1}{C_t y_n} \left[\frac{2 b_g}{3 b} (k_n^{3/2} - k_v^{3/2}) + 2 a_g (k_n^{1/2} - k_v^{1/2}) + a a_g \lambda_v^n \right] \tag{20}$$

a and b are the counterparts, used in the integration of k-equation, of a_g and b_g and λ_v^n a function of k in the near wall control volume :

$$\lambda_v^n = \begin{cases} \dfrac{2}{\sqrt{-a}} \left(tg^{-1} \left(\dfrac{k_n}{-a} \right)^{1/2} - tg^{-1} \left(\dfrac{k_v}{-a} \right)^{1/2} \right) & \text{if } a < 0 \\[4mm] \dfrac{1}{\sqrt{a}} \ln \left(\dfrac{k_n^{1/2} - a^{1/2}}{k_n^{1/2} + a^{1/2}} \dfrac{k_v^{1/2} + a^{1/2}}{k_v^{1/2} - a^{1/2}} \right) & \text{if } a > 0 \end{cases} \tag{21}$$

The χ-equation is replaced by the following algebraic relation obtained from local equilibrium hypothesis :

$$\chi = \frac{1}{C_t} \frac{k_P^{1/2} g_P}{y_P} \tag{22}$$

The integration of the set of equations developed here above is performed by using the finite volume method, with staggered nonuniform grid arrangement. The power law scheme is used to discretize these equations. The discretized equations are solved by the classical Thomas algorithm in conjunction with a line-by-line method and a block correction procedure. The coupling of different variables is managed by the SIMPLER algorithm of Patankar (14).

APPLICATIONS

To assess the capabilities of the present model to reproduce experimental facts, the thermal boundary layer development on a heated flat plate is studied.

The first case is concerned with the simultaneous development of the hydrodynamical and thermal boundary layers on a flat plate at fixed temperature. The air, flowing at a Reynolds number per unit length of 1.41×10^6, is used as fluid medium ($Pr = 0.71$). Computation is performed on a grid consisting of 40x27 control volumes in the downstream and transversal directions, respectively. The experimental data are those given in (7).

Comparison of the predicted skin friction coefficient ($\tau_w / (\rho U_0^2 / 2)$) and the Stanton number ($q_w / (\rho c_p U_0 (T_w - T_0))$) are shown in Fig.2. As we see, the predictions are in good agreement with experiments over the entire domain of interest. Fig.3 compares the predicted time scale ratio R with the experimental results and , here also, a good agreement is achieved.

In the second case, a thermal boundary layer, which develops differently from its hydrodynamical counterpart on a flat plate with step change in wall heat flux, is calculated. The same working fluid and grid are used and the Reynolds number per unit length is of the order of 6.03×10^5. The experimental data are those of Antonia et al (15) who provide extensive measurements about the thermal fields.

The predictions for the mean and fluctuating temperatures are shown in Figs.4 and 5, respectively. For this case, also, the agreement between predictions and experiments is acceptable. The same discrepancy for the mean temperature profile in the wall region was obtained with a second order model by Samaraweera (16).

CONCLUSIONS

The examples considered in the preceding section show the capabilities of the present four-equation turbulence model to predict real flow situations. Further improvement and extension are necessary to take into account specific characters of turbulent convective heat transfer phenomena such as strong anisotropy for instance. The model can still greatly benefit from further validation to achieve full confidence. Unfortunately, this point is still waiting for detailed experiments in practical geometries (recirculating flows, ...) becoming available.

REFERENCES

1. Arpaci V.S. and Larsen P.S., 1984 *Convection Heat Transfer*, Prentice-Hall, New-Jersey.
2. Newman G.R., Launder B.E. And Lumley J.L., 1981 J. Fluid Mech. **111** pp.217-232.
3. Elghobashi S.E. and Launder B.E., 1983 Phys. Fluids **26** pp.2415-2419.
4. Chung M.K. and Sung H.J., 1984 Int. J. Heat Mass Transfer **27** pp.2387-2395.
5. Jones W.P. and Musonge P., 1988 Phys. Fluids **31** pp.3289-3604.
6. Malin M.R. and Younis B.A., 1990 Int. J. Heat Mass Transfer **33** pp.2247-2264.
7. Nagano Y. and Kim C., 1988 ASME - J. Heat Transfer **110** pp.583-589.
8. Yoshizawa A., 1988 J. Fluid Mech. **195** pp.541-555.
9. Launder B.E. and Spalding D.B., 1974 Comput. Meths. Appl. Mech. Engrg. **3** pp.269-289.
10. Reynolds A.J., 1975 Int. J. Heat Mass Transfer **18** pp.1055-1068.
11. Launder B.E., 1976 Chap.6 in *Turbulence* Bradshaw P. (ed.), Springer-Verlag, Berlin, pp.231-287.
12. Chieng C.C. and Launder B.E., 1980 Numer. Heat Transfer **7** pp.59-75.
13. Kader B.A., 1981 Int. J. Heat Mass Transfer **24** pp.1541-1544.
14. Patankar S.V., 1980 *Numerical Heat Transfer and Fluid Flow*, Hemisphere & McGraw-Hill, Washington D.C.
15. Antonia R.A., Danh H.Q. and Prabhu A., 1977 J. Fluid Mech. **80** pp.153-177.
16. Samaraweera D.S.A., 1978 *Turbulent Heat Transport in Two and Three-Dimensional Temperature Fields*, Ph.D. Thesis, University of London.

NOMENCLATURE

C Model constant
c_p Specific heat [J/kg°C]
g Variance of temperature fluctuations [°C²]
k Turbulent kinetic energy [m²/s²]
P Mean pressure [Pa]
Pr Prandtl number
q Heat flux [W/m²]
R Time scale ratio
Re Reynolds number
T Mean temperature [°C]
T_q Friction temperature [°C]
t Fluctuating temperature [°C]
U Mean velocity [m/s]
u Fluctuating velocity [m/s]
x Coordinate
y Distance to wall [m]

Greek letters
α Heat diffusivity [Pa s]
δ Boundary layer thickness [m]
ϵ Rate of dissipation of k [m²/s³]
μ Viscosity [Pa s]
ρ Density [kg/m³]
τ Shear stress [Pa]
χ Rate of dissipation of g [°C²/s]

Subscripts
o   Reference
i, j Coordinate directions
n North face of control volume
t Turbulent
v Viscous sublayer edge
w Wall

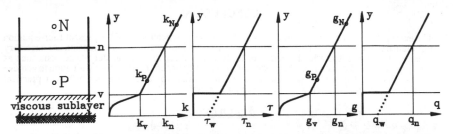

Fig.1 Two-layer model of wall-adjacent zone.

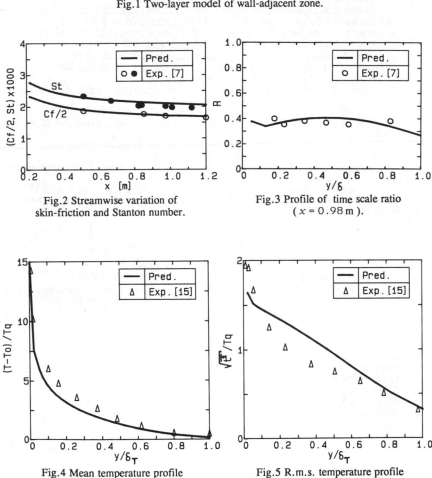

Fig.2 Streamwise variation of
skin-friction and Stanton number.

Fig.3 Profile of time scale ratio
($x = 0.98$ m).

Fig.4 Mean temperature profile
($x/\delta_0 = 42.9$).

Fig.5 R.m.s. temperature profile
($x/\delta_0 = 42.9$).

COUPLING EFFECTS IN ELECTROHYDRODYNAMIC HEAT TRANSFER IN A CHANNEL FLOW

A. Ould El Moctar+, H. Peerhossaini+, P. Le Peurian* and J.P. Bardon+

This paper describes the basic principles and physical modelling of systems based on electrohydrodynamic heat transfer, generally referred to as ohmic heating. The technique consists of heat generation and transfer in liquids by passage of mains frequency electric current through the liquid itself. Preliminary numerical results are presented and an experimental appratus designed and constructed for verification of these results is described .

1. Introduction

Over heating of poor conducting liquids adjacent to the hot surfaces in heat exchangers can not only cause burning or degrading of the liquids, but it is at the origin of the heat exchanger fouling. Furtheremore , in classical hot surface heating method, the heating capacity of the system is restricted by the available surface area of the heat exchanger.

An alternative method to obviate the basic difficulties of heating by heat conduction from a hot surface heat exchanger is the dierct resistance heating or ohmic heating. It consists of causing an electric current to pass directly in the flowable media between pairs of electrodes.This technique can be applied only to the liquids containing free ions, hence can become electric conductors. Different ions present in the liquid start to move once the liquid is under the influence of an alternative electric field. This permanent mouvement of ions is accompanied by a heat generation in the liquid bulk that can be formulated by the Joule law. In this manner electrohydrodynamic volume heating offers a major advantage in comparison to the conventional heating methods for viscous, poor thermal conducting liquids or particulates. In situ and laboratory experiments have proved the efficiency of this heating method.

+ *Laboratoire de Thermocinétique URA de CNRS No. 869, ISITEM, La Chantrerie, C.P. 3023, 44087, Nantes - France*
* *EDF/DER Département A.D.E., Centre des Renardières, B.P. No. 1 77250, Moret sur Loing - France*

From the fundamental point of view ohmic heating of liquids is a complex physaical problem in which a strong interaction of thermal, hydraulic and electric phenomena can be observed. In fact the thermphysical properties of the liquid in flow are functions of temperature, while the temperature field in its turn depends on the residence time in the system (in which the fluid is under the effect of the electrical field), i.e. in the flow field. In a channel flow configuration in which there exists a velocity difference between the fluid flowing in the core and the fluid flowing close to the wall, the resultant temperature gradient can generate a free convection flow superposed on the main stream. The effect of natural convection appears as an accelerated field adjacent to the electrodes .

2. A physical model

In the framework of this program we have developed a physical model which takes into account the above mentioned interactions. The basic equations consist of the continuity equation, the Navier-Stokes equation, the equation of conservation of energy, and the equation of electric potential applied to this specific complex problem [1 & 2]. For an incompressible fluid they can be written as follows:

$$\text{div } V = 0$$
$$\rho \frac{DV}{Dt} = -\text{grad}p - \rho(T)g + \text{div}(\mu(T) \text{ grad}V) \tag{1}$$
$$\rho(T)c_p \frac{DT}{Dt} = \text{div}(\lambda(T)\text{grad}T) + \sigma(T)(\text{grad}U)^2$$
$$\text{div}(-\sigma(T)\text{grad}U) = 0$$

Dimensional analysis of the equations (1) reveals that the system is governed by four non-dimensional numbers:

$$A = \frac{H}{l}$$ channel aspect ratio

$$Re = \frac{\rho V_m l}{\mu}$$ Reynolds number

$$Pe = \frac{V_m . l}{a}$$ Peclet number

$$Ri = \frac{Gr}{Re^2} = \frac{\beta . g . U . I}{\rho . c_p . V_m^3}$$ Richardson number

This system of coupled equations is then numerically solved. The numerical code used for this purpose is a control volume code, called Ulysse, developed in the Laboratoire National d'Hydraulique (LNH) of EDF. It is essentially designed to solve the Navier-Stokes equation with or without heat transfer. We have supplimented this code to take into account the electric potentiel equation and then we have solved the resulting simultanous system of eqautions [2&3].

For the first approach, we have considered a fully develped two dimensional channel (Poiseuille) flow preceded by a diffuser and followed by a simple nozzle.

3. Numerical results

In the above parameter space, a physical state is identified by four parameter values. By a systematic variation of the parameters one can generate a multitude of thermoelectrohydrodynamic regimes. Using such a procedure, it is possible to single out the effect of each parameter on the resulting thermal and hydrodynamic fields, and therefore finding an optimal combination of parameters for the best heat transfer results. Here however, we restrict ourselves to a single set of parameters in order to demonstrate the generic results and using them to point out the basic physical phenomena encountered in electrohydrodynamic heat transfer.

Let us consider a physical state identified by the following set of parameters:

A = 2.5
Re = 500
Pe = 4290
Ri = 1.3

The flow enters the electrohydrodynamic domain with a fully developed Poiseuille velocity profile and a uniform temperature profile. Figure 1 shows the velocity profiles at the entrance and exit sections of the domain. We notice that the parabolic velocity profile at the entrance has been considerably deformed as it has passed through the electric field. It is accelerated at the regions close to the electrodes and decelerated in the central zone of the channel. Profile deformation is a manifestation of the strong coupling between hydrodynamic, thermal and electric fields. In the absence of such coupling the velocity profile at the exit section would have been a parabolic one similar to that of the entrance. In fact low speed flow close to the electrodes (in the parabolic velocity profile) causes a longer residence time and therefore, the fluid particles undergo a longer electric conduction effect as they pass in the field. Its direct effect would be more resistance heating close to the electrodes which causes a free convection effect in this region. It is manifested by an acceleration on the velocity perofile. In the central zone of the channel the flow undergoes an opposite effect. To satisfy the continuity condition, flow acceleration close to the electrodes is compensated by flow deceleration in the middle of the channel. Such a velocity redistribution entrains a more uniform temperature distribution at the exit section of the electrohydrodynamic domain. In Figure 2 we have shown the temperature profiles at the exit section of the domain. Profile "a" corresponds to the parabolic velocity profile all through the domain, while profile "b " corresponds to the deformed velocity profile. This result constitutes one of the advantages of heaters designed based on ohmic heating.

An interesting question rises as to the variation of the temperature difference between the central and close to the electrode regions ΔT_{pc}, as a function of the distance from the entrance to the thermoelectrohydrodynamic domain. An equally interesting aspect is the spatial evolution of V_{ax}, the velocity at the channel center, as fluid particles pass through the electric field. In Figure 3 we have plotted ΔT_{pc} and V_{ax} versus Y, the coordinate along the channel axis. As it can be noticed from the Figur 3, ΔT_{pc} and V_{ax} show an ondulating behavior in the streamwise direction which gradually gets attenuated. It can also be seen that the spatial oscillation of ΔT_{pc} and V_{ax} are out of phase. To verify this behavior we have computed the velocity profiles at several positions downstream of the entrance to the electrohydrodynamic domian. They are superposed with the entrance parabolic velocity profile in Figure 4. This figure clearly shows that the phenomenon of flow acceleration close to the electrodes and the consequent flow deceleration at the channel center is repeated back and forth as fluid particles have passed through the channel. This behavior can be interpreted based on the same mechanism which is at the origin of velocity profile deformation. In this procedure it is natural to observe that spatial oscillations of ΔT_{pc} and V_{ax} be out of phase. In fact it shows the strong coupling between the hydrodynamic and heat transfer phenomena. The phase lag, as an indication of the time response of the system, is a function of fluid velocity, electric power and of course thermophysical properties of the fluid, or any nondimensional combination of them.

On the other hand attenuation of the oscillations which is an indication of temperature and velocity uniformity at the exit section of the system is a function of the dimensionless parameters discussed before. In order to quatify the temperature uniformity of the heated fluid at the system outlet, we have used standard deviation σ_{tp} of the temperature profile to define a uniformity criterion $C_{tp} = \sigma_{tp} / (T_s - T_e)$. In this definition, T_s and T_e are outlet and inlet temperatures respectively and σ_{tp} is defined as $\sigma_{tp} = \dfrac{\sum \Delta x_i . T_i^2}{\sum \Delta x_i} - T^2$. We have then computed and plotted the values of the criterion as a function of Ri, Pe, Re and A [5]. The results show that C_{tp} decreases with Ri, Re, and A and it increases with Pe. However the decrease of C_{tp} with Ri is much larger than the variation of this parameter with other dimensionless numbers. This observation implies that free convection effect is dominating in this process.

References

1) Jones,T.B. (1979) : *Electrohydrodynamically enhanced heat transfer in liquids - A Review,* Advances in heat transfer,**14**,107-148

2) Ould El Moctar, A.; Peerhossaini, H. ; Le Peurian, P. ; Bardon, J.P.(1991): *Procédé de chauffage volumique d'un liquide par conduction électrique directe; Principe et Modélisation,* Revue Générale de Thermique, **360**, 747-756

3) Ould El Moctar, A.; Peerhossaini, H. ; Le Peurian, P. ; Bardon, J.P.(1992): *Ohmic heating of complex fluids* (submitted for publication)

4) Peerhossaini, H. and Wesfreid,J.E. (1988): *On the inner structure of streamwise Görtler roles,* Int. J. Heat and Fluid Flow,9,12-18

5) Ould El Moctar, A. (1992): *Etude des phenomènes physiques couplés lors du chauffage volumique d'un liquide ionique en écoulement par conduction electrique directe,* Ph.D. Thesis, ISITEM, University of Nantes.

Nomenclature

A	: Channel aspect ratio (length to width ratio)	
a	: mean thermal diffusivity of fluid	$(m^2 s^{-1})$
c_p	: constant presure specific heat	$(J kg^{-1} K^{-1})$
D/Dt	: total derivative	
div	: divergence	
g	: gravitation acceleration	$(m s^{-2})$
Gr	: Grashof number	
grad	: gradient operator	
H	: electrode length	(m)
I	: current intensity	(A)
l	: inter-electrode distance	(m)
p	: presure	(Pa)
Pe	: Peclet number	
Pr	: Prandtl number	
Re	: Reynolds number	
Ri	: Richardson number	
t	: time	(s)
T	: temperature	(K)
u	: spanwise velocity component	$(m s^{-1})$
U	: electric potential	(V)
U_0	: potential difference	(V)
v	: streamwise velocity component	$(m s^{-1})$
V	: velocity vector	$(m s^{-1})$
V_m	: mean streamwise velocity	$(m s^{-1})$

GREEK SYMBOLES:

β	: expansion coefficient of fluid at constant presure	(K^{-1})
Δ	: Laplacian operator	
λ	: thermal conductivity	$(W m^{-1} K^{-1})$
μ	: viscosity	$(kg m^{-1} s^{-1})$
ν	: kinematic viscosity	$(m^2 s^{-1})$
ρ	: specific mass	$(kg m^{-3})$
σ	: electric conductivity	$(S m^{-1})$

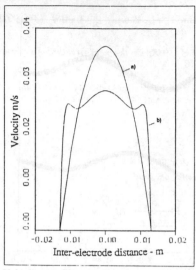

Fig. 1: Velocity profiles at the entrance (a) and exit (b) sections of the thermoelectrohydrodynamic domain

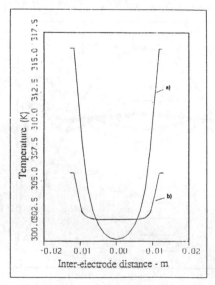

Fig. 2: Temperature profiles at the exit section of the thermoelectrohydrodynamic domain: (a) corresponds to a parabolic velocity profile all through of the domain, (b) corresponds to a deformed velocity profile through the domain

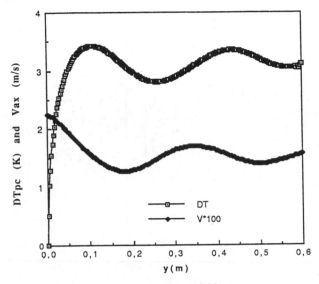

Fig. 3: Spatial evolution of ΔT_{pc} and V_{ax}

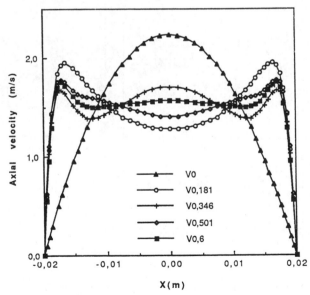

Fig. 4: Velocity profile at the entrance V0, and 4 downstream positions in the thermoelectrohydrodynamic domain

MEASUREMENT AND MODELLING OF FOULING ON FINNED TUBES

N.F. GLEN[*], S. FLYNN[*], J.M. GRILLOT[**] and P. MERCIER[**]
* NEL, East Kilbride, Glasgow, G75 0QU, UK.
** GRETh, Centre d'Etudes Nucléaires de Grenoble, 38041 Grenoble, France.

This paper describes a study of the deposition and removal of weakly dispersed particles at high mass concentrations on finned tubes. Experimental tests were performed on a heat exchanger test section consisting of a row of ten tubes, with two adjacent columns of uncooled half section tubes to simulate a complete array. Tests have been carried out for different particle sizes, flow rates and total mass of particulate injected. Thermal fouling resistances have been determined from these data. Local conditions of the flow (velocity and temperature) in the test section have been obtained from a computational fluid dynamic model and predictions of deposit formation have been made using a model which incorporates the concept of deposit retention probability.

INTRODUCTION

Extended surface exchangers offer many advantages over plain tube exchangers for gas-side applications but their use at present is usually limited to clean environments, due to uncertainties about the effects of fouling on their performance. A better understanding of fouling mechanisms would therefore allow the design and operation of more efficient heat exchangers. This paper describes recent work aimed at obtaining improved predictions for thermophoretically enhanced deposition onto extended surfaces.

THEORY

For particulate deposition, mass deposition rates can be calculated in terms of a collection efficiency, η, defined as the fraction of the free stream particle mass flow (through the projected area of the body) which actually strikes the body. For each mechanism η_i is a function of a number variables such as particle diameter and density, and gas velocity. Thus

$$\eta_i = \Sigma \, f(d_p, \rho_p, u_g, \ldots) \tag{1}$$

where the summation is over the full particle size range to be modelled, assuming all the mechanisms are independent. The total collection efficiency η_{tot} is then given by

$$\eta_{tot} = 1 - \Pi \, (1 - \eta_i) \tag{2}$$

where the product is over all the active mechanisms. The mass deposition rate can then be calculated from

$$\dot{M} = \eta_{tot} \, [\, \rho_g \, u_g \, \omega \, A_{proj}]. \tag{3}$$

A previous model by Glen and Howarth [1,2] based on the Israel and Rosner [3] correlation for inertial impaction has been extended to incorporate the Gökoglu and Rosner (4) mass transfer correlation approach for small particle diffusion / thermophoresis.

The theoretical expressions for collection efficiency for various mechanisms all assume unit sticking probability, ie. all material which reaches a surface remains there. In practice this is not the case, since particles may bounce, or deposit may be removed by incoming particles or fluid shear. At present there are no *a priori* methods for calculating instantaneous sticking probabilities and it is necessary to use a deposit retention probability, P_r. Equation 3 can be rearranged and used to calculate a collection efficiency η_x from experimental data, which is the net result of all the deposition mechanisms and incorporates the effects of deposit removal. The deposit retention probability is thus given by

$$P_r = \eta_x / \eta_{tot} .$$ (4)

Values of P_r can then be correlated with relevant system parameters such as gas and surface temperatures, gas velocity or material properties. The correlations can then be used to modify theoretical collection efficiency calculations for other conditions. The calculations described above have been encoded as a Fortran 77 program, FOULPRED.

EXPERIMENTAL

Experimental tests were performed on a heat exchanger test section installed in the GAZPAR test loop at GRETh, described by Grillot (5), and shown in figure 1. The test section is an exchanger with a row of ten tubes, with two adjacent columns of uncooled half section tubes to simulate a complete array. The test section parameters are given in table 1.

Particles are produced by an aerosol generator which uses an ultrasonic pulverisation technique, as described by Bouland and Bourbigot (6). The vibration of a piezoelectric ceramic plate submerged in a salt solution produces an ultrasonic fountain at the liquid/gas interface. The resultant uniformly sized droplet mist is dried to produce near-monodisperse particles.

Fluid temperatures are measured at the inlets and outlets on the gas and liquid sides of the test section. The gas-side pressure drop across the test section is also measured. Particle concentration and size distribution are measured at the inlet to the test section using an Andersen cascade impactor. The aerosol flow rate is also logged and at the end of each run the test section is carefully dismantled, photographed and the central row deposit mass determined.

Local conditions of the flow (velocity and temperature) in the test section have been obtained from a computational model, the TRIO code, described by Mercier and Villand (7). This package is based on 3-D statistically averaged equations for mass, momentum and energy. Turbulent fluxes are modelled with turbulent diffusivities derived from the k-ϵ model. The local conservation equations are spatially discretized according to the finite volume approach.

RESULTS AND DISCUSSION

The experimental results from the GAZPAR tests are given in tables 2 and 3. The particle size distributions, converted from aerodynamic equivalent to geometric diameters, are shown in figure 2. Although there is a small tail of larger particles in the isothermal tests, the total mass contained in particles greater than about $7\mu m$ is less than one per cent, so impactive effects will be negligible.

The TRIO runs were done with a non-regular refined mesh of $43 \times 11 \times 22$ cells. In the mid-plane, between the fins, the flow separates behind the cylinder as expected and a recirculation zone occupies a large space between the tubes. In the fin plane the vortex occupies a much shorter space between the fins and the impaction point of the flow on the second fin occurs at a lower angle and generates a 3-D behaviour of the flow. These results are in excellent agreement with the experimental results, as shown in figure 3, a sketch of a typical deposit.

The experimental data were processed with FOULPRED and the results are given in table 4. Figure 4 shows the experimental and calculated deposit masses as a function of test duration. From figure 4 it can be seen that the runs with heat transfer are tending towards an asymptotic value, probably due to shear removal. The isothermal deposits have not built up enough to reach this limit. The predicted values are much higher, indicating that the most of the material which reaches the surface does not stick.

Figure 5 shows the experimental and predicted collection efficiencies respectively. This clearly shows that the experimental collection efficiency peaks after about 55 hours. The theoretical collection efficiencies show no time dependence since the calculations do not include any removal mechanisms. However, although the absolute values of the theoretical calculations are wrong, it is worth noting that the for any given time, the ratio of the theoretical collection efficiency with heat transfer to that without is similar to the corresponding ratio of the experimental collection efficiencies. This suggests that the calculation for thermophoretic enhancement to small particle deposition is correct.

The core of the problem therefore remains the retention probability, shown in figure 6. More detailed calculations of the flow around the deposit as it builds, using the TRIO package, would allow explicit deposit removal mechanisms to be modelled. However, it is not yet possible to calculate instantaneous sticking probabilities, so it will remain necessary to determine experimental retention probabilities and correlate these with relevant system parameters to allow valid predictions to be made.

CONCLUSIONS

A series of measurements have been made on the deposition and removal of weakly dispersed micronic particles on a finned heat exchanger section under controlled conditions. The TRIO computational fluid dynamic code was used to calculate local flow conditions through the test section and the results were in excellent agreement with the observed deposit patterns. The FOULPRED code was used to determine collection efficiencies and deposit retention probabilities. More detailed calculations of the deposition process are clearly necessary to extend the range of validity of the modelling process.

Acknowledgements

This research is financed in part by the Commission of the European Communities within the frame of the Rational Use of Energy category of the JOULE R & D Programme. It is published with the approval of the Chief Executive, NEL.

References

1 Glen N.F. and Howarth J.H., 1988. Second UK National Conference on Heat Transfer, 1, I.Mech.E., p. 401-420.

2 Glen N.F. and Howarth J.H., 1989. In 'Deposition from Combustion Gases'. IOP Short Meetings Series No 23, 1989, Institute of Physics, p. 33-48.
3 Israel R. and Rosner D.E., 1983. Aerosol Sci. and Technol., 2, 45-51.
4 Gökoglu S.A. and Rosner D.E., 1984. Int. J. Heat Mass Transfer, 27(5), 639-646.
5 Grillot J.M., 1989. Thèse de docteur en génie des procédés. Institut National Polytechnique de Grenoble.
6 Bouland D. and Bourbigot Y., 1981. Rapport Interne SPT No 260, CEA/DTP/SPIN/LPA.
7 Mercier P. and Villand M., 1991. Eurotherm Seminar No 18, Hamburg.

Nomenclature

A	area	m^2
C	concentration	$kg\ m^{-3}$ at 273.15K
d	diameter	m
M	mass	kg
$\dot{M}$	mass deposition rate	$kg\ s^{-1}$
P	probability	
T	temperature	K
t	time	s
u	velocity	$m\ s^{-1}$
ρ	density	$kg\ m^{-3}$
ω	mass fraction	
η	collection efficiency	

Subscripts

g	gas	s	surface
i	i^{th} mechanism	r	retention
p	particle	tot	total theoretical
proj	projected / frontal	x	experimental

Tables

TABLE 1 - Test section characteristics

Tubes	Internal diameter	18 mm
	External diameter	20 mm
Bundle	Height	80 mm
	Width	40 mm
	Length	400 mm
	Longitudinal pitch	40 mm
	Transverse pitch	40 mm
	External area (central row)	83,441 mm^2
	Area enhancement factor	3.3
	Flow minimum cross section	1,472 mm^2
	Fins per metre	100

TABLE 2 - Experimental conditions

		With heat transfer	Isothermal
Particles	Aerodynamic mass median diameter (μm)	1.8	2.20
	Geometric standard deviation	1.5	1.68
Aerosol	Mass flow rate (kg h^{-1})	25.05	30.4
	Inlet temperature (°C)	185.0	80.0
	Velocity, based on minimum cross section		
	at inlet temperature	6.14	5.77
	at mean temperature	5.67	5.77
Water	Temperature (°C)	40.0	80.0

TABLE 3 - Experimental results

Run	t	C_p	u_g	T_g	T_s	M_x
	(s)	(10^{-3} kg m^{-3})	(ms^{-1})	(°C)	(°C)	(10^{-3} kg)
C4	24192	0.179	5.695	146.0	41.5	0.45
C2	85032	0.151	5.705	148.0	44.0	1.38
F	118440	0.188	5.767	150.0	42.5	2.50
C3	182098	0.171	5.713	148.0	40.0	4.26
G	212760	0.193	5.780	151.0	44.0	5.70
H	328320	0.183	5.785	152.0	44.0	7.36
C5	576360	0.176	5.757	151.0	42.0	9.28
IA1	37999	0.161	5.772	81.1	80.9	0.251
IA5	103189	0.142	5.771	82.4	80.7	0.925
IA3	173742	0.151	5.794	82.6	82.6	2.64
IA4	237933	0.151	5.773	81.5	80.0	3.42

TABLE 4 - Results from FOULPRED

Run	t	M_x	M_{tot}	η_x	η_{tot}	P_r
	(s)	(10^{-3} kg)	(10^{-3} kg)			
C4	24192	0.45	9.4	0.00350	0.0728	0.048
C2	85032	1.38	7.5	0.00363	0.0723	0.050
F	118440	2.50	9.2	0.00377	0.0742	0.051
C3	182098	4.26	9.1	0.00462	0.0749	0.062
G	212760	5.70	0.2	0.00466	0.0737	0.063
H	328320	7.36	132.7	0.00412	0.0743	0.055
C5	576360	9.28	225.9	0.00308	0.0751	0.041
IA1	37999	0.251	6.0	0.00115	0.0274	0.042
IA5	103189	0.925	4.5	0.00178	0.0279	0.064
IA3	173742	2.64	5.5	0.00283	0.0273	0.103
IA4	237933	3.42	5.5	0.00268	0.0278	0.096

Figures

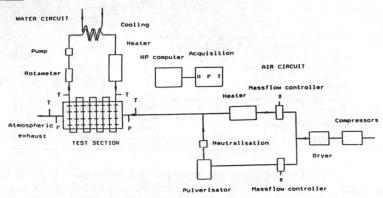

Fig. 1 GAZPAR test loop

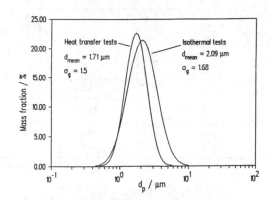

Fig. 2 Experimental particle size distributions

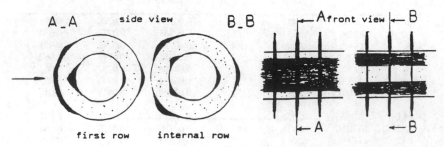

Fig. 3 Sketch of typical deposit pattern

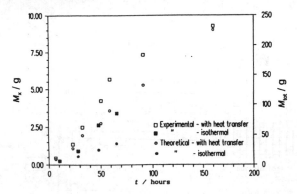

Fig. 4 Experimental and calculated deposit mass against time

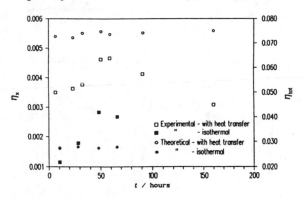

Fig. 5 Experimental and theoretical collection efficiency against time

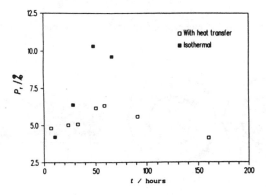

Fig. 6 Deposit retention probability against time

PLANT DEMONSTRATION OF A TECHNIQUE TO MEASURE LOCAL FOULING RESISTANCE

D C THOMPSON[1] AND J BRIDGWATER [2]*

Heat transfer coefficients can vary considerably with
position within an exchanger and with time.   Exchangers
may need periodic cleaning and it is very often difficult
to be sure how effective this process has been.   A device
was built to permit the local fouling resistance on the
tube side of an exchanger to be measured as a function of
position, the measurements being taken on a reboiler from a
urea plant when taken off-line for the purposes of
cleaning.   The initial fouling was shown to be
non-uniform.   The efficacy of cleaning was determined
before the exchanger was put back on line.   The cleaning
process was found to be significantly more effective in
some regions than others, being most so in regions of high
thermal resistance.

Introduction

When fouling arises, it sometimes happens that the foulant is seen clearly
and it is possible to assess how efficient a cleaning process is by visual
inspection alone.   In others the foulant layer may be hard to see and this
will be important if thin layers of high thermal resistance are formed.
With such systems it can be hard to be at all sure if any cleaning process
has been successful.

Such issues were of considerable importance in a reboiler used on a 1000
te/day plant for the manufacture of urea.   The synthesis reactions yields a
mixture of urea, water and ammonium carbamate together with excess ammonia.
The mixture was fed to a high pressure decomposer unit at 155° C to effect
separation and break down the carbamate to ammonia and carbon dioxide.
This is shown in Fig. 1.   The process stream passes to the base of a single
pass heat exchanger being fed to the tube side that is mounted vertically.
The process product mixture, being a mixture of gas and liquid, is removed
from the top.   Steam is fed to the top of the shell side and removed from
the base, the level being controlled automatically.   When the reboiler was
clean, the shell pressure was 8 bara.   In its most fouled condition, the
shell-side pressure rose to 12 bara, since the steam, supply pressure was 13
bara.   The performance of the decomposer then limited the throughput of the
plant.

(1) ICI Fertilizers, P O Box 1, Billingham, Cleveland, TS23 1LE
(2) School of Chemical Engineering, University of Birmingham,
Edgbaston, Birmingham B15 2TT

*   Address for correspondence

The reboiler on this unit was fouling significantly over a period of months, limiting overall plant potential (Fig. 2). Measurements of the overall heat transfer coefficient revealed that it lay below the design value, was decreasing with time and was only marginally improved by in-situ heat exchanger cleaning techniques. After 700 days a new reboiler was installed and it was cleaned *in situ* after a further 400 days and the overall heat transfer coefficient returned to its initial value. However it then declined rapidly such that the first reboiler, since subjected to off-line cleaning, was reinstalled. The scheme of operation was thus to hold a reboiler on line, replace it, and then carry out off-line cleaning. Thus while various techniques had been tested for the removal of the fouling layer, because of its tenacious nature no technique had been formed which could be completed during the duration of a normal plant shut-down.

The deposit formed was tough and hard to remove. The precise position in which the deposit was laid down and the efficiency of the cleaning method used both needed to be known. It would have been too easy to put back into service a reboiler that had been cleaned inadequately. The deposit contained ferric oxide as a main ingredient with minor amounts of chromium, lead, and titanium with traces of other metals. When a tube is cleaned by some off-line method, the improvement could not be observed with, for example, an optical technique because a highly polished surface is left by the cleaning, the "polished" fouling layer looking like clean metal. Here we describe a thermal method for determining the local fouling resistance in the tubes of the reboiler when it was taken out of service prior to cleaning and when it had been subject to a cleaning process.

Procedure

When a reboiler was removed from the plant for cleaning, it was laid on its side with one end being slightly higher than the other and the end plates were removed. There were 2799 tubes operating in parallel. These were of 14.4 mm internal diameter (19.1 mm outer diameter) set in a triangular pattern with centres 25.4 mm apart. The tubes were 4.9 m in length.

The principle used is to determine the local heat flux in a region of a tube and, with a knowledge of the local temperature difference applied between the shell side and the tube side, to calculate a local overall heat transfer coefficient. In order to know the temperature on the shell side, that side is connected to an L.P. steam supply and the condensate exit is vented via a steam trap to atmosphere. The steam flow is such that condensate is being removed at the steam supply pressure and the temperature on the shell side may be said with some confidence to be given by the saturation temperature at the steam pressure used and to be the same throughout the exchanger.

It is the tube side temperature distribution that needs particular effort (Fig. 3). The principle is to feed water at a known flowrate and temperature to a single tube and to measure the bulk water temperature as a function of position. Water is fed to a tube through a pipe A that ends in a flange G in which is seated a compressible ring. This is forced into the metal around the test tube, the force being obtained from expanding rubber insert D fitting into three tube holes adjacent. Water flow was measured by rotameter B (0 - 1,000 l/h) and controlled by a valve C on the water inlet to the measuring system.

At the other end of the tube under test, a similar fitting, again using three adjacent tube holes, is used to fit a pipe flush to the outlet. The warmed water flows up out of T-piece H to a drain. At the very end of tube is fitted a disc H with a gland at the centre that permits a 5 m long hollow tube K of outer diameter 6 mm containing a thermocouple L mounted at its tip. This thermocouple can be moved in and out, thereby allowing the temperature to be measured as a function of axial position. From marks on the outside of the small tube, the position of the tube tip in the tube was known. This end of the heat exchanger is a trifle higher than the other only to ensure that gas is flushed from the tube readily on starting up. The tube in which the thermocouple was mounted was hollow with the thermocouple wires passing along its axis. The thermocouple was sealed into the end using an epoxy resin that prevented ingress of water. Near the tip, vanes were fitted that served to hold the thermocouple in the centre of the tube.

The temperature is measured at a number of axial positions. From the change in bulk liquid temperature, the heat transferred can be found. From the area of the tube wall and the difference in temperatures (steam side temperature - average bulk liquid temperature), the overall heat transfer coefficient over that length of the tube is calculated. Temperature readings were taken at intervals of 500 mm. Steam side temperatures were measured by carrying out a dry run with the probe, steam side temperatures then being those of the probe in the absence of heat flux. Thus one thermocouple was used for all the measurements. The equipment could be operated with the stands of the measuring device M either at the water outlet (as shown) or on the water inlet with the T-piece H and seal J being moved to the water inlet. The whole system was first built and tested in the laboratories at the University of Birmingham and was then moved to the plant at ICI.

Results & Discussion
_____

In the work reported here, studies were carried out on each of twelve tubes of the reboiler. Experiments were on some occasions reported at four water flow rates. The reboiler was examined after it was taken out of service from the process and again after a contractor had carried out his cleaning routine on the tube bundle.

The measurements taken from the twelve tubes were averaged and the average heat transfer coefficient measured with a flow rate of 600 l/h are reported as a function of position in Fig. 4. Heat transfer occurs most readily at the process stream inlet, but the transfer coefficient decreases to a third in the lower part of the exchanger, that corresponding approximately to the existence of liquid condensate on the shell side during normal operation. Further up the overall heat transfer coefficient rises significantly, though not reaching the value at the process side and finally diminishes sharply near the top.

Figure 5 shows the overall heat transfer coefficient as a function of position, measured at four different water flow rates for a cleaned tube. These are obtained on just one tube; the information gained using 300 l/h cannot be compared easily with the rest as the water boiled and a two phase steam-water mixture spilled out of the drainpipe. Part of the fluctuation in the relationship between overall heat transfer coefficient and position arises from the need to take temperature at two axial distances and to use the difference to calculate the increase in axial enthalpy flux which were typically 2°C. Short scale perturbations are thus not necessarily a process effect. Averaging pairs of adjacent readings eliminates this effect to a fair extent (Fig. 6).

When the average overall heat transfer coefficient before cleaning is calculated, we find

Flow (*l*/h)	Average overall heat transfer coefficient (W/m^2K)
300	2150
400	1860
500	1540
600	1980

We do not find the expected relationship of heat transfer coefficient increasing with flow rate. As already remarked, boiling recurred at 300 *l*/h and it is likely that vapour bubble development was present also at 400 *l*/h.

When the data from the rig were compared for a particular reboiler with plant operation on line, we found the following values of average overall heat transfer coefficient (W/m^2K)

	On line	Test rig
Fouled reboiler	1020	980
Cleaned reboiler	1850[*]	2060

The data denoted by [*] is an underestimate since an assessment was carried out after 20 days of plant operation. The indications are that the rig gives reliable guidance on the process performance to be anticipated.

Let us now compare the data from Figs. 4 and 6 and recall the existence of a steam condensate layer on the shell side in normal operation. High transfer coefficients occur both at the process liquid inlet or when the tubes are contact with steam, as in the top two-thirds of the reboiler. The overall local heat transfer coefficient U measured by the present technique is a minimum at about 1½ - 2 m from the base, the approximate position of the condensate-steam interface in normal operation. This seems to promote the deposition of foulant but no clear chemical or physical explanation could be adduced.

The cleaning process, one of blasting with a fine grit or nut shell from the top, caused an improvement in coefficient along the exchanger but was particularly effective in the region most prone to fouling.

Conclusion

The method used here can be applied to equipment that is off-line and can be a useful way of determining the manner in which fouling resistance is distributed throughout a heat exchanger.   It can also be used to understand cleaning procedures and assessing the efficacy thereof.   It obtains local values of an overall heat transfer coefficient and inferences from deposit thickness using thermal conductivities are not necessary.   It would be possible also to develop a robot to carry out measurements in an operating exchanger.   Indeed perhaps one can argue with some force that in view of the critical nature of fouling in determining the performance of heat exchange equipment such a development is now overdue.

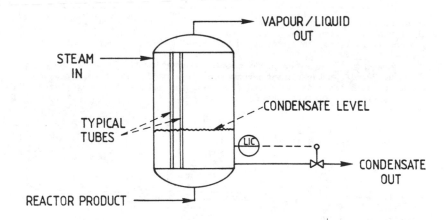

Figure 1   Reboiler used as a high pressure decomposer

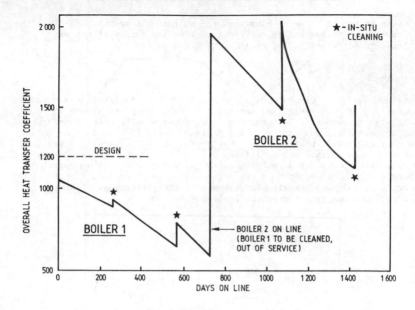

Figure 2   Reboiler performance;   average overall heat transfer
            coefficient vs time

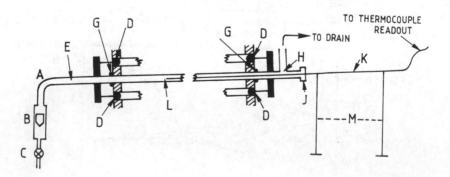

Figure 3   Equipment for measuring local overall heat transfer coefficient
            Note that the water exit is slightly higher than the water
            inlet

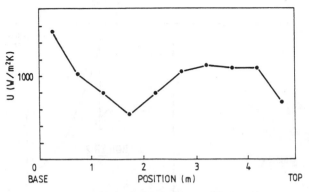

Figure 4   Local overall heat transfer coefficient U as a function of
position prior to cleaning

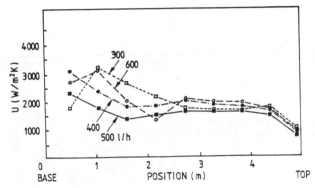

Figure 5   Local overall heat transfer coefficient U as a function of
position after cleaning

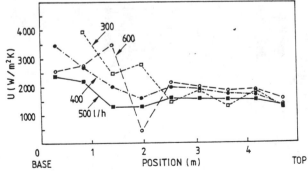

Figure 6   Local overall heat transfer coefficient U as a function of
position after cleaning.   Smoothed data

MECHANISM AND SOLVENT EFFECTS IN CHEMICAL REACTION FOULING

D.I. Wilson & A.P. Watkinson
Department of Chemical Engineering, University of British Columbia, 2216 Main Mall,
Vancouver , B.C., Canada V6T 1Z4

Chemical reaction fouling involving the autoxidation of alkenes has been modelled using indene in solution with a mineral oil, kerosene, tetralin and trichlorobenzene. The fouling mechanism is intricately connected to the formation of foulant precursors, which occurs via a free radical mechanism based on peroxy radicals. Solvent effects are important in the chemical reaction rate, the solubility relationships of reaction products and the actual reaction pathway. Evidence is presented to show that the deposit undergoes aging on the heat transfer surface.

INTRODUCTION

Chemical reaction fouling was classified by Epstein (1) as the formation of deposits on heat transfer surfaces where the surface itself is not a reactant. The fouling precursors can be generated by polymerisation, pyrolysis or oxidative reactions. The nature of the chemical reaction influences the conditions under which fouling occurs and the form of the fouling resistance - time behaviour. This paper describes part of an ongoing study of fouling in which organic liquids are heated under conditions such that autocatalytic oxidation (autoxidation) reactions occur.

Deposit formation due to hydrocarbon oxidation was studied by Taylor and colleagues in relation to the formation of gums in jet fuels undergoing evaporation. The tendency to form deposits was related to hydrocarbon chain length, branching and composition, and the presence of olefins, heteroatomic species and metallic ion catalysts. Taylor (2) studied the deposit formation tendency of selected alkenes in a relatively inert solvent, dodecane. Mayo and Lan (3) studied the dissolved gum and deposit formation in model solutions of specific alkenes in solvents which were inert to autoxidation, and found that the tendency to form deposits was strongly related to the structure and reactivity of the alkene.

The autoxidation of alkenes has been extensively studied in the chemical literature and Mayo and Lan have recently applied this knowledge to explain their results. Autoxidation involves a free radical mechanism where the hydroperoxy radical, ROO; is the dominant species. The simplified kinetic scheme appears as;

Initiation	$\rightarrow$ ROO$\cdot$			[1]
Propagation	ROO$\cdot$ + RH	$\xrightarrow{kh}$ ROOH + R$\cdot$	(hydrogen abstraction)	[2]
	ROO$\cdot$ + RH	$\xrightarrow{ka}$ ROOR$\cdot$ [$\equiv$ R$\cdot$]	(addition)	[3]
	R$\cdot$ + O_2	$\rightarrow$ ROO$\cdot$		[4]
Termination	2 ROO$\cdot$	$\rightarrow$ non radical products		[5]

It can be seen that addition leads to the formation of dimers, trimers and polyperoxide species.

Compounds which favour addition include alkenes and particularly those with conjugated or cyclic structures which reduce the activation energy of the addition step. Long chain terminal alkenes also show increased tendency to undergo addition reactions with oxygen. Styrene, methylstyrene, cyclopentene, dicyclopentadiene, indene, and hexadec-1-ene are all known to form polyperoxides under such conditions (4). Mayo and Lan concluded that gum and deposit formation are related to the tendency to form such polymeric oxidated species.

Asomaning and Watkinson (5) investigated the fouling of heat transfer surfaces by model solutions of alkenes in kerosene under oxygenated conditions at high heat fluxes. Compounds such as oct-1-ene and dec-1-ene which were known to react mainly via hydrogen abstraction (and thus produce hydroperoxides, epoxides and carbonyl cleavage products) were found to produce little fouling, whereas indene, hexadec-1-ene and dicyclopentadiene generated significant amounts of deposit. Identical experiments run without oxygen produced much less deposit, confirming the role of oxygen in the fouling mechanism. Autoxidation of the olefin was thought to produce insoluble gum precursors which would then be deposited. Little data was available, however, to relate the observations to the known chemistry and particularly the role of the solvent in this solubility-linked process.

This study focuses on the fouling of solutions of indene (C_9H_8) in four solvents. Indene was chosen because of its tendency to form polyperoxides (Russell (6)) with relatively little epoxide or formaldehyde. The latter is known to interfere with chemical analysis and also change the autoxidation mechanism by increasing the rate of hydroperoxide decomposition. The solvents were selected to provide sensible heat transfer at the heat fluxes and surface temperatures involved (200 kW/m²; 180°C) whilst reflecting different solvent properties:
(1)   tetralin [$C_{12}H_{12}$] an aromatic cyclic alkane known to undergo autoxidation;
(2)   trichlorobenzene [$C_6H_3Cl_3$] an aromatic, polar solvent;
(3)   Paraflex mineral oil - a medium molecular mass paraffinic oil with little aromatic content;
(4)   kerosene - a commercial product which is mostly aliphatic

EXPERIMENTAL

Figure 1 shows the experimental fouling loop, which recirculates liquid from the holding tank through flow measuring devices (orifice plates and rotameters) and the heat transfer probe. The apparatus has been described in detail previously (7). This study used an annular probe manufactured by Heat Transfer Research Inc. A 10 cm. long electrically heated section is located on the annulus core downstream of flow entry effects and heat transfer is measured using thermocouples in the liquid and buried in the probe surface. Constant heat flux and turbulent flow conditions were maintained throughout each experiment so that the fouling resistance, R_f, could be calculated from the expression;

$$R_f = \frac{1}{U} - \frac{1}{U_o} = \frac{(T_s - T_b)}{\dot{q}} - \frac{(T_{s,o} - T_{b,o})}{\dot{q}_o} \quad \ldots [6]$$

where surface temperatures were corrected for heat flux effects. Bulk temperatures in the tank were maintained at 80°C using water cooling coils and the tank pressure was maintained at 37 kPa using system air.

All initial fouling runs were performed with bulk temperatures of 80°C and heat fluxes adjusted to give an initial surface temperature of 180°C. Table I shows how the Reynolds and Prandtl numbers varied with the choice of solvent. The tank was charged with solvent and was presaturated with air before an experiment and brought up to 80°C using the tank heating coils. Indene (Aldrich Chem. Co., 90 wt% +) was added, overpressure restored, power applied to the probe and the system run until deposition produced surface temperatures approaching the probe's safety limit (300-320°C). Heating was then stopped and the probe removed, inspected and any deposit removed for further analysis. Liquid samples were taken regularly and analysed for chemical activity.

Cleaning between experiments was performed using successive rinses of clean solvent and solvent/acetone mixtures and gave sufficiently good experimental reproducability.

The solid deposit was sent for elemental analysis and its chemical activity studied using a Fourier Transform Infra Red (FTIR) spectrophotometer. Analysis of autoxidation solutions is quite difficult due to the interaction of the chemical species present and the lability of the polyperoxides formed. Liquid samples were titrated immediately to give hydroperoxide concentrations using ASTM method E298, in which iodine liberated from sodium iodide at 60°C by the hydroperoxide is titrated against 0.1 mol/L sodium thiosulphate. Peroxide linkages present in polyperoxides cannot be determined by simple procedures. Some liquid samples were also analysed by FTIR. A simple gravimetric procedure was used to estimate the proportion of high molecular weight material in solution; this material was precipitated using hexane and the precipitate filtered out, dried and weighed. Gas chromatography (GC) was used to monitor indene concentration in tetralin, trichlorobenzene and the Paraflex oil. This analysis also showed that the impurities present in the indene were indene derivatives remaining from its manufacture.

RESULTS AND DISCUSSION

The pure solvents were run for two days at the maximum flow rate and surface temperatures of 180°C to check that fouling by the solvent was negligible. The kerosene and Paraflex oil gave no change in heat transfer coefficient and a small increase in peroxide number. By contrast, the peroxide analysis for tetralin shown in Figure 2 and the tetralin concentrations obtained from GC analysis showed that tetralin had undergone significant autoxidation to its hydroperoxide and the secondary product, tetralone, as described in the literature. No fouling was observed, however. Trichlorobenzene did not prove to be inert, producing an unusual black solid containing chlorine. Trichlorobenzene was thus abandoned as a solvent.

A solution of indene [0.74 mol/L] in tetralin did not produce any deposit even though GC analysis showed that both indene and tetralin were being consumed throughout the run. No fouling was observed in all the experiments involving tetralin. The peroxide analysis results are shown in Figure 2. This figure includes results from experiments using hexadec-1-ene, which also yields polyperoxides in autoxidation reactions. It can be seen that the autoxidation of tetralin is being suppressed by the added alkene, which was observed for cumene in tetralin by Russell (8). Here the indene and tetralin are undergoing cooxidation, which breaks up the chain reaction producing indene polyperoxide and also modifies the tetralin kinetics. Thus tetralin acts as an inhibitor and prevents the formation of foulant precursor. A somewhat similar result was found by Taylor (2) who showed that the presence of aromatics reduced deposition from dodecane. These results confirm that prediction of fouling in hydrocarbon mixtures from experiments on single pure components will be very difficult since the reactions of free radicals in mixtures are so complex.

Four concentrations of indene in kerosene (0.84, 0.74, 0.4 and 0.15 mol/L) produced significant fouling deposits and showed very similar behaviour. Figures 3 and 4 show the fouling resistance and peroxide concentration variation with time for solutions of 0.4 mol/L indene in kerosene and in Paraflex. The time axes on these plots have been corrected for the induction period for initiation of autoxidation, given by the time at which the hydroperoxide levels start to increase. The induction period was shorter at higher concentrations, increasing from 20 hours for 0.84 mol/L to 95 hours for 0.15 mol/L. The fouling resistance curve follows an accelerating profile which begins after the hydroperoxide concentration reaches a maximum. The hydroperoxide concentration with time shows sequential features in common with prior autoxidation research, i.e. induction period, linear increase, acceleration, maximum concentration, decrease in concentration. However, the data does not fit profiles given in the literature, such as that predicted for the unimolecular dissociation of hydroperoxide to peroxy radicals in equation [7].

$$\frac{d[ROOH]}{dt} = k_h [indene][ROOH]^{0.5} - k_d[ROOH] \qquad \dots [7]$$

Unimolecular decomposition of hydroperoxides becomes bimolecular at higher concentrations which complicates fitting data to any single kinetic scheme. Predicted kinetic profiles feature maxima which increase with the initial concentration of indene. The similarity of maxima observed in the experiments (around 13 mmol hydroperoxide per litre) suggests that there is a solubility limit for the indene products. These peaks coincided with the start of the accelerated fouling period and were accompanied by visible traces of orange gum in the solution; more gum was seen as reaction proceeded.

Equation [7] does give a relatively simple initial rate expression and this was fitted to the data obtained. A similar relationship was also found by Russell (9). Figure 5 shows the initial rates of hydroperoxide formation found for the experiments in kerosene and Paraflex. The rate is proportional to the initial indene concentration and gives an intercept which can be attributed to the reaction in the solvent itself. Figure 5 also shows that the propagation constant, k_h, varies slightly between the solvents. Van Sickle et al. (10) found that the rate constant for cyclopentene oxidation fell as the solvent polarity decreased in the order acetontrile > benzene > cyclohexane. Since the kerosene is more aromatic than the Paraflex, a larger value of k_h is expected.

Four concentrations of indene in Paraflex were studied (0.15, 0.41, 0.68, 0.71 mol/L). Induction periods were generally shorter than in kerosene but again depended on the initial indene concentration.The fouling resistance profile again shows accelerated fouling but the peroxide profiles show a more gradual rise to a maximum, at which point fouling becomes significant. The GC analysis shows that indene consumption does not become significant until this gradual rise period, after which the indene consumption can be fitted reasonably well to a pseudo first order rate expression. This simplified approach was used by Norton and Drayer (11) in their work on hexadecene autoxidation. Table II shows the values obtained for the rate constant. The value is found to lie within the bounds of experimental error.

Indene is seen to undergo an accelerated fouling rate with time in both solvents. Since surface conditions are maintained nearly constant by the constant heat flux and flow rate conditions, the acceleration must be due to an increase in the concentration of fouling precursor. The results above suggest that this occurs once the polymeric species formed have exceeded the solubility limit of the solvent. As reaction continues, the mean mass and number of polyperoxide molecules will increase and thus increase the concentration of fouling precursor. The levels of hexane-insoluble material were indeed found to increase significantly after the peroxide values had peaked and fouling had started. The role of this insoluble material is currently under investigation.

Figure 6 shows the fouling resistance - time results for 0.41 mol/L indene in Paraflex at surface temperatures of 180°C, 210°C and 240°C. These results show a distinct transition from a linear fouling rate regime to an accelerated one, the transition occuring under conditions where the peroxide concentration had begun to flatten out and indene reaction rate increase significantly.

Infra red analysis of the solutions showed the presence of hydroperoxides and large concentrations of carbonyl compounds, particularly when fouling had started. FTIR analysis of gum isolated from the solution showed it to contain hydroperoxy, carbonyl and indene (aromatic and alkene) activity. The secondary products of the autoxidation reactions can include alcohols, aldehydes, esters, ketones and ethers; Ueno et al. (12) reported finding indanone, phthalic acid and other cleavage products of the poly- and hydro-peroxides in their autoxidation mixtures.

The solid deposit found on the probes was a chunky, brown material on the top surface and a harder, darker material next to the probe wall. Optical microscope examination of the deposit surface showed it to be composed of small, irregular particles. The thickness of deposit depended on the local surface temperature, which was not uniform along the probe. Table III lists the results of C:H:O analysis based on the indene molecule. Most deposits have reduced oxygen compared to the polyperoxide $[C_9H_8O_2]$ which can be explained by subsequent 'aging' reactions occuring on the hot probe wall. The reduced oxygen content is especially noticeable for cases where the deposits were sampled from locations at different temperatures. Deposits at the hot end of the probe showed less oxygen than those at the cooler end. Deposits from near the hot probe/deposit interface showed less oxygen than those near the cooler deposit/fluid interface. This aging hypothesis is also supported by the FTIR spectra for the deposits which show a complex picture of aromatic and carbonyl peaks, derived from rearrangements of the peroxide linkages. Further work is needed to quantify the aging process.

CONCLUSIONS

Fouling experiments have been carried out using indene in four air-saturated solvents chosen to reflect a range of polarity and aromaticity. Under oxygenated conditions, trichlorobenzene solvent was unstable and led to significant fouling in the absence of indene. Tetralin, which undergoes autoxidation itself, acted as an inhibitor to indene polyperoxide formation, and no thermal fouling was observed. With both kerosene and Paraflex solvents, heavy fouling occurred once the hydroperoxide concentration in the mixture reached a critical level. This threshold condition for fouling seems to be related to the solubility limit of the polymeric fouling precursors. The fouling resistance exhibited an accelerating rate behaviour with time. The effects of indene concentration and surface temperature were established. Deposit analysis showed that the polyperoxides undergo aging reactions which reduce the oxygen content of the deposit as the surface temperature is increased.

ACKNOWLEDGEMENTS

The ongoing support of the Natural Sciences and Engineering Research Council of Canada to A.P.W. and the continuing support of D.I.W. by the University of British Columbia are gratefully acknowledged. The assistance of Dr. G. Zhang and the support of the Argonne National Laboratories is acknowledged.

REFERENCES

1.   Epstein,N, (1983) Heat Transfer Engineering, 4(1), 43-56
2.   Taylor WF, (1969) Ind. Eng. Chem. Prod. Res. Dev., 8(4), 375-380
3.   Mayo FR and Lan BL, (1986) Ind. Eng. Chem. Prod. Res. Dev., 25, 333-348
4.   Scott G, (1965) Atmospheric Oxidation and Antioxidants Ch.2, Elsevier, Amsterdam
5.   Asomaning S and Watkinson AP, (1990) AIChemE. Conf. Proceedings, San Diego'90
6.   Russell GA, (1956) J.Am Chem.Soc., 78, 1035-1040
7.   Fetissoff PE, Watkinson AP and Epstein N, (1982) Proc. 7th Intl. Heat Transfer Conf.,
       Vol .6, 391-396
8.   Russell GA, (1955) J.Am.Chem.Soc., 77, 4583-4590
9.   Russell GA, (1956) J.Am.Chem.Soc., 78, 1041-1044
10. Van Sickle DE, Mayo FR and Arluck RM, (1965) J.Am.Chem.Soc., 87(21), 4832-87
11. Norton CJ and Drayer DE (1968) in ACS 'Advances in Chemistry' Ser. 75, 78-92
12. Ueno Y, Kotetsu N and Yoshida T, (1974) Aromatikkusu, Japan, 26(3), 117-20

Nomenclature			Subscripts	
k	rate constant	$(L/mol)^{0.5}/s$ or $1/s$	a	addition
q	heat flux	W/m^2	b	bulk
[ROOH]	hydroperoxide concn.	mol/L	d	decomposition
R_f	fouling resistance	m^2K/W	h	hydrogen abstraction
t	time	s	o	start of experiment
T	temperature	K	s	surface
U	overall h.t.c.	W/m^2K		

TABLE I - Flow conditions and physical properties of solvents used in fouling runs

Solvent	Kinematic viscosity 10^{-6} m^2/s	Re	Pr	Solvent Nature	Source
tetralin	0.931	21 500	12.5	aromatic, non polar	Aldrich Chem. Co.
trichlorobenzene	0.803	24 000	12.1	aromatic, polar	Aldrich Chem.Co.
kerosene	0.898	14 000	14.1	mixture, non polar	Imperial Oil Co.
Paraflex HT10	3.630	2 900	47.7	aliphatic, non polar	Petro-Canada

Properties calculated at 80°C

TABLE II - Pseudo-first order rate constants for indene reaction in Paraflex oil

Indene Concentration mol/L	Pseudo-first order Rate Constant hr^{-1}	Regression Fit R^2
0.680	0.0346	0.957
0.402	0.0337	0.937
0.147	0.0405	0.986
0.738	0.0387	0.953
0.407	0.0354	0.981
0.407	0.0375	0.900

TABLE III - Elemental Analysis of Indene Derived Materials

Indene Concentration mol /L	Solvent	Surface Temperature °C	Elemental Analysis	Origin of Material
0.74	kerosene	180	$C_9 H_{7.55} O_{1.65}$	Top of deposit
0.40	kerosene	180	$C_9 H_{7.27} O_{1.30}$	Top of deposit
0.15	Paraflex	180	$C_9 H_{12} \ O_{1.23}$	Top of deposit
0.68	Paraflex	180	$C_9 H_{9.6} O_{0.56}$	Top of deposit
0.71	Paraflex	180	$C_9 H_{8.4} O_{1.47}$	Top of deposit
0.41	Paraflex	180	$C_9 H_{9.67} O_{0.97}$	Probe wall - cool
		180	$C_9 H_{10.1} O_{1.16}$	Top of deposit
		180	$C_9 H_{13.3} O_{0.56}$	Probe wall - hot
0.41	Paraflex	210	$C_9 H_{10.9} O_{0.8}$	Top of deposit
0.41	Paraflex	240	$C_9 H_{9.8} O_{1.77}$	Soln. precipitate
		240	$C_9 H_{8.98} O_{0.87}$	Top of deposit
		240	$C_9 H_{8.97} O_{0.75}$	Probe wall - hot

Indene Polyperoxide $(C_9 H_8 O_2)_n$

Figure 1   **Apparatus used in fouling experiments**

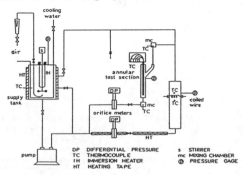

Figure 2 Hydroperoxide Concentrations in tetralin

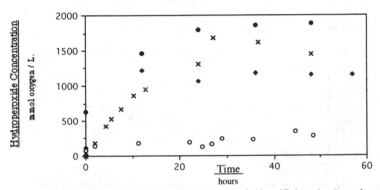

• = tetralin; o = 0.5 mol/L indene in tetralin; x = 0.46 mol/L hexadec-1-ene in tetralin
♦ = 0.26 mol/L hexadec-1-ene in tetralin

Figure 3 Fouling Resistance vs Time   [indene] = 0.4 mol/L

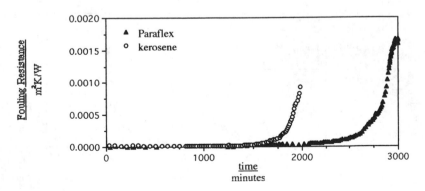

Figure 4 Hydroperoxide Concentration vs. Time   [indene] = 0.4 mol/L

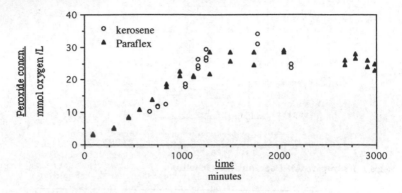

Figure 5 Initial Rate of Hydroperoxide Formation vs. [indene]

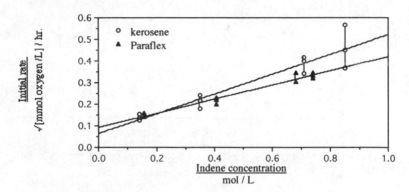

Figure 6 Effect of Surface Temperature on Fouling Resistance.

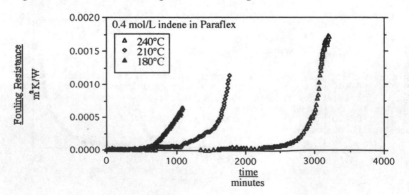

THE EFFECT OF SYSTEM PARAMETERS ON THE FOULING PERFORMANCE OF HEAT EXCHANGERS

J E Hesselgreaves
Energy and Environment Centre, NEL, East Kilbride, Glasgow G75 0QU

In a heat exchanger system in a fouling environment, it is known that the thermal performance is affected by system parameters, and in particular the head/flow characteristic of the circulating pump or fan.

This paper describes a simple analysis which examines the relationship between the pump or fan characteristic and the performance of basic heat exchanger surfaces with a given uniform layer of fouling material or frost. The surface velocity is shown to be strongly dependent on a dimensionless parameter dependent on the gradient of the pump or fan characteristic and the system resistance curve.

Some consequences for design to minimise the effects of fouling or frosting are given.

INTRODUCTION

In a heat exchanger system in a fouling environment, it is known that the thermal performance is affected by such system parameters as the gradient of the circulating pump or fan characteristic.

The nature of this relationship is examined in this paper. For simplification, the layer of fouling material or frost is assumed to be uniform on the surface, and the mechanism of its deposition is not considered.

The two basic heat exchanger configurations treated in this paper are those of the parallel circular tube and the parallel flat plate. In the former case typical practical examples are in shell-and-tube heat exchangers, and in water tube and fire tube boilers. Fully developed laminar flows and fully developed (rough) turbulent flows are treated for this case. Parallel flat plates are found in tube-in-plate type air conditioning and refrigeration heat exchangers. The low operating Reynolds number common in such applications means that the flow is frequently laminar, and only this flow mechanism is treated in this case.

BASIC ANALYTICAL MODEL

The object of this analysis is to predict the changes in flow and in mean flow velocity through a heat exchanger as a result of a given build-up of fouling layer. This layer, by reducing the available cross-sectional area for flow, increases the resistance and causes a reduction in flow which is dependent on the gradient of the circulating fan or pump characteristic, as shown schematically in Fig. 1.

The simplifying assumptions made in the analysis are as follows:

a   the flows are idealised to fully developed turbulent (rough) or laminar, b   the fouling or frost layer is uniformly distributed on the surface, and has a surface roughness the same as that of the clean surface,

c   the system consists simply of the heat exchanger surfaces and the circulating pump or fan (in one case a residual circuit resistance is allowed for), and

d   the shape of the pump or fan characteristic is idealised to linear over the range of interest.

SUMMARY OF ANALYTICAL RESULTS

The main analysis for each case is presented in Appendix I. Three controlling dimensionless parameters are defined:

i   A parameter describing the fouling thickness in terms of the characteristic surface dimension. This is defined as

d^+ = effective diameter (with fouling)/clean diameter (for tubes)

b^+ = effective plate spacing (with fouling)/clean spacing (for parallel plate surfaces).

ii   A parameter, P, formed by

$$P = P_1/P_2$$

where $P_1 = \dfrac{\text{change in pump head from clean condition}}{\text{change in flow from clean condition}} = \dfrac{\Delta H}{\Delta Q}.$

= average gradient of pump characteristic

and $P_2 = \dfrac{\text{reference (clean) pump head}}{\text{reference (clean) flow}} = \dfrac{H_R}{Q_R}.$

iii   A parameter, R_{CR}, formed by

$$R_{CR} = R_C/R_R$$

where R_C = resistance of circuit excluding fouled tubing, and

R_R = resistance of tubes in clean condition.

The volume flowrate and mean throughflow velocity in the pipe or duct are now expressed in terms of these parameters, for each of the four flow conditions.

Pipe Flow, Fully Turbulent, without Circuit Resistance

The volume flowrate relative to its reference value is given by

$$\frac{\dot{V}}{\dot{V}_R} = d^{+5} \frac{P^2 + 4P/d^{+5}) + 4/d^{+5})^{\frac{1}{2}} - P}{2} \tag{1}$$

The mean velocity in the pipe, relative to its reference value, is given by

$$\frac{u}{u_R} = d^{+3} \frac{(P^2 + 4P/d^{+5} + 4/d^{+5})^{\frac{1}{2}} - P}{2} \tag{2}$$

Pipe Flow, Fully Turbulent, with Circuit Resistance

$$\frac{\dot{V}}{\dot{V}_R} = \frac{\{P^2 + 4A(P + 1)\}^{\frac{1}{2}} - P}{2A},$$ (3)

where $\qquad A = \dfrac{1/d^{+5} + R_{CR}}{1 + R_{CR}},$

and $\qquad \dfrac{u}{u_R} = \dfrac{V}{V_R \ d^{+2}}.$ (4)

(Note that when the circuit resistance is zero, these equations reduce to equations (1) and (2) above.)

Pipe Flow, Laminar, without Circuit Resistance

$$\frac{\dot{V}}{\dot{V}_R} = \frac{1 + P}{1/d^{+4} + P};$$ (5)

$$\frac{u}{u_R} = \frac{1 + P}{1/d^{+4} + P} \ \frac{1}{d^{+2}}.$$ (6)

Parallel Plates, Laminar Flow

$$\frac{\dot{V}}{\dot{V}_R} = \frac{1 + P}{1/b^{+3} + P};$$ (7)

$$\frac{u}{u_R} = \frac{1}{b^+} \ \frac{1 + P}{1/b^{+3} + P}.$$ (8)

DISCUSSION OF ANALYTICAL RESULTS

General Features of Results

Figures 2-4 show that the volume flowrate in a fouled system invariably falls with increasing fouling thickness, but the rate of fall decreases as the parameter P increases.

In some fouling environments, the throughflow velocity has a controlling influence on the rate of fouling. Mean velocities are accordingly also shown in Figs 2-4. For values of P above about 1.0-2.0, an increase in velocity is initially observed as the fouling layer grows, but eventually a rapid fall ensues as the available gap closes up.

When the resistance of the remainder of the circuit is significant, the effect of the controlling parameter R_{CR} on the results is very similar to that of P, as is shown in Fig. 5.

Effect on Heat Transfer

For a given heat exchanger with, for example, fixed flow conditions on the non-fouling side, several factors interact. In the calculation of N_{Tu}, (Appendix II) each of the components change as the fouling layer increases.

The net result is an increase in N_{Tu}, giving a closer temperature approach of the streams. In terms of total heat transferred, which is important in, among others, refrigeration applications, this closer approach is offset by the lower mass flow. Preliminary calculations, shown graphically in Fig. 6, indicate that for circular tubes in turbulent and laminar flow, the heat flowrate decreases with increased fouling thickness. The rate of fall is lowest the higher the value of P, and is higher the higher the 'clean' value of N_{Tu}.

For parallel plates in laminar flow, there is an initial increase in heat flowrate with fouling thickness, for moderate values of P (eg 4). This increase is smaller the higher the initial N_{Tu}. With further increase in fouling thickness the heat flowrate falls. The heat flowrate thus exhibits similar behaviour to that of mean velocity, although it is N_{Tu} sensitive.

Effect on Fouling Rates

If the 'clean' throughflow velocity is close to, but below the critical fouling velocity of some applications, then the build-up of a fouling layer with a suitably high value of P will cause the 'fouled' throughflow velocity to exceed this value, and the fouling would then stop. This process might explain the asymptotic fouling observed in some cases.

CONCLUSIONS

a   The volume flowrate through the passages of a fouled or frosted heat exchanger is strongly dependent on a parameter P, which is a function of the operating flowrate, pressure drop, and the gradient of the circulating pump or fan characteristics, and is readily calculable at the design stage of a system.

To reduce the effect of fouling on a heat exchanger, a high value of P is desirable. This can be achieved in two ways:

i   by design of the heat exchanger and circuitry for high flows and low pressure drops, and

ii   by choosing circulating pumps and fans with a high characteristic gradient.

(By way of example, for an industrial boiler feed application, the value of P at the best efficiency point (efficiency ≈ 0.85) of a pump was 2.4. At a higher flow, such that the efficiency was 0.75, the value of P was 7.0).

A further reduction in effect is caused by a finite circuit resistance, but use of this mechanism is not recommended because of increased pumping costs.

b   The effect of P on mean velocity is more marked than on flowrate, and for typical values of P an initial increase of velocity with fouling thickness is observed. If the fouling rate (or more properly the removal rate) is strongly dependent on velocity, a possible mechanism for asymptotic fouling is indicated.

c   Preliminary calculations indicate that for circular tube flows, total heat flowrate reduces monotonically with fouling thickness, whereas for flat parallel plates, an initial increase in heat flowrate is possible.

ACKNOWLEDGEMENT

This work was supported in part by the Department of Energy through an IEA (International Energy Agency) programme on studies of the performance of practical extended surface systems.

NOTATION

A   Surface area

b   Plate spacing

c_p Specific heat at constant pressure

d   Diameter

f   Darcy friction factor

g_n Gravitational acceleration

H   Head

ΔH Difference in head

L   Length of tube or plate

N_{Tu} Number of thermal units

P   Parameter defined by equation (I.8)

P_1 Defined by equation (I.5)

P_2 Defined by equation (I.6)

$\dot{Q}$ Heat flowrate

R   Resistance

R   As suffix, denotes reference conditions

ΔT Temperature difference

u   Mean through flow velocity

$\dot{V}$ Volume flowrate

α Surface heat transfer coefficient

ε Effectiveness

ρ Fluid density

APPENDIX I

ANALYSIS OF FLOW IN SYSTEM

Fully Turbulent (Rough) Pipe Flow, Zero Circuit Resistance

We treat here a basic circular tube of a given length, representing an element of a heat exchanger. The rough surface case is chosen because in practice most pipes are rough, and because it represents a theoretical extreme, the other being fully laminar flow.

The pressure drop through the pipe in single-phase can be expressed by

$$H = 4f \; \frac{L}{d} \; \frac{u_m^2}{2g_n} \qquad (I.1)$$

where f is the Darcy friction factor, which is constant in this case.

Alternatively, we may write, in terms of the volume flow $\dot{V}$,

$$H = R\dot{V}^2 \qquad (I.2)$$

where

$$R = \frac{f}{\pi^2 d^4} \; \frac{L}{4} \qquad (I.3)$$

where R is the circuit resistance.

Equation (I.3) shows that the resistance is inversely proportional to the fifth power of diameter. If the tube is fouled uniformly so that a clean diameter d_R is reduced to an effective diameter d (with the same roughness) then the resistance related to reference conditions (suffix R) is given simply by

$$\frac{R}{R_R} = \left(\frac{d_R}{d}\right)^5 = \frac{1}{d^{+5}} \qquad (I.4)$$

where d^+ is the effective dimensionless diameter.

Referring now to Fig. 1, an increased resistance interacting with the circulating pump characteristics gives a reduced flow, V, and an increased head, H, from the reference conditions V_R and H_R.

We now put

$$\frac{\Delta H}{\Delta \dot{V}} = \frac{H - H_R}{\dot{V}_R - \dot{V}} = P_1 \qquad (I.5)$$

where P_1 is the average gradient of the pump characteristic over the flow range.

We also put

$$\frac{H_R}{\dot{V}_R} = P_2. \qquad (I.6)$$

Expanding equation (I.5) in terms of (I.3) and (I.4), we obtain a quadratic equation for V in terms of V_R m P_1, P_2 and d^+. This is solved to give

$$\frac{\dot{V}}{\dot{V}_R} = \frac{d^{+5}}{2} \{ (P^2 + 4P/d^{+5} + 4/s^{+5})^{\frac{1}{2}} - P\} \tag{I.7}$$

where
$$P = \frac{P_1}{P_2}. \tag{I.8}$$

The ratio of mean velocity to its reference value then follows naturally as

$$\frac{u}{u_R} = \frac{d^{+3}}{2} \{ (P^2 + 4P/d^{+5} + 4/d^{+5})^{\frac{1}{2}} - P\}. \tag{I.9}$$

Fully Turbulent Pipe Flow, Finite Circuit Resistance

We now generalise the above treatment to the case in which the tube system is only part of a circuit which we assume is unfouled and has a constant resistance R_C.

Equation (I.2) above is now replaced by

$$H = R_T V^2$$

where
$$R_T = R + R_C$$

$$= \frac{R_R}{d^{+5}} + R_C \text{ (from equation (I.4) above).} \tag{I.10}$$

By substitution in equation (I.5) and after some algebraic manipulation we obtain

$$\frac{\dot{V}}{\dot{V}_R} = \frac{\{P^2 + 4A(P + 1)\}^{\frac{1}{2}} - P}{2A}, \tag{I.11}$$

where
$$A = \left(\frac{1/d^{+5} + R_C/R_R}{1 + R_C/R_R}\right). \tag{I.12}$$

In the limiting case when R_C/R_R is zero, this reduces to equation (I.7). The mean velocity through the tubes is, correspondingly,

$$\frac{u}{u_R} = \frac{\dot{V}}{\dot{V}_R} \frac{1}{d^{+2}}. \tag{I.13}$$

Fully Developed Laminar Pipe Flow

In this case the Darcy friction factor f is inversely proportional to Reynolds number, which easily yields the relationship

$$H = R\dot{V}, \qquad (I.14)$$

where
$$R = \frac{R_R}{d^{+4}}. \qquad (I.15)$$

Substitution as before gives a linear equation for $\dot{V}$ in terms of $\dot{V}_R$, d^+ and P, such that

$$\frac{\dot{V}}{\dot{V}_R} = \frac{1 + P}{1/d^{+4} + P}, \qquad (I.16)$$

and the corresponding velocity is given by equation (I.13).

Fully Developed Laminar Flow between Parallel Plates

This is the basic geometry characteristic of the tube-in-plate heat exchangers commonly used in air conditioning and refrigeration systems. Typical Reynolds numbers based on plate spacing are in the region 500 < Re < 2000, which is normally laminar.

It is assumed that a uniform fouling or frost layer exists which reduces the plate spacing from the reference (clean) value b_R to an effective value b.

As for case I.3 above, the friction factor is inversely proportional to Reynolds number, giving equation (I.14) again, but where R is given now by

$$R = \frac{R_R}{b^{+3}} \qquad (I.17)$$

where
$$b^+ = \frac{b}{b_R}. \qquad (I.18)$$

A straightforward analysis then gives the volume flowrate

$$\frac{\dot{V}}{\dot{V}_R} = \frac{1 + P}{1/b^{+3} + P}, \qquad (I.19)$$

and the mean velocity is given by

$$\frac{u}{u_R} = \frac{1}{b^+} \frac{1 + P}{1/b^{+3} + P}. \qquad (I.20)$$

APPENDIX II

Preliminary Thermal Analysis

We now investigate the effect of the above flow analysis on the thermal performance, using a simple N_{Tu} method. For convenience, the analysis is limited to the limiting case in which the surface (including fouling layer) temperature is constant. It is also assumed that the effects of fluid density changes are negligible.

In this case the effectiveness of the exchanger surface is given by

$$\varepsilon = 1 - \exp(- N_{Tu}). \tag{II.1}$$

The heat flowrate $\dot{Q} = \varepsilon \rho \dot{V} c_p \Delta T, \tag{II.2}$

where ΔT is the temperature difference between the flow at inlet and the wall (or fouling layer) surface, which with the above assumption remains constant during the fouling process.

The N_{Tu} is given by

$$N_{Tu} = \frac{\alpha A}{\rho \dot{V} c_p} \tag{II.3}$$

where α is the surface heat transfer coefficient
(overall heat transfer coefficient for this case), and

A is the effective surface area.

The heat transfer coefficient is directly related to the Nusselt number, which is constant for fully developed laminar flow in pipes and between parallel plates, and approximately proportional to Reynolds number for fully developed turbulent (rough) pipe flow.

Using these relationships, we obtain

for turbulent pipe flow,

$$N_{Tu} = \frac{N_{Tu,R}}{d^+}; \tag{II.4}$$

for laminar pipe flow,

$$N_{Tu} = N_{Tu,R} \frac{\dot{V}_R}{\dot{V}}; \tag{II.5}$$

and for laminar flow between parallel plates,

$$N_{Tu} = N_{Tu,R} \frac{\dot{V}_R}{\dot{V}} \frac{1}{b^+}; \tag{II.6}$$

where $N_{Tu,R}$ is the reference (clean) value of N_{Tu}, for each case. Using these relationships, for selected values of $N_{Tu,R}$, it is possible to compare values of heat flowrate Q for the fouled and unfouled cases, by solving equation (II.2).

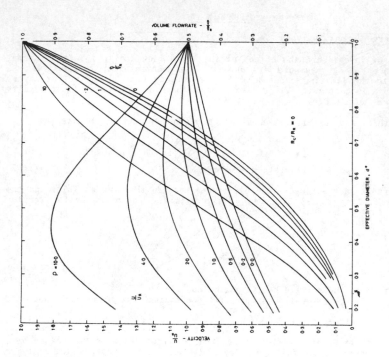

FIG 2   VARIATION OF VOLUME FLOWRATE AND VELOCITY WITH EFFECTIVE PIPE DIAMETER (TURBULENT FLOW)

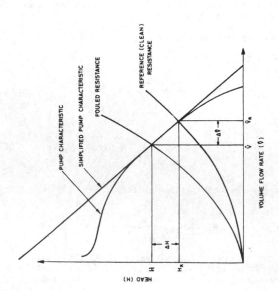

FIG 1   PUMP CHARACTERISTIC / SYSTEM RELATIONSHIP

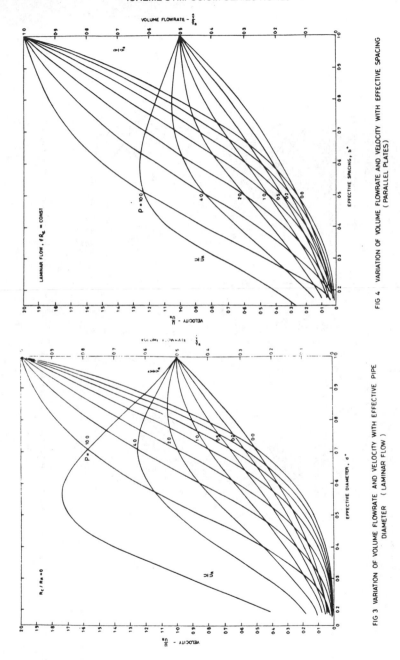

FIG 4   VARIATION OF VOLUME FLOWRATE AND VELOCITY WITH EFFECTIVE SPACING
(PARALLEL PLATES)

FIG 3   VARIATION OF VOLUME FLOWRATE AND VELOCITY WITH EFFECTIVE PIPE
DIAMETER   (LAMINAR FLOW)

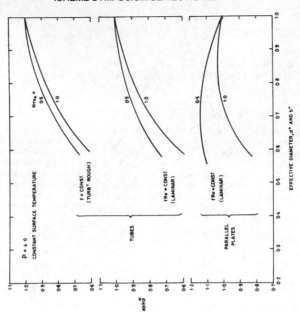

FIG 6 VARIATION OF HEAT FLOW RATE WITH EFFECTIVE DIAMETER

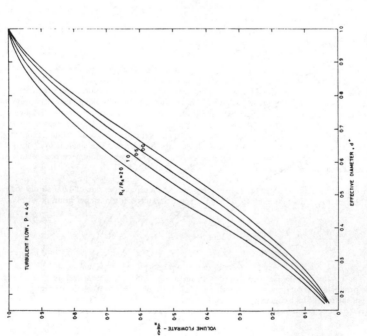

FIG 5 VARIATION OF VOLUME FLOWRATE WITH EFFECTIVE DIAMETER AND CIRCUIT RESISTANCE

A STUDY OF FOULING FROM KRAFT PULP BLACK LIQUOR

C.A. Branch and H.M. Müller-Steinhagen
Department of Chemical and Materials Engineering,The University of Auckland, Auckland, New Zealand

Heat transfer and heat transfer fouling was measured for New Zealand Forest Products Kraft pulp black liquor as a function of surface temperature, bulk temperature, velocity and liquor solids concentration. For technical reasons, most experiments have been performed in an annular test section with a heated stainless steel core. However, some results are reported for a PTFE-coated heater and for a plate heat exchanger. The 'clean' heat transfer coefficients have been compared with the prediction of correlations for convective and boiling heat transfer. The deposits have been analyzed with respect to appearance, composition and process conditions for which they were obtained.

1. INTRODUCTION

1.1 The Kraft Process

The Kraft process is the most common process to produce high strength paper from wood chips. Wood chips are digested for approximately three hours at 180°C and at elevated pressure using a solution of dissolved sodium hydroxide, sodium carbonate and sodium sulphide, the so-called white liquor. Lignin, starches, sugars, resins and fatty acids are dissolved, leaving wood pulp fibres and the spent cooking chemicals (weak black liquor). The weak black liquor is concentrated in multiple effect evaporators to either strong black liquor (45-50% solids) or heavy black liquor (65-70%). After adding make-up sodium sulphate, the concentrated liquor is sprayed into the chemical recovery furnace, where steam is generated by combustion of the organic component and chemicals are recovered from the smelt beneath the furnace. The smelt is then dissolved in water, to form green liquor, and undergoes various treatments to convert sodium carbonate to sodium hydroxide and to remove insoluble material.

The evaporators play a vital part in the recovery cycle. At a production rate of 1200 air dry tonnes of pulp per day, the multiple effect evaporation demand is approximately 350 tonnes per hour, or 40 MW.

1.2 Scope of Present Work

Recently, New Zealand Forest Products Kinleith pulp and paper mill modernised their black liquor evaporators. The new evaporator set consists of 9 steam heated falling film evaporators with corrugated surfaces. The production of high concentration firing strength (> 65% solids) black liquor brought with it a severe fouling problem. The high concentration (No. 1) effect is designed to control fouling on the heat transfer surfaces. There are three sub-effects which have a variable flow pattern to allow for regular washing cycles.

The liquor flow pattern is switched at regular intervals to ensure the heating elements are continually washed with incoming intermediate liquor.

Although fouling was considered in the design phase of the evaporators, the problem can, at certain operation periods, cause great reductions in evaporator capacity. Under ideal conditions, the evaporator is designed to handle 9 to 12 switches per day. When fouling can no longer be controlled by switching, weak black liquor washes are used. Fig. 1 shows plant data for August 1990, indicating an escalation of fouling in the second half of the month.

Many aspects of fouling from black liquor have been reported in the literature [1]. However, no systematic study of the effect of pertinent process variables on the fouling from firing strength Kraft black liquor could be found. The purpose of this study was to investigate fouling phenomena under forced convective and boiling heat transfer conditions which closely model the actual conditions found in industrial evaporators.

2. TEST EQUIPMENT AND EXPERIMENTAL PROCEDURE

The test rig designed and used for the fouling experiments is described elsewhere [2]. It consists of a closed, temperature conrolled flow loop with an electrically heated annular test section.

Concentrated black liquor was obtained from New Zealand Forest Product's Kinleith mill in 60 litre -100 litre batches. All results shown in each of the following diagrams were obtained from the same sample.

The fouling resistance could be calculated from the change in heat transfer coefficient with time

$$R_f(t) = \frac{1}{\alpha(t)} - \frac{1}{\alpha_o} \tag{1}$$

where α_o is the initial 'clean' heat transfer coefficient. The heat transfer coefficient was calculated from

$$\alpha = \frac{\dot{Q}}{A(T_w - T_b)} \tag{2}$$

The measured clean heat transfer coefficients for forced convective and subcooled flow boiling heat transfer have been compared with the predictions of correlations from the literature for liquor solids concentrations ranging from 0% to 65% [3]. Excellent agreement was obtained using the Gnielinski equation [4] for convective heat transfer and the Gorenflo [5] correlation for boiling heat transfer. A compilation of correlations for the physical properties of black liquor is given in [6].

3. EXPERIMENTAL RESULTS

3.1 Plant Data

Fig. 2 shows the fouling resistance as a function of time measured at the Kinleith Mill No. 1 evaporator. Fouling has a characteristic delay period, followed by a period of severe fouling, after which the rate of fouling decreases again. The heat exchanger was switched and cleaned with weak black liquor after 200 minutes of operation.

3.1 Laboratory Data

The following effects of process parameters on the fouling rate has been found from numerous experiments with the experimental test rig [2]:

Effect of solids concentration   Figure 3 shows a comparison of the fouling curves for three liquor solids concentrations. The measured trends agree very well with the plant data shown in Fig. 4. The difference between the fouling rates of the 65% liquor and the 60% liquor is approximately two orders of magnitude. No fouling was observed for the 55 % liquor or any lower concentration. The delay time is also strongly affected by the liquor solids concentration.

Effect of heat transfer surface temperature   The fouling rate increases considerably with increasing temperature, while the delay time decreases with increasing surface temperature. If the delay time is assumed to be governed by a zeroth order rate equation the effects of surface temperature may be expressed by the following correlation:

$$\frac{1}{t_d} = 2.444 \cdot 10^{20} \exp\left(\frac{-1.647 \cdot 10^5}{R \cdot T_w}\right) \tag{3}$$

Effect of flow velocity   For constant surface and bulk temperatures, the fouling rate remained essentially constant with flow velocity, which is characteristic for chemical reaction and adhesion controlled fouling. The delay time is a function of the velocity, showing a maximum for a velocity around 50 cm/s.

Effect of bulk temperature   For constant surface temperature and flow velocity the rate of fouling increases with decreasing bulk temperature. The delay time remains almost constant with variation in bulk temperature. This suggests that the delay time is a function of the surface conditions.

3.3 Appearance And Composition Of Deposits

The deposit typically had a smooth appearance and was strongly bonded to the heating surface. With increasing heat flux, the deposit became more porous, with what appeared to be black liquor in the pores immediately adjacent to the heating surface. For all experiments the deposit was completely soluble in water, thus indicating a carbonate-sulphate type scale. It was noticed that high heat flux fouling experiments generally gave highly porous scales.

For experiments with a long delay time, scanning electron micrographs show a uniform crystalline structure with the growth of the deposit from nucleation sites. If deposition occurs without a pronounced delay time, a much less defined crystal structure is obtained.

The elemental composition of a deposit was obtained with an EDAX 9100 energy-dispersive X-ray spectrometer system. The major elements identified are Sodium, Potassium, Sulphur and Silicon. A peak near the silicon and sulphur positions could indicate aluminium. Calcium may also be present although the sample did not show it clearly.

Using X-ray diffraction, the following components have been identified in the deposit: Na_2CO_3, $Na_2S \cdot H_2O$, Na_2S_2, $Na_6CO_3(SO_4)_2$ (Burkeite), $K_2Ca(CO_3)_2$ and $CaCO_3$. Therefore, the deposits consisted mainly of burkeite and calcium carbonate with quantities of the liquor itself. The fact that the liquor is present in the deposit is not surprising since most deposits were to some extent porous and the liquor was generally found in the pores.

3.4 Mitigation Of Fouling

Chemical treatment Two commercial liquid blends of polymeric antifoulant were investigated. Recommended dosage for once through processes is 25 ppm. Based on the recommended dosage, a 65 % solids liquor which caused considerable fouling was tested with each of the chemical additives.

Figure 4 shows a comparison of fouling from untreated and S62-2 treated liquor. It was found that the recommended dosage only marginally reduced the fouling; the severe characteristic fouling rate was still observed in the final stages. Increasing the dosage from 25 ppm to 50 ppm made no significant improvement. Large improvements were noted for the 100 ppm and 200 ppm runs. Severe fouling was delayed until nearly 7 times the untreated fouling time. In the case of the 200 ppm treatment no fouling escalation was observed for the 52 hour duration of the run.

The chemical treatment with the second chemical, S89-2, began with the recommended dosage of 25 ppm. Severe fouling was observed earlier than in the untreated case for this treatment. The same trend was observed for further increases in concentration.

Modification of heat exchanger Fouling runs were performed with a Teflon-coated heat transfer surface. Contrariwise to the stainless steel surface, the fouling curve for Teflon showed a strong saw-tooth character. Once the fouling grows to a resistance of around 0.15 m^2K/kW the deposit sloughs. The average period of removal is estimated at 15 minutes. After about 300 minutes the sloughing diminished and an excursion into severe fouling occurred. This suggests that the continuous growth and removal of deposit may wear down the Teflon coating.

A number of fouling experiments have been performed with a plate and frame heat exchanger for a 68% solids black liquor which rapidly fouled the annular test section for comparable heat fluxes [7]. Over an experimental period of 24 hours, no fouling has been observed in the plate heat exchanger, for two different surface temperatures and flow velocities.

4. CONCLUSIONS

Fouling experiments with Kraft black liquor closely reproduced the fouling curves found from plant measurements. The fouling rates increase considerably with increasing liquor solids concentration and heat transfer surface temperature. The fouling rate was insensitive to variations in velocity, at least over the observed range of flow rates. Contrariwise, there is an effect of flow velocity on the delay time before fouling starts.

Chemical treatment of high concentration liquors can reduce the fouling potential, however, the required additive concentration has to be quite high. The application of additives is highly specific to a given liquor composition and increased fouling may be produced with unsuitable additives.

Surface coating with Teflon can reduce the overall fouling rate, as can the application of plate and frame heat exchangers.

5. REFERENCES

1. Frederick W.J., and Grace T.M.: Scaling in Alkaline Spent Pulping Liquor Evaporators. Proceedings, Intl. Conf. on Fouling of Heat Transfer Equipment, p. 587, (1979).

2. Branch, C.A. and Müller-Steinhagen, H.M.: Fouling During Heat Transfer to Kraft Pulp Black Liquor,Part I: Experimental Results. submit. for publ. in Can. J. Chem. Eng., 1991

3. Branch, C.A., Müller-Steinhagen, H.M. and Jamialahmadi, M.: Convective and Subcooled Flow Boiling Heat Transfer to Kraft Pulp Black Liquor. submit. for publ. in APPITA Journal, 1991

4. Gnielinski, V.: Wärmeübertragung in Rohren. VDI-Wärmeatlas, VDI-Verlag, Düsseldorf (1986).

5. Gorenflo, D.: Behältersieden. VDI Wärmeatlas, Sect. Ha, ed. 5, (1988).

6. Branch, C.A. and Müller-Steinhagen, H.M.: Physical Properties Of Kraft Black Liquor APPITA J. Vol.44, No. 5, pp. 339-341 (1991)

7. Branch, C.A., Müller-Steinhagen, H. and Seyfried, F.: Heat Transfer to Kraft Black Liquor in Plate Heat Exchangers. APPITA J. Vol. 44, No. 4, pp.270-272 (1991)

6. LIST OF SYMBOLS

A	heat transfer surface area, m^2
$\dot{Q}$	heat flow rate, W
R	universal gas constant, 8.314 J/mol $\cdot$ K
R_f	fouling resistance, m^2K/kW
t_d	delay time, min
T_b	bulk temperature, °C
T_w	heat transfer surface temperature, °C
v_{av}	average flow velocity, m/s
α	heat transfer coefficient, $W/m^2 \cdot K$
α_o	clean heat transfer coefficient, $W/m^2 \cdot K$

7. ACKNOWLEDGEMENTS

The authors are indebted to PAPRO NZ and to New Zealand Forest Products Ltd. for continuous support of their investigations. The assistance of Heat Transfer Research Inc. and Sachtleben GmbH is gratefully acknowledged.

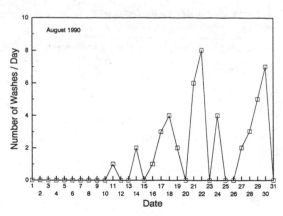

Fig. 1   No. 1 effect weak black liquor washes per day

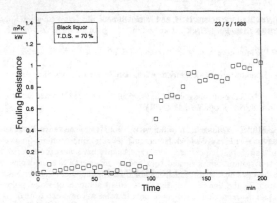

Fig. 2   No. 1 effect fouling resistance versus time

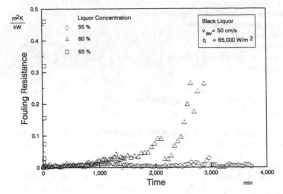

Fig. 3   Effect of liquor concentration on the fouling rates

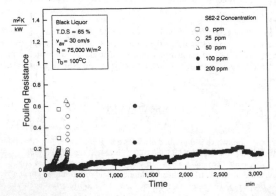

Fig. 4   Fouling resistance versus time for varying S62-2 concentrations

CONVECTIVE AND NUCLEATE BOILING HEAT TRANSFER TO WOOD PULP FIBRE SUSPENSIONS.

J. Middis, H.M. Müller-Steinhagen, and G.G. Duffy
Department of Chemical and Materials Engineering, The University of Auckland, Auckland, New Zealand

Heat transfer coefficients to flowing wood pulp fibre suspensions have been measured for a wide range of parameters such as pulp concentration, flow velocity, heat flux and bulk temperature. The experiments have been performed in an annular test section with a heated core and in an externally heated pipe. Both, forced convective heat transfer and subcooled flow boiling heat transfer have been investigated. In the annular test section, a small addition of wood pulp fibres resulted in a moderate improvement of heat transfer as compared to that for pure water. In the heated pipe, where higher flow velocities and pulp concentrations could be investigated, a more complex relationship between these parameters and the heat transfer coefficient exists, with both heat transfer enhancement and heat transfer reduction being observed.

INTRODUCTION

In the paper and associated wood processing industries, wood fibres are often handled as a suspension, with water as the suspending fluid. The fibres are hollow cylindrical structures with length to diameter ratios between 30 and 120 (Figure 1). The ability of the fibres, when in suspension, to entangle and form flocs and coherent networks results in a suspension with a unique flow behaviour.

For any given pulp fibre suspension, the flow mechanism depends strongly on the velocity. At low velocities the flow tends to be plug-like, and the frictional pressure drop of the suspension is higher than that of water. At higher velocities, a water (or fibre/water) annulus forms at the pipe wall and a pressure drop lower than that of water can be observed. The latter phenomenon is called drag reduction and it has been studied in some detail by Duffy et al [1,2].

Heat transfer to a moving fluid is a result of both conduction at the wall and convection in the fluid. For turbulent flow, the addition of pulp fibres modifies the turbulence structure and the momentum transfer within the fluid. These changes in the rate of momentum transfer cause a variation in the friction factor. Since momentum and heat transfer are inter-related (Reynolds [3]), it can be expected that fibres would also modify the convective heat transfer.

TEST EQUIPMENT

Both rigs had a similar flow loop, which is described elsewhere by Branch et al [4]. The diference was that one test section was an internally-heated annulus and the other an externally-heated pipe.

The first test section consisted of an electrically heated stainless steel rod mounted concentrically inside a larger pipe to give an annular flow area. The important dimensions of the annular test section are given in Table 1.

Table 1: Dimensions of annular test section

Length	100 mm
Heater rod diameter	10.7 mm
Annulus outside diameter	25.4 mm
Hydraulic diameter	14.7 mm
Hydrodynamic entry length	298.6 mm
Thermal entry length	82.6 mm

In the second test rig, the suspension flowed through an electrically heated pipe. This rig was equipped with a considerably larger pump, allowing bulk flow velocities up to 10 m/s even for high fibre concentrations. The main dimensions of this test unit are given in Table 2.

Table 2: Dimensions of heated pipe

Length	250 mm
Inside diameter	25.3 mm
Hydrodynamic entry length	2.0 m
Thermal entry length	143 mm

In both test heaters, the wall temperature was measured using four thermocouples which were positioned below the surface of the heater. The heat transfer surface temperature was calculated from these measurements by deducting the temperature drop between thermocouple location and heater surface from the measured temperature:

$$T_w = T_{tc} - \frac{\dot{q}}{\lambda / x}$$ (1)

λ is the thermal conductivity of the heater material, and x is the distance between thermocouple location and outside surface. Inlet and outlet bulk temperatures were measured with thermocouples placed in the flow upstream and downstream of the test section. The local heat transfer coefficient was calculated using:

$$\alpha = \frac{\dot{q}}{T_s - T_b}$$ (2)

where the bulk temperature was taken to be a position-weighted average value of the inlet and outlet temperatures.

EXPERIMENTAL PROCEDURE

The fibres used in the experiments were bleached Kraft pulp fibres with an average length of about 3 mm and an average diameter of 35 μm, obtained from New Zealand Forest Products Kinleith Mill. The fibres were soaked in water for a minimum of 18 hours, before being added to the test rig, to absorb water and swell. The sheets of fibres were then hand torn and dispersed in a commercial blender before being added to the tank. The solution was circulated through the test rig for thirty minutes to fully distribute the fibres before any data were collected. For each pulp concentration (consistency), the heat transfer coefficient was measured as a function of either the bulk velocity, the heat flux or the bulk temperature. Consistency measurements were taken at the beginning and end of each run and the average value taken as the overall consistency. The beginning and end consistencies were always in agreement, with differences less than 5%.

The range of experimental parameters is shown in Table 3.

Table 3: Range of parameters

Bulk velocity	annulus	$0.2 < u$ (m/s) < 2.5
	pipe	$0.5 < u$ (m/s) < 10
Pulp concentration	annulus	$0.0 < C$ (%) < 0.8
	pipe	$0.0 < C$ (%) < 3.8
Heat flux	annulus	$5000 < q$ (W/m^2) < 600000
	pipe	$5000 < q$ (W/m^2) < 100000
Bulk temperature	annulus	$45 < T_b$ (°C) < 80
	pipe	$28 < T_b$ (°C) < 32

RESULTS AND DISCUSSION

Figure 2 shows the heat transfer coefficient as a function of the heat flux and the fibre concentration. The curve consists of two parts; a forced convective heat transfer regime where the heat transfer coefficient is independent of the heat flux, and a subcooled nucleate boiling regime where the heat transfer coefficient increases with increasing heat flux. Unfortunately, there is quite some scatter in the data because the fibres tended to block the flow control valve for this low flow velocity, causing variations in the actual bulk velocity. The error was highest for small temperature differences between wall and bulk (i.e. low heat fluxes) at low flow rates. Despite the scatter of data, this figure shows that the effect of wood pulp fibre concentration is much more pronounced for the convective heat transfer region than for the boiling region. Two mechanisms are responsible for this effect: i) the micro-convection caused by bubble formation and detachment is more important than the macro-convection caused by the turbulent eddies; and ii) the vapour bubbles are pushing the fibres away from the heated surface, giving a liquid-only layer next to the heated surface which gives the suspension the same boiling heat transfer coefficient as water at the same conditions.

For all fibre concentrations, the convective heat transfer coefficient increased with increasing flow velocity. For the experiments shown in Figure 2, the presence of fibres always reduced the convective heat transfer coefficients below the values of pure water at the same heat flux. This effect was only found for very low flow velocities. For higher flow velocities, enhancement and reduction of heat transfer was observed, depending on flow velocity and fibre concentration. Figure 3 presents the heat transfer coefficient as a function of the heat flux for a velocity of 1 m/s. Within the investigated range of heat fluxes, only the convective heat transfer regime can be observed. As the pulp concentration increases from 0% - 0.2%, heat transfer enhancement of up to 10% occurs. Above this concentration, the heat transfer coefficients decrease until they once more reach the water values. Similar results were obtained for experiments at 0.6 m/s, 1.75 m/s, and 2.5 m/s, with a maximum increase in heat transfer at approximately 0.2% pulp consistency.

Figures 4a) and 4b) show the effect of bulk temperature on the heat transfer coefficient. The heat transfer coefficient increases with the bulk temperature due to the change in physical properties, particularly the viscosity, of the water. Comparison of the two figures shows that the addition of pulp fibres increases this trend.

To extend the range of flow velocities up to 10 m/s and the range of pulp consistencies up to 3.3%, the test rig with the tubular heater was designed. It was limited to measurements in the forced convective heat transfer regime. Figure 5 shows the heat transfer coefficient as a function of the bulk velocity for a range of pulp concentrations. At high concentrations and high velocities, the suspension heat transfer coefficient lies below the water curve. Measurements of the pressure drop showed that in this region the friction factor was also reduced. By comparison with earlier work of Duffy *et al* [2] it was deduced that at these velocities a turbulent fibre/water annulus was present. Thus it appears that the fibre/water annulus with its turbulence-damping effect reduces both the heat transfer and momentum transfer rates. At lower velocities and higher fibre concentrations, a region of enhanced heat transfer occurs. This is a result of the suspension flow being plug-like. Between these to limits lies a complicated transitional region. Here, pressure drop and heat transfer show similar, although not identical trends.

Because of the differences in the design of the two test rigs, it is difficult to make any direct comparison. The inside diameter of the heated pipe was much larger than the length of the fibres; whereas in the annular test section, hydraulic diameter and fibre length were of the same order of magnitude. It has been shown by Duffy and Lee [5] that the mechanism of drag reduction is a turbulent core effect. Thus, if the formation of this core is restricted by the dimensions of the flow channel, this may have a considerable effect. Additionally, the velocity profile in the tubular test section was fully developed, whilst it was still developing in the annular heater. Nevertheless, both sets of results show some interesting features. The annular data show that the addition of a small amount of wood pulp fibres to a fluid will have (at least) no detrimental effect on the heat transfer performance. The pipe flow data show the effect of these fibres for concentrations more commonly encountered in industry.

CONCLUSIONS

Experiments with Kraft pulp fibre suspensions in annular flow have shown heat transfer enhancement over a wide range of bulk velocities. The increase in heat transfer coefficient (10%) and the amount of pulp added (0.2%) were both quite small.

The heat transfer coefficient was independent of the heat flux in the convective heat transfer regime, and independent of the pulp concentration in the boiling heat transfer regime. The fibres increased the influence of the bulk temperature on the heat transfer coefficient.

Experiments with the same fibres in pipe flow (at much higher concentration and velocity), showed that the heat transfer enhancement is a function of both, flow velocity and fibre concentration.

ACKNOWLEDGEMENTS

The authors would like to acknowledge the financial assistance of PAPRO (NZ) Ltd.

NOMENCLATURE

C	pulp consistency, %	Subscripts
q	heat flux, W/m^2	
Re	Reynolds number	b
T	temperature, °C	tc
u	bulk velocity, m/s	w
x	thermocouple to wall distance, m	
α	heat transfer coefficient, W/m^2K	
λ	thermal conductivity, W/mK	

Subscripts

b bulk
tc thermocouple
w wall

REFERENCES

[1]   Duffy G.G.; A Study of the Flow Properties of New Zealand Wood Pulp Suspensions.; PhD Thesis, (1972); University of Auckland.

[2]   Duffy, G.G., Titchener, A.L., Lee, P.F.W., and Moller, K.; The Mechanisms of Flow of Pulp Suspensions in Pipes; Appita (March 1976); 25, 5, pp 363 - 370.

[3]   Reynolds, O.; Scientific Papers (1901), 1; Cambridge Univ. Press, London & New York.

[4]   Branch, C.A., Müller-Steinhagen, H.M. and Jamialahmadi, M.; Convective and Subcooled Flow Boiling Heat Transfer to Kraft Pulp Black Liquor. submit. for publ. in TAPPI Journal, 1991

[5]   Duffy, G.G. and Lee, P.F.W.; Drag Reduction in the Turbulent Flow of Wood Pulp Suspensions; Appita (January 1978); 31, 4, pp. 280-286.

Figure 1: Kraft wood pulp fibres.

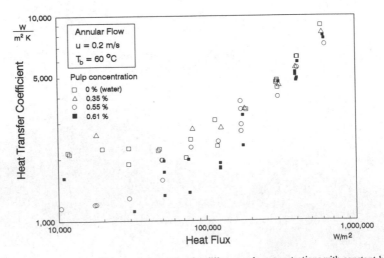

Figure 2: Pulp heat transfer coefficient versus heat flux for different pulp concentrations with constant bulk velocity (0.2 m/s).

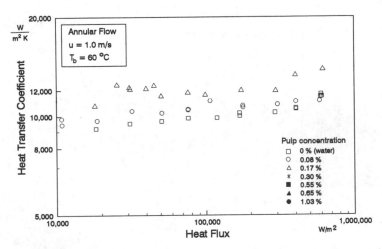

Figure 3: Pulp heat transfer coefficient versus heat flux for different pulp concentrations with constant bulk velocity (1.0 m/s).

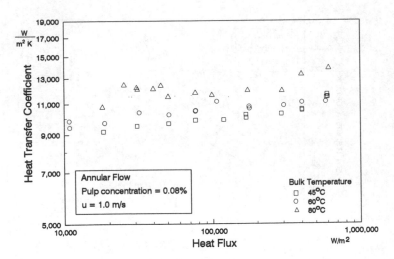

Figure 4a: Pulp heat transfer coefficient versus heat flux for different bulk temperatures with constant pulp concentration (0.08%).

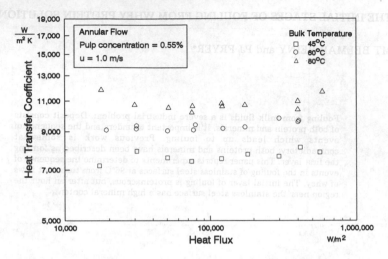

Figure 4b: Pulp heat transfer coefficient versus heat flux for different bulk temperatures with constant pulp concentration (0.55%).

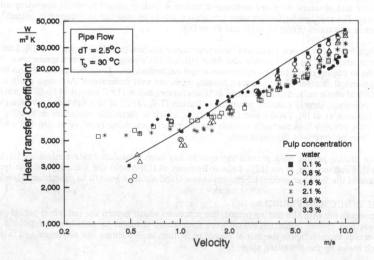

Figure 5: Pulp heat transfer coefficient versus velocity for a range of pulp concentrations.

THE INITIAL STAGES OF FOULING FROM WHEY PROTEIN SOLUTIONS

MT BELMAR-BEINY and PJ FRYER*

Fouling from milk fluids is a severe industrial problem. Deposit consists of both protein and minerals. It is important to understand the sequence of events which leads up to fouling. Previous work is somewhat contradictory; both proteins and minerals have been described as forming the first layer. This paper reports experiments to determine the sequence of events in the fouling of stainless steel surfaces at 96°C from turbulent flows of whey. The initial layer of fouling is proteinaceous, but after an hour the region near the stainless steel surface has a high mineral content.

FOULING IN MILK PROCESSING

The formation of unwanted fouling deposit within process plant is a common and severe problem in the food industry. Fouling is a kinetic process; there may be an *induction period*, in which heat transfer and pressure drop are unchanged, before a *fouling period* in which operating efficiency is reduced. If a rate law for fouling were available it would be possible to determine optimal operating cycles or run times (Fryer et al [1] and Fryer [2]).

Fouling from milk-based fluids has been thoroughly studied (see Hallstrom et al [3], Lund et al [4], Kessler and Lund [5] and Lalande and Rene [6]), but little progress has been made towards reducing fouling in commercial plant by other than empirical techniques. The composition of the deposit, and the chemical changes which occur on heating milk, are well understood. Although protein makes up only 3% of whole milk, deposits formed at temperatures below 110°C consist of 50-60% protein and 30-35% minerals, largely calcium phosphate (Burton [7,8]). Half of this protein is ß-lactoglobulin (ß-lg) (Lalande et al [9], Tissier and Lalande [10]), which is thermally unstable; on heating above 70-74°C the protein first partially unfolds (denaturation), exposing reactive sulphydryl (-SH) groups, and then polymerises (aggregation).

In the fouling period, bulk protein aggregation has been considered rate-controlling [6], Gotham et al [11], Paterson and Fryer [12]). Belmar-Beiny et al [13] model the fouling period in terms of the volume of the fluid hot enough for protein to react, and obtain a good fit to experimental data.

THE INDUCTION PERIOD

By comparison with the fouling period, the processes which govern the induction period are not well understood. If the induction period were understood it might be possible to extend it. It is thus important to investigate the initial events in fouling, to determine the sequence and rate of events which make up the induction stage.

*Department of Chemical Engineering, University of Cambridge
Pembroke Street, Cambridge CB2 3RA

Experiments to examine the initial stages of fouling are not easy. In situ measurement techniques such as ellipsometry are difficult at high temperatures and cannot be used with stainless steel. The adsorption of ß-lg onto idealised surfaces has been thoroughly studied (Arnebrandt et al [14]) but as commercial heat transfer surfaces are rough and of complex metallurgy, the relevance is uncertain. Surface characterization requires different techniques (e.g. electron microprobe analysis (EMPA), X ray photoelectron spectroscopy (XPS)) as a single technique is very unlikely to give a meaningful solution. However surface analysis techniques are both time consuming and expensive; it is therefore necessary to ensure that the results gained balance the cost and the time expended.

Several contradictory studies have been made of the initial events in milk fouling. Dupeyrat et al [15] and Yoon and Lund [16] suggest that although deposition of the first layer of fouling is surface-dependent, the surface has no subsequent influence, but McGuire and Swartzel [17] consider deposition rates to be strong functions of surface properties. Burton [7] stated that the desolubilisation of minerals, particularly calcium phosphate, was the primary stage of deposition. In contrast, Baier [18] and Delsing and Hiddinck [19] concluded that protein was the most important constituent of the earliest fouling layer, although adsorption was strongly favoured by the presence of cations such as calcium. Tissier and Lalande [10] found a mineral-rich sublayer, about 0.02 µm thick after 1 minute, extending eventually to 15µm, formed adjacent to the surface. It was concluded that this layer formed first, and suggested that the onset of fouling required the presence of minerals. Daufin et al [20] concluded that, for whey and milk, proteins were the first species to adhere; but considered that fouling was linked to the presence of calcium phosphate so that the inhibition of calcium phosphate formation will reduce fouling.

Both protein and minerals may be involved in the early stages in fouling, but the sequence of events is unclear. Gotham [21] carried out an extensive study of fouling from whey protein concentrates (WPC); after an hour, a thick, largely protein, deposit is formed. Figure 1 shows the lower regions of a WPC deposit, made up of protein aggregates about 0.2 - 0.7 µm in diameter; no distinct layer of the type observed by Tissier and Lalande [10] is seen. X ray microanalysis of deposit shows that its mineral content is much greater near the surface, as found by [10]. This paper reports experiments to study the initial stages of fouling of WPC solutions from turbulent flow onto stainless steel surfaces.

MATERIALS AND METHODS
Preparation of test fluid Whey protein concentrate (80% protein) was weighed to give 1% final protein concentration and then dissolved in distilled water.

Fouling rig and operation. A simple rig was constructed as shown in Figure 2; an electric kettle was adapted and used as a simple heat exchanger. A 25 cm long stainless steel tube (AISI 321, 6.35 mm o.d., 0.006" thick) was placed inside the heat exchanger. Test fluid inlet and outlet temperatures were measured by thermocouples. Stainless steel tubes were cleaned for 30 minutes at 45°C with a 0.1% (v/v) Lipsol detergent and thoroughly rinsed with distilled water. The test fluid was preheated in a water bath and then added to the reservoir placed 60 cm above the inlet tube. Fluid flowed through the tube by gravity in turbulent flow (Reynolds numbers ca. 15 000): after the time required the system was cooled with distilled water to quench any further reactions. The test fluid inlet temperature was varied between 64 and 73°C, and experimental runs lasted between 4 and 210 seconds. A constant water temperature of 96°C was maintained in the heat exchanger.

Analytical methods. Fouled tubes were removed, dried at room temperature and cut into small sections for Scanning Electron Microscopy (SEM), EMPA, XPS and contact angle measurements. For SEM work, the specimen was gold sputtered and analysed in an SEM model JEOL JSM-820. For EMPA mapping and EMPA the specimen was carbon sputtered and analysed on a CAMECA SX50 Electron Microprobe fitted with a LINK analytical ED system (AN 1000) at 20kV and 30nA. Contact angle measurements were done using a telescope and protractor eyepiece from Ealing Electro Optics (Watford, UK) Small distilled water drops were delivered from a vertical micro syringe (Model E-100, Ramé- Hart, Inc.) onto horizontally clamped test surfaces, and the drop profile was then viewed. XPS analysis was done in an AEI - ES 200.

RESULTS
Contact angle measurements for different inlet temperatures and contact times were carried out. The contact angle for clean stainless steel is 49°; after only a few seconds of contact, the angle increases significantly, (69° after 4 seconds with whey) showing that the surface has become more hydrophobic. After this, as the contact time increases, the contact angle decreases, so the surface becomes more hydrophilic.

Figure 3 shows an SEM of a surface fouled with WPC for 120s (Temperatures: inlet 73°C, outlet 75°C). Figure 4 shows an SEM of a fouled surface with WPC for 150s under the same temperature conditions. The surface is now fully covered with protein aggregates which are heterogeneously distributed across the surface. This protein layer was identified with EMPA through the presence of its sulphur content; elemental maps for sulphur, calcium, titanium iron and chromium showed that both calcium and sulphur were found together. XPS analysis (Figure 5) confirmed the EMPA results; proteinaceous material was identified through the carbon, nitrogen and sulphur peaks.

Figure 6 (a) and (b) shows XPS spectra obtained for unfouled and fouled stainless steel with a contact time of 4 seconds. The layer was identified as proteinaceous material (nitrogen and carbon peaks). This result confirms results of contact angle measurements that at 4 seconds contact time the protein has already adsorbed to the surface. This layer of protein is not thick enough to show in the SEM; this only occurs at longer contact times and higher temperatures as shown in figure 4. At a fluid inlet temperature of 68°C and 210s contact time the protein layer was not yet seen under the SEM.

Calcium was detected with EMPA at 150s contact time (Temperatures: inlet 73°C, outlet 75°C) as shown in Figure 5, though it was not detected in any other specimen. Phosphorus was never found. In all specimens studied, XPS analysis found no calcium or phosphorus.

DISCUSSION

The above results demonstrate clearly that protein is the first species adsorbed here. The contact angle measurements can be explained by the bilayer model of protein adsorption proposed by Arnebrandt et al [22]. They suggest that for a hydrophilic surface (clean metal surfaces are usually hydrophilic) the first layer of protein was irreversibly bound to the metal surface through polar aminoacid chains of the protein. Such strong ionic binding leads to an unfolding of the protein with exposure of hydrophobic loops into the aqueous solution. In the case of a negative surface charge and protein molecules with a negative net charge it is expected that the upper layer will tend to orient its positive residues downwards and the negatively charged side chains outwards from the surface. The protein molecules are attached to the first layer through hydrophobic and/or ionic forces and the whole layer is highly hydrated. At some point protein aggregates formed in the bulk begin to adhere to the surface.

X ray mapping shows that at 150s contact time (Temperatures: inlet 73°C and outlet 75°C) sulphur is found over all the fouled surface, and is in higher concentrations where protein aggregates are found in clusters. Calcium seems to be bound to the protein, confirming previous results (Mulvihill and Donovan [23]) about the sensitivity of ß lactoglobulin to calcium aggregation. Calcium phosphate is not formed in the time scale of these experiments. EMPA and X ray mapping using EMP, is a useful technique once the concentration of elements sought is within the detection limits of the machine. To register a signal the deposit has to be thick enough to stop the electrons beam going through the stainless steel. In this case, below 150 seconds no useful data was obtained.

XPS analysis proved to be a more useful technique as it analyses layers of about 50nm. However it has the disadvantage that it analyses over about one square centimetre, so the elements to be analysed need to be above a certain concentration to be detected. Calcium was found in one specimen by EMPA but not by XPS, the reason being that probably the element was found closer to the stainless steel. One way of solving this problem is to bombard the surface with Argon for a specific time and then do the analysis. In this way an elemental composition as a function of depth can be obtained; bombardment progressively exposes regions further and further into the deposit.

The final fouling deposit, after more than one hour contact time, contains a mineral rich layer near the surface. It is not clear yet whether this mineral rich layer is deposited on top of the initial protein layers, or whether phosphorus diffuses through the deposit to the surface during later fouling. Between about 120 and 150 seconds, protein aggregates formed in the bulk and containing bound calcium, become adsorbed to the surface. These aggregates will increase the deposit thickness and lower heat transfer efficiency. It is probable that the point were they begin to deposit constitutes the end of the induction period.

In future work an elemental composition as a function of depth will be sought using XPS for the experimental conditions used in this work and for higher temperatures. Auger mapping might be more useful than EMP mapping, as it analyses depths of about 10nm.

CONCLUSIONS

Experiments have been carried out to examine the first layer of fouling from WPC solutions. The first layer of deposit, formed in the first 4 seconds of contact between fluid and surface, appears to be a homogeneous protein layer . No minerals were found at this time. After 150 seconds of contact the surface is covered with protein aggregates bound to calcium. These protein aggregates were seen under SEM, and their composition identified with EMP mapping and XPS. It seems that at this fluid inlet temperature (73°C) the induction period ends between 120 and 150 seconds. Contact angle measurements suggest that the bilayer model developed by Arnebrandt et al [13] to describe ß-lactoglobulin adsorption to chromium surfaces can be applied to the industrial problem of complex milk fluids and stainless steel surfaces.

ACKNOWLEDGEMENTS

Financial support for this work was provided by AFRC. We are also grateful for the assistance of T Blesser and Dr SJB Reed of the Earth Sciences Department, A Moss of the Materials Science Department, Cambridge, and Dr N Owen, University College London.

REFERENCES

1. Fryer, PJ, Hobin, PJ and Mawer, SP. *Can. Journal of Chem. Eng.*, **66**, 558, (1988).
2. Fryer, PJ. *Proceedings of the 2nd UK Nat. Heat Transfer Con.*,Vol I, IMech E, (1988).
3. Hallstrom, B, Tragardh, C and Lund, DB. (eds) *Fundamentals and Applications of Surface Phenomena associated with Fouling and Cleaning in Food Processing*, Univ. of Lund, Sweden. (1981).
4. Lund, DB, Sandu, C and Plett, C. (eds)*Fouling and Cleaning in Food Processing*, Univ. of Madison, USA. (1985)
5. Kessler, HG and Lund, DB. (eds) *Fouling and Cleaning in Food Processing*, Univ. of Munich, FRG. (1989)
6. Lalande, M and Rene, F. in *Fouling Science and Technology*, (ed. Melo, L, Bott, TR and Bernardo, CA,), NATO ASI E-145, Kluwer, Amsterdam, 557-573. (1988)
7. Burton, H. *J.Dairy Res.*, **35**, 317-330. (1968)
8. Burton, H. *UHT Processing of Milk and Milk Products*, Elsevier, London. (1988)
9. Lalande, M, Tissier, JP and Corrieu,G. *Biotech.Prog.*, **1**, 131-139, (1985).
10. Tissier, JP and Lalande, M. *Biotech.Prog.*, **2**, 218-229, (1986).
11. Gotham, SM, Fryer, PJ and Pritchard, AM. In ref [5] , 1-13 (1989).
12. Paterson, WR, and Fryer, PJ. *Chem.Eng.Sci.*, **43**, 1714-1717, (1988).
13. Belmar-Beiny, MT, Gotham, SM, Fryer, PJ and Pritchard, in press *J.Fd.Eng.*, 1992..
14. Arnebrandt, T, Barton, K and Nylander, T. *J.Colloid Interface Sci.*, **119**, 383 (1987).
15. Dupeyrat, M, Labbe, JP, Michel, F, Billoudet, F and Daufin, G. *Le Lait*, **67**, 465-486, (1987).
16. Yoon, J and Lund, DB. In ref [5] 59-80 (1989).
17. McGuire, J and Swartzel, KR. *J. Food Proc. Pres.* , **13**, 145-160. (1989).
18. Baier, RB. (1981) In ref [3], pp 168-189, 1991.
19. Delsing, BMA, and Hiddinck, J. *Neth. Milk and Dairy J.*, **37**, 139-148, (1983)
20. Daufin, G, Labbe, JP, Quemerais, A, Brule, G, Michel, F and Roignant, M. *Le Lait*, **67**, 139-364, (1987).
21. Gotham, SM. PhD thesis, Cambridge University, (1990).
22. Arnebrandt, T, Ivarsson, B, Larsson, K, Lundstrom, I, Nylander, T. *Progr. Colloid & Polymer Sci.*, **70**, 62-66, (1985)
23. Mulvihill, DM and Donovan, M. *Irish Journal of Food Science and Technology*, **11**, 43-75, (1987).

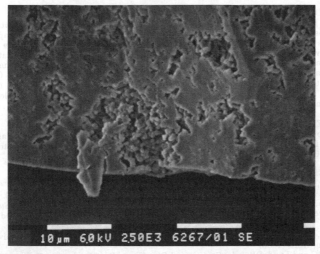

Figure 1. Deposit on stainless steel (AISI 316), 1.0% Protein (WPC 35%), Protein inlet temperature 73°C and outlet temperature 83°C, Oil inlet temperature 97°C and outlet temperature 95°C.

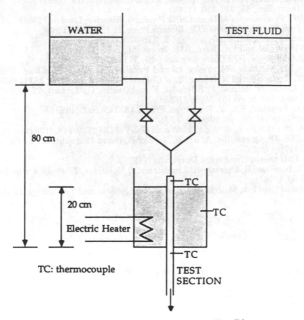

Figure 2. Schematic Diagram of Fouling Rig

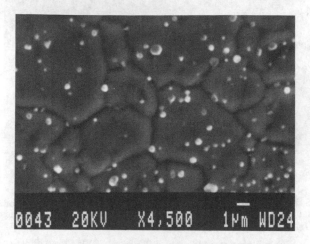

Figure 3. Fouled stainless steel surface (AISI 321) with WPC for 120 seconds
(Temperatures: inlet 73°C and outlet 75°C, wall 96°C)

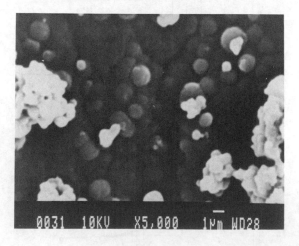

Figure 4. Fouled stainless steel surface (AISI 321) with WPC for 150 seconds
(Temperatures: inlet 73°C and outlet 75°C, wall 96°C)

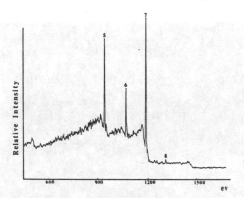

Figure 5. XPS spectra on a fouled stainless steel surface (AISI 321) with WPC for 150 seconds (Temperatures: inlet 73°C and outlet 75°C, wall 96°C) with indication of main characteristic peaks: $5 = O_{1s}$, $6 = N_{1s}$, $7 = C_{1s}$, $8 = S_{2p}$

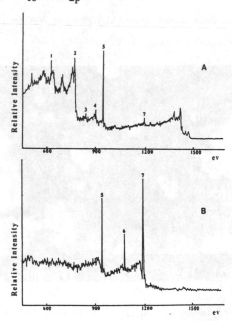

Figure 6. XPS spectra on a clean stainless steel surface (AISI 321) (a) and a fouled stainless steel surface (AISI 321) (b) with WPC for 4 seconds (Temperatures: inlet 73°C and outlet 75°C, wall 96°C) with indication of main characteristic peaks:

$1 = Fe_{Auger}$, $2 = Fe_{2P}$, $3 = Ni_{Auger}$, $4 = Cr_{2P}$, $5 = O_{1s}$, $6 = N_{1s}$, $7 = C_{1s}$.

CHARACTERISATION OF INCINERATOR FOULING

J.H. Howarth *, P. Séguin **, P. Tabaries **, and O.Osborn ***
* NEL, East Kilbride, Glasgow, G75 0QU
** CNIM, BP208, 83507 la Seyne-sur-Mer, CEDEX, France
*** Formerly School of Chemical Engineering, University of Birmingham, UK.

This paper describes aspects of a study of gas-side fouling in a refuse incineration plant. The chemical composition of deposits and fly ash are examined, including SEM/EDAX, XRD and DTA data. The results are used to quantify the active deposition mechanisms. The results indicate that 20–30% of gas-side deposition in incineration plant occurs in the form of non-impactive deposition, including condensation-type phenomena and enhanced inertial impaction of calcium-rich particles.

INTRODUCTION

Studies on the modelling of gas-side fouling have been under way for several years [1], including work on refuse firing, diesel exhaust, pulverised coal [2] and stoker coal firing. The method developed involves theoretical calculation of particle transport to the heat transfer surface taking all active mechanisms into account. Using the measured particulate concentration, the overall theoretical collection efficiency is calculated. Mass deposition rates are measured experimentally, which are used in combination with particulate concentration data to calculate experimental collection efficiencies. The ratio of these two collection efficiencies provides a set of residence probabilities [3] for particulate fouling. These residence probabilities comprise the effects of instantaneous particle retention, and of deposit removal. The P values so obtained are then used in a predictive model of the deposition process.

The above approach requires, however, an quantifying the different particles attributable to different mechanisms or groups of mechanisms. This paper reports the development of a method deployed to quantify the contribution made to deposit mass by non-particulate mechanisms of deposition based on measurements from a refuse incineration plant at Toulon, France. Clearly, if the elemental composition of different size fractions of gas borne particulate are known, along with the composition of deposits, any difference between the two can be ascribed to either disproportionate deposition from particular species, or to non-particulate deposition.

EXPERIMENTAL

An air cooled stainless steel tubular probe with detachable semi-cylindrical sample pieces was used to obtain samples of fouling deposits from a refuse incineration plant at Toulon, France. The thermocoupled sample pieces were oriented at 90° to gas flow normal to gas flow, allowing exposure to the flow of flue gases. The deposit mass was determined by the change in mass from the clean condition.

Sample piece metal temperatures (T) were measured with stainless steel sheathed K-type thermocouples inserted in axially drilled wells. Gas stream temperature (T) was measured with similar thermocouples, and gas stream velocity was measured with a pitot tube. Particulate concentrations were measured with a modified British Coal Utilisation Research Association BCURA cyclone and

CHARACTERISATION OF INCINERATOR FOULING

J.H. Howarth[*], P. Seguin[**], F. Tabaries[**] and G Osborn[***]
* NEL, East Kilbride, Glasgow, G75 0QU
** CNIM, BP208, 83507 la Seyne-sur-Mer, CEDEX, France
*** Formerly School of Chemical Engineering, University of Birmingham, Birmingham, UK.

This paper describes aspects of a study of gas-side fouling in a refuse incineration plant. The chemical composition of deposits and fly ash are examined, including SEM/EDAX, XRD and DTA data. The results are used to quantify the active deposition mechanisms. The results indicate that 20-70% of gas-side deposition in incineration plant occurs in the form of non-impactive deposition, including condensation-type phenomena and enhanced inertial impaction of calcium-rich particles

INTRODUCTION

Studies on the modelling of gas side fouling have been under way for several years (1), including work on refuse firing, diesel exhaust, pulverised coal (2) and stoker coal firing. The method developed involves theoretical calculation of particle transport to the heat transfer surface taking all active mechanisms into account. Using the measured particulate concentration, the overall theoretical collection efficiency is calculated. Mass deposition rates are measured experimentally, which are used, in combination with particulate concentration data, to calculate experimental collection efficiencies. The ratio of these two collection efficiencies provides overall residence probabilities (P) for particulate fouling. These residence probabilities combine the effects of instantaneous particle retention, and of deposit removal. The P values so obtained are then used in a predictive model of the deposition process.

The above approach relies however, on quantifying the deposit fractions attributable to different mechanisms or groups of mechanisms. This paper reports the development of a methodology for quantifying the contribution made to deposit mass by non-particulate mechanisms of deposition, based on measurements from a refuse incineration plant at Toulon, France. Clearly, if the chemical composition of different size fractions of gas-borne particulate are known, along with the composition of deposits, any difference between the two can be ascribed to either disproportionate deposition from particulate species, or to non-particulate deposition.

EXPERIMENTAL

An air cooled stainless steel tubular probe with detachable semi-cylindrical sample pieces was used to obtain samples of fouling deposits from a refuse incineration plant at Toulon, France. The thermocoupled sample pieces were oriented so as to present both upstream and downstream surfaces to the flow of flue gases. The deposit mass was determined by the change in mass from the clean condition.

Sample piece metal temperatures (T_s) were measured with stainless steel sheathed K-type thermocouples inserted in axially drilled wells. Gas stream temperature (T_g) was measured with similar thermocouples, and gas stream velocity was measured with a pitot tube. Particulate concentrations were measured with a modified British Coal Utilisation Research Association (BCURA) cyclone and filter configured for external sampling.

The samples of deposit and gas-borne particulate were examined using a scanning electron microscope (SEM). Chemical compositions were determined using the X-ray microprobe technique either directly with intact fouled sample pieces or using samples of deposit removed from the shells. Further investigations were made using X-ray diffraction (XRD), and differential thermal analysis with thermo-gravimetry (DTA). Details of these techniques have been published previously (3).

RESULTS

The mass gain test data, including velocity, particulate concentration, mass gain rate and temperatures are given in TABLE 1. These data are used to calculate experimental collection efficiencies, defined as the ratio of deposition flux to mass flux. Collection efficiency is plotted against T_g in FIGURE 1. A typical DTA/TG trace is shown in FIGURE 2. The composition of deposits and of BCURA cyclone and filter particulate samples are given in TABLE 2 and FIGURE 3(A-E).

DISCUSSION

The relationship between experimental collection efficiency and T_g, given in FIGURE 1, suggests a probable change in the deposition process at about 700°C. This relates to the endotherms in FIGURE 2, which show that melting behaviour occurs at similar temperatures. The following discussion tries to reach an understanding of these processes and to quantify them for future modelling studies.

The deposition processes operative in gas-side fouling can be grouped as a first approximation into impactive and non-impactive processes. Non-impactive processes will include small particle thermophoresis and diffusion processes (both Brownian and eddy), vapour condensation and gas/solid reactions. The solids collected by the BCURA cyclone impactor are greater than 4 μm in size, and are considered to represent the material available for impaction in the gas stream; the filter mass represents that part of the particulate mass available for non-impactive deposition.

The contributions to the deposit mass made by these processes may be approximately quantified using the data for chemical composition of the different solid samples collected. The observations which form the basis for these calculations are as follows: (see FIGURE 3)

- The alumino-silicate content in the deposits is always much lower than in the BCURA cyclone dust;

- Conversely, S, Cl and K are consistently much higher in the deposits than in the cyclone dust.

- The Ca content of the cyclone dust and the upstream deposits are generally very similar, whereas they are much lower in the downstream deposits.

- The filter material consists of solidified lumps of condensed salts. These are rich in S, Cl and K. They contain very little Ca or alumino-silicate.

The deductions made from these observations are given below.

a)   >4μm dust forms only a part of the deposits, both upstream and downstream. Therefore impactive processes account for only a part of the overall deposition.

b)   The deposits therefore contain material deposited by processes other than simple inertial impaction. This fraction of the deposits can be determined by calculation of the enrichment of each component relative to the composition of the cyclone dust, assuming that SiO_2 in the deposit arrives only by impaction.

c)   The source of this enrichment could be:

> non-impactive particulate mechanisms;
> vapour deposition;
> neutralisation-type reactions between SO_2 and HCl, and bases in the deposit;
> enhanced P values for certain types of particle in the $>4\mu m$ dust which are high in the enriched components.

d)   The absence of Ca-rich small particles in the filter analyses suggests that the Ca enrichment in the upstream deposits is due to enhanced impaction or retention of Ca from the $>4\mu m$ dust. The lack of Ca enrichment in the downstream deposits supports this idea, since *simple* inertial impaction will not be active at this surface.

The causes of enhanced retention of Ca-rich particles on the upstream surface of the deposits is less easily explained. It may result from the presence of soft or molten Ca-rich particles in flight, or from enhanced retention due to shape factors. XRD investigation of the deposits has identified several phases. They possibly include a $Ca_xO_yCl_z$ phase, which may be involved in these capture processes.

e)   Other components besides Ca are enriched in both the upstream and downstream deposits. The presence of formerly liquid material in the filter samples suggests that condensable material could account for at least part of the non-impactive enrichments. K, S and Cl dominate the filter material, and this correlates well with the remaining components of the enrichment. It seems therefore that the enrichment is from two sources, namely enhanced inertial impaction and some other non-impactive process. These non-impactive process are likely to be operative downstream as well as upstream. In the absence therefore of impactive Ca enrichment at the downstream surface, Ca should be depleted in downstream deposits relative to the Ca in the cyclone dust, due to the addition of these other components. This is indeed the case.

f)   A part of the enriched deposit fraction is therefore definitely not impactive and is labelled "condensation" for the present. The most suitable driving force for modelling such processes is $(T_g-T_s)/T_s$, which is shown in FIGURE 4 to give the expected physical trend. It is estimated that between 0.4 and 5.2 ppm KCl would be needed in the furnace to account for this "condensation", assuming 100% capture efficiency.

Further work will permit better quantification of the processes described, and also give information on the rate of gas-solid reactions between the deposits and gaseous species.

CONCLUSION

Deposition in refuse incineration has been quantified, and shown to have a relationship to T_g, with an increase in deposition rate at approximately 700°C, which accords well with the observed melting behaviour of the deposited material.

The deposition process has been shown to result from a complex set of processes, including inertial impaction, enhanced retention of inertially impacted particles rich in Ca, small particle deposition and "condensation" deposition.

Acknowledgements

This research is financed in part by the Commission of the European Communities within the frame of the Rational Use of Energy category of the JOULE R & D Programme. It is published with the approval of the Chief Executive, NEL.

References

1   Glen N.F. and Howarth J.H., 1988. 2nd National UK Heat Transfer Conference,1, 401-420.

2   Glen N.F. and Howarth J.H., 1989, In "Deposition from combustion gases". IOP Short Meetings Series No 23. Institute of Physics, 33-48.

3   Ewart W.R., 1988. 2nd National UK Heat Transfer Conference,1, 421-432.

TABLE 1 - Mass gain test data

Test name	Temperatures			Time	Deposit mass			Dust concentration	Velocity
	gas	metal u/s*	d/s#		d/s	u/s	total		
	°C	°C	°C	hours	mg	mg	mg	kgm^{-3}	ms^{-1}
T-1-01	473	332	316	5.11	99.2	64.5	163.7	0.390	11
T-1-02	490	403	402	4.30	169.2	130.0	299.2	0.381	11
T-1-03	498	415	410	6.86	256.0	171.3	427.3	0.377	11
T-2-01	721	435	430	4.84	1003.4	283.2	1286.6	0.350	15
T-2-02	737	389	354	3.68	1212.5	177.1	1389.6	0.344	15
T-2-03	738	432	381	1.27	194.9	90.4	285.3	0.344	15
T-2-04	738	415	352	1.21	321.3	62.3	383.6	0.344	15
T-2-05	725	521	505	1.95	678.6	103.6	782.2	0.348	15
T-2-06	719	320	252	4.42	1297.2	187.1	1484.3	0.350	15
T-2-07	661	331	249	1.51	139.7	51.9	191.6	0.372	15
T-2-08	732	526	479	7.77	3793.5	478.4	4271.9	0.346	15
T-3-01	326	142	131	2.09	30.4	30.6	61.0	0.724	10
T-3-02	320	180	182	4.25	55.3	46.3	101.6	0.731	10
T-4-01	393	182	145	2.55	39.6	35.4	75.0	0.210	10
T-4-02	284	188	194	2.14	14.1	11.5	25.6	0.440	10
T-4-03	392	290	255	4.14	45.8	37.7	83.5	0.210	10
T-5-01	747	564	527	2.92	1847.0	573.8	2420.8	0.819	14
T-5-02	789	580	597	4.91	2814.4	533.8	3348.2	0.787	14
T-5-03	835	601	593	0.39	202.9	78.2	281.1	0.754	14
T-5-04	818	632	636	2.00	1217.7	227.5	1445.2	0.766	14
T-6-01	472	364	358	2.15	61.5	26.7	88.2		

* u/s = upstream   # d/s = downstream

TABLE 2 - Chemical composition of deposits and particulate material (% oxides)

Test	Na	Mg	Al	Si	P	S	Cl	K	Ca	Ti	Cr	Fe	Ni	Zn	Pb
Upstream deposits															
T-2-01	1	0	0	2	0	16	20	9	39	0	0	8	0	3	1
T-2-02	1	0	3	5	0	13	27	11	30	0	0	7	0	2	0
T-2-03	0	0	3	8	0	16	24	10	27	0	0	8	0	3	0
T-2-04	0	0	10	7	0	10	25	9	32	0	0	7	0	2	0
T-2-06	0	0	3	5	0	11	25	14	37	0	0	2	0	3	0
T-2-07	1	0	3	3	0	10	30	21	21	0	0	3	0	4	3
T-2-08	0	0	3	6	0	16	18	9	44	0	0	2	0	2	0
T-3-01	2	0	3	7	0	13	28	18	23	0	0	3	0	3	1
T-4-01	4	0	2	5	0	10	30	22	12	0	0	3	0	6	5
T-4-02	0	0	5	11	0	7	25	21	23	0	0	3	0	4	0
T-4-03	8	0	2	4	0	21	23	26	8	0	0	4	0	3	2
T-5-01	0	0	1	6	0	15	19	9	45	0	0	2	0	2	0
T-5-02	0	0	3	7	0	16	18	8	45	0	0	2	0	1	0
T-5-03	0	0	6	7	0	13	25	10	34	0	0	3	0	1	0
T-5-04	0	0	3	5	2	19	18	8	41	0	0	3	0	1	0
Average			3	6		14	24	14	31					3	1
Downstream deposits															
T-2-01	0	0	0	5	0	25	24	13	8	0	0	10	0	8	7
T-2-02	1	0	0	4	0	12	30	18	12	0	0	8	0	6	7
T-2-03	0	0	0	6	0	18	30	12	14	0	0	13	0	5	2
T-5-02	0	0	3	6	0	24	28	16	17	0	0	4	0	2	0
Average			1	5		20	28	15	13					5	4
BCURA cyclone															
T-1	1	3	13	24	2	5	6	4	34	2		4		2	
T-2	1	3	13	24	2	6	5	5	29	2		7		4	
T-3	1	3	13	22	2	5	7	5	36	3		3		3	
T-4	1	3	12	21	2	7	8	5	35	2		3		3	
T-5	1	3	13	22	2	5	3	2	45	2		2		2	
BCURA filter															
T-2(02)	3	0	2	3	0	12	25	32	2	0	1	19			
T-2(03)	4	0	4	7	0	11	24	29	2	0	1	17			

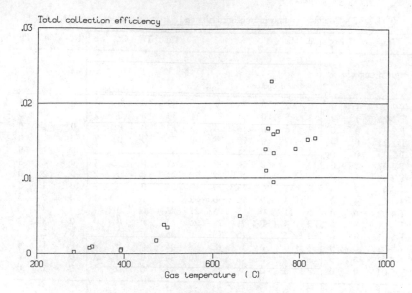

Fig. 1 Total collection efficiency versus gas temperature

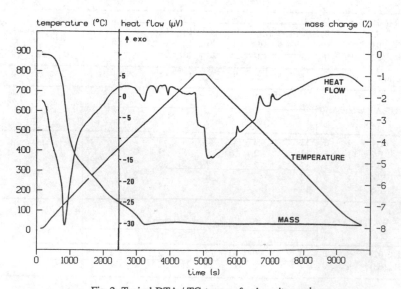

Fig. 2 Typical DTA / TG traces of a deposit sample

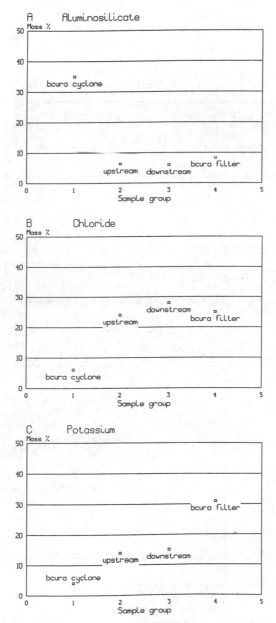

Fig. 3 Average composition of different solid sample groups

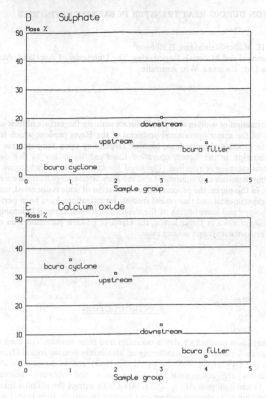

Fig. 3 (cont.) Average composition of different solid sample groups

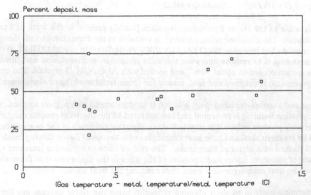

Fig. 4 Condensation upstream deposit mass versus temperature driving force

SCALE FORMATION DURING HEAT TRANSFER IN BAUXITE REFINERIES

M. Jamialahmadi, H. Müller-Steinhagen, B Robson[1]
Department of Chemical and Materials Engineering, The University of Auckland, Auckland, New Zealand
[1]Alcoa of Australia Ltd., Kwinana, WA, Australia

The formation of sodium aluminium silicate scale on the surface of heat exchangers is one of the major operational problems of the Bayer process, which is generally used to produce alumina from bauxite. Depending upon temperatures and silica concentration in the liquor, operation times can be as low as five days before chemical cleaning is required. The need for stand-by heat exchangers, costs for cleaning chemicals and equipment and the associated energy losses are a major cost factor. In this paper, the process related formation of scale is discussed, using results from experimental and theoretical studies. Investigations have been performed on morphology and composition of the deposit, thermodynamic relationships for the solubility of silica in Bayer liquor, the effect of process parameters on clean heat transfer coefficients and scaling rates.

1. INTRODUCTION

The majority of aluminium produced today is manufactured from bauxite. The term bauxite refers to an ore or mixture of minerals formed by the weathering of aluminium-bearing rocks. Hydrated alumina occurs in bauxite as gibbsite ($Al_2O_3 \cdot 3H_2O$), bohmite ($\gamma\ Al_2O_3 \cdot H_2O$) and as diaspore ($\alpha\ Al_2O_3 \cdot H_2O$). The main impurities of bauxites are compounds of silicon, iron and titanium. Silicon occurs as kaolinite $Al_2O_3 \cdot 2SiO_2 \cdot 2H_2O$ and halloysite $Al_2O_3 \cdot 2SiO_2 \cdot 4H_2O$. To extract the alumina from the bauxite, the Bayer process is used. The bauxite is blended and ground to particles usually finer than 1 mm and then digested in a strong caustic soda solution between 150°C and 180°C. The digestion reaction is:

$$Al_2O_3 \cdot 3H_2O + 2NaOH \rightarrow 2NaAlO_2 + 4H_2O \tag{1}$$

During the digestion in the Bayer liquor, reactive silica (usually present in the form of kaolinite) as well as alumina will dissolve. The dissolved silica, however, is unstable in the Bayer liquor and, therefore, precipitates to an equilibrium level as insoluble $Na_2O \cdot Al_2O_3 \cdot 2SiO_2$ or desilication product (DSP). With the alumina in solution, the next step is to remove and wash the solid residue by sedimentation and filtration. The diluted clean solution is then cooled to about 70°C, and seeded with $Al_2O_3 \cdot 3H_2O$ crystals. The coarser fractions of the resulting precipitate is calcined and sold, mainly for commercial smelting to aluminium.

The cool, dilute and alumina-depleted spent solution is supersaturated with respect to silica. This spent liquor from the precipitation building is re-heated and concentrated to the digestion process conditions in a counter-current evaporation process before it is recycled to the digestion process. In the evaporator, the spent liquor is first heated by steam in shell and tube heat exchangers with the spent liquor flowing inside the tubes. Then the solution is flashed in a series of flash tanks. The rate of silica precipitation increases exponentially with the temperature of the spent liquor and a portion of the silica in the liquor reacts to form desilication product (DSP) scale on the heat exchanger tube walls, thus reducing the heat transfer rates.

The major operating costs in modern alumina refineries using the Bayer process are the costs of energy. Therefore, regular cleaning of heat exchangers is required, often in less than bi-weekly intervals. The scale is removed from the fouled tubes with inhibited sulphuric acid solutions. The acid solutions attack the heater tubes as well as the scale on the tube walls, requiring frequent tube replacement.

The serious effect of scale formation on the performance of heat exchangers in the Bayer process has long been appreciated by all who are involved with the aluminium industry. It may be safely said that scale formation is the most important factor in design and operation of the evaporation building of the Bayer process. A thorough understanding of the fouling kinetics and the effects of operational parameters on the fouling behaviour of silica is required to improve operation and design of the shell and tube heat exchangers which are extensively used. Therefore, extensive experimental and theoretical investigations have been performed by Alcoa of Australia Ltd. in co-operation with the Department of Chemical and Materials Engineering, University of Auckland.

2. EXPERIMENTAL EQUIPMENT AND PROCEDURE

2.1 Test Rig

The test apparatus used for the present investigations is described in detail, elsewhere [1]. It consists of a closed, temperature controlled flow loop with two parallel test sections. Three types of test heaters have been used:

i) an electrically heated cylindrical heating rod, mounted concentrically within the surrounding pipe. The dimensions of the test section are: diameter of heating rod = 10.7 mm, annular gap = 14.73 mm, length of heating rod = 400 mm, length of heated section = 100 mm

ii) a similar design, but with 2.5 mm x 2.5 mm stainless steel wire clippings fluidized in the annular space, to simulate a fluidized bed heat exchanger [2]. Modifications have been made to the design to avoid particle carry-over.

iii) a 25 mm steel pipe with an electrical resistance heater soldered to the pipe outside. The heated length is 160 mm.

2.2 Experimental Procedure and Data Reduction

Spent Bayer liquor has been obtained in 200 litre drums from an Alcoa of Australia refinery.

The range of the experimental parameters is shown in Table 1.

Flow velocity	0.1 m/s < u < 1.5 m/s
Heat flux density	$10000 \ W/m^2 < \dot{q} < 400000 \ W/m^2$
Bulk temperature	$30°C < T_b < 110°C$
System pressure	105 kPa < p < 350 kPa

Table 1 Range of Parameters

The fouling resistance on the liquor side is calculated from

$$R_f = \frac{1}{\alpha} - \frac{1}{\alpha_o} \qquad (2)$$

with α_o being the heat transfer coefficient at the beginning (i.e. t=0) and α being the coefficient at some later time when fouling has taken place. The local heat transfer coefficient is defined as:

$$\alpha = \frac{\dot{q}}{T_w - T_b} \tag{3}$$

3. EXPERIMENTAL RESULTS AND DISCUSSION

3.1 Appearance of Deposits

The tough and adherent deposit has a characteristic rippled surface, which has also been observed for fouling from geothermal waters [3].

3.2 Fouling Resistance

Figs. 1a) - 1c) show the measured fouling resistances as a function of time and heat flux. The scatter in these Figures is partly caused by the complexity of the deposition process itself, the rippled nature of the deposit and some minor fluctuations in power supply and bulk temperature. The initial increase in heat transfer is thought to be due to the increased roughness of the rippled silica deposit. A rippled surface increases the turbulence level in the flow zone near the surface of the heating element and, therefore, improves heat transfer until the insulating effect of the growing deposit becomes dominant.

After the initial improvement at the beginning of the experiments, the experimental results for silica concentrations of 0.6 g/l and 1.2 g/l show an almost linear increase with respect to time. A linear relationship is generally characteristic of tough, hard, adherent deposits and indicates that the deposition rate is constant and that there is no removal. These are also the characteristics of the silica scale observed at the end of each run on the surface of the heating element. The scale could not be removed solely by acid wash, therefore, abrasive paper was used.

Those fouling runs which were performed at a silica concentration of 1.6 g/l do not conform to the almost linear fouling curve discussed above. The fouling rate shows a strongly decreasing behaviour after an initially linear period. This type of fouling curve is characteristic for particulate fouling, where strong removal mechanisms are present. The formation of particles for this concentration must occur by bulk precipitation.

The extent of the reduction in heat transfer coefficient with time and the mechanism of the fouling are strongly affected by the concentration of dissolved silica and the temperature driving force between the heat transfer surface and bulk fluid. Maximum reduction in the heat transfer coefficient was obtained

- for the highest silica concentration

- the maximum temperature difference between heat transfer surface and fluid bulk

- the lowest liquid flow rate.

3.3 Fluidized Bed Test Section

To investigate the possible application of fluidized bed heat exchangers, experiments have been performed with a stationary fluidized bed using stainless steel wire clippings as fluidized particles. A typical experiment is shown in Fig. 3, comparing the performance of the fluidized bed test section with that of the parallel empty test section of identical geometry. The heat transfer coefficient in the empty test section drops rapidly within the first 300 minutes of the experiment, due to the formation of sodium aluminium silicate on the heat transfer surface. Due to the abrasive action of the steel particles, the fluidized bed test section indicated hardly any fouling-related drop in heat transfer. This result was confirmed by visual inspection of the two heaters, after the experiment: the heating rod installed in the fluidized bed was essentially clean and shiny, while the other heating rod was covered by tough, adherent silica scale which had to be removed mechanically. An attempt to remove the deposit by installing the fouled heater in the fluidized bed did not show any beneficial results within two days of operation.

4. MODELLING OF DSP SCALE FORMATION

The precipitation reaction of the DSP scale on the heat transfer surfaces can be expressed by the following reaction:

$$2Na_2SiO_3 + 2NaAlO_2 + 2H_2O \rightarrow DSP\downarrow + 4NaOH \quad with \quad \Delta H_r < 0 \tag{4}$$

For typical Bayer liquors, the rate equation of the above reaction is of second order with respect to the degree of silica supersaturation driving force:

$$\dot{n} = \frac{dc_s}{dt} = -K_s(c_b - c_s^*)^2 \tag{5}$$

The kinetic constant K_s is a function of temperature. The silica solubility c_s^* can be calculated as a function of temperature, free caustic soda and alumina concentration of the liquor. From a mass balance around a differential pipe segment one obtains:

$$\frac{d_t}{d_i} = \exp\left(\frac{\dot{n}}{2\rho_s}t\right) \tag{6}$$

with d_i being the clean (initial) pipe diameter and d_t being the pipe diameter after a certain time of operation. Assuming constant thermal conductivity and density of the deposit, the fouling resistance is related to the mass of deposit per unit heat transfer area according to eq. (7):

$$R_f = \frac{m}{\rho_s \lambda_s} \approx \frac{d_t - d_i}{2\lambda_s} \tag{7}$$

Combining equations (6) and (7):

$$R_f = \frac{d_i}{4\lambda_s}\left(1 - \exp(\frac{\dot{n}}{\rho_s}t)\right) \tag{8}$$

The kinetic constant K_s was obtained by curve-fitting the experimental data for $R_f > 0$ m^{2}K/W:

$$K_s = \exp\left(10.09727 - \frac{6164.036}{T}\right) \tag{9}$$

Fig. 4 shows that equations (5)-(9) predict the measured fouling rates satisfactorily. The model can not be used to predict the effects of surface roughness on the initial part of the fouling resistance vs. time curves. For the silica concentration of 1.6 g/l, the measured fouling resistances are considerably higher than the predicted values, indicating that particulate fouling occurs in addition to chemical reaction fouling.

5. CONCLUSIONS

Scaling rates from Bayer process liquor with silica concentrations below or equal to 1.2 g/l increase with increasing heat transfer surface temperature and silica concentration, but are almost independent of flow velocity. For silica concentrations above and equal to 1.6 g/l, particulate deposition contributes to the fouling process, causing the fouling curve to shift from linear to asymptotic shape. The asymptotic fouling resistance decreases with increasing flow velocity.

A model has been formulated for the prediction of deposition rates from first principles. The agreement between measured and predicted fouling resistances is excellent, as long as the scale formation is caused by chemical reaction at the heat transfer surface, i.e. SiO_2 concentration is below or equal to 1.2 g/l

Scale formation in our laboratory apparatus could be prevented by installing a fluidized bed.

6. SYMBOLS

c_b	silica concentration, kg/m^3
$c_{s_\bullet}$	silica solubility, kg/m^3
c_s	silica solubility at surface temperature, kg/m^3
d_i	tube inside diameter, m
d_t	diameter at time t, m
K_s	reaction constant, $m^3/kg/min$
m	mass of deposit per unit area of the clean tube, kg/m^2
$\dot{n}$	fouling rate, $kg/m^3/min$
q	heat flux, W/m
R_f	fouling resistance, $m^2 \cdot K/W$
t	time, min
T_b	bulk temperature, K
T_w	wall temperature, K
U	overall heat transfer coefficient, W/m^2K
α	heat transfer coefficient, W/m2K
ΔH_r	heat of reaction, J/mol
ρ_s	density of the deposit kg/m^3
λ_s	thermal conductivity of the deposit, W/m/K
ε	voidage

7. REFERENCES

1. Jamialahmadi, M. and Müller-Steinhagen, H.M.: Convective and Subcooled Boiling Heat Transfer to BAYER Process Liquor. Light Metals 1992, pp. 141-150.

2. Klaren ,D.G.: The fluid bed heat exchanger: Principles and modes of operation and heat transfer results under severe fouling conditions. Fouling Prev. Res. Dig., vol.5, No.1, March 1983.

3. Bott, T.R., Gudmundsson, J.S.: Rippled Silica Deposits in Heat Exchanger Tubes. Proc. 6th Int. Heat Transfer Conf., 4, 373, Hemisphere Publ. Corp., 1978.

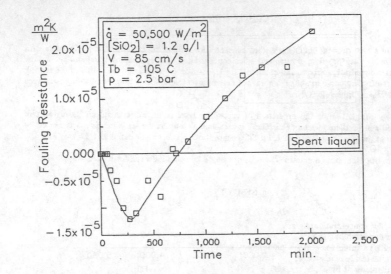

Figure 1a)   Fouling resistance as a function of time for a silica concentration of 1.2 g/l.
$\dot{q} = 50,500 \ W/m^2$. $T_s = 116°C$

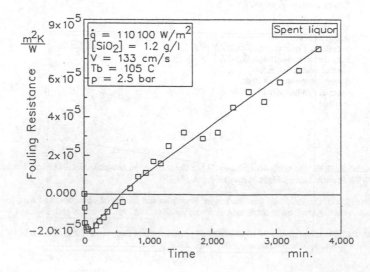

Figure 1b)   Fouling resistance as a function of time for a silica concentration of 1.2 g/l.
$\dot{q} = 110,100 \ W/m^2$. $T_s = 121°C$

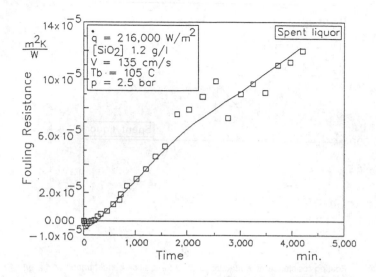

Figure 1c)   Fouling resistance as a function of time for a silica concentration of 1.2 g/l.
q = 216,000 W/m². T_s = 136°C

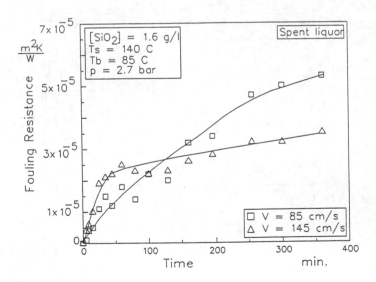

Figure 2   Effect of liquor flow velocity on the fouling behaviour for a silica concentration of
1.6 g/l

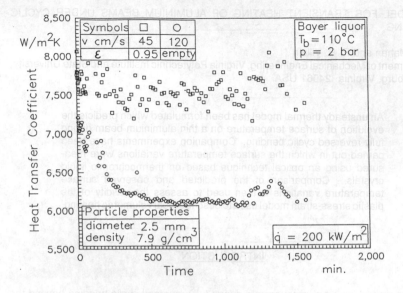

Figure 3   Heat transfer fouling in the fluidized bed section and in the empty test section

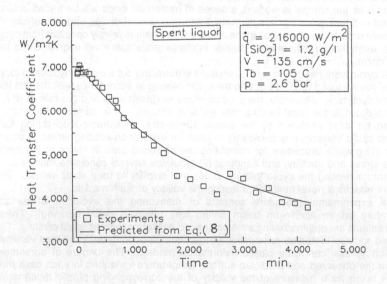

Figure 4   Measured and predicted heat transfer coefficients for a silica concentration of 1.2 g/l.

A MODEL FOR TRANSIENT HEATING OF ALUMINIUM BEAMS UNDER CYCLIC BENDING

J. R. Mahan and F. R. Villain
Department of Mechanical Engineering, Virginia Polytechnic Institute and State University, Blacksburg, Virginia 24061 USA.

An unsteady thermal model has been formulated which predicts the evolution of surface temperature on a thin aluminium beam during fully reversed cyclic bending. Companion experiments have been carried out in which the surface temperature variations were measured using an optical technique based on thermochromic liquid crystals. Comparison of the predicted and observed surface temperature variations is then used to assess the validity of the plastic stress-strain model used to describe local volumetric heating.

INTRODUCTION

The net work done on a metal sample during fully reversed cyclic bending beyond the elastic limit can be divided into two parts: that which permanently damages the sample, and that which goes into raising its temperature. If a plot is made of local stress versus local strain as the sample is worked, a series of hysteresis loops will be traced out, as illustrated in figure 1(a). For a sufficiently limited number of cycles and for sufficiently small strain amplitudes, the shape of these loops will remain relatively constant from cycle to cycle, even though their relative position in stress-strain space will migrate along the fatigue curve.

The cyclic work per unit volume associated with tracing out a closed hysteresis loop, given by the enclosed area, is equal to the cyclic heating in accordance with the first law of thermodynamics. Therefore, the product of the enclosed area and the rate at which it is traced out is the local heating rate per unit volume in the sample. In order to calculate the area enclosed by the plastic stress-strain hysteresis loop during fully reversed cyclic bending, it is necessary to have an expression relating stress to strain. Sandor (1) gives a procedure for computing the enclosed area in terms of the cyclic fracture stress and ductility, and Landgraf (2) tabulates several candidate models from the literature relating the cyclic fracture stress and ductility to their static values. The paper introduces a novel method for testing the validity of such models.

The experimental procedure consists of measuring the evolution of surface temperature on an aluminium beam during fully reversed cyclic bending. These measurements are obtained using a method based on thermochromic liquid crystals. The observed surface temperature variation is then compared with the predicted variation based on several candidate plastic deformation models. The degree of agreement between the observed and predicted surface temperature variations in each case may then be taken as a measure of the validity of the corresponding plastic deformation model. In the present paper the technique is applied to a single aluminium alloy (6061-T6) which has well documented mechanical and thermal properties.

ANALYSIS

The system to be analyzed consists of a flat beam of thickness $2b$, width w, and free length L, which is rigidly clamped at one end ($x = 0$). Initially the beam is in thermal equilibrium with the surrounding air at T_∞, but at time $t = 0$, its free end ($x = L$) is subjected to a sinusoidal transverse displacement. This results in local heating in the beam and a subsequent increase in local temperature.

The differential equation describing the evolution of temperature on the surface of the beam may be derived by considering an energy balance on a differential volume element of length dx, thickness $2b$, and width w. The result is a version of the unsteady "fin" equation modified to include a local volumetric source term $q(x)$,

$$k \frac{\partial^2 \theta(x,t)}{\partial x^2} - \frac{h}{b} \theta(x,t) + q(x) = \rho c \frac{\partial \theta(x,t)}{\partial t} , \qquad (1)$$

where $\theta(x,t) = T(x,t) - T_\infty$. The assumption of one-dimensional heat conduction implied by equation (1) is consistent with a thin beam, a high thermal conductivity, and a low thermal conductance at the surface/air interface, i.e., $hb/k \ll 1$. In solving equation (1) it is assumed that the clamped end is insulated, that the beam is initially at the temperature of the surroundings, and that heat is lost by convection from the beam surfaces, including the free end; that is,

$$\partial\theta(0,t)/\partial x = \theta(x,0) = 0, \text{ and } \partial\theta(L,t)/\partial x = h\theta(L,t)/k .$$

Once the source term $q(x)$ has been specified, the standard Crank-Nicolson finite difference method is used to solve equation (1) subject to these initial and boundary conditions.

Equation (1) requires that the variation of local heating be known as a function of position x along the beam. While the thermoelastic effect does produce a local cyclic temperature oscillation which may in principle lead to heat dissipation, this effect will be negligible compared to plastic working of the sample if the cyclic frequency of the fully reversed bending is sufficiently high. In this case the thermoelastic contribution to the local temperature variation will appear as an oscillation superimposed on the relatively slowly evolving local temperature increase provoked by plastic working. Then the associated heat conducted out of a given volume element due to the thermoelastic effect during one half-cycle will be balanced by a like amount of heat conducted in during the next half-cycle, so that only a cyclic *plastic* stress-strain model is needed to define q(x).

According to Sandor (1), if only the plastic (nonlinear) part of the strain is considered, the hysteresis loop corresponding to one cycle appears as shown in figure 1(b). The area enclosed by this loop is identical to that enclosed by a stress-strain loop which includes elastic strain, because this latter contributes no *net* area. Then the local volumetric work performed during one cycle is

$$W(x,y) = 2 \int_0^{\Delta\varepsilon_p(x,y)} \sigma(x,y) d\varepsilon_p(x,y) , \qquad (2)$$

where the symbols are defined in figure 1(b). According to Sandor's model the nonlinear part of the stress in figure 1(b) and equation (2) may be written

$$\sigma(x,y) = K [\varepsilon_p(x,y)]^n , \qquad (3)$$

where K and n are material constants. In writing equation (2) and the subsequent equations it is recognized that this work is performed uniformly throughout the width (z-coordinate) of the beam. Substitution of equation (3) into equation (2) and integration yields

$$W(x,y) = \frac{2}{n+1} \, \sigma_a(x,y) \, \Delta\varepsilon_p(x,y) \ . \tag{4}$$

This result, formulated for static loading, can be extended to the case of cyclic loading by replacing the static stress amplitude σ_a and the static plastic strain range $\Delta\varepsilon_p$ with their corresponding cyclic values. Then indicating the cyclic quantities with primes (') and multiplying equation (4) by N, the number of bending cycles per unit time, the local volumetric heating rate is given by

$$Q(x,y) = \frac{2N}{n'+1} \, \sigma_a'(x,y) \, \Delta\varepsilon_p'(x,y) \ . \tag{5}$$

Equation (3) can also be used to relate the cyclic stress amplitude and the cyclic plastic strain range to the cyclic fracture ductility and the cyclic fracture stress,

$$\Delta\varepsilon_p'(x,y) = \varepsilon_f' \, [\sigma_a'(x,y)/\sigma_f']^{1/n'} \ , \tag{6}$$

so that the local volumetric heating rate is finally given by

$$Q(x,y) = \frac{2N}{n'+1} \, \sigma_a'(x,y) \, \varepsilon_f' \, [\sigma_a'(x,y)/\sigma_f']^{1/n'} \ . \tag{7}$$

The values of the cyclic quantities might reasonably be expected to change as the sample is worked; thus, the values used should be thought of as mean values over the duration of the experiment.

In the experiment the free end of the cantilever beam is driven sinusoidally at a specified displacement amplitude rather than with a specified force amplitude. Therefore, in order to evaluate the local volumetric heat dissipation rate using equation (7), the cyclic stress amplitude must be related to the displacement amplitude.

The local stress in a cantilevel beam loaded at its free end by a force F is

$$\sigma(x,y) = \frac{My}{I} = \frac{F(L-x)y}{I} \ , \tag{8}$$

and the corresponding elastic deflection at $x = L$ would be

$$v(L) = - FL^3/3EI \ . \tag{9}$$

The importance of plastic effects diminishes with distance from the clamped end such that plastic effects contribute to the overall behavior of the beam only near the clamped end. Therefore, the actual deflection of the beam for most practical situations is well estimated by equation (9) even in the presence of local plastic deformation. Then eliminating the force F between equations (8) and (9) and replacing $\sigma(x,y)$ by $\sigma_a'(x,y)$ and $v(L)$ by $v_a(L)$, the amplitude of the sinusoidal deflection at $x = L$, there results

$$\sigma_a'(x,y) = - \frac{3EI v_a(L)(L-x)}{L^3} \, y \ . \tag{10}$$

The heat source term in equation (1) can now be obtained by substituting equation (10) into equation (7) and integrating the result over the beam thickness. When this is done there results

$$q(x) = \frac{2N}{n'+1} \frac{1}{2+1/n'} \frac{\varepsilon_f'}{(\sigma_f')^{1/n'}} \frac{b}{2} [3bEv_a(L)(L-x)/2L^3]^{1+1/n'} . \tag{11}$$

EXPERIMENT

The test apparatus consists of a variable-speed electric motor mounted on a baseplate. Also mounted on the baseplate is a steel clamp whose jaws are faced with bakelite to provide thermal insulation. This clamp holds rigidly one end of the sample to be tested. An eccentric on the motor shaft provides oscillatory motion to a rigid link whose pivoted end is attached to the "free" end of the test sample. This arrangement provides fully reversed cyclic bending of the sample at a fixed amplitude and frequency.

The evolution of the beam surface temperature is monitored using an optical technique, based on thermochromic liquid crystals, perfected by Hippensteele and his coworkers (3) at NASA's Lewis Research Center. Briefly, the technique exploits the fact that a surface coated with a thin layer of cholesteric liquid crystals and illuminated by white light will reflect light whose colour depends on the local surface temperature. The variation of wavelength with temperature is highly nonlinear such that certain wavelengths can be reflected in only very narrow temperature ranges. In the case of the liquid crystals used in the current work, the yellow colour band and the black/red transition, because of their relative sharpness, are especially sensitive indicators of temperature. The entire upper surface of the beam is coated with a thin layer of liquid crystals, and three 0.25-mm diameter copper-constantan thermocouples are spot welded on the centerline of the upper surface 4.3 (0.17 in.), 32.5 (1.28 in.) and 89.8 mm (3.54 in.) from the clamped end. The thermocouples are used for in situ calibration of the liquid crystals.

After the aluminium beam has been installed in the test apparatus, the drive motor and a microcomputer-driven data acquisition system are then started simultaneously. The data acquisition system records the thermocouple readings at time intervals ranging from two to five seconds and signals the experimenter to trip the camera used to record the instantaneous colour pattern of light reflected from the liquid crystal layer on the upper surface of the beam. The motor speed is measured using a stroboscopic tachometer.

RESULTS

The measured and predicted temperature profiles for a typical experimental run are shown in figure 2. In this case a 6061-T6 mill-finish aluminium beam 2.0 mm (0.080 in.) thick, 101.6 mm (4.0 in.) wide and 106.7 mm (4.2 in.) long was subjected to a 10 mm (0.39 in.) deflection amplitude at a nominal frequency of 15 Hz. The mass density, thermal conductivity, and specific heat of the sample were taken to be 2715.1 kg/m^3, 365.0 W/m·°C and 962.0 W/kg·°C, respectively. The modulus of elasticity was taken to be 68.94 GPa, the static true fracture (σ_f) 61.6MPa, and the static true ductility (ε_f) 0.43. These are handbook values for 6061-T6 mill-finish aluminium.

The nominal heat transfer coefficient was independently estimated to be 6.0

$W/m^2 \cdot °C$, although a value of 16.5 $W/m^2 \cdot °C$ gives the best agreement between theory and experiment. The heat transfer coefficient would be expected to vary somewhat along the beam because the local amplitude of the beam motion increases from the clamped end to the free end. However, figure 2 shows that the results are not very sensitive to heat transfer coefficient, and so no attempt has been made to consider this variation. Sandor's suggested value of 0.11 was used for the cyclic strain hardening exponent n'. A parametric study showed that the model is very sensitive to the value assumed for this exponent. It turns out that very simple models could be used for the cyclic true fracture stress and the cyclic true ductility: for the former Halford and Morrow (4) suggest $\sigma_f' = \sigma_f$, and for the latter Morrow (5) suggests $\varepsilon_f' = \varepsilon_f$. Several other more complex models for these quantities were obtained from the literature and tested. However, they either produced worse agreement between theory and experiment or else no significant improvement was obtained using them. The second set of curves (dashed lines) in figure 2 represent improved agreement between theory and experiment obtained by adjusting the heat transfer coefficient and nominal frequency within the uncertainties of their values.

CONCLUSION

An unsteady thermal model has been formulated, based on plastic deformation models from the literature, capable of predicting the evolution of temperature along an aluminium beam undergoing fully reversed cyclic bending. The technique introduced in this paper can be thought of as a first step in developing a method for evaluating candidate plastic deformation models. The technique is relatively insensitive to the heat transfer coefficient, whose value is not well known, but is very sensitive for determining the cyclic strain hardening coefficient.

ACKNOWLEDGEMENTS

The authors are indebted to the Structural Acoustics Branch of the Acoustics Division at NASA's Langley Research Center for their partial support of this work under NASA Contract NAS1-18461-3. A debt of gratitude is also owed to Hallcrest Manufacturing Company for its generosity in providing the liquid crystals used in this work.

REFERENCES

1. Sandor, B. I., 1972, Fundamentals of Cyclic Stress and Strain, pp. 58-61, The University of Wisconsin Press, Madison.
2. Landgraf, R. W., 1970, ASTM STP 467, pp. 17 and 20.
3. Hippensteele, S. A., Russell, L. M., and Stepka, F. S., 1983, ASME Tran., J. of Heat Trans., 105, pp. 184-189.
4. Halford, G. R., Morrow, JoDean, 1962, ASTM Proc., 62, pp. 695-707.
5. Morrow, JoDean, 1968, Fatigue Design Handbook, Chapter 3.2, Society of Automotive Engineers, New York.

NOMENCLATURE

b	Beam half-thickness (m)	$q(x)$	Heat source term (W/m^2)
c	Specific heat (W/kg•°C)	T	Temperature (°C)
E	Elastic modulus (GPa)	t	Time (s)
F	Force (N)	v	y-deflection (m)
h	Heat transfer coefficient (W/m^2•°C)	x	Axial coordinate (m)
I	Moment of Inertia (m^4)	y	Thickness coordinate (m)
k	Thermal conductivity (W/m•°C)	z	Width coordinate (m)
L	Length of beam (m)	Greek	
M	Moment (N•m)	ε	Strain (-)
N	Bending frequency (Hz)	θ	$T(x,y) - T_\infty$ (°C)
n	Strain hardening exponent (-)	ρ	Mass density (kg/m^3)
Q	Local volumetric heating (W/m^3)	σ	Stress (MPa)

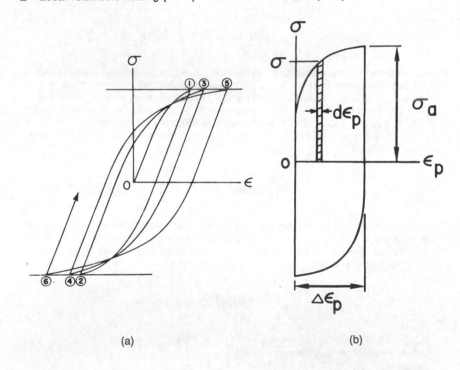

(a)   (b)

Figure 1. (a) Typical Stress-Strain for Cyclic Fully Reversed Bending with Fixed Strain Limits, and (b) Plastic Part of the Stress-Strain Hysteresis Loop (Reproduced from Sandor (1)).

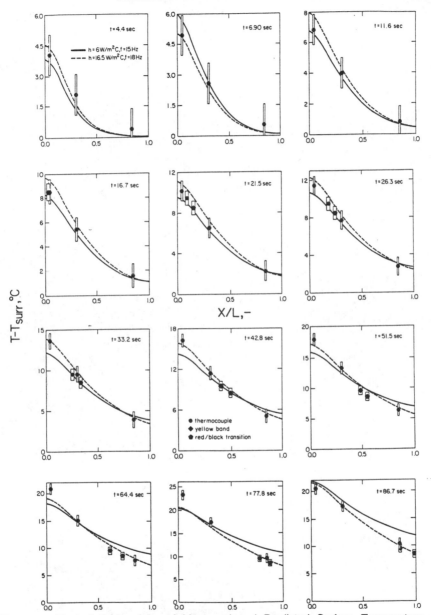

Figure 2. Typical Comparison of Measured and Predicted Surface Temperature Distributions ($n' = 0.11$, $\sigma_f' = \sigma_f$, and $\varepsilon_f' = \varepsilon_f$).

STEADY AND TRANSIENT CONDUCTION IN COMPOSITE MATERIALS

G.C.J. Bart
Applied Physics Department, Delft University of Technology, The Netherlands

The models that exist in literature to predict the apparent thermal conductivity for steady conduction in a composite material are not as such applicable in the case of transient conduction. It will be showed here when the analytical solutions for the diffusion equation in a homogeneous medium can be applied and also bounds for the transient response with lumped models will be given.

INTRODUCTION

Thermal conductivity and transient heat transfer in a solid composite material are of interest in many applications. Two different spatial composition distributions within cubical cells have been studied. In the first model, named the cubical face model, one of the phases is continuously distributed the other is distributed as equal sized solitary cubes. In the other model, named the cubical edge model, both phases are continuously distributed. Figure 1 shows these cases.

STEADY CONDUCTION MODELS

In the literature a great number of models for the steady conduction of a two phase composite exist [2,3,4]. Steady heat conduction is described by the Laplace equation

$$\nabla . \lambda \nabla T = 0 \tag{1}$$

together with the appropriate boundary conditions. The full solution of the Laplace equation can be approximated by a control volume finite difference method [1].
It is possible to obtain bounds for the possible apparent thermal conductivities for an arbitrary distribution of the two materials 1 and 2 with fractions f_1 and $f_2 = 1 - f_1$ and thermal conductivities λ_1 and λ_2 from the weighted harmonic mean:

$$1/\lambda_{app} = f_1/\lambda_1 + f_2/\lambda_2 \tag{2}$$

found by a series connection of the two materials, and the weighted arithmetic mean:

$$\lambda_{app} = f_1 \lambda_1 + f_2 \lambda_2 \tag{3}$$

found by a parallel connection of the two materials. These upper and lower limits are known as Wiener [2] bounds.
    If the overall behaviour of the two-phase composite is isotropic the possible apparent thermal conductivity range is given by the Hashin-Shtrikman [3] bounds:

$$\lambda_{app} = \lambda_2 + f_1 / \{ 1/(\lambda_1 - \lambda_2) + f_2/(3 \lambda_2) \} \tag{4}$$

respectively

$$\lambda_{app} = \lambda_1 + f_2 / \{ 1/(\lambda_2 - \lambda_1) + f_1/(3 \lambda_1) \} . \tag{5}$$

Because of the extra constraint these limits are narrower than those given by the harmonic and arithmetic mean.

If the geometry of the composite material is such that all boundaries between the two materials are exactly parallel or exactly perpendicular to the main heat flow direction, as is the case for the cubical face geometry and the cubical edge geometry depicted in figure 1, it is possible to obtain bounds for the apparent thermal conductivity by adding adiabatic planes parallel to the main heat flow or isothermal planes perpendicular to the main heat flow. In that case a network of thermal conductances in series and parallel is obtained. The apparent thermal conductivity of this network can be calculated. By using adiabates a lower bound, and by using isotherms an upper bound for the apparent thermal conductivity is found. This technique has been described by Crane and Vachon [4] who apply it for granular materials.

With this model the real spatial distribution of the phases of the composite material is exploited, so, generally speaking, closer bounds than with the former models should be obtained.

Comparison of steady conduction models

The results for our two spatial distributions for the different bounds and the numerical finite difference solution are given in figures 2 and 3 for the case $\lambda_2/\lambda_1 = 10^3$. It can be concluded in these cases that the Wiener bounds and Hashin-Shtrikman bounds are not capable predicting even the order of magnitude of the apparent thermal conductivity. Yet one of the Hashin-Shtrikman results, derived for a spherical geometry can serve as a good estimator for the cubical face model.

The Crane-Vachon models can give useful estimations of the thermal conductivity in many cases, especially when one of the two fractions is rather small.

As the bounds given with the Crane-Vachon models do not always lie within those given by the Hashin-Shtrikman model the narrowest bounds for the cubical structures are found with a combination of these two models.

TRANSIENT CONDUCTION MODELS

The description of transient heat conduction in composite materials will be restricted to the cubical face and the cubical edge geometry that have already been described. An example of such a material is given in figure 4. Due to symmetry there is no heat flow between neighbouring cells perpendicular to the main heat flow.

Each elementary cell with volume V is composed from different materials and is characterized by an equivalent steady thermal conductivity λ_{app} as known from the numerical solution of the Laplace equation and by an effective volumetric heat capacity $(\rho c_p)_e$, that in the case of two materials with volume fractions f_1 and f_2 respectively is given by:

$$(\rho\, c_p)_e = f_1\, (\rho\, c_p)_1 + f_2\, (\rho\, c_p)_2 \ . \tag{6}$$

For the detailed description of the transient heat conduction of such a composite material one always needs a numerical solver for the partial differential equation with position dependent thermal conductivity λ and volumetric heat capacity $\rho\, c_p$. For each material, the diffusion equation:

$$\rho\, c_p\ \partial T/\partial t = \nabla\, \lambda\, .\, \nabla T \tag{7}$$

holds with the appropriate boundary conditions. These can be reduced into a set of linear algebraic equations for a numerical solution. The control volume finite differences method [1] is an example of such a conversion.

In a few limiting cases, e.g. short times or scale of the inhomogeneities small compared to the overall size, a description of the transient heat conduction can be obtained by using existing analytical solutions for homogeneous media. Further, simplified models of the transient heat conduction can be made with lumped models.

Short time approximation

For a composite medium the penetration theory [5] can be used to find a short time solution for differential equation 7. In this case at $x = 0$ the composite material has an area S_1 characterized by thermal conductivity λ_1 and diffusivity κ_1 and an area S_2 characterized by λ_2 and κ_2. Supposing that the two heat flows going into respectively material 1 and 2 are independent, then, at $t = 0$ with a temperature step $(T_w - T_o)$ at $x = 0$, the total heat flow at time t will be given by:

$$Q = (S_1 \lambda_1 / \sqrt{\pi \kappa_1 t} + S_2 \lambda_2 / \sqrt{\pi \kappa_2 t}) (T_w - T_o) . \qquad (8)$$

This approximation is valid as long as the two penetration depths $\delta_1 = \sqrt{\pi \kappa_1 t}$ and $\delta_2 = \sqrt{\pi \kappa_2 t}$ do not reach boundaries between the two materials. Also, starting from the borderline between areas S_1 and S_2, the uniform heat flux at $x = 0$ in both materials will be affected. The size of this effect perpendicular to both the main heat flow and the borderline is of the order δ_1 and δ_2 in the two materials respectively. As a consequence the result of equation 8 can no longer be used if the product of $(\delta_1 + \delta_2)$ and the length l_b of the borderline is not small compared to the total area $(S_1 + S_2)$.

Homogeneous medium approximation

When the characteristic size of the elements of the composite medium becomes very small compared to the overall size L of the composite domain and the direction of the local temperature gradient has become constant (and consequently isothermal planes remain isothermal), the composite can be treated like an homogeneous medium with a thermal conductivity that equals the steady apparent thermal conductivity λ_{app} from the numerical solution of the Laplace equation and an effective volumetric heat capacity as given in equation 6. Different solutions can be found in Carslaw and Jaeger [5].

If the two media have characteristic sizes d_1 and d_2, the local temperature gradient will become constant if:

$$Fo_1 = \lambda_1 t / \{(\rho c_p)_1 d_1^2 \} \quad > 1 \qquad (9)$$

and

$$Fo_2 = \lambda_2 t / \{(\rho c_p)_2 d_2^2 \} \quad > 1 . \qquad (10)$$

For short times the homogeneous medium approximation can not be used. However it can be an useful approximation for long times if

$$Fo_L = \lambda_{app} t / \{(\rho c_p)_e L^2\} < \min (Fo_1, Fo_2) , \qquad (11)$$

when the heat transfer process in the direction of the main heat flow is not disturbed by local heat transfer phenomena. The accuracy of this approximation will become better the smaller Fo_L becomes compared to Fo_1 and Fo_2.

Lumped RC-models

The transient heat transfer of a material that has been built up from a number of identical rectangular cells has to be described. (See figure 4.)

When describing the transient heat conduction of such a composite material use can be made of a lumped model built with thermal resistances and heat capacities. In these lumped models the overall heat balance is preserved by taking care that the total heat capacity of the system and its model are equal.

These models are used to obtain limits for the heat content of the composite material. This material will be represented by a number of elementary building blocks (as given in figure 5) that, connected together, form networks as given in figure 6. As input a step function is used. The transient behaviour of this network is described by a number of first order differential equations that can be written in matrix form. The solution is composed with help of the eigenvalues and eigenvectors of this matrix [6]. Building block 'a' is used to obtain a lower limit of extracted heat (slower response than reality) and block 'b' for an upper limit (faster response).

Two different methods of obtaining RC-models have been used.

In the first method the lower limit of apparent thermal conductivity as given by Hashin and Shtrikman together with building block 'a' have been applied to obtain a slower response and the Hashin and Shtrikman upper limit with building block 'b' to get a faster response.

With the second method, following the Crane-Vachon approach, isotherms were added perpendicular to the main heat flow to make the problem one dimensional. Then the response could be described with a number of building blocks 'b' using the real thermal conductivity of that part. This again yields a network as given in figure 6, here to obtain a faster response. Also adiabates parallel to the main heat flow were added and combined with building block 'a'. Then a few networks as in figure 6 are obtained, the responses of which should be added to obtain a complete slower response.

Comparison of transient conduction models

For simulations on the two spatial distributions given in figure 1, four of these octants or elementary cells have been used in series connection. The thermal conductivity ratio of the two materials has been chosen $\lambda_2/\lambda_1 = 10^3$ and the heat capacity ratio by $(\rho\ c_p)_2/(\rho\ c_p)_1 = 10^3$, giving the materials the same thermal diffusivity. For the cubical face geometry material 1 can be found in the faces and for the cubical edge geometry material 2 is found along the edges (solid line part in figure 1). Fraction $f_1 = f_2 = 0.5$ has been chosen. The results are given in figures 7 and 8 respectively.

In figure 7 it can be seen that the step response obtained with a numerical finite difference method, the best available approximation to the exact solution, is bounded by the faster response of the Crane-Vachon approach and the slower response using Hashin-Shtrikman. Also it can be seen that for $Fo > 0.5$ the homogeneous medium description seems adequate. In that range equations 9, 10 and 11 are obeyed. The part of the curve where the penetration theory is valid falls outside the range of the graph.

In figure 8 the bounds are not close to the numerical solution. Here for $Fo < 3.10^{-3}$ the numerical solution fits nicely on that given by the penetration theory and for $Fo > 0.2$ the analytical solution for an homogeneous medium fits well with the numerical results.

CONCLUSIONS

With these examples it has been showed that a homogeneous medium approximation of the transient heat conduction in a composite material, using the apparent thermal conductivity obtained with a steady numerical model and the effective volumetric heat capacity, only gives satisfactory results under the restrictions given by equations 9, 10 and 11. The Fourier numbers mentioned above are not generally valid, they for example depend on the number of elementary cells changing the overall dimension L, the material properties and characteristic sizes d_1 and d_2. The penetration theory can be used to obtain the (very) first part of the response.

Bounds obtained with lumped models can sometimes be useful as estimates. If the range obtained with these bounds is too wide a numerical solver should be used.

References
1. Patankar, S.V., (1980), Numerical Heat Transfer and Fluid Flow, Hemisphere Publ. Corp., New York.
2. Wiener, O., (1912), Abh. Sachs. Akad. Wiss. Leipzig Math.-Naturwiss. Kl. 32, 509.
3. Hashin, Z. and Shtrikman, S., (1962), J. of Appl. Physics, 33, 10, 3125-3131.
4. Crane, R.A. and Vachon, R.I., (1977), Int. J. Heat Mass Transfer, 20, 711-723.
5. Carslaw, H.S., and Jaeger, J., (1959) Conduction of heat in solids, 2nd. ed., Oxford, Clarendon Press.
6. Smith, B.T., Boyle, J.M., Dongarra, J.J., Garbow, B.S., Ikebe, Y., Klema, V.C. and Moler, C.B., (1976), Lecture Notes in Computer Science 6, Matrix Eigensystem Routines - EISPACK Guide, Springer Verlag, Berlin.

Nomenclature

c_p	specific heat		δ	penetration depth
C	heat capacity		κ	thermal diffusion
f	fraction		ρ	density
Fo	Fourier number			
g	fraction			
l	length		app	apparent
Q	extracted heat		b	border
R	thermal resistance		e	effective
S	area		L	total length
t	time		o	initial
T	temperature		r	relative
x	position coordinate		w	wall

Figure 1. One octant of cubical face geometry (left) and one octant of cubical edge geometry (right).

Figure 2. Upper and lower limit of the apparent thermal conductivity ratio λ_{app}/λ_1 for the cubical face geometry according to the Wiener model (———), the Hashin-Shtrikman model (– – –) and the Crane-Vachon model (....) compared with a numerical finite differences solution (o o o). Fraction f_1 of thermal conductivity λ_1 is found in the inner cube.

Figure 3. Upper and lower limit of the apparent thermal conductivity ratio λ_{app}/λ_1 for the cubical edge geometry according to the Wiener model (———), the Hashin-Shtrikman model (– – –) and the Crane-Vachon model (....) compared with a numerical finite differences solution (o o o).

Figure 4. Two dimensional example of non homogeneous material composed of identical rectangular cells. One of the materials is shaded with ▨. The adiabates due to symmetry are indicated with an a. The direction of the main heat flow is indicated with ⇒.

Figure 5. Building blocks for one dimensional conduction simulation.

Figure 6. General RC-system to be solved.

Figure 7. Transient response (relative extracted heat vs dimensionless time) for the cubical face geometry with four octants in series and $f_1 = f_2 = 0.5$. (———) numerical solution, (– – – –) analytical solution homogeneous medium, (——— ———) bounds from the Hashin-Shtrikman approach, (——— – ———) bounds according the Crane-Vachon approach.

Figure 8. Idem figure 7 for the cubical edge geometry. (.....) penetration theory.

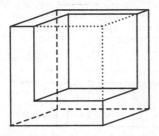
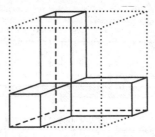

Figure 1

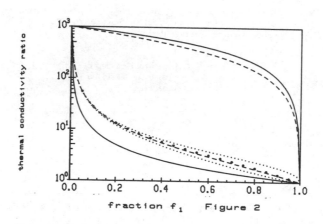

fraction f_1 Figure 2

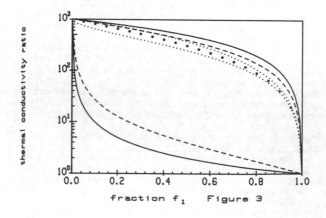

fraction f_1 Figure 3

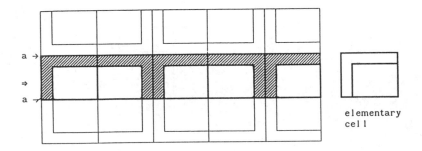

Figure 4

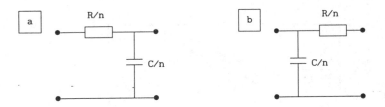

Figure 5

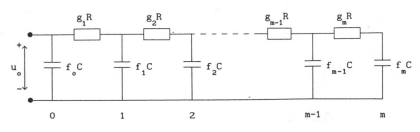

Figure 6

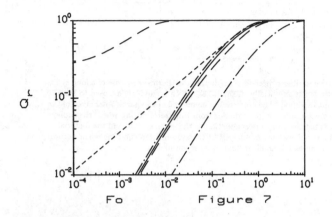

Figure 7

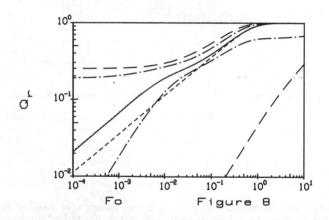

Figure 8

TRANSIENT CONDUCTION IN A LONG CYLINDER WITH ONSET OF NATURAL CONVECTION

T.W.Davies, U.Krempels and M.A.Patrick

School of Engineering, Exeter University, Exeter EX4 4QF, UK.

This paper describes a method of predicting the thermal response of a long cylinder, suspended in an initially stagnant gas, which is suddenly exposed to heating by isotropic radiation. The prediction is composed of two parts; an analysis of the temperature evolution in the cylinder from time zero, during which the radiant heat flux is balanced by pure conduction to the ambient fluid and the material of the cylinder, followed by an analysis of the system from the time when natural convection causes the ambient fluid to be set in motion.

INTRODUCTION

Transient conduction in cylinders immersed in a fluid is of fundamental importance in the analysis of the physics and chemistry of industrial processes involving the heat treatment of fibres. In modelling the chemical and physical transformations which may occur as fibres pass through critical heat transfer zones the starting point is the thermal response of the fibre as it enters, for example, a flame or furnace environment. If the temperature at each point within a fibre can be predicted as a function of time then predictions may also be made about the progress of thermally sensitive events such as chemical reaction or phase change.

In most previous analyses, that have included the effect of convection at the surface of the fibre, the convective heat transfer has been computed using empirical correlations for the fully developed Nusselt number at the surface of a cylinder for either forced or natural convection. In rapid heating processes the application of fully developed convection correlations in the early stages may be unjustified for certain combinations of process parameters and fibre/fluid properties. In particular, when a significant proportion of the energy transfer to the fibre takes place within a time before the development of convection, errors in the predicted temperature evolution of the fibre may result from the use of steady state Nusselt number correlations.

This paper presents an analytical method of predicting the thermal response of an isothermal fibre, in a stationary gas, which is suddenly exposed to isotropic thermal radiation. The analysis first describes events leading up to the time when fully developed surface convection conditions may be assumed.

PROBLEM DESCRIPTION

At time $t = 0$ the cylinder and the ambient fluid are both at the same uniform temperature T_i when the cylinder is suddenly exposed to a circumferentially uniform radiant flux from a source at a temperature T_R. As the surface of the cylinder is heated by the radiant flux and the surface temperature T_s rises, concentric conduction waves propagate inwards towards the centre of the cylinder and outwards into the surrounding stationary fluid as illustrated in Figure 1.

The dimensionless positions of the conduction fronts in the solid and fluid at elapsed time t are denoted by δ_1 and δ_2 respectively. When the Rayleigh number of the system reaches a critical value of about 1000

(Ref.(1)) natural convection will be initiated and eventually the cylinder will reach an equilibrium temperature at which the rate of radiant energy absorption at the surface is balanced by the rate of energy loss by convection. The problem has two phases separated by a critical Rayleigh number, (Ra_{crit}), and the solution to each must be matched at the time when Ra_{crit} is reached.

SOLUTION TECHNIQUE

The method of solution of both phases of the problem is based on the heat balance integral technique which is especially suitable for non linear diffusion problems (see Refs. (2) and (3)). For the early pure conduction phase, heat balances are written for the thermal waves spreading through the solid and fluid and also at the surface of the cylinder. These balances produce three simultaneous ordinary differential equations (for T_s, δ_1 and δ_2) which may be solved using the Runge-Kutta method. For the post-Ra_{crit} phase of the problem where heat transfer in the fluid is calculated from a steady-state natural convection correlation the thermal response of the cylinder is available directly from the integrated heat balance on the solid conduction wave.

In either the pre- or post-convection phase solutions two possibilities must be anticipated which arise because of the change in the boundary conditions of the problem at the moment when the conduction wave in the solid penetrates to the centre of the cylinder. Depending on the particular values of the physical properties and temperatures of the system under study it is possible, for example, for penetration of the solid conduction front to occur before or after the ambient fluid boundary layer reaches Ra_{crit}. If penetration occurs during the pure conduction phase in the fluid then the first part of the solution will have two components which are matched at the time of penetration. Similarly if penetration occurs after Ra_{crit} has been reached and convection initiated then the second part of the solution will now have two components which are matched at the penetration time.

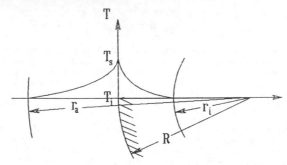

Fig.1   Definition of problem geometry and variables

Solution for pre-convection phase

The cylinder is assumed to be of infinite length, homogeneous material, opaque to thermal radiation, constant thermal properties, uniform initial temperature T_i and surrounded by a fluid with semi-infinite expanse, constant thermal properties, which is transparent to thermal radiation and is of uniform initial temperature T_i. For the pre-convection period the heat transfer inside the body and inside the ambient fluid is governed by the same heat diffusion equation:

$$\frac{\partial T}{\partial t} = \alpha \nabla^2 T \tag{1}$$

Using cylindrical coordinates and introducing non dimensional variables for the time and the temperature we obtain

$$r\frac{\partial \theta}{\partial Fo} = R^2 \frac{\partial}{\partial r}\left(r\frac{\partial \theta}{\partial r}\right) \qquad for\ the\ cylinder\ (0 \leq r \leq R) \tag{2}$$

$$r\frac{\partial \theta}{\partial Fo_f} = R^2 \frac{\partial}{\partial r}\left(r\frac{\partial \theta}{\partial r}\right) \qquad for\ the\ ambient\ fluid\ (r \geq R) \tag{3}$$

where : $\quad \theta = T/T_R \quad , \quad Fo = \alpha_s t/R^2 \quad , \quad Fo_f = (1/\zeta)Fo \quad$ are non dimensional temperature, time for the cylinder and time for the fluid respectively and $\quad 1/\zeta = \alpha_f/\alpha_s \quad$.

In the analysis which follows the equations for the cylinder and the ambient fluid are treated separately.

Pre-penetration period

The first step in the integral technique is to integrate equation (2) over the width of the heated region. For the thermal wave in the solid cylinder:

$$\int_{r_i}^{R} r \frac{\partial \theta}{\partial Fo} dr = \int_{r_i}^{R} R^2 \frac{\partial}{\partial r} \left(r \frac{\partial \theta}{\partial r} \right) dr \tag{4}$$

Using the differentiation rule of Leibnitz and integrating:

$$\frac{d}{dFo} \int_{r_i}^{R} r\theta \, dr + r_i \theta_i \frac{dr_i}{dFo} = R^3 \left(\frac{\partial \theta}{\partial r} \right)_{r=R} \tag{5}$$

Up to this point the calculation is exact. To evaluate the integral a function for the temperature profile in the thermal boundary layer must be introduced. For this purpose the temperature distribution across the layer is assumed to be of the form:

$$\theta = a_0 + a_1 \xi + a_2 \xi^2$$

where : $\quad \xi = (R-r)/(R-r_i) \quad$ is non-dimensional radial position. With the boundary conditions :

- on the surface $\xi = 0$, $(r = R)$: $\quad \theta = \theta_s$

- on the conduction front $\xi = 1$, $(r = r_i)$ $\quad \theta = \theta_i$ and $\quad (\partial\theta/\partial\xi)_{\xi=1} = 0$

the coefficients a_i may be evaluated to give the temperature profile:

$$\theta = \theta_s + 2(\theta_i - \theta_s)\xi - (\theta_i - \theta_s)\xi^2 \tag{6}$$

Introducing equation (6) into equation (5) and using the non dimensional thermal layer thickness δ_1, evaluation of each term produces the following ordinary differential equation for the surface temperature and conduction front position as a function of elapsed heating time:

$$\left[\frac{1}{3}(\theta_s - \theta_i) + \frac{1}{6}\delta_1(-\theta_s + \theta_i)\right] \frac{d\delta_1}{dFo} + \left[\frac{1}{12}\delta_1(4 - \delta_1)\right] \frac{d\theta_s}{dFo} = 2\frac{\theta_s - \theta_i}{\delta_1} \tag{7}$$

A second differential equation is obtained in the same way by integrating equation (3) over the width of the thermal boundary layer in the fluid:

$$\left[\frac{1}{3}(\theta_s - \theta_i) + \frac{1}{6}\delta_2(\theta_s - \theta_i)\right] \frac{d\delta_2}{dFo} + \left[\frac{1}{12}\delta_2(4 + \delta_2)\right] \frac{d\theta_s}{dFo} = 2\frac{\theta_s - \theta_i}{\zeta \delta_2} \tag{8}$$

The set of three equations is completed by the differential heat balance equation on the surface which is written as :

$$k_s \left(\frac{\partial T}{\partial r}\right)_{r=R} - k_f \left(\frac{\partial T}{\partial r}\right)_{r=R} = \sigma \epsilon (T_R^4 - T_s^4) \tag{9}$$

Introducing non dimensional variables for the temperature, length and the Radiation number $Rn = \dfrac{\sigma \epsilon T_R^3 R}{k_s}$ equation (9) may be rearranged to give:

$$\delta_1 = \delta_2 \frac{2(\theta_s - \theta_i)}{\delta_2 \, Rn \, (\theta_R^4 - \theta_s^4) - 2\lambda(\theta_s - \theta_i)} \tag{10}$$

where : $\lambda = k_f/k_s$

To summarise, the set of three differential equations (7),(8) and (10) describe the temporal variation of the surface temperature and conduction front positions during the pre-penetration phase for pure conductive cooling.

Post-penetration period

After the thermal boundary layer reaches the middle of the cylinder a new variable describing the mid-temperature T_m (in non dimensional form $\theta_m = T_m/T_R$) must be introduced instead of the thermal layer thickness in the cylinder which is now fixed. Equation (2) is integrated over the fixed heated range ($0 \leq r \leq R$). The temperature profile in the cylinder is changed, because of the new boundary conditions, to:

$$\theta = \theta_s + 2(\theta_m - \theta_s)\delta - (\theta_m - \theta_s)\delta^2 \tag{11}$$

with $\delta = (R - r)/R$; and the temperature gradient at the surface ($\delta = 0$) inside the body becomes

$$(\partial\theta/\partial\delta)_{\delta=0} = 2(\theta_m - \theta_s) \tag{12}$$

Similar manipulation of the basic balance equations as was used for the pre-penetration phase is now applied to the equations for the heat balances in the solid, in the fluid and on the surface. The corresponding set of differential equations describing the post- penetration phase so produced is:

$$\frac{d\theta_s}{dFo} = \frac{8(\theta_s - \theta_m) + 12\dfrac{\lambda}{\zeta}\dfrac{1}{\delta_2^3}\dfrac{\theta_s - \theta_i}{2 + \delta_2}}{2 + 2\,Rn\,\theta_s^3 + \dfrac{\lambda}{\delta_2}\left(1 + \dfrac{1}{2}\dfrac{4 + \delta_2}{2 + \delta_2}\right)} \tag{13}$$

$$\frac{d\theta_m}{dFo} = 8(\theta_s - \theta_m) - \frac{d\theta_s}{dFo} \tag{14}$$

$$\frac{d\delta_2}{dFo} = \frac{2\dfrac{\theta_s - \theta_i}{\zeta\,\delta_2} - \dfrac{1}{12}\delta_2(4 + \delta_2)\dfrac{d\theta_s}{dFo}}{\dfrac{1}{6}(\theta_s - \theta_i)(2 + \delta_2)} \tag{15}$$

Starting conditions

To solve the set of equations (7),(8),(10) a set of starting values for $(\theta_s - \theta_i)$, δ_1 and δ_2 not equal to zero is required. These starting conditions are calculated using linearised equations to describe the heating problem at times very close to zero. Equations (7), (8) and (10) are linearised for very small values of non dimensional time Fo by introducing the approximations:

$$4 - \delta_1 \approx 4, \quad and \quad 4 + \delta_2 \approx 4$$

$$\theta_R^4 - 4\theta_s^3\theta_i + 3\theta_s^4 \approx \theta_R^4 - 4\theta_i^4 + 3\theta_i^4 = \theta_R^4 - \theta_i^4$$

$$\delta_1(\theta_i - \theta_s) \approx 0, \quad and \quad \delta_2(\theta_s - \theta_i) \approx 0$$

Further, taking $(\theta_s - \theta_i)$, δ_1, and δ_2 to be proportional to $\sqrt{Fo}$ leads to expressions for the starting values:

$$\theta_s = \frac{Rn(\theta_R^4 - \theta_i^4)}{2\left(\sqrt{\frac{1}{6}} + \lambda\sqrt{\frac{\zeta}{6}}\right)}\sqrt{Fo_0} + \theta_i \tag{16}$$

$$\delta_2 = \sqrt{6/\zeta}\sqrt{Fo_0} \tag{17}$$

The third initial value δ_1 is calculated from the heat balance equation (10) using θ_s and δ_2 from equations (16) and (17).

Surface heat loss by convection

As in the heating phase with pure conductive loss to the ambient fluid, the transient heat conduction in the solid cylinder is described by equation (2). The difference from the previous analysis is embodied in the equation for the heat balance on the surface, which now has the form:

$$k_s \left(\frac{\partial T}{\partial r} \right)_{r=R} = \sigma \epsilon (T_R^4 - T_s^4) - h (T_s - T_i) \tag{18}$$

Pre-penetration period $(0 < r_i < R)$

As the mechanism of convective cooling on the surface in no way interferes with the heating mechanism inside the cylinder equation (7) still describes the transient heat transfer in the cylinder during the pre-penetration period. Introducing the non dimensional temperature θ and length δ the surface heat balance equation (18) is more conveniently written using the non dimensional groups:

Radiation number $\quad Rn = \dfrac{\sigma \epsilon R T_R^3}{k_s}, \quad$ and $\quad$ Biot number $\quad Bi = \dfrac{h R}{k_s}.$

This leads to the equation

$$- (\partial \theta / \partial \delta)_{\delta=0} = Rn (\theta_R^4 - \theta_s^4) - Bi (\theta_s - \theta_i) \tag{19}$$

From equation (6), noting that $\xi = \delta / \delta_1$

$$(\partial \theta / \partial \delta)_{\delta=0} = 2((\theta_i - \theta_s)/\delta_1)$$

After substitution in equation (19) we can solve for δ_1:

$$\delta_1 = \frac{2 (\theta_s - \theta_i)}{Rn (\theta_R^4 - \theta_s^4) - Bi (\theta_s - \theta_i)} \tag{20}$$

This expression may be differentiated with respect to non-dimensional time to yield:

$$\frac{d \delta_1}{d Fo} = \frac{2 Rn (\theta_R^4 + 3 \theta_s^4 - 4 \theta_s^3 \theta_i)}{\left[Rn (\theta_R^4 - \theta_s^4) - Bi (\theta_s - \theta_i) \right]^2} \frac{d \theta_s}{d Fo} \tag{21}$$

Substituting for δ_1 and $d \delta_1 / d Fo$ in equation (7) and integrating gives:

$$Fo_2 = Fo_1 + \frac{1}{3} \int_{\theta_{s1}}^{\theta_{s2}} \left[\frac{2 (\theta_s - \theta_i)}{G^2} + \frac{(\theta_s - \theta_i) H - (\theta_s - \theta_i)^2}{G^3} - \frac{(\theta_s - \theta_i)^2 H}{G^4} \right] d \theta_s \tag{22}$$

where

$$H = 2 Rn (\theta_R^4 + 3 \theta_s^4 - 4 \theta_s^3 \theta_i)$$
$$G = Rn (\theta_R^4 - \theta_s^4) - Bi (\theta_s - \theta_i)$$

Now the time history of the surface temperature is easily obtained by performing this integration over the time period Fo_1 to Fo_2 where the surface temperature rises from $\theta_s(Fo_1) = \theta_{s1}$ to $\theta_s(Fo_2) = \theta_{s2}$

The integral can be evaluated by any convenient numerical method. In this study a general purpose integrator (NAG routine D01AJF) was used. The heat transfer coefficient h is calculated according to an appropriate correlation for natural convection, depending on whether the cylinder is vertical or horizontal, and will be a function of $(T_s - T_i)$. The starting values for the integration come from the results of the pre-convection and pre-penetration calculation.

$$Fo_1 = Fo_{crit} = Fo(Ra_\delta = Ra_{\delta,crit})$$

$$\theta_{s1} = \theta_{s,crit} = \theta_s(Ra_\delta = Ra_{\delta,crit})$$

The second dependent variable, δ_1, the thermal layer thickness in the cylinder, is obtained from equation (20) completing all data necessary to describe the heating state of the cylinder.

Post-penetration period

As in the pre-penetration period we need two equations to solve the problem. The first equation is derived from the heat diffusion equation in the solid body. Because there is no difference from the post-penetration period of the pre-convection phase, equation (14) is also valid in this section of the heating process. The temperature gradient at the surface is given by equation (12) so that we may write for the mid-temperature:

$$\theta_m = \frac{2\theta_s + (\partial\theta/\partial\delta)_{\delta=0}}{2} \tag{23}$$

The temperature gradient at the surface is given by equation (19) as in the pre-penetration period. Thus, differentiating equation (23) with respect to non dimensional time, we obtain the time variation of the mid-temperature:

$$\frac{d\theta_m}{dFo} = \left(1 + \frac{1}{2}Rn\,4\,\theta_s^3 + \frac{1}{2}Bi\right)\frac{d\theta_s}{dFo} \tag{24}$$

Substituting this term in equation (14), separating the variables and integrating leads to:

$$Fo_2 = Fo_1 + \frac{1}{2}\int_{\theta_{s1}}^{\theta_{s2}}\frac{1 + Rn\,\theta_s^3 + \frac{1}{4}Bi}{G}\,d\theta_s \tag{25}$$

By integration, any chosen time Fo_2 when the surface temperature reaches the value $\theta_s(Fo_2) = \theta_{s2}$ can be calculated. The lower integration limits Fo_1 and θ_{s1} are (i) the penetration time Fo^* and the surface temperature at this time $\theta_s(Fo^*) = \theta_s^*$ obtained from a pre-penetration and convection calculation or (ii) the values Fo_{crit} and the surface temperature at this time $\theta_s(Fo_{crit}) = \theta_{s,crit}$ obtained from a post-penetration and pre-convection calculation when the Rayleigh number reaches the critical value.

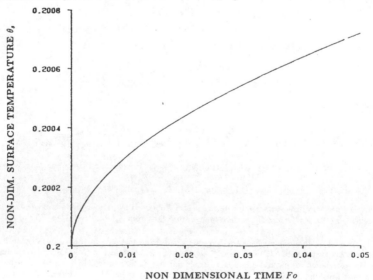

Fig.2 Development of non dimensional surface temperature θ_s versus non dimensional time Fo for a cylinder during the early conductive-cooling, pre-penetration period.

The same numerical integrator as in the pre-penetration period is used to solve equation (25). The mid-temperature can be calculated easily from equation (23).

RESULTS

Figure 2 shows sample results for the change of surface temperature with time for a long glass fibre of diameter 1.5 μm with initial surface temperature 300K when immersed in initially still air and exposed to radiation from a perfectly emitting surface at a temperature of 1500K. The surface absorptivity of the glass is assumed to be 0.835 which corresponds to a radiation number of 0.0024. The early heating period shown ($Fo \leq 0.05$) represents a real time of about 21 μs for this case and corresponds to an initial surface heating rate of about 6×10^4 K/s. Shortage of space prevents inclusion of further results and a proper parametrical study will be left to a subsequent paper in which the transition to natural convection will be featured.

CONCLUSIONS

1. The heat balance integral technique has been used to construct a set of predictive equations for the thermal response of a long cylinder exposed to sudden radiant heating with boundary conditions which depend on conduction to the surrounding gas.

2. The analysis provides a study of the thermal response during the initial heating period before convection effects are established but allows for later transition to natural convection when the conduction layer has become thick enough to be unstable.

NOTATION

		Greek	
Bi	Biot number (hR/k_s)	α	thermal diffusivity
c	heat capacity	δ	position, $(R-r)/R$
Fo	Fourier number	δ_1	$(R-r_i)/R$
h	heat transfer coefft.	δ_2	$(r_a-R)/R$
k	thermal conductivity	ϵ	emissivity of solid
Ra	Rayleigh number	θ	non dimensional temp., (T/T_R)
Ra_{crit}	...critical	λ	conductivity ratio, (k_f/k_s)
Rn	Radiation number	ξ	$(R-r)/(R-r_1)$
r	radial position	ρ	density
R	radius of cylinder	σ	Stefan-Boltzmann constant
t	time from start	ζ	diffusivity ratio, (α_s/α_f)
T	temperature		
T_R	...of radiation source		

REFERENCES

[1] **Patrick, M.A.** and **Wragg, A.A.**, Int.J.Heat Mass Transfer, 1975, Vol 18, pp 1397-1407.

[2] **Goodman, T.R.**, Advances in Heat Transfer, 1964, Vol 1, Academic Press, New York, pp 51-122.

[3] **Davies, T.W.**, Applied Math. Modelling, 1988, Vol 12, pp 429-433.

TRANSIENT CONDUCTION IN A PLATE WITH SINUSOIDALLY VARYING SURFACE HEAT FLUX

T.W.Davies, M.A.Patrick and M. Scholtes

School of Engineering, Exeter University, Exeter EX4 4QF, UK.

This paper describes an approximate analytical solution to the problem of transient heat conduction in an initially isothermal plate one side of which is suddenly exposed to a time-dependent radiant heat flux. As this side of the plate heats up it begins to lose heat to its surroundings by a combination of natural convection and re-radiation, making the problem highly non-linear in temperature. Once the heating effect has penetrated to the other side of the plate energy is also lost to the environment from this face by a combination of natural convection and radiation. The thermal response of the entire plate is predicted using an analysis based on the heat balance integral technique.

INTRODUCTION

There exist a group of geometrically simple transient conduction problems which are nonlinear in temperature and therefore are difficult to analyse. The heat balance integral (HBI) technique of Goodman [1] provides a useful starting point for the construction of approximate but relatively simple and accurate solutions to such problems. The problem under consideration here is that of a large plate, initially isothermal and at the same temperature as its environment, one side of which is suddenly exposed to a time–dependent (actually sinusoidal) radiant heat flux. Examples might be that of a solar collector, or a plate rotating in a radiation field. The time–dependent temperature distribution in the plate is approximated by a polynomial whose coefficients are time–dependent and can be determined using the boundary conditions of the problem and an integrated form of the heat conduction equation. The thermal history of the plate is predicted in two stages; the temperature response up to that time at which the thermal wave propagating from the front surface reaches the back face (called the penetration time), and the temperature response thereafter. The two solutions are matched at the penetration time.

PROBLEM DEFINITION

For conduction in a one-dimensional plate (i.e., infinite in y- and z-directions) the heat conduction equation may be written as:

$$\frac{\partial T}{\partial t} = \alpha \frac{\partial^2 T}{\partial x^2}, \qquad (1) \qquad \text{or, nondimensionally,} \qquad \frac{\partial \theta}{\partial F} = \frac{\partial^2 \theta}{\partial X^2} \qquad (2)$$

Boundary Conditions

At both bounding surfaces of a vertical plate (i.e., $X = 0$ for the 'front' and $X = 2$ for the 'back' surface) the heat transfer to or from the plate environment may be equated to the conductive flux at the boundary surfaces. For convective cooling of the front surface of the plate ($X = 0$),

$$k_s \left[\frac{\partial T}{\partial x} \right]_{X=0,conv.} = h \left(T_a - T_s \right), \text{or, nondimensionally,} \left[\frac{\partial \theta}{\partial X} \right]_{X=0,conv.} = Bi \left(\beta - 1 \right) \qquad (3)$$

The physical properties ν, α and k are assumed to be constant but for the natural convection processes considered here the Biot number is a function of temperature. For turbulent free convection at a vertical plate it may be shown [2] that:

$$Bi = 0.102 L \frac{k_f}{k_s} \left[\frac{g \left(\beta - 1 \right)}{\nu \alpha_f \left(\beta + 1 \right)} \right]^{\frac{1}{3}}. \qquad (4)$$

Equation (3) can now be written as:

$$\left[\frac{\partial \theta}{\partial X}\right]_{X=0,conv.} = B\left(\frac{\beta-1}{\beta+1}\right)^{\frac{1}{3}}(\beta-1). \tag{5}$$

All these equations assume the steady state and are not necessarily valid for transient convective heat transfer. In fact, for highly transient processes the heat transfer exceeds quasi-steady state values for rising surface temperatures, and is less for falling surface temperatures. Sparrow and Gregg [3] conclude that it is justified to neglect these effects for small values of ρ and c_p (i.e., for gases). If the front surface of the plate ($X = 0$) is assumed to be receiving heat by radiation the intensity of which is a (sinusoidal) function of time, $\dot{q}(F)$, and also loses heat by radiation by exchange with its surroundings then the radiant component of the surface heat balance may be written as:

$$k_s\left[\frac{\partial T}{\partial x}\right]_{X=0,rad.} = \underbrace{-\dot{q}(F)}_{heating} + \underbrace{\sigma\epsilon\left(T_s^4 - T_{rsink}^4\right)}_{cooling}.$$

Before the heating process starts, the plate is at equilibrium with the environment and a radiation background temperature is assumed to exist, such that $T_{rsink} = T_i = T_a = const.$ The surface heat balance is nondimensionlised using T_a:

$$\left[\frac{\partial \theta}{\partial X}\right]_{X=0,rad.} = -\frac{L}{k_s T_a}\dot{q}(F) + \frac{L\sigma\epsilon T_a^3}{k_s}\left(\beta^4 - 1\right), or, \quad \left[\frac{\partial \theta}{\partial X}\right]_{X=0,rad.} = -Q\dot{q}(F) + R\left(\beta^4 - 1\right). \tag{6}$$

For the case when the front surface experiences a time–dependent incident heat flux which then leads to combined radiant and convective cooling the instantaneous surface heat balance is extended to:

$$\left[\frac{\partial \theta}{\partial X}\right]_{X=0,comb.} = B\left(\frac{\beta-1}{\beta+1}\right)^{\frac{1}{3}}(\beta-1) + R\left(\beta^4 - 1\right) - Q\dot{q}(F) \tag{7}$$

The essence of the HBI method is that the form of the approximating polynomial used for the temperature distribution in the solid is established using the boundary conditions of the problem. At the beginning of the transient conduction process the boundary conditions at the leading edge of the thermal wave propagating into the plate from the front surface help to establish the solution to this phase of the heating process, referred to here as the pre–penetration period.

Pre-Penetration Period

For the initial heating period, where the thermal wave propagates as if in a semi–infinite solid, it is assumed that the temperature distribution is of quadratic form:

$$\theta = a_0 + a_1 X + a_2 X^2 \tag{8}$$

with time dependent coefficients $a = f(F)$ and the boundary conditions:

$$X = \delta^* \quad \Rightarrow \quad \theta = \gamma \quad and \quad (\partial\theta/\partial X) = 0$$
$$X = 0 \quad \Rightarrow \quad \theta = \beta$$

using these conditions with equation (8) gives:

$$a_0 = \beta, \qquad a_2 = \frac{-a_1^2}{4(\gamma-\beta)}, \qquad \delta^* = \frac{2(\gamma-\beta)}{a_1}, \tag{9}$$

The general heat conduction equation (2) is integrated over the expanding region occupied by the thermal wave, i.e., $0 \leq X \leq \delta^*$:

$$\int_{X=0}^{\delta^*} \frac{\partial\theta}{\partial F}dX = \left[\frac{\partial\theta}{\partial X}\right]_{X=\delta^*} - \left[\frac{\partial\theta}{\partial X}\right]_{X=0}. \tag{10}$$

Using the Leibnitz rule for the differentiation of an integral:

$$\frac{d}{dx}\int_{a(x)}^{b(x)} f(x,y)\,dy = \int_{a(x)}^{b(x)} \frac{\partial f}{\partial x}dy + f(b,x)\frac{db}{dx} - f(a,x)\frac{da}{dx}$$

the heat balance of equation (10) may be expanded as:

$$\frac{d}{dF}\int_{X=0}^{X=\delta^*}\theta(F,X)\,dX = \int_{X=0}^{X=\delta^*}\frac{\partial\theta(F,X)}{\partial F}\,dX + \theta(\delta^*,F)\frac{d\delta^*}{dF} - \underbrace{\theta(0,F)\frac{d0}{dF}}_{=0}. \tag{11}$$

The heat balance on an advancing thermal wave may thus be written as:

$$\frac{d}{dF}\int_{X=0}^{X=\delta^*}\theta(F,X)\,dX - \theta(\delta^*,F)\frac{d\delta^*}{dF} = \left[\frac{\partial\theta}{\partial X}\right]_{X=\delta^*} - \left[\frac{\partial\theta}{\partial X}\right]_{X=0} \tag{12}$$

Substitution of the approximating temperature profile (equation (8)) into equation (12) and integration gives:

$$\frac{d}{dF}\left[a_0\delta^* + a_1\frac{\delta^{*2}}{2} + a_2\frac{\delta^{*3}}{3}\right] - \left[a_0 + a_1\delta^* + a_2\delta^{*2}\right]\frac{d\delta^*}{dF} = \left[\frac{\partial\theta}{\partial X}\right]_{X=\delta^*} - \left[\frac{\partial\theta}{\partial X}\right]_{X=0}$$

which simplifies to

$$\delta^*\frac{da_0}{dF} + \frac{\delta^{*2}}{2}\frac{da_1}{dF} + \frac{\delta^{*3}}{3}\frac{da_2}{dF} = \left[\frac{\partial\theta}{\partial X}\right]_{X=\delta^*} - \left[\frac{\partial\theta}{\partial X}\right]_{X=0}. \tag{13}$$

Using Equations (9) and the temperature gradients at either side of the thermal wave:

$$[\partial\theta/\partial X]_{X=\delta^*} = 0 \qquad [\partial\theta/\partial X]_{X=0} = a_1$$

enables equation (13) to be rewritten as:

$$\frac{2(\gamma-\beta)}{a_1}\frac{d\beta}{dF} + \frac{2(\gamma-\beta)^2}{a_1^2}\frac{da_1}{dF} + \frac{8}{3}\frac{(\gamma-\beta)^3}{a_1^3}\frac{da_2}{dF} = -a_1. \tag{14}$$

Differentiation of $a_2 = -a_1^2/4(\gamma-\beta)$ with respect to F and substitution into equation (14) yields:

$$\frac{4}{3}\frac{(\gamma-\beta)}{a_1^2}\frac{d\beta}{dF} + \frac{2}{3}\frac{(\gamma-\beta)^2}{a_1^3}\frac{da_1}{dF} + 1 = 0. \tag{15}$$

Using the identity

$$\frac{(\gamma-\beta)^2}{a_1^3}\frac{da_1}{dF} = -\frac{1}{2}\frac{d}{dF}\left[\frac{(\gamma-\beta)^2}{a_1^2}\right] - \frac{(\gamma-\beta)}{a_1^2}\frac{d\beta}{dF}$$

equation (15) may be rewritten as a general from of the HBI equation, where only the insertion of the expression for the surface heat flux, a_1, is now needed to complete the solution:

$$\frac{2}{3}\int_1^\beta\frac{(\gamma-\beta)}{a_1^2}\,d\beta - \frac{1}{3}\left[\frac{(\gamma-\beta)}{a_1}\right]^2 = -\int_0^F dF^\dagger = -F \tag{16}$$

For the case under consideration equations (7) and (8) allow the front surface heat balance to be written as:

$$a_1 = \left[\frac{\partial\theta}{\partial X}\right]_{X=0} = B\left(\frac{\beta-1}{\beta+1}\right)^{\frac{1}{3}}(\beta-1) + R(\beta^4-1) - Q\dot{q}(F) \tag{17}$$

The HBI for this problem can then be finally rewritten as:

$$F = \frac{1}{3}\left[\frac{1-\beta}{B\left(\frac{\beta-1}{\beta+1}\right)^{\frac{1}{3}}(\beta-1) + R(\beta^4-1) - Q\dot{q}(F)}\right]^2$$
$$-\frac{2}{3}\int_1^\beta\frac{1-\beta}{\left[B\left(\frac{\beta-1}{\beta+1}\right)^{\frac{1}{3}}(\beta-1) + R(\beta^4-1) - Q\dot{q}(F)\right]^2}\,d\beta \tag{18}$$

Using equation (18) the time F at which the surface temperature β reaches a specified value can be easily evaluated by numerical integration. The corresponding temperature profiles may then be calculated using equations (9). Once the leading edge of the thermal wave reaches the back face of the plate the boundary conditions of the problem need revision, and the solution enters a second phase referred to as the post–penetration period.

Penetration Time, F^*

The solution for the temperature distribution in the plate derived from the pre–penetration analysis must match that derived from the post–penetration analysis at the moment of penetration. The penetration time is defined as the moment when the thermal front reaches the back of the plate, i.e., when: $\delta^* = 2.0$ (because the plate is of thickness 2L) and where, from equation (9) with $\gamma = 1$:

$$\delta^* = \frac{2(1-\beta)}{a_1} = \frac{2(1-\beta)}{B\left(\frac{\beta-1}{\beta+1}\right)^{\frac{1}{3}}(\beta-1) + R(\beta^4 - 1) - Q\dot{q}(F)} \tag{19}$$

The penetration front temperature β^* may be evaluated when $\delta^* = 2.0$.

Post-Penetration Period

When the thermal disturbance has reached the back face of the plate, this marks the end of the 'semi–infinite' behaviour of the problem and the boundary conditions on the heat balance need to be redefined. A second polynomial is chosen to approximate the temperature profile in the post– penetration period:

$$\theta = b_0 + b_1 X + b_2 X^2 \tag{20}$$

the derivative of which is :

$$\frac{\partial \theta}{\partial X} = b_1 + 2b_2 X$$

Using the following four boundary conditions allows the coefficients of equation (20) to be evaluated as:

1. $X = 0 \Rightarrow \theta(F, 0) = b_0 = \beta$ $\tag{21}$

2. $X = 0 \Rightarrow \dfrac{\partial \theta(F, 0)}{\partial X} = b_1 = -B\left(\dfrac{1-\beta}{1+\beta}\right)^{\frac{1}{3}}(1-\beta) - R(1 - \beta^4)$ $\tag{22}$

3. $X = 2 \Rightarrow \theta(F, 2) = \eta = b_0 + 2b_1 + 4b_2$ $\tag{23}$

4. $X = 2 \Rightarrow \dfrac{\partial \theta(F, 2)}{\partial X} = b_1 + 4b_2 = -B\left(\dfrac{\eta-\gamma}{\eta+\gamma}\right)^{\frac{1}{3}}(\eta-\gamma) - R(\eta^4 - \gamma^4)$ $\tag{24}$

These relationships can also be used to establish the following link between the temperatures on the front (β) and back (η) faces of the plate:

$$\eta + B\frac{(\eta-1)^{\frac{1}{3}}}{(\eta+1)^{\frac{1}{3}}} + R\eta^4 = \beta - Q\dot{q}(F) + B\frac{(\beta-1)^{\frac{1}{3}}}{(\beta+1)^{\frac{1}{3}}} + R(\beta^4 - 1) + R \tag{25}$$

For a chosen β value the expression on the RHS of equation (25) can be evaluated, and the equation can be solved numerically for the only remaining variable, η.

The heat balance integral for the plate in the post–penetration period, where $\delta^* = 2 = $const. may be written (using the Leibnitz rule) as:

$$\frac{d}{dF}\int_{X=0}^{X=2} \theta(F, X)\,dX = \int_{X=0}^{X=2} \frac{\partial \theta(F, X)}{\partial F}\,dX + \underbrace{\theta(2, F)\frac{d2}{dF}}_{=0} - \underbrace{\theta(0, F)\frac{d0}{dF}}_{=0}$$

which simplifies to:

$$\frac{d}{dF}\int_{X=0}^{X=2} \theta(F, X)\,dX = \left[\frac{\partial \theta}{\partial X}\right]_{X=2} - \left[\frac{\partial \theta}{\partial X}\right]_{X=0}$$

$$\frac{d}{dF} \int_{X=0}^{X=2} \left(b_0 + b_1 X + b_2 X^2\right) dX = \left[\frac{\partial \theta}{\partial X}\right]_{X=2} - \left[\frac{\partial \theta}{\partial X}\right]_{X=0}$$

$$\Rightarrow \frac{d}{dF} \left[b_0 X + \frac{1}{2}b_1 X^2 + \frac{1}{3}b_2 X^3\right]_{X=0}^{X=2} = b_1 + 4b_2 - b_1$$

$$\Rightarrow \frac{d}{dF} \left(2b_0 + 2b_1 + \frac{8}{3}b_2\right) - (0) = 4b_2$$

$$\Rightarrow 2\frac{db_0}{dF} + 2\frac{db_1}{dF} + \frac{8}{3}\frac{db_2}{dF} = 4b_2.$$

Using the equations (21) and (23)

$$\Rightarrow 2\frac{d\beta}{dF} + \frac{d\eta}{dF} - \frac{d\beta}{dF} - 4\frac{db_2}{dF} + \frac{8}{3}\frac{db_2}{dF} = 4b_2$$

$$\Rightarrow \frac{d\beta}{dF} + \frac{d\eta}{dF} - \frac{4}{3}\frac{db_2}{dF} = 4b_2$$

leads to the general form of the HBI for the post–penetration period $F > F^*$:

$$\int_{\beta^*}^{\beta} \frac{d\beta}{4b_2} + \int_{\eta=\gamma}^{\eta} \frac{d\eta}{4b_2} - \frac{1}{3}\int_{b_2^*}^{b_2} \frac{db_2}{b_2} = \int_{F^*}^{F} dF^\dagger.$$

Integration yields the following expression for $F > F^*$, which gives the time–temperature dependence in the post–penetration period of the transient heating process:

$$F = F^* + \int_{\beta^*}^{\beta} \frac{d\beta}{Q\dot{q}(F) - B\frac{(\beta-1)^{\frac{4}{3}}}{(\beta+1)^{\frac{1}{3}}} - R(\beta^4 - 1) - B\frac{(\eta-1)^{\frac{4}{3}}}{(\eta+1)^{\frac{1}{3}}} - R(\eta^4 - 1)}$$

$$+ \int_{\eta=\gamma}^{\eta} \frac{d\eta}{Q\dot{q}(F) - B\frac{(\beta-1)^{\frac{4}{3}}}{(\beta+1)^{\frac{1}{3}}} - R(\beta^4 - 1) - B\frac{(\eta-1)^{\frac{4}{3}}}{(\eta+1)^{\frac{1}{3}}} - R(\eta^4 - 1)} \tag{26}$$

$$- \frac{1}{3}\ln \left[\frac{Q\dot{q}(F) - B\frac{(\beta-1)^{\frac{4}{3}}}{(\beta+1)^{\frac{1}{3}}} - R(\beta^4 - 1) - B\frac{(\eta-1)^{\frac{4}{3}}}{(\eta+1)^{\frac{1}{3}}} - R(\eta^4 - 1)}{Q\dot{q}(F^*) - B\frac{(\beta^*-1)^{\frac{4}{3}}}{(\beta^*+1)^{\frac{1}{3}}} - R(\beta^{*4} - 1)}\right]$$

For given convection and radiation conditions and physical properties for the fluid and the solid, equation (26) is the desired relation between time and front and back surface temperatures. With equation (25) giving the back surface temperature as a function of the front surface temperature, the corresponding time after penetration can now be calculated for any chosen value of the front surface temperature $\beta > \beta^*$.

Expressing equation (26) in the form:

$$F = F^* + F_1 + F_2 + F_3. \tag{27}$$

then in the integral F_1 the dependence between $d\beta$ and the expression in the denominator as a function of β is taken into consideration during a numerical integration. However the relationship between $d\beta$ and η is much more complicated, as shown by equation (25). Unfortunately η as function of β cannot be introduced into equation (26).

Therefore the numerical solution of F_1 is split into a number of finite integrations over small differences of β, allowing readjustment of the value of η as a function of the last value of β. The quantity η is then considered as constant over this small finite integration sub–interval, making F_1 over this sub–interval a function of β only. The same approximation has to be used for F_2 with β being a function of η. In order to investigate the accuracy of this approach an explicit finite difference solution for the same problem was produced. . The plate was divided into ten strips and a time increment of 0.008 was employed, giving a value for the stability criterion of 5. A comparison between the front and back surface temperatures predicted by the HBI method and this numerical method is given in Table 1, for the initial heating phase of the problem, from which it will be seen that the two methods are generally in good agreement except at the very beginning of the heating process. The discrepancies, due in part to the 'lumping' used in the finite difference scheme, can be reduced by using a Hermite polynomial to approximate the temperature distribution as shown by Figure 1.

HBI			explicit fin. diff.			rel. diff.[%]	
F	β	η	F	β	η	front	back
0.0088	0.300	0.200	0.008	0.248	0.200	17.3	0.0
0.0437	0.400	0.200	0.040	0.340	0.200	15.0	0.0
0.1250	0.500	0.200	0.128	0.455	0.200	9.0	0.0
0.2963	0.600	0.200	0.296	0.562	0.202	6.3	1.0
0.5333	0.672	0.200	0.536	0.644	0.221	4.2	10.5
0.6138	0.683	0.219	0.616	0.663	0.231	2.9	5.5
0.7050	0.696	0.238	0.704	0.681	0.243	2.2	2.1
0.9063	0.722	0.275	0.904	0.714	0.272	1.1	1.1
1.1388	0.748	0.311	1.136	0.745	0.304	0.4	2.3
1.4150	0.774	0.346	1.416	0.772	0.339	0.3	2.0
1.7606	0.800	0.380	1.760	0.798	0.374	0.3	1.6
2.0363	0.816	0.402	2.040	0.814	0.397	0.2	1.2
2.3906	0.832	0.424	2.392	0.831	0.419	0.1	1.2
2.8744	0.848	0.445	2.872	0.847	0.441	0.1	0.8
3.6850	0.864	0.466	3.696	0.864	0.465	0.0	0.2
7.0081	0.880	0.487	7.008	0.880	0.488	0.0	0.2
7.8619	0.880	0.488	7.864	0.881	0.489	0.1	0.2
10.5563	0.8815	0.4895	10.560	0.8816	0.4896	0.0	0.0
12.9700	0.8816	0.4897	12.968	0.8816	0.4897	0.0	0.0

Table 1 *Comparison of HBI solution with numerical solution for $R = 0.4$, $B = 0.8$, $\gamma = 0.2$, $\Delta X = 0.2$ and $\Delta F = 0.008$ i.e.,stability criterion $= 5$.*

Heat Flux Oscillation

A sinusoidal, time-dependent, radiative heat flux on the front surface of the vertical plate is assumed. The boundary condition on the front surface is radiative heating countered by free convective cooling and radiant exchange with the background radiation sink. On the rear surface the plate is cooled by free convection and radiation to the same constant radiation sink temperature. The ambient fluid temperature is considered as constant and not affected by the cooling of the plate.

This model can for example be used for rotating plates exposed to a thermal radiation source or a plate exposed to oscillatory heating, where the ambient fluid temperature stays constant. The radiative heat flux on the front surface of the plate is arbitrarily chosen to be $Q\dot{q}(F) = 1.6 - 1.5\cos\left(2R^2F\right)$ The values of $Q\dot{q}(F)$ are oscillating between $Q\dot{q}_{min} = 0.1$ and $Q\dot{q}_{max} = 3.1$. The other parameters are set at $R_{front} = R_{back} = 0.4$, $B_{front} = B_{back} = 0.8$, $\theta_a = \theta(F = 0) = \gamma = 1.$. The ambient temperature, which is used to render the temperatures dimensionless, is equal to the uniform initial temperature. Figure 2 shows the predicted response of the plate surface temperatures to the sinusoidal driving function. The parameters of this particular example are such that a steady state temperature distribution is reached at the peak front surface heat flux which leads to the rear surface maintaining a constant temperature until the effects of falling surface heat flux penetrates to the back face. During the cooling phase the back surface again becomes isothermal during the period after a second steady state has been reached and before the effects of the next heating phase propagate through the plate.

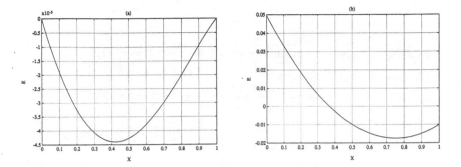

Figure 1 Relative error E in (a) θ throughout the plate and (b) $d\theta/dX$ throughout the plate at the steady state and using a Hermite polynomial approximation of the temperature profile. $E = 100 \times (HBI - A)/A$, where HBI is the value predicted by the present integral technique and A is the corresponding value predicted from the analytical solution given by Carslaw and Jaeger [4].

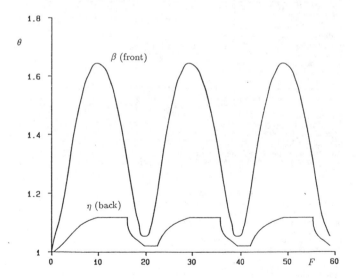

Figure 2 Dimensionless temperature responses of the front and back faces of a plate, the front face of which is exposed to a sinusoidal variation in incident heat flux of the form $Q\dot{q}(F) = 1.6 - 1.5\cos(2R^2F)$. Both surfaces lose heat to the environment by a combination of natural convection and radiant exchange, with the problem parameters set at R = 0.4, B = 0.8, $\theta_a = \gamma = 1$.

CONCLUSIONS

The HBI method has been used to develop predictive equations for the temperature response of a plate with time–dependent, non–symmetrical, nonlinear boundary conditions. Numerical results are obtained by a process of numerical integration over successive sub–intervals of elapsed time, during which a linear dependence of surface temperature response on the exciting function is assumed. The solutions appear to be stable, particularly at small elapsed times, and in reasonable agreement with an explicit finite difference solution.

REFERENCES

[1] **Goodman, T. R.**,Advances in Heat Transfer, 1964, Vol 1 Academic Press, *New York,* pp 51–122.

[2] **McAdams, W. H.**,Heat Transmission, 3rd edition , 1954, McGraw-Hill, *New York,* pp 165–183.

[3] **Sparrow,E.M.**and **Gregg,J.L.** Int. Jnl. Heat Mass Transfer, 1960, Vol. 82C No. 3, pp 258–260.

[4] **Carslaw,H.S.** and **Jaeger,J.C.**Conduction of Heat in Solids, 2nd edition, 1959, Oxford Univ. Press, pp 149–150.

NOMENCLATURE

B	constant in eqn (5)
$Bi = hL/k_s$	Biot number
$F = (\alpha/L^2)\,t$	Fourier number (dimensionless time)
F^*	F at penetration (penetration time)
L	half value of thickness of plate
$Nu = hl/k_f$	Nusselt number
$Q = L/k_s T_a$	heat flux parameter
$R = L\sigma\epsilon T_a^3/k_s$	Radiation number
T	temperature
T_a	ambiant fluid temperature
T_i	initial solid temperature at $t = 0$
T_r	radiation source temperature
$T_{r sink}$	radiation sink temperature
T_s	surface temperature
$X = x/L$	dimensionless value of position x
c_p	specific heat
h	heat transfer coefficient
k_f	thermal conductivity of the fluid
k_s	thermal conductivity of the solid
l	length of the surface
$\dot{q}$	rate of heat flux per unit surface area
t	time
$\alpha = k/\rho c_p$	thermal diffusivity
β	dimensionless front surface temperature
β^*	front surface temperature at penetration time
γ	dimensionless initial temperature
δ^*	dimensionless position of conduction front
ϵ	radiative emissivity of surface
η	dimensionless back surface temperature
θ	dimensionless temperature
$\theta_{i,a,r,\ldots}$	dimensionless expressions of $T_{i,a,r,\ldots}$
κ	dimensionless radiation sink temperature
λ	dimensionless radiation source temperature
ν	kinematic viscosity of fluid
ρ	density
σ	Stefan Boltzmann constant $5.669 \cdot 10^{-8}\left[\frac{W}{m^2 K^4}\right]$

ESTIMATION OF TRANSPORT COEFFICIENTS IN PACKED BEDS

E.A. Foumeny[*], C. McGreavy[*] and H. Pahlevanzadeh[*]
J.A.A. Castro[**], A.E. Rodrigues[***]

[*]Department of Chemical Engineering, The University of Leeds, U.K.
[**]Department of Chemical Engineering, University of Coimbra, Portugal.
[***]Department of Chemical Engineering, University of Oporto, Portugal.

A descriptive mathematical model of a packed bed heat exchanger has been employed to analyse a series of data extracted from an experimental rig. Computational techniques have been exploited to estimate and subsequently correlate the extracted values of the associated thermal coefficients. The results reveal that, unlike published design data, the transport parameters are also influenced by the physical dimensionality of the packing matrix.

INTRODUCTION

Design of any heat transfer equipment requires accurate knowledge of the related transport parameters. The important parameters associated with packed bed systems are the effective radial thermal conductivity, k_{er}, and the wall heat transfer coefficient, h_w. These coefficients have been studied by many investigators (Yagi and Wakao, 1959; Zehner and Schluender, 1970; De Wasch and Froment, 1972; Cybulski and Kawecki, 1972; Dixon and Cresswell, 1979) and numerous correlations have been proposed for the prediction of the parameters. A comprehensive review of the correlations has been given by Stankiewicz (1989) and Foumeny and Ma (1991). In general, the proposed design data for packed bed systems falls into two distinctive categories. One is based on plug flow idealization, i.e. single value, while the other is associated with local flow conditions.

Most of the available models of packed beds provide thermal predictions which differ from one another. This is mainly attributed to the contradictory nature of the published design data. This clearly signifies the importance of a set of well founded and generalized design information which could be used with a reasonable degree of confidence. Rectification of this certainly necessitates a descriptive mathematical representation of the physical reality as well as a series of experimental data. Suitable optimization techniques may then be used for the treatment of the observed data for the extraction of the values of the appropriate thermal coefficients.

As far as this work is concerned, a two-dimensional non-plug flow model of packed beds has been used to analyse a series of steady state data, obtained from a packed bed rig. Values of the wall coefficient and effective radial conductivity have been obtained for a variety of physical and operating conditions. The intention of this contribution is to describe the mathematical treatment of a series of experimental data together with the employed estimation strategies.

MATHEMATICAL MODEL

The mathematical equations describing the heat exchange between the packing matrix and its environment can be expressed by the following equations:

$$\text{Fluid:} \quad \frac{\partial T_s}{\partial \theta} = -u(r) \frac{\partial T_f}{\partial x} - \frac{6h(r)(1-\epsilon(r))}{\rho_f C_f d_p \epsilon(r)} (T_f - T_s) + \frac{1}{\rho_f C_f r \epsilon(r)} \frac{\partial}{\partial r} \left[r k_{er}(r) \frac{\partial T_f}{\partial r} \right]$$

$$\text{Solid:} \quad \frac{\partial T_s}{\partial \theta} = \frac{6h(r)}{\rho_s C_s d_p} (T_f - T_s)$$

$$\theta \leq 0 \quad T_f = T_s = T_i \quad \text{and} \quad x = 0 \quad T_f = T_{fi}$$

$$r = 0 \quad \frac{\partial T_f}{\partial r} = 0 \quad \text{and } r = R \quad -k_{er}(r) \frac{\partial T_f}{\partial r} = h_w (T_f - T_w)$$

These equations represent a non-plug flow system, and contain variables and parameters, ϵ, u, h and k_{er}, that are radial position dependent. For a plug flow model, these properties become independent of position. The model equations are solved numerically using finite difference schemes and details of these together with the employed flow and structural non-uniformities can be found elsewhere (Pahlevanzadeh, 1990). It is worth noting that the stated model is a transient one which can be used for the treatment of transient as well as steady state data.

EXPERIMENTAL INFORMATION

The experimental packed bed heat exchanger and the related operational procedures have been described in detail by Chalbi (1984). The experimental set-up consisted of three sections: the fixed bed heat exchanger, the air and saturated steam feeds, and sensors for the measurement and control. The heat exchanger section was constructed from a tube of stainless steel material, 0with a fixed length of 1.5 m. Several tubes with varying internal diameter, 1.02×10^{-2} - 5.57×10^{-2}, were used. The beds were packed with Alumina particles, $d_p = 3.1 \times 10^{-3}$ m. The mean bed voidage calculated on the basis of the mass of the dry packed particles and the particle density of 1230, kg/m^3, as given in Table 1. The packed section was contained inside a well insulated shell.

TABLE 1: Data of Chalbi's Packed Bed

Tube Diameter $d_t \times 10^2 (m)$	d_t/d_p	Mean Voidage ϵ
5.57	17.97	0.376
3.00	9.68	0.399
2.33	7.52	0.419
1.78	5.74	0.424

Air was used as process fluid which at 8 bar was expanded and dried before entering the packed bed at a temperature of 17°C with the flowrate being measured at the outlet of the bed. The bed container was heated at constant temperature. For the purpose the saturated steam was generated at a regulated pressure of 1.016 bar with its condensing temperature at 107°C. The measured quantities were the air flowrate and the radial temperature profiles at different axial positions, 0.25 m, 0.5 m, 0.75 m and 1.25 m from the bed entrance. At each cross-section, the radial temperature profile at steady state was established by means of thermocouples which were distributed at given radial positions, with one being located 3×10^{-3} m away from the wall and the rest spaced at 5×10^{-3} m internal. Table 2 lists the observed temperature profile of a typical run. Further details may be found elsewhere (Chalbi, et al. 1986; Pahlevanzadeh, 1990).

TABLE 2: Experimental Results for a Typical Run, $d_t/d_p = 17.97$

Radial Axial	$r_1=0.0$	$r_2=0.005$	$r_3=0.01$	$r_4=0.015$	$r_5=0.025$	$r_6=0.025$
	Run S1: G = 0.5 x 10^{-3} kg/s, Re_p = 34.4					
Y1 = 0.25	61.0	60.0	62.0	65.0	72.0	81.0
Y2 = 0.50	90.0	90.0	92.0	92.0	98.0	98.0
Y3 = 0.75	101.0	101.0	103.0	102.0	104.0	104.0
Y4 = 1.25	106.0	106.0	107.0	106.0	106.0	106.0

ESTIMATION STRATEGY

The two-dimensional non-plug flow model of packed beds has been employed to analyze the reported experimental data of Chalbi (1984). The descriptive mathematical model of the system and the observed information, together with a suitable optimization algorithm are utilized to extract the best estimates of the desired parameters. The optimization algorithm allows the minimum of the sum of squares of residuals to be found. The procedure is based on a combination of Gauss-Newton and steepest-descent method. This algorithm is due to Marquardt (1963) and has proved to be very powerful and efficient convergence with relatively poor starting guesses of the unknown coefficients. A least squares objective function is used. In this method, the Gauss-Newton normal equations are modified by adding a factor λ,

$$(\underline{A}^t\underline{A} + \lambda\underline{I}) \Delta\hat{\underline{A}} = \underline{A}^t (Y - \ddot{Y})$$

where $\underline{I}$ is the identity matrix. Thus λ is added to each term of the main diagonal of the $\underline{A}^t\underline{A}$ matrix.

The least squares objective function used here can be presented as:

$$obj = \sum_{j=1}^{Na} \sum_{i=1}^{Nr} (T_{fexp}(r_i,y_j) - T_{fcal}(r_i,y_j,h_w,k_{er}))^2$$

where

Na, Nr = number of axial and radial positions, respectively

$T_{fexp}(r_i,y_j)$ = the temperature measured at the position (r_i,y_j) with r and y denoting radial and axial positions, respectively

$T_{fcal}(r_i,y_j,h_w,k_{er})$ = the calculated fluid temperature at position (r_i,y_j) by solving the model equations based on h_w and k_{er}

For optimizing the occurrence of multiple local optima the estimation programme was run several times with different initial value of the parameters, i.e. h_w and k_{er}, in order to check that the estimated values were based on the global optimum and not local.

RESULTS AND DISCUSSION

Four sets of experimental runs based on fixed particle diameter, d_p, and varying tube diameter, $d_t = 0.0557$ m, 0.03 m, 0.0233 m and 0.0178 m, have been chosen from Chalbi's work for the estimation of the two thermal coefficients, h_w and k_{er}. Each set is based on six different mass flowrates. For each experimental run, four attempts based on different initial values of the parameters, h_w and k_{er}, were performed in order to establish the optimum values. The best estimates of the thermal coefficients for the different beds are summarized in Table 3.

In addition, the experimental data of six runs of one bed, $d_t/d_p = 17.97$, together with a two-dimensional plug flow model have been used to extract the best estimates of the two coefficients, h_w and k_{er}, so that a direct comparison could be made between rival models of the system. The results of this study are compared with those obtained by Chalbi et al. (1986), Table 4. The estimated results extracted from the two-dimensional non-plug flow model demonstrate that both coefficients, h_w and k_{er}, decrease with a decrease in the mean Reynolds number, Re_p. This is a consequence of the lateral mixing. The results also reveal that both parameters are dependent upon d_t/d_p.

Comparison between the estimated results based on the non-plug flow model, Table 3, with those obtained by Chalbi et al. (1986) using a plug flow model, exhibits noticeable difference between the corresponding estimates. For relatively large beds, e.g. $d_t/d_p = 17.97$, the difference between the corresponding values of the two estimates is not very significant while for small beds, e.g. $d_t/d_p = 5.74$, the discrepancy was found to be quite important. This is caused by the pronounced anisotropic nature of the bed structure.

Inspection of Table 4 indicates that the difference between the corresponding estimated coefficients, h_w and k_{er}, of this work and that of Chalbi et al. (1986) to be as much as 8.4%. This discrepancy is mostly attributed to the nature of the employed model. The model used here is a heterogeneous, $T_f \neq T_s$, while the one developed by Chalbi et al. (1986) is of pseudohomogeneous, $T_f = T_s$, nature. In reality, there is a temperature gradient between the two phases.

In a further attempt, the estimated thermal coefficients, Table 3, have been used to generate correlations which will allow h_w and k_{er} to be predicted. Most of the previous investigators expressed the wall heat transfer coefficient and the effective radial thermal conductivity in terms of just the flow conditions, i.e. Reynolds number. However, it is now clear that these design parameters are also influenced by the physical dimensionality of the system, d_t/d_p, as well as the packing Reynolds number, Re_p.

In order to find suitable correlations for the desired parameters a range of mathematical expressions have been used. In each case the coefficients of the expressions were estimated by minimizing the sum of squares of residuals on the estimated values of h_w and k_{er} by means of Marquardt's method (1963). The established correlations for the wall heat transfer coefficient and the effective radial thermal conductivity for a system of packed bed heat exchanger with air as the flowing fluid and Alumina as packing (for $d_t/d_p = 5-18$) are as follows:

$$h_w = 354.3 \, d_p/d_t + 13.5 \, (d_p/d_t)^{0.642} \, Re_p^{0.685}$$

$$K_{er} = K_{er}^o + 0.0047 \, (1 + 24. \, d_p/d_t)^{-1} \, Re_p^{1.067}$$

where the static effective radial conductivity, $k_{er}^o = 0.95 \, k_f(1-\varepsilon)/(0.034+k_f/k_s)$, contribution is based on a previous correlation (Yagi and Kunii, 1975).

TABLE 3: Estimates of Heat Transfer Parameters Extracted from a Non-Plug Flow Model for Different Beds

d_t/d_p	Run No.	Flowrate x 10^3 (kg/s)	Re_p	h_w (W/m^{2}K)	k_{er} (W/m K)
17.97	S1	0.5	34.4	25.98	0.249
	S2	1.6	110.1	37.89	0.449
	S3	2.6	178.9	55.42	0.582
	S4	3.7	254.8	73.53	0.769
	S5	4.9	337.2	94.30	0.938
	S6	8.0	550.4	140.99	1.508
9.68	T1	0.5	122.0	73.21	0.893
	T2	1.4	332.0	106.22	1.682
	T3	2.3	552.0	231.65	2.200
	T4	3.9	961.0	314.72	3.537
	T5	4.9	1194.0	374.43	3.868
	T6	9.3	2262.0	629.82	6.186
7.52	U1	0.5	196.6	142.75	0.985
	U2	1.57	629.0	456.77	1.352
	U3	2.6	1022.0	574.54	2.150
	U4	3.7	1454.0	674.11	2.992
	U5	4.9	1929.0	762.99	3.987
	U6	8.0	3145.0	1119.42	6.299
5.74	V1	0.5	348.0	405.22	0.98
	V2	1.0	742.0	489.07	1.399
	V3	2.0	1400.0	740.73	2.145
	V4	2.6	1724.0	800.24	2.969
	V5	3.7	2579.0	865.33	4.523
	V6	4.5	3139.0	953.94	5.342

TABLE 4: Comparison Between Estimated Values of This and Published Work, Chalbi et al., (plug flow model), $d_t/d_p = 17.97$

Run No.	Flowratex10^{-3} kg/s	Re_p	h_w			k_{er}		
			Chalbi et al.	This Work	e%	Chalbi et	This al.	e% Work
SP1	0.5	34.4	17.3	19.0	8.5	0.23	0.22	1.3
SP2	1.6	110.1	36.5	38.7	5.8	0.42	0.43	1.9
SP3	2.6	178.9	56.0	56.2	0.4	0.57	0.63	8.4
SP4	3.7	254.6	75.0	74.5	0.7	0.76	0.81	6.1
SP5	4.9	337.2	97.6	95.4	2.3	0.94	1.01	6.3
SP6	8.0	550.4	144.1	142.5	1.1	1.52	1.60	5.5

Comparison between the various values associated with k_{er} obtained by rival approaches, e.g. present, Chalbi et al. (1986), De Wasch and Froment (1971), and Yagi and Kunii (1957), for varying diameter ratios, d_t/d_p, revealed interesting trends. Inspection of the results highlighted that at $d_t/d_p = 17.97$ the value of k_{er} predicted from correlation presented in this work is much higher than the others, while for low d_t/d_p the discrepancy becomes insignificant. In addition, the present wall heat transfer coefficient correlation is compared with those obtained by Beek (1962), Caldwell (1968), Li and Finlayson (1977), Dixon (1978) and Chalbi et al. (1986) and the results indicate that the wall heat transfer coefficient, h_w, increases with increasing value of the flow parameter, Re_p. From a comparison point of view, the present data on h_w are higher than the corresponding published data for bed with $d_t/d_p = 7.52$ with a reversed pattern for bed with $d_t/d_p = 17.97$. This dependency between the observed wall heat transfer coefficient and the bed structural parameter, d_t/d_p, is evident from all the data. Typical results are being given in Figure 1.

CONCLUDING REMARKS

A two-dimensional heat transfer model of packed beds has been used to analyze the experimental data reported by Chalbi (1984). For this, the descriptive mathematical model of the system and the experimental data together with a suitable optimization algorithm were used to find the best estimates of the desired parameters. The estimated data were found to be influenced by the physical dimensionality of the system, d_t/d_p, as well as Re_p. The influence of d_t/d_p on the two thermal parameters experienced to be appreciable. The established data have been used to generate suitable correlations and comparison between published and presented data showed noticeable differences between the two, particularly for beds with higher tube to particle diameter ratios.

NOMENCLATURE

C	heat capacity (J/kg K)		$G, \overline{G}$	flowrates, (kg/s, kg/m^{2}s)
d	diameter (m)		ϵ	voidage (-)
h	convective coefficients (W/m^{2}K)		ρ	density (kg/m^3)
k	thermal conductivity (W/mK)			
r	radial co-ordinate		Subscripts	
R	tube radius (m)		er	effective radial
Re	Reynolds number, $G\, d_p/\mu$		f,s	fluid, solid
T	temperature		i	initial
u	velocity (m/s)		p,t	particle, tube
θ	time (s)		w	wall

REFERENCES

Beek, J., (1962). "Design of packed catalytic reactors", Adv. Chem. Eng. 3, p. 203.

Caldwell, A.D., (1968). "Wall heat transfer coefficient with gas flow through packed beds", Chem. Eng. Sci. 23, p. 393.

Chalbi, M., (1984). Ph.D. Thesis, University of Compiegne, France.

Chalbi, M., Castro, J.A., Rodrigues, A.E., Zoulalian, A., (1987). "Heat transfer parameters in fixed bed exchangers", The Chem. Eng. Journal 34, p. 89.

Cybulski, A. and Kawechi, W., (1972). "Analysis of methods of the measurement of the transfer coefficients in packed beds". Inz. Chem. 2 (2), p. 343.

De Wasch, A.P. and Froment, G.F., (1971). "Heat transfer in packed beds", Chem. Eng. Sci. 27, p. 567.

Dixon, A.G., (1978). "Heat transfer in packed beds of low tube/particle diameter ratio", Ph.D. Thesis, University of Edinburgh, Scotland.

Dixon, A.G. and Cresswell, D.L., (1979). "Theoretical prediction of effective heat transfer parameters in packed beds". A.I.Ch.E.J. 25, p. 663.

Foumeny, E.A. and Ma, J., (1991) in Heat Exchange Engineering by Foumeny, E.A. and Heggs, P.J. (Editors), Vol. 1, Chapter 11, Simon and Schuster International.

Marquardt, D. W., (1963). "Estimation of non-linear parameters". J. Soc. Ind. Appl. Math. 11(2), p. 431.

Pahlevanzadeh, H., (1990). "Heat transfer studies in packed bed systems", Ph.D. Thesis, University of Leeds.

Stankiewicz, A., (1989). "Advances in modelling and design of multitubular fixed-bed reactors". Chem. Eng. Technol. 12, p. 113.

Yagi, S. and Kunii, D., (1960). "Studies of heat transfer near wall surfaces in packed beds". A.I.Ch.E.J. 6, p. 97.

Yagi, S. and Wakao, N., (1959). "Heat and mass transfer from wall to fluid in packed beds", A.I.Ch.E.J. 3, p. 373.

Zehner, P. and Schluender, E.U., (1970). "Waermeleitfahigkeit von schuettunger bei massigen temperatuern". Chem. Eng. Technol. 42, p. 933.

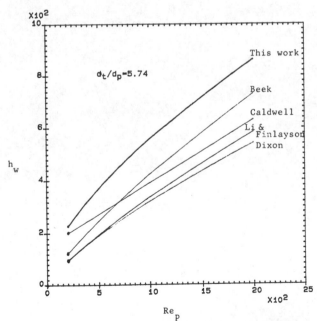

Figure 1:   Comparison between wall heat transfer coefficients of rival correlations

HEAT TRANSFER TO MOVING FIBRED BLADES IN PACKED BEDS

R.M. Davies and A.R. Thit

British Gas plc, Midlands Research Station, Solihull, West Midlands, B91 2JW

The measurement of heat transfer coefficients at the
surface of moving stirrer blades in packed beds is
described. The use of an existing theory to analyse the
results of these measurements, made in laboratory
apparatus, in the design of components for a
commercial coal gasification reactor is explained.

INTRODUCTION

British Coal's continuous slagging gasifier, developed by British
Gas, may slagging (BGC/Lurgi) process, developed by Lurgi and Scott [1,2] for the
production of a medium calorific value gas which can be used as fuel gas and
synthesis gases as well as to generate electricity via a Gasification-Combined
Cycle scheme. The Slagging Gasifier, shown in Fig.1, is a counter-current fixed
bed process operated at pressures of up to 30 bar. Coal is fed into the top of
the reactor via a lock-hopper system. The gasifier is water-cooled and
refractory lined. The bottom part of the coal bed is heat and is kept clean
and oxygen are injected and high velocity through tuyeres. A zone of extreme
turbulence known as a raceway is formed in front of the tuyere where
temperatures of residual char take place. Temperatures in excess of 2000 degC
are reached, ensuring that the mineral matter in the coal forms a molten slag
which collects in a hearth and is discharged periodically into water.
Gasification and devolatilisation reactions occur in the main part of the coal
bed above the raceway. The product gas leaves the reactor via an offtake above
the top of the bed. The process was developed at the British Gas Westfield
Development Centre in Scotland over a number of years and has been demonstrated
there on a 500 tonne/day plant.

The gasifier is designed to handle a range of fuels, including bituminous coals
of caking rank. The properties of these coals are such that the gas system
swells and cakes as they are heated. A multi-bladed, water-cooled rotating
stirrer is used to maintain the free-flowing fuel down the gasifier. The
blades of this stirrer are subject to thermal stresses and to abrasion by the
coal particles. In order to be able to design the blades for a long lifetime it
is advantageous to be able to calculate their surface temperatures and the heat
fluxes through them. This requires a knowledge of the heat transfer coefficient
at the blade surfaces. This paper describes how these coefficients have been
determined under gasification conditions.

HEAT TRANSFER TO MOVING STIRRER BLADES IN PACKED BEDS

R.M. Davies and A.R. Tait
British Gas plc., Midlands Research Station, Solihull, West Midlands

The measurement of heat transfer coefficients at the surface of moving stirrer blades in packed beds is described. The use of an existing theory to apply the results of these measurements, made in laboratory apparatus, to the design of components for a commercial coal gasification process is explained.

INTRODUCTION

British Gas plc., in conjunction with Lurgi GmbH, has developed the British Gas/Lurgi Slagging (BGL) Gasifier, as described by Lacey and Scott (1), for the production of a medium calorific value gas which can be used as fuel gas and synthesis gas as well as to generate electricity via a Gasification-Combined Cycle scheme. The Slagging Gasifier, shown in Fig.1, is a counter-current fixed bed process operated at pressures of up to 70bar. Coal is fed into the top of the reactor via a lock-hopper system. The gasifier is water-cooled and refractory lined. The hottest part of the fuel bed is near its base where steam and oxygen are injected at high velocity through tuyeres. A zone of extreme turbulence known as a 'raceway' is formed in front of the tuyeres where combustion of residual char takes place. Temperatures in excess of 2000 deg.C are reached, ensuring that the mineral matter in the coal forms a molten slag which collects in a hearth and is discharged automatically into water. Gasification and devolatilisation reactions occur in the main part of the coal bed above the raceway. The product gas leaves the reactor via an offtake above the top of the bed. The process was developed at the British Gas Westfield Development Centre in Scotland over a number of years and has been demonstrated there on a 500 ton/day plant.

The gasifier is designed to handle a range of fuels, including bituminous coals of varying rank. The properties of these coals are such that they may soften, swell and cake as they are heated. A multi-bladed, water-cooled rotating stirrer is used to maintain the free flow of the fuel down the gasifier. The blades of this stirrer are subject to thermal stresses and to abrasion by the fuel particles. In order to be able to design the blades for a long lifetime it is advantageous to be able to calculate their surface temperatures and the heat fluxes through them. This requires a knowledge of the heat transfer coefficient at the blade surface. This paper describes how these coefficients have been determined under gasification conditions.

THEORY

The most direct way of determining the heat transfer coefficients would have been to carry out temperature and heat flux measurements in the gasifier itself. However, this was not possible due to the severity of the operating conditions. The alternative was to take a theoretical approach but a survey of the literature revealed little previous work on the subject of heat transfer to surfaces immersed in packed beds and even less on stirred systems. Schlunder (2) reviewed the subject and he provided a method which was thought to be suitable for analysing heat transfer data for this type of system. The heat transfer in the gasifier takes place by a combination of conduction, convection and radiation, with both gas and solids phases taking part; however, Schlunder's analysis applies to the particle and gas conductive and particle convective terms but excludes gas convection and radiation terms. These have to be calculated separately by empirical means. The approach advocated by Schlunder is based on a transient analysis of heat transfer between a stationary packed bed and an immersed surface. The time-dependent coefficient is defined as :

$$h(t) = \frac{Q(t)}{Tw(t) - Tmean(t)} \quad W/m^2/K \tag{1}$$

In general the temperature of the bed near the wall (Ts) will lie somewhere between Tw and Tmean and hence a maximum coefficient can be defined as :

$$hmax(t) = \frac{Q(t)}{Tw(t) - Ts(t)} \quad W/m^2/K \tag{2}$$

This approach can be applied to a a steady moving bed by utilising the concept of contact time. If h is the limiting case of a moving bed with infinite contact time then hmax will be the limiting case for zero contact time. When applied to the case of a rotating stirrer in a packed bed the contact time becomes inversely proportional to the stirrer speed. The real stirrer in the gasifier should, therefore, approach a maximum coefficient at a high stirrer speed. Schlunder further suggests that the coeficient, h, at any speed should be given by the relationship :

$$\frac{h}{hmax} = \frac{1}{1 + \sqrt{(\pi/4.Ntherm.Nmech)}} \tag{3}$$

In this expression Ntherm is a dimensionless parameter which is a function of the thermal properties of the bed while Nmech is an empirical stirrer effectiveness parameter which is a function of blade geometry (i.e. its shape but not its size).

$$Ntherm = \frac{hmax^2}{K.\rho.Cp.speed} \tag{4}$$

The overall effective bed conductivity, K, can be calculated by the method of Wakao and Kaguei (3). In a high temperature, radiation dominated system such as the gasifier it should also be possible to calculate hmax from :

$$hmax = 4.\sigma.Tmean^3 \quad W/m^2/K \tag{5}$$

However, before equations (3), (4) and (5) could be applied to the gasifer it was necessary to test the theory and to determine empirically the correct value of Nmech. Since Nmech should only be a function of the shape of the blade it was decided to employ a laboratory scale rig to obtain suitable validation and design data.

APPARATUS

A cut-away diagram of the rig is shown in Fig. 2. The main vessel was 0.5m in diameter and 1.0m high. It was filled with a static bed of particles through which hot air could be blown from the bottom. The wall of the vessel was cooled by six copper cooling coils and the vessel was instrumented with heat flux meters and thermocouples placed at different heights. The blades themselves were also water-cooled and fully instrumented and were fitted to a water-cooled shaft which was driven by a variable-speed motor. Details of the blade and shaft are shown in Fig. 3. Separate metered cooling water flows were deliverd to the six coils, the blades and the stirrer shaft. The instrumentation on the rig consisted mainly of temperature and flow measuring devices. These were divided between static and rotating sensors, the data from which were collected by two different logging systems. The static data presented no unusual problems but the data from the blade and shaft was dealt with in a novel way. Taking temperatures from a rotating system usually involves passing low voltage thermocouple signals across rotating slip-rings. This can lead to large errors in the measured temperatures because of electrical noise. This was overcome by mounting a datalogger directly on top of the rotating shaft, so that it rotated with it, and then by passing a digital signal through a slip-ring system, which also carried mains power to the data logger. All the data from both loggers was then collected on a personal computer for further analysis.

EXPERIMENTS

The procedures for validating Schlunder's theory and for determining Nmech were the same. The air flow, stirrer speed and air inlet temperature were set and the rig was left to reach steady state. This was judged to have been achieved when the air outlet temperature became constant. The time required to reach steady state varied from 1 to 2.5 hrs depending on the air flowrate used, the lower flows taking the longer time. At each steady state a number of samples of data from each instrument were taken.

A matrix of experiments were carried out in which the air flows were varied from .001 to .003 $s.m^3/s$, the air temperature from 160 deg.C to 240 deg.C and the stirrer speed from 0 to 5 rpm. Three different types of packing, .007, .015 and .02 m diameter alumina spheres, .01 m diameter steatite spheres and gravel of mean size .01 m were used. Two different types of blades, a simple cylindrical design as indicated in Fig. 3 and a shaped blade more typical of the gasifier were fitted to the shaft.

The point of these experiments was to calculate the heat transfer coefficient at the blade surface. This was given by :

$$h = \frac{Q}{Tw - Tmean} \quad W/m^2/K \tag{6}$$

The heat flux per unit area was given by the heat flux meters embedded in the blades, the surface temperature Tw was given by the blade thermocouples and the air temperature Tmean was given by a heat balance over the relevant part of the packed bed. An overall heat balance was also carried out as a check on the validity of the experiments.

Since Schlunder's theory applies only to the contributions of bed conduction and solids convection to the heat transfer coefficient, account had to be taken of the gas convection which occurs in practice. The gas convective coefficient was measured in the experiments by turning off the air supply and measuring the transient heat flux to the blade as the bed cooled down. This was used to estimate the conductive and solids convective contribution at the point of cut-off and hence, by subtraction, the gas convective term.

RESULTS

A typical result of a 'cut-off' experiment is shown in Fig. 4. Although the heat flux drops quite quickly there is no sharp cut-off. This is because of the thermal inertia of the blade itself. However, once the rate of cooling becomes linear the data can be regressed linearly, as indicated, to give the conductive and solids convective heat load. In this case the gas convective term represented almost two thirds of the heat load. When this was converted into a heat transfer coefficient it could be subtracted from the overall value given by equation (6). Fig. 5a indicates a typical set of values of the overall coefficient for one point on the shaped blade. There is a significant amount of scatter. However, when the gas convective term was removed, as shown in Fig. 5b, the scatter was reduced significantly and it can be seen that hmax was approached above a stirrer speed of about 1 rpm. Figs. 5c and 5d show plots of h/hmax against Ntherm and 1/h against √Ntherm. In the former h/hmax tends towards unity as Ntherm tends towards zero, as equation (3) predicts. In the latter the data falls on a straight line, as would also be predicted by equation (3). Plots such as 5d can be used to determine the value of Nmech from the gradient which is equal to $\sqrt{(\pi/4.Nmech)}$/hmax. When Nmech was determined for the different packings and blade sections it was found that there was a significant variation around the surface of the blade and that lower values were obtained in gravel than in uniform spheres. Typical results for the shaped blade were 0.61 when using the spheres and 0.24 for the gravel.

CONCLUSIONS

The results presented in Figs. 5c and 5d confirm that Schlunder's theory can be used to analyse heat transfer to moving surfaces within packed beds. The theory proposes a mixing parameter, Nmech, and equation (6) indicates that the lower the value of Nmech the more closely the heat transfer coefficient will approach its maximum. This means that for a greater degree of mixing there will be a lower sensitivity to stirrer speed. This analysis in combination with the experimental apparatus described here provides a viable method for assesing the effectiveness of different designs of stirrer blade and for estimating the heat flux to them in the real, reacting situation. This can be done by calculating a value of hmax from equation (5), calculating Ntherm from the appropriate physical propery data via equation (4), and then by inserting these values and the measured value of Nmech as given by the experiments into equation (3). A gas convective coefficient would then have to be calculated and added to give an overall coefficient. It is true to say that without these experiments it would not be possible to obtain a value of the overall coefficient at gasifier conditions and hence it would not be possible to provide vital design data.

ACKNOWLEDGEMENT

This paper is published with the permission of British Gas plc.

REFERENCES

1. Lacey, J.A. and Scott, J.E., 1985, Proceedings of the IGU 16th World Gas Conference, Munich.

2. Schlunder, E., 1982, International Heat Transfer Conference, Munich.

3. Wakao, N. and Kaguei, S., 1982, Heat and mass transfer in Packed Beds, Gordon and Breach, New York.

NOMENCLATURE

h, h(t)	steady state and transient heat transfer coefficients	$W/m^2/K$
hmax, hmax(t)	maximum value of heat transfer coefficients	$W/m^2/K$
Q, Q(t)	steady state and transient heat flux to immersed surface	W/m^2
Tw, Tw(t)	steady state and transient surface temperature	K
Tmean, Tmean(t)	steady state and transient mean bed temperature	K
Ts, Ts(t)	steady state and transient bed temperature close to surface	K
Ntherm, Nmech	dimensionless parameters	
K	effective bed thermal conductivity	$W/m/K$
ρ	bed bulk density	Kg/m^3
Cp	bed specific heat capacity	$J/Kg/K$
speed	blade rotational speed	sec^{-1}
σ	Stefan-Boltzman constant	5.67×10^{-8} $W/m^2/K$

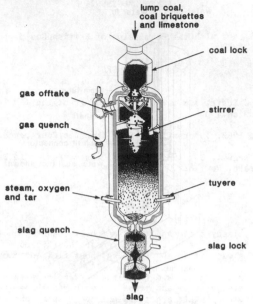

FIG. 1. THE BGL GASIFIER

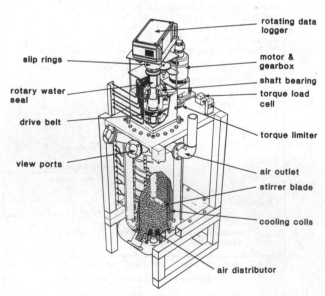

FIG. 2. THE STIRRER HEAT TRANSFER RIG

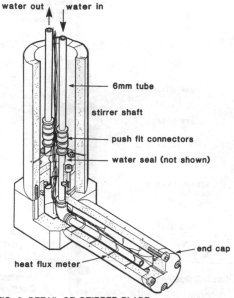

FIG. 3. DETAIL OF STIRRER BLADE

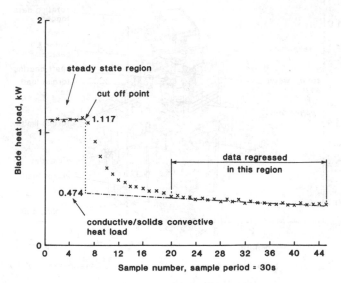

FIG. 4. TYPICAL CUT-OFF EXPERIMENT

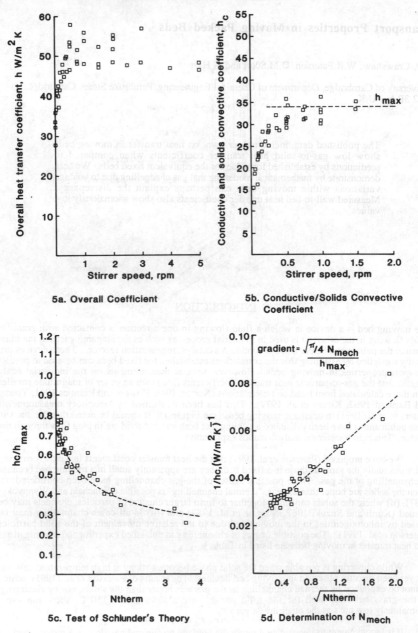

5a. Overall Coefficient

5b. Conductive/Solids Convective Coefficient

5c. Test of Schlunder's Theory

5d. Determination of N_{mech}

FIG. 5. RESULTS FOR SHAPED STIRRER BLADE

Transport Properties in Moving Packed Beds

J. P. Crawshaw, W.R.Paterson, D.M.Scott and G.Hart

University of Cambridge, Department of Chemical Engineering, Pembroke Street, Cambridge.
CB2 3RA   U.K.

The published data, including our own, on heat transfer in moving beds show low gas-to-solid heat transfer coefficients when compared to predictions by established formulae for the equivalent fixed beds. We here demonstrate by mathematical modelling that gas channelling due to voidage variations within moving beds can perhaps explain the discrepancy. Measured wall-to-bed heat transfer coefficients also show anomalously low values.

INTRODUCTION

The moving bed is a device in which a fluid flowing in one direction is contacted with granular solids flowing in the other; it is used in industrial processes such as the slagging gasifier, the blast furnace, the pebble bed heater and the continuous catalyst regeneration reactor. These devices are usually modelled by assuming that the correlations established for fixed beds can be used to predict transport properties in moving beds. However, several measurements on the industrial scale suggest that the gas-to-particle heat transfer coefficients (htc's) are an order of magnitude smaller than those calculated from fixed bed correlations (Norton 1946, Bowers and Reintjes 1961, Young and Barnard 1962, Kitaev et al. 1977). This has been confirmed by laboratory measurements (Paterson et al. 1991) in adiabatic moving beds (see Figure 1). It should be noted that the htc's in these publications have been calculated assuming that both gas and solid are in plug flow through the reactor. This heat transfer anomaly demands explanation.

We have proposed (Paterson et al. 1991) that the heat transfer coefficients in a slowly moving bed are actually the same as those in a fixed bed: they are apparently small in a moving bed because of channelling of the gas. Three possible causes of the gas channelling have been considered: (i) when the solids are hotter than the gas, "thermal channelling" is possible (Wonchala and Wynnyckyj 1987), (ii) where the solids can sinter together to form larger clusters of particles, channels may be formed (Kunii and Suzuki 1967, Norgate et al. 1982) and (iii) in all cases channelling may be caused by inhomogenities in the moving bed due to the relative movement of the solid particles (Paterson et al. 1991). The possible causes of channelling in published experimental investigations into heat transfer in moving beds are listed in Table 1.

While clustering is possible when the solid phase becomes sticky at high temperature, this has not yet been directly observed in moving bed devices. The results of Paterson et al. (1991), which cannot be explained by thermal channelling as the gas was hotter than the solids, nor by clustering, as the glass ballotini used for the solid phase were always below 150°C, show the same anomalously low htc's as the other investigators'.

It is inconvenient to investigate directly the voidage profile and gas flow in a moving bed, as the flow of solids makes design of the apparatus difficult. Fortunately pressure drop measurements (Paterson et al. 1992) suggest that the structure of a "frozen" bed (which has been moving and has then been stopped) is the same as the structure of the corresponding moving bed. Information about the structure of a moving bed may therefore be gained by (a) flooding the corresponding frozen bed with a solidifying material, cutting the resulting solid into sections and studying them visually, and (b) making gas residence time distribution measurements in the corresponding frozen bed.

The consequences of gas channelling due to voidage inhomogenities have been investigated by simple mathematical modelling. This demonstrates that the observed discrepancies in gas-to-particle htc's could be caused by plausibly small variations in bed voidage. The model and some results are presented below.

GAS FLOW MALDISTRIBUTION MODELLING

The model of Schlunder (1977) and Martin (1978) of gas flow through a non-uniformly packed bed is now extended to a counter current moving bed system. The simplest model of the voidage variation is a sharp, step change in the voidage. The bed is assumed to have one voidage, ε_2, in the core and another lower voidage, ε_1, in the annulus near the wall. The pressure gradient, $\Delta P/L$, will be the same for both regions leading to a different gas superficial velocity, u_1 and u_2, in each region. The pressure gradient for each region may be described by the Ergun equation

$$\frac{\Delta P}{L} = \frac{150(1 - \varepsilon_i)^2 \mu u_i}{\varepsilon_i^3 d_p^2} + \frac{1.75(1 - \varepsilon_i)\rho u_i^2}{\varepsilon_i^3 d_p} \qquad i = 1,2 \qquad (1)$$

While there is some question as to the validity of this equation with the coefficients 150 and 1.75 for moving beds (Paterson et al. 1992), it was considered adequate given the preliminary nature of the model. The uniform inlet superficial gas velocity, u, is given by

$$u = (1 - \varphi)u_1 + \varphi u_2 \qquad (2)$$

where φ is the fraction of the bed cross-sectional area in region 2. We set $\Delta P_1 = \Delta P_2$ then equations (1) and (2) comprise 3 equations in 3 unknowns ΔP, u_1 and $u_{.2}$. By fixing ΔP, u_1 and $u_{.2}$ may be calculated.

The main assumptions in developing the thermal model are:

      (i) in each zone the gas is in plug flow with no cross flow
      (ii) the solids are in plug flow throughout
      (iii) each zone is adiabatic, i.e. there is no heat flow across the zone boundaries
      (iv) the particles are small enough for temperature profiles within
          the solid to be ignored
      (v) the physical properties are constant

The gas-to-solid heat transfer coefficient, h, is calculated from the temperatures of gas, T_g, and solids, T_s, at some point, x, within the bed and at the solids inlet, $x = 0$:

$$h = \frac{Q_g}{A \, \Delta T_{LM}} \qquad (3)$$

where

$$Q_g = G_g c_g (T_{g,x} - T_{g,o}) \qquad (4)$$

$$A = \frac{6(1 - e) \times \text{volume of bed}}{d_p} \quad (5)$$

$$\Delta T_{LM} = \frac{(T_{g,x} - T_{s,x}) - (T_{g,o} - T_{s,o})}{\ln\left(\dfrac{T_{g\,x} - T_{s\,x}}{T_{g\,o} - T_{s\,o}}\right)} \quad (6)$$

and

$c_{g,s}$ - heat capacity of gas , solids respectively

$G_{g,s}$ - mass flux of gas , solids respectively

The main aim of the model is to calculate the voidage variations and area fractions (φ) for the two zones necessary to account for the observed heat transfer coefficients. Temperatures are often measured at the centre line of the bed, that is within zone 2 of our model, while solids temperatures, which are difficult to measure within the moving bed, have been calculated by heat balance using the solids inlet temperature and the gas temperatures for the two positions:

$$T_{s,x} = \frac{G_g c_g (T_{g,x} - T_{g,o})}{G_s c_s} + T_{s,o} \quad (7)$$

This heat balance is valid only for plug flow of gas and solids in the whole bed; where channelling occurs it is a source of error in determining h.

The model may be applied by first assigning values for $\varepsilon_1, \varepsilon_2, \varphi, G_g, G_s$ and the bed geometry, then using equations (1) and (2) to determine the velocities in each region. Then typical experimental values are assigned to $T_{s,o}$ and $T_{g,o}$ and the temperature profiles in each zone are then calculated using the htc's for each zone estimated from the fixed bed correlation of Wakao and Kaguei (1982). Finally an apparent htc is calculated from equations (3) to (7) using the temperatures in the bed core (region 2) and the overall gas and solid flow rates.

For parameters typical of the experimental work of Paterson et al. (1991), the results of the modelling are shown in Figure 2. The apparent heat transfer coefficient is very sensitive to bypassing and falls rapidly from the uniform bed value as φ is reduced below 1. For a voidage variation of only 10% between the core and the edge of the bed, the htc becomes a factor of 10 smaller than that predicted by the fixed bed correlation when $\varphi = 0.7$.

The model cannot readily be tested against the other workers results, because they reported insufficient experimental detail, leaving us unclear as to the means and location of the temperature measurements.

WALL HEAT TRANSFER COEFFICIENTS

The wall heat transfer coefficient (U_w) has also been measured in a steam jacketed moving bed. The apparatus consisted of the adiabatic rig described by Paterson et al. (1991) with the insulation removed and a steam jacket installed. Heat flux gages were installed within the steam jacket on the outside of the tube containing the moving bed in order to measure both the steam temperature and the heat flux into the bed. The air temperature at the bed wall was also measured and U_w was evaluated from the temperature differe nce and the heat flux.

Figure 3 shows a comparison of the measured U_w to those calculated by the correlation of Colledge (1985) for the equivalent fixed beds, that is those of equal mean voidage, bed radius and Reynolds number. Again the established correlation for fixed beds predicts heat transfer coefficients much larger than those measured in moving beds. This too is qualitatively consistent with the gas having channelled preferentially to the bed core, leaving a flow deficit in the outer region, consequently reducing the rate of heat transfer at the wall.

CONCLUSIONS

The published data on heat transfer in moving beds show low htc's when compared to predictions by established formula for the equivalent static packed beds. The results of Paterson et al. (1991) cannot be explained by appealing to either thermal channelling or to clustering. However, the results of a simple, preliminary model indicate that voidage variations of about 10% between a core which occupies 70% of the bed cross-sectional area and an annulus close to the wall can account for the observed behaviour of the htc's. While the model is crude, the voidage variations it demands are modest and this indicates that the proposed channelling mechanism is feasible.

Work is in progress to measure directly the voidages and residence time distributions of frozen beds so that experimental evidence for the existance of gas flow maldistribution may be presented.

ACKNOWLEDGMENT

We are grateful to British Gas and SERC for financial support.

REFERENCES

Bowers, T.G., and Reintjes, H. "A review of fluid to particle heat transfer in packed and moving beds". Chem. Eng. Prog. Symp. Ser. 1961, v.57, no. 32 , pp. 69 - 74

Colledge, R.A. "Heat and mass transfer at the wall of a packed bed at high Reynolds numbers " Ph.D. Thesis, University of Cambridge, 1985

Kitaev, B.I., Yaroshenko, Yu.G., and Suchkov, V.D. "Heat Exchange in Shaft Furnaces", pg. 136 , 1st. edition , Pergamon Press , 1977

Kunii, D. and Suzuki, M. "Particle to fluid heat and mass transfer in packed beds of fine particles" Int. J. Heat and Mass Transfer 1967, 10, pp. 845 - 852

Martin, H. "Low peclet number particle to fluid heat and mass transfer in packed beds" Chem. Eng. Sci. 1978 , 33 , pp. 913 - 919

Norgate, T.E, Batterham, R.J., Thurlby, J.A., and Povey, P.C. "Mixing effects in large scale packed beds" Chemeca (Tenth Australian Chemical Engineering Conference, 1982, pp. 83 - 87

Norton, C.L. "The pebble bed heater - a new heat transfer apparatus" J. Am. Cer. Soc. 1946 , v. 29, part 7 , pp. 187 - 193

Paterson, W. R., Hart, G. and Scott, D. M. "Heat Transfer in Moving Packed Beds" Proceedings of the 1991 I. Chem. E. Research Event, Queens' College, Cambridge. 315-316

Paterson, W. R., Crawshaw, J. P., Hart,G., Parker, S. R., Scott, D. M. and Young, J. P. "Pressure Drop in Countercurrent Flow Through Moving Particulate Beds" Chemical Engineering Research and Design (accepted for publication 1992)

Schlunder, E.U. "On the mechanism of mass transfer in heterogeneous systems - in particular in fixed beds, fluidized beds and on bubble trays" Chem. Eng. Sci. 1977, 32, pp. 845 - 851

Wakao, N. and Kaguei, S. "Heat and Mass Transfer in Packed Beds" Gordon and Breach, Science Pub. Inc. 1982 , pg. 293

Wonchala, E.P., and Wynnyckyj, J.R. "The phenomenon of thermal channelling in counter - current gas solid heat exchangers" Can. J. Chem. Eng. 1987 , 65 , pp. 736 - 743

Young, P.A., and Barnard, D.A. "The cooling of sinter", in Agglomeration ed. William A. Knepper, Interscience Publishers , 1962

Table 1 Possible causes of gas flow maldistribution

Investigator	Date	Hot phase	$d_p/(mm)$	Clustering
Norton	1946	Solid	8 to 12	No
Bowers and Reintjes	1961	Gas	8 to 38	Possible
Young and Barnard	1962	Solid	?	Possible
Kiatev et al.	1967	Gas	?	Possible
Paterson et al.	1991	Gas	3	No

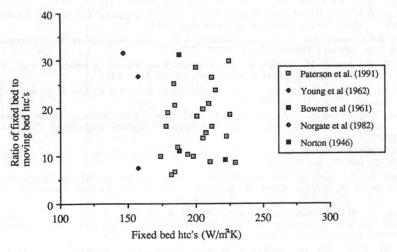

Figure 1: Ratio of fixed bed to moving bed gas-particle htc's versus fixed bed htc's

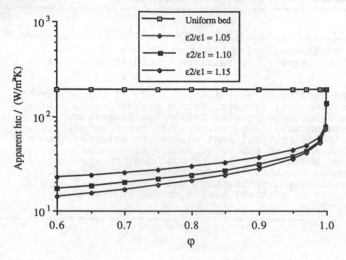

Figure 2: Influence of bypassing on apparent htc (ε_l = 0.37)

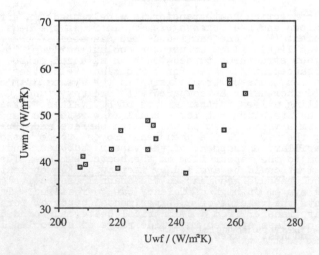

Figure 3: Wall heat transfer coefficients predicted for fixed beds, Uwf, versus those measured in equivalent moving beds, Uwm

FORCED CONVECTION COOLING OF A SURFACE EMBEDDED IN POROUS STRUCTURE

V. V. Kuznetsov, S. B. Chikov
Institute of Thermophysics, Russian Academy of Sciences,
Prospect Lavrentyev 1, Novosibirsk 630090, Russia

The results of experimental study of forced convection heat transfer from a heater inside a liquid-saturated porous medium are presented. The
data on wall temperature, heat transfer coefficients and critical heat fluxes are obtained in a
wide range of subcooled liquid flow rates. It is
also shown that the porous medium application
causes the heat transfer enhancement in the single-phase convection mode and allows to get higher
values of critical heat flux due to flow rate increasing.

INTRODUCTION

The forced convection heat transfer is widely used in different
cooling systems and heat exchangers. To increase the heat transfer coefficient (HTC) one can reduce the diameter of the channel
where cooling liquid flows or use the coolant boiling. Application of porous structures turned to be an advanced method of heat
transfer enhancement /1/.Even for low coolant velocities the porous structures cause powerful mixing in the system of microchan-
nels and HTC increasing as compared with ordinary tube. Once the
coolant boiling arises further growth of HTC values takes place
in the porous structure but for a small flow rate the vapour bubbles can form a vapour film on the heated surface and thus sharply reducing the HTC /2/. The phenomena proves to be associated
with the capillary entrapment of the vapour bubbles and their
amalgamation to the vapour film on the heated surface. The capillary entrapment could be avoided by increasing the coolant flow
rate and the pressure gradient in the channel respectively. Thus
the cooling system of high efficiency can be realized. This report presents the results of an experimental study of forced convection heat transfer from a cylindrical heater surrounded with a
packed bed of glass spheres in a wide range of subcooled liquid
flow rates and heat fluxes.

EXPERIMENTAL APPARATUS AND PROCEDURES

The experimental apparatus employed in this investigation is the
plant consisted of reservoir, test section and condenser.The test
section shown in Fig. 1 is a vertical stainless steel tube of 6mm
external diameter 1 with a wall thickness of 1mm and 400 mm long

mounted in a thick-wall fluoroplastic tubing of 28 mm inner diameter 2 and then in stainless steel container. For high flow rates the pressure gradients caused by gravity are negligible   as
compared with the frictional losses and the results obtained for
vertical and horizontal channels should be the same. The channel
was filled with a uniformly packed bed of glass spheres 3 of 1.3
mm mean diameter, which were carefully placed into the test section to ensure uniformity in the structure of porous matrix. The
measured core porosity of the liquid-saturated packed spheres is
0.37. The heating of central stainless steel tube is provided by
electrical current, its capacity is measured carefully to   estimate the heat flux value. The temperature along the heater   and
in the porous bed is defined with 12 chromel-copel thermocouples
of 0.8 mm diameter: 6 of them are embedded in the wall of   the
heated tube 4 and the others are placed in the core of the   bed
to define the coolant temperature in the channel. The experimental accuracy of the thermocouples is 0.5K.

The inlet pressure and the pressure drop are measured   with
industrial electrical manometers with a factory calibrated accuracy of 2 percent full scale. An industrial rotameter₃ measures
the flow rate which has a range of 0.5 → 2.5 x 10⁻⁴ m³/s and a
factory calibrated accuracy of 1 percent full scale.

A computer-based data aquisition system is employed to record
accurately and swiftly the signals from every measurement   channel and to carry their initial processing. Measurements are carried out once steady-state hydrodynamic and thermal conditions
are reached. The fluid properties (e.g. density, specific heat,
velocity, thermal conductivity, etc.) are based on the inlet temperature and pressure and are calculated using regression curve
fits. The standard error of the fluid property estimates is negligible.

THEORETICAL ANALYSIS OF FORCED SINGLE-PHASE CONVECTION

Consider the single-phase convection heat transfer in porous medium for the configuration shown in Fig.1. The pressure drop in
 porous medium proves to be mainly associated with shear   stress
in porous bulk and one can believe that in scales much more than
the pore diameter the average velocity is constant   across the
flow section. Therefore, for the steady flow of uncompressible
liquid in porous medium the heat transfer equation can be writt-
en as

$$C_p \dot{m} \frac{\partial T}{\partial z} = \lambda_{ef} \left(\frac{\partial^2 T}{\partial z^2} + \frac{1}{r} \frac{\partial T}{\partial r} \right)$$

where z-coordinate is settled along the channel axis and r
is settled along the radius, C_p is the specific heat of   liquid
and λ_{ef} is the effective thermal conductivity defined in /3/ as

$$\frac{\lambda_{ef}}{\lambda_\ell} = 1.47 + 0.1 (Re_\ell)(Pr_\ell)$$

The boundary conditions on the heated surface when $r = R_1$
must express the high thermal resistance near the wall associated with molecular mechanism of heat transfer. In this case

$$-\lambda_{ef}\left(\frac{\partial T}{\partial r}\right)_{r=R_1+\delta_T} = \alpha_w(T_w - T_1) = \dot{q} \; ,$$

where α_w is wall HTC, δ_T is thermal layer thickness, and T_1 is the liquid temperature outside the thermal layer.

Usially, $\delta_T \ll R_1$ and the liquid temperature when $r = R_1$ can be determined from (1) without taking into account the wall HTC and T_1 can be defined from (3).

As shown in /3/ α_w can be estimated from

$$Nu_w = \frac{\alpha_w d_e}{\lambda_{ef}} = 0.09(Re_e)^{0.8}(P_{T_e})^{0.33}$$

where $d_e = 4\varepsilon d/6(1-\varepsilon)$ is the equivalent pore diameter, and the thermal layer thickness can be estimated as $\delta_T \approx \lambda_{ef}/\alpha_w$. The outer wall of the channel is insulated and

$$\lambda_{ef}\left(\frac{\partial T}{\partial r}\right) = 0$$

when $r = R_2$.

The solution of (1) with boundary conditions (3) and (5) is made in /4/ and looks like

$$T_\ell(z,r) = T_{in} + \frac{\dot{q}R_1}{\lambda_{ef}}\left\{\frac{1}{(B^2-1)}\left[2F_0 + \frac{2(r^*)^2-1}{4} - B^2\left(\ell_n\frac{r^*}{B} - \frac{\ell_n B}{(B^2-1)} + \frac{3}{4}\right)\right] + \right.$$

$$+ \sum_{n=0}^{\infty} \frac{\pi \, J_1(\mu_n) \, J_1(\mu_n B)}{\mu_n[(J_1(\mu_n))^2 - (J_1(\mu_n B))^2]} \left[J_0(\mu_n r^*)Y_1(\mu_n B) - \right.$$

$$\left. - Y_0(\mu_n)J_1(\mu_n B)\right] exp(-F_0 \mu_n)\bigg\}$$

where $F_0 = az/R_1^2$, $a = \lambda_{ef}/\dot{m}c_p$, $r^* = r/R_1$, $B = R_2/R_1$, and μ_n are the roots of equation

$$J_1(\mu_n)Y_1(\mu_n B) = Y_1(\mu_n)J_1(\mu_n B)$$

Considering the total HTC as

$$\alpha_\Sigma = \dot{q}/(T_w - T_b)$$

where T_b is average bulk temperature of the liquid, from (6) it can be obtained finally

$$\alpha_\Sigma = \frac{\dot{q}}{\left[(T_\ell(R_1) - T_b) + \frac{\dot{q}}{\alpha_w}\right]}$$

DISCUSSION OF EXPERIMENTAL RESULTS

In this section the main experimental results of the forced convection with and without boiling are compared with the results
of the theoretical consideration of single-phase forced convection and known data for the RC318 pool boiling.
   The data on the difference between wall and mixed-mean fluid temperatures are shown in Fig.2 for subcooled flow boiling in
the inlet and outlet cross-sections when the inlet ΔT_{sub} = 30K
and $\dot{m}$ = 125 kg/m^{2}s. For high heat fluxes the points illustrated
for porous and hollow channels coincide and are in good accordance with the curves 1 and 2 for RC318 pool boiling /5/

$$\alpha_{pool} = \frac{\dot{q}}{T_w - T_{sat}} = 30.9 \frac{[P_c/[P_0]]^{1/4}[\dot{q}/[W/m^2]]^{3/4}}{[T_c/[K]]^{7/8} M_r^{1/8}} (0.14 + 2.2 \bar{p}_r) \; [W/m^2 K]$$

which is typical for developed subcooled boiling in channels.The
main difference between the values of wall temperatures in porous and in hollow channels is observed for small heat fluxes in
single-phase convection mode. The points for porous channel are
in good accordance with the curve 3 provided by (6) and the poi-
ntsfor hollow one agree with the line 4 which represents the known equation /6/ for single-phase forced convection heat transfer for internally heated annuli:

$$Nu = 0.86 \left(\frac{R_2}{R_1}\right)^{-0.16} \frac{f(Re_\ell - 10^3) Pr_\ell}{8[1 + 12.7\sqrt{\frac{f}{8}}(Pr_\ell^{2/3} - 1)]}$$

where

$$f = [1.82 \cdot lg(Re_\ell) - 1.64]^{-2}$$

   High pressure gradients in porous media cause the reduction
of the bubbles detachment diameter. For this mass velocity it is
significantly less than the pore diameter and capillary entrapment does not take place, so fully developed subcooled boiling
arises just after the first bubbles leave the heated surface.
   Flow rate influence on the wall temperature vs heat flux
value is shown in Fig. 3 when the inlet ΔT_{sub} = 30K. With the
flow rate growth the wall temperature for single-phase convection in porous channel reduces and the CHF increases.Once the flow
rate influence on the CHF in hollow channel is weak, one can see
that for the small coolant velocity the CHF for porous channel
is lower significantly than for hollow channel, while for high
velocity the CHF in porous channel takes place even later than
that in hollow one. The wall temperature values obtained from the
experiment for both velocities are in good accordance with values calculated by (6) for single-phase convection (lines 1 and 2)
and by (9) when boiling arises (line 3). It is also important

that in transcritical heat transfer in porous structure the wall temperature does not exceed 200°C for the heat fluxes close to CHF.
    There are data on the HTC for porous channel fitted in Fig. 4 for three mass velocities when the inlet subcooling ΔT_{sub}=30K The HTC for single-phase convection depend significantly on the flow rate but display their independence of the heat flux. In case of boiling the curves for all velocities coincide and the HTC values increase with the heat flux growth untill the crisis takes place.

CONCLUSIONS

It is shown that the HTC in channel filled with porous structure are high enough even if the porous bed heat conductivity is comparatively small. The main differences between the heat transfer rate for porous and hollow channel occur in single-phase convection mode. The single-phase convection HTC for cylindrical heat exchanger in channel filled with a porous bed were calculated and their values are compared with the experimental results for the RC318 various velocities.
    Once the developed subcooled nucleate boiling arises, the wall temperature quantities for porous channel prove to be similar to that in hollow one untill the CHF being reached. The hysteresis was not observed because of small bubbles detachment diameter, so the developed subcooled boiling onset can be defined at the moment when the wall temperature in single-phase convection mode approaches the value corresponding to the boiling curve.
    The CHF in porous channel depends essentially on the coolant flow rate. Thus, when the flow rate is m = 50 kg/m^{2}s the CHF in hollow channel exceeds considerably the CHF in porous one. The difference reduces with the flow rate being increased and for the rates more than m = 200 kg/m^{2}s the CHF values for porous channel occur even higher than those for hollow one. It is also important that the crisis in porous channel is of "soft" character as compared with the crisis in the channel without porous filling.

NOMENCLATURE

$\dot{q}$ heat flux, W/m^2
$\dot{m}$ mass velocity, kg/m^2s
$u = \dot{m}/\rho_\ell$ flow velocity, m/s
d diameter of a sphere, m
L test section length, m
S cross-section area, m^2
$z = \bar{Z}/L$ dimensionless coordinate
ε porosity
T temperature, K
P pressure, Pa
ρ density, kg/m^3
λ thermal conductivity, $W/m\,K$
γ viscosity, $Pa\,s$
c_p heat capacity, J/K

α heat transfer coefficient, W/m^2K
M_r relative molecular mass

Indexes:

b   = bulk
l   = liquid
w   = wall
c   = critical
r   = reduced
e   = equivalent
ef = effective
sat = saturation
sub = subcooling

Dimensionless groups:

$Re = 2(R_2 - R_1)\dot{m}/\gamma_\ell$ = Reynolds number
$Re_e = 2\dot{m}d/3(1-\varepsilon)\gamma_\ell$ = effective Reynolds number
$Pr = C_p\gamma/\lambda$ = Prandtl number
$Nu = 2(R_2 - R_1)\alpha/\lambda$ = Nusselt number

REFERENCES

1. Koh J. C. Y., Colony R., 1974, ASME Journal of Heat Transfer, vol. 96.
2. Parmentier E.M., 1979, Int. J. Heat Mass Transfer, vol. 22, p. 849-855.
3. Jagi S., Kunii D., Endo K., 1964, Int. J. Heat Mass Transfer, vol. 7, p. 333.
4. Luikov A. V., 1967, Heat Conductivity, Moscow, High School.
5. Danilova G. N., Kupriyanova A. V., 1970, Heat Transfer Sov. Res., vol. 2(2), p. 79-83.
6. Armstrong Robert C., Bauer R., Bergles Arthur E. et al., 1983, Heat Exchanger Design Handbook, vol. 2, "Fluid Mechanics and Heat Transfer", Hemisphere Pub. Corp.

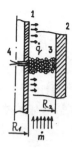

Fig. 1. Scheme of the test section

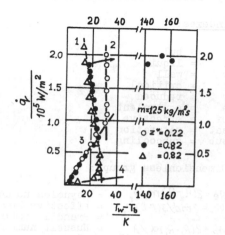

Fig. 2. Heat flux vs wall temperature: O,●-for porous channel, Δ - for hollow channel

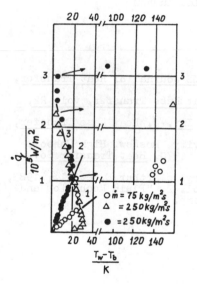

Fig. 3. Heat flux vs wall temperature: O,●-for porous channel, Δ - for hollow channel

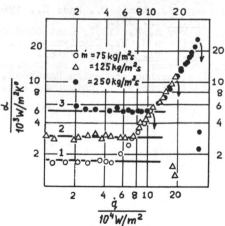

Fig. 4. HTC vs heat flux for porous channel

FREE CONVECTION IN AN INCLINED CAVITY WITH A PARTIALLY HEATED WALL AND PARTLY FILLED WITH A POROUS MEDIUM

P. H. Oosthuizen and J. T. Paul
Queen's University, Kingston, Ontario, CANADA K7L 3N6

Two-dimensional natural convective flow in an inclined square cavity with part of one wall heated to a uniform temperature and the opposite wall uniformly cooled to a lower temperature and with the remaining walls adiabatic has been considered. The cavity is partly filled with a fluid and partly filled with a porous medium, which is saturated with the same fluid, the fluid layer being separated from the porous medium layer by an impermeable partition which offers no resistance to heat transfer. This partition is, in general, inclined at an angle to the walls of the cavity. The governing equations written in terms of stream function and vorticity have been expressed in dimensionless form. The dimensionless equations, subject to the boundary conditions, have been solved using a finite element procedure. Results have then been obtained for a range of values of the governing parameters and the effect of these parameters on the heat transfer rate across the cavity has been studied.

INTRODUCTION

The situation considered in the present study is shown in Figure 1. It involves two-dimensional natural convective flow in an inclined cavity with part of one wall AB heated to a uniform temperature, T'_H, and the opposite wall DE uniformly cooled to a temperature, T'_C, which is less than T'_H, and with the remaining walls BCD and AFE adiabatic. The cavity is partly filled with a fluid and partly filled with a porous medium, which is saturated with the same fluid, the fluid layer being separated from the porous medium layer by an impermeable partition, CF, which offers no resistance to heat transfer. This partition is, in general, inclined at an angle to the walls of the cavity.

The situation under consideration is an approximate model of some situations that occur in building practice. Previous studies of vertical partitioned cavities partly filled with a porous medium have indicated that the heat transfer rate across the cavity is very significantly reduced as compared to that for an unpartitioned fluid filled cavity. The main aim of the present study was to determine how this reduction in the heat transfer rate is effected by the position and size of the heated wall section when the wall is only partly heated. The results have application in situations where the heat transfer across a cavity has to be reduced but cost considerations require that as little insulation material as possible be used.

The flow and heat transfer in cavities that are partly filled with a fluid and partly filled with a porous medium have been considered by a number of workers both for the case where there is no barrier between the layers and for the case where there is an impermeable barrier between the layers. Typical of these studies are those of Poulikakos and Bejan (1), Lauriat and Mesguich (2), Beckermann et al (3), Oosthuizen and Paul (4), Tong and Subramanin (5), Tong et al (6) and Song and Viskanta (7). The present study differs from these in that inclined cavity with a partly heated wall and with an inclined barrier has been considered.

GOVERNING EQUATIONS AND SOLUTION PROCEDURE

It has been assumed that the flow is steady, laminar and two-dimensional and that fluid properties are constant except for the density change with temperature which gives rise to the buoyancy forces, this being treated using the Boussinesq approach. It has also been assumed that, in the porous medium, the inertia term in the momentum equation is negligible. The usual Darcy assumptions have been adopted in the porous layer, except that the viscous shear stress term, i.e. the Brinkman term, has been retained although the inertia term has been neglected.

The solution has been obtained in terms of the stream function and vorticity defined, as usual, by:

$$u' = \frac{\partial \psi'}{\partial y'} \quad , \qquad v' = -\frac{\partial \psi'}{\partial x'} \quad , \qquad \omega' = \frac{\partial v'}{\partial x'} - \frac{\partial u'}{\partial y'} \tag{1}$$

The prime $('\ )$ denotes a dimensional quantity. In the porous layer, the velocity is, of course, the superficial or Darcian mean velocity.

The following dimensionless variables have then been defined:

$$\psi = \psi'/\alpha_f \quad , \qquad \omega = \omega' W'^2/\alpha_f \quad , \qquad T = (T - T'_C)/(T'_H - T'_C) \tag{2}$$

where $\alpha_f = \lambda_f/\rho_f c_f$ and where the subscript f denotes fluid properties. The cold wall temperature, T'_C, has been taken as the reference temperature. The coordinate system used is shown in Figure 2.

In terms of these dimensionless variables, the governing equations for the porous medium are:

$$\frac{\partial^2 \psi}{\partial x^2} + \frac{\partial^2 \psi}{\partial y^2} = -\omega \tag{3}$$

$$\left(\frac{\nu}{\nu_f}\right)\left(\frac{\partial^2 \omega}{\partial x^2} + \frac{\partial^2 \omega}{\partial y^2}\right) - \frac{\omega}{Da} = -Ra\left(\frac{\partial T}{\partial x}\cos\phi + \frac{\partial T}{\partial y}\sin\phi\right) \tag{4}$$

$$\frac{\partial^2 T}{\partial x^2} + \frac{\partial^2 T}{\partial y^2} - \left(\frac{\alpha}{\alpha_f}\right)\left(\frac{\partial \psi}{\partial y}\frac{\partial T}{\partial x} - \frac{\partial \psi}{\partial x}\frac{\partial T}{\partial y}\right) = 0 \tag{5}$$

where $\alpha = \lambda/\rho_f c_f$ and ϕ is the angle of inclination of the cavity.

Similarly, the dimensionless governing equations for the fluid layer are:

$$\frac{\partial^2 \psi}{\partial x^2} + \frac{\partial^2 \psi}{\partial y^2} = -\omega \tag{6}$$

$$\left(\frac{\partial^2 \omega}{\partial x^2} + \frac{\partial^2 \omega}{\partial y^2}\right) - \frac{1}{Pr}\left(\frac{\partial \psi}{\partial y}\frac{\partial \omega}{\partial x} - \frac{\partial \psi}{\partial x}\frac{\partial \omega}{\partial y}\right)$$

$$= -Ra\left(\frac{\partial T}{\partial x}\cos\phi + \frac{\partial T}{\partial y}\sin\phi\right) \tag{7}$$

$$\frac{\partial^2 T}{\partial x^2} + \frac{\partial^2 T}{\partial y^2} - \left(\frac{\partial \psi}{\partial y}\frac{\partial T}{\partial x} - \frac{\partial \psi}{\partial x}\frac{\partial T}{\partial y}\right) = 0 \tag{8}$$

where $Ra = \beta g(T'_H - T'_C)\omega'^3/\nu\alpha$.

The boundary conditions on the solution are:

On all walls: $\psi = 0, \quad \dfrac{\partial \psi}{\partial n} = 0$

At $x = 0$ (on AB in Fig. 1): $T = 1$

At $x = 1$: $T = 0$

On remaining wall segments: $\dfrac{\partial T}{\partial n} = 0$

n is the coordinate measured normal to the wall surface considered. On the impermeable partitions between the layers, which, by assumption, offers no resistance to heat transfer, the following conditions apply:

$$\psi = 0, \quad \frac{\partial \psi}{\partial n} = 0$$

$$\frac{\partial T}{\partial n}\Big|_i = \frac{\partial T}{\partial n}\Big|_{i+1} \left(\frac{\alpha_{i+1}}{\alpha_i} \right)$$

where the subscripts i and $i+1$ refer to conditions on the two sides of the partition.

The above dimensionless equations, subject to the boundary conditions, have been solved using the finite element procedure. The solutions for the porous medium and fluid layers are obtained simultaneously using the matching conditions across the impermeable partitions. Nodal points were selected to lie along the porous medium and the plain fluid regions so that the elements were either entirely in the porous medium or entirely in the plain fluid. The numerical procedure adopted has previously been successfully used to study several cavity flow problems involving either pure fluids or porous media or both e.g. see Oosthuizen and Paul (4).

RESULTS

The solution has the following parameters (see Figure 2):

- The Rayleigh number, Ra
- The Darcy number, Da
- The Prandtl number, Pr
- The angle of inclination of the cavity, θ
- The position of the partition, B
- The angle of inclination of the partition as defined by Δ
- The size of the heated wall section, S
- The position of the heated wall section, c
- The diffusivity ratio, α_p / α_f
- The viscosity ratio, ν_p / ν_f

In the present study, the diffusivity ratio has been taken as 1. This does not correspond exactly to any real physical situation but is closely satisfied in many cases. The viscosity ratio has also been taken as 1 which appears, on the basis of available experimental results, to be a reasonably good assumption. It is implicit in the use of these assumptions that the porous medium is saturated with the same fluid as that which occupies the fluid-filled layers.

With the diffusivity and viscosity ratios taken as 1, the governing parameters reduce to the Rayleigh number, the Darcy number, the Prandtl number and the geometrical arrangement. Results have only been obtained for a Prandtl number of 0.7.

The main results considered here are the mean heat transfer rate across the cavity. The heat transfer rates have been expressed in terms of mean Nusselt numbers based on the full cavity width, W' and the overall temperature difference, $(T'_H - T'_c)$.

Typical variations of Nu with barrier position for a vertical cavity and a vertical barrier for a fixed Ra, Da and c and various heated section sizes is shown in Figure 3. It must be emphasised that the mean heat transfer rate in the Nusselt number is per unit area of the full cavity wall. As a result, the Nusselt number decreases with dimensionless heated section size. It will be seen from the results given in Figure 3 that for all heated section sizes the mean heat transfer rate decreases rapidly with increasing B until the barrier is approximately in the middle of the cavity. For larger values of B the barrier position has relatively little effect on the heat transfer rate. The results given in Figures 4 and 5 illustrate the effect of the position of the heated element up the wall on the mean Nusselt number. In all cases, the heat transfer rate is highest when the element is approximately centred on the wall. The effect of inclining the barrier on the mean heat transfer rate is illustrated by the results given in Figure 6. It will be seen that it is only when the heated element is placed near the bottom of the wall that inclination of barrier has a significant effect on the mean heat transfer rate and this is only true when B is relatively small. Although the results presented are for the same values of Ra and Da and for a vertical cavity, the effect of changes in these parameters has been investigated and found not to significantly alter any of the conclusions drawn from the results.

CONCLUSIONS

For fixed Rayleigh number, fixed Darcy number and fixed heated element size, the most important factors determining the mean heat transfer rate are the barrier position and the element position. However provided the barrier is beyond the centre of the cavity, its position has little effect on the mean heat transfer rate. The heat transfer rate is lowest when the heated element is placed near the top or the bottom of the wall.

ACKNOWLEDGEMENTS

This work was supported by the Natural Sciences and Engineering Research Council of Canada.

REFERENCES

1. Poulikakos, D. and Bejan, A., 1983, Int. J. Heat Mass Transfer , 26 , pp. 1805-1813.
2. Lauriat, F. and Mesguich, F., 1984, ASME Paper 84-WA/HT-101.
3. Beckermann, C., Ramadhyani, S. and Viskanta, R., 1986, Natural Convection in Porous Media, ASME HTD-56 , pp. 1-12.
4. Oosthuizen, P.H. and Paul. J.T., 1986, Natural Convection in Porous Media, ASME HTD-56 , pp. 75-84.
5. Tong, T.W. and Subramaniam, E., 1986, Int. J. Heat and Fluid Flow , 7 , pp. 3-10.
6. Tong, T.W., Faruque, M.A., Orangi, S. and Sathe, S.B., 1986, Natural Convection in Porous Media, ASME HTD-56 , pp. 85-93.
7. Song, M. and Viskanta, R., 1991, Heat Transfer in Enclosures, ASME HTD-177 , pp. 1-12.

NOMENCLATURE

c	$=$	specific heat
Da	$=$	Darcy number
g	$=$	gravitational acceleration
K	$=$	permeability of porous medium
λ	$=$	thermal conductivity
Nu	$=$	mean Nusselt number based on W'
n	$=$	$B\, n' \,/\, w'$
n'	$=$	coordinate measured normal to surface
Pr	$=$	Prandtl number
p'	$=$	pressure
Ra	$=$	Rayleigh number based on W'
T	$=$	dimensionless temperature
T'	$=$	temperature
T'_H	$=$	temperature of hot wall
T'_c	$=$	temperature of cold wall
u'	$=$	velocity component in x' direction
v'	$=$	velocity component in y' direction

W' = width of cavity
x = dimensionless x' coordinate
x' = horizontal coordinate position
y = dimensionless y' coordinate
y' = vertical coordinate position
β = coefficient of thermal expansion
Δ = displacement of barrier
ν = kinematic viscosity
ρ = density
ϕ = angle of inclination of cavity
ψ = dimensionless stream function
ψ' = stream function
ω = dimensionless vorticity
ω' = vorticity

Subscripts

f = fluid properties

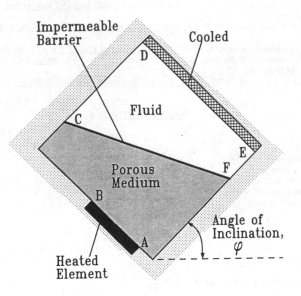

Figure 1.   Flow configurations considered

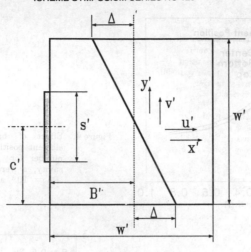

Figure 2.   Coordinate system used

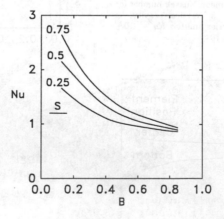

Figure 3.   Effect of barrier position and element size on mean Nusselt number for centrally positioned heated element, a vertical cavity, a vertical barrier and for $Ra = 10^5$ and $Da = 10^{-4}$

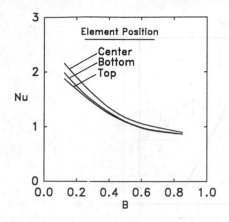

Figure 4. Effect of barrier position and element position on mean Nusselt number for $s = 0.5$, a vertical cavity, a vertical barrier and for $Ra = 10^5$ and $Da = 10^{-4}$

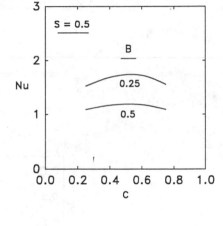

Figure 5. Effect of barrier and element position on mean Nusselt number for $s = 0.5$, a vertical cavity, a vertical barrier and for $Ra = 10^5$ and $Da = 10^{-4}$

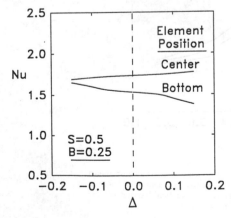

Figure 6. Effect of barrier displacement on mean Nusselt number for $s = 0.5$, $B = 0.25$, a vertical cavity and for $Ra = 10^5$ and $Da = 10^{-4}$

INVESTIGATION ON STEAM CONDENSATION FRONT PROPAGATION IN POROUS MEDIA

M. H. Shi and Y. C. Hu

Southeast University, Nanjing, 210018 China

The steam condensation front velocity in a semi—infinite horizontal porous layer was investigated theoretically and experimentally. Based on the volumetric average method, a physical model for steam condensing flow in one—dimensional porous media initially saturated with water when receiving a sudden inflow of steam was proposed. A general expression for the steam condensation front velocity was obtained. The particular forms for different cases of lateral heat losses were deduced. Experiments were conducted in a stainless steel tube filled by glass beads as porous media. Reasonable agreements were found between predicted and experimental data.

INTRODUCTION

Transient multiphase flow through porous media with phase change is of importance in a wide variety of current applications, particularly in petroleum industry. Thermal oil recovery has been considered as the most useful method in enhancing oil recovery. Steam injection into oil field produces a condensing flow which heats the oil sands and reduces crude oil viscosity. Therefore the determination of steam flow and condensation front propagation in porous media are of primary importance to clarification of the mechanism of oil thermal recovery and energy saving of the process.

Steam condensing flows in porous media have been studied by many researchers. Udell et al(1, 2) analyzed one—dimensional unsteady condensing flows in porous media at lateral adiabatical conditions. Neuman (3) considered the effect of lateral heat loss but assumed constant condensation front propagation velocity. Nilson (4) studied one—dimensional transient flow of hot dry vapor injected into a cold initially dry solid matrix using the generalized Darcy law. Ni (5) proposed a approximate method to calculate the constant codensation front propagation velocity for engineering thermal oil recovery design. All these previous investigation are instractive but both analytical and experimental work considering the lateral heat loss in the transient condensing flow remains inadequate.

In this work a model of one—dimensional steam condensing flow with lateral heat loss in porous media initially saturated with water when receiving sudden inflow of steam was developed. By use of integration method, a general expression for the velocity of the moving steam condensation front was obtained.

ANALYTICAL FORMULATION

Consider a semi—infinite horizontal porous layer subjected to injection of steam flow from left side at a constant mass flow rate m_{in} and quality α. The porous media is intially saturated with water at the temperature far below the saturation temperature under the system pressure. The porous medium is homogeneous and incompressible. The steam flows in the cold solid matrix, losing its energy and forming condensate which flows cocurrently with the vapor. Therefore steady propagation of the steam condensation front can be observed in the medium and the medium can be divided into two distinct regions, vapor—liquid two phase region and liquid region, as shown in Fig. 1, where $X_o(\tau)$ is the location of the condensation front at time τ.

To simplify the problem the following assumptions are made:

1. The flow is one—dimensional and horizontal, the lateral heat loss does not influence the main flow direction.

2. Temperature gradient along the flow direction within the two—phase region is sufficiently small, so heat conduction can be neglected.

3. The pressure gradients in liquid and vapor phases are equal.

4. At the pore level, thermodynamic equilibrium is expected to exist between the vapor, liquid and solid phases.

5. The thermal properties for liquid, vapor and solid are constant, the volumetric parameters are used everywhere.

With these assumptions, the governing equations for heat and mass transfer in a control volume of infinitesimal length dx within two—phase region are:

Continuous equation

$$\frac{\partial}{\partial \tau}(\varepsilon s_L \rho_L + \varepsilon s_v \rho_v) + \frac{\partial}{\partial x}(\rho_L u_L + \rho_v u_v) = 0 \tag{1}$$

Energy equation

$$\frac{\partial}{\partial \tau}(\varepsilon s_L \rho_L h_L + \varepsilon s_v \rho_v h_v + (1 - \varepsilon)\rho_s h_s) + \frac{\partial}{\partial x}(\rho_L h_L u_L + \rho_v h_v u_v) + \frac{q_f P}{A} = 0 \tag{2}$$

Where $s_L + s_v = 1$.

Integrating equation (1) and (2) from inlet $(X_o = 0)$ to $X_o(\tau)$, one obtains

$$m_w - m_{in} = \frac{d}{d\tau}\int_o^{x_o}(\rho_L s_L + \rho_v s_v)\varepsilon dx = \varepsilon(\rho_L - \rho_v)\frac{d}{d\tau}\int_o^{x_o}s_L dx + \varepsilon \rho_v \frac{dX_o}{d\tau} \tag{3}$$

$$\{[h_L(1 - \alpha) + h_v \alpha]m_{in} - h_L m_w\} = \frac{d}{d\tau}\int_o^{x_o}(\rho_L h_L s_L + \rho_v h_v s_v)\varepsilon dx +$$

$$\frac{d}{d\tau}\int_o^{x_o}(1 - \varepsilon)\rho_s h_s dx + \int_o^{x_o}\frac{q_f P}{A}dx \tag{4}$$

Where $q_f(x, \tau)$ is the lateral heat flux at x and τ.

Substituting equation (3) to equation (4) and eliminating m_w, one obtains:

$$\alpha m_{in}(h_v - h_L) - 2\varepsilon \rho_L h_L \frac{d}{d\tau}\int_o^{x_o}s_L dx + \varepsilon \rho_v(h_L + h_v)\frac{d}{d\tau}\int_o^{x_o}s_L dx$$

$$- \varepsilon \rho_v(h_L + h_v)\frac{dX_o}{d\tau} - (1 - \varepsilon)\rho_s h_s \frac{dX_o}{d\tau} - \frac{P}{A}\int_o^{x_o}q_f dx = 0 \tag{5}$$

Defining $\bar{s}_L$ as the average liquid saturation in the two—phase region at time τ and $\bar{q}_f$ as the average lateral heat flux, thus

$$\bar{s}_L = \frac{1}{X_o(\tau)} \int_o^{x_o} s_L dx, \tag{6}$$

$$\bar{q}_f = \frac{1}{X_o(\tau)} \int_o^{x_o} q_f dx, \tag{7}$$

Substituting equaiton (6) and (7) into equation (5), yields

$$(R_1 \bar{s}_L - R_2) \frac{dX_o}{d(\tau)} + (R_1 \frac{d\bar{s}_L}{d(\tau)} X_o(\tau) - \frac{P}{A} \bar{q}_f X_o(\tau)) + R_3 m_{in} = 0 \tag{8}$$

where

$$R_1 = \varepsilon(\rho_v(h_v + h_L) - 2\rho_L h_L), \ R_2 = \varepsilon\rho_v(h_v + h_L) + (1 - \varepsilon)\rho_s h_s, \ R_3 = (h_v - h_L)\alpha.$$

Setting

$$G(\tau) = (R_1 \frac{d\bar{s}_{Lo}}{d\tau} - \frac{P}{A} \bar{q}_f)/(R_1 \bar{s}_L - R_2),$$

$$F(\tau) = - R_3 m_{in}/(R_1 \bar{s}_L - R_2)$$

thus equation (8) becomes

$$\frac{dX_o}{d\tau} + G(\tau)X_o = F(\tau) \tag{9}$$

This is a nonlinear ordinary differential equation, its general solution is

$$X_o(\tau) = e^{-\int_o^\tau G(\tau)d\tau}(\int_o^\tau F(\tau)e^{-\int_o^\tau G(\tau)d\tau}d\tau + c) \tag{10}$$

From initial condition $\tau=0$, $X_o=0$, one gets $C=0$. Thus the general expression of the condensation front propagation velocity can be obtained as

$$V_f(\tau) = \frac{dX_o}{d\tau} = F(\tau) - G(\tau)e^{-\int_o^\tau G(\tau)d\tau}(\int_o^\tau F(\tau)e^{-\int_o^\tau G(\tau)d\tau}d\tau) \tag{11}$$

Obviously, for given thermal properties of fluid and porous material the velocity V_f mainly depends on the inlet mass flow rate m_{in}, lateral heat flux q_f, and liquid saturation $\bar{s}_L$. Here we emphasize to analyze the effects of lateral heat loss to the velocity V_f.

1. Lateral adiabatical case, $\bar{q}_f=0$

This is a ideal situation. The experiments (2) show that in this case the average liquid saturation $\bar{s}_L$ in the two—phase region far from the moving front can be considered as constant, that is $\bar{s}_L(\tau)=s_c=$ constant.

From equation (11), one can get the simplified form of V_f as

$$V_f = \frac{m_{in}\alpha(h_v - h_L)}{2\rho_L h_L s_c \varepsilon + \rho_v(h_L + h_v)(1 - s_c)\varepsilon + (1 - \varepsilon)\rho_s h_s} \tag{12}$$

From equation (12), one can simply calculate the condensation front velocity V_f without knowing the temperature distribution in the liquid region. The results also show that steam condensation front velocity is independent of the time, and for a given inlet mass flow rate, the velocity is a constant. This agrees with the experimental result reported in reference (2).

2. Constant lateral heat flux, $\bar{q}_f = C$

It could happen in the case of lateral heat loss by thermal convection to the surrounding. Usually the lateral heat loss is small by comparison with condensation, we can still assume constant liquid saturation in two—phase region far from the moving front and one—dimensional flow. In this case parameters G and F are independent of the time, thus the equation (11) becomes

$$V_f = Fe^{-Gt} = \frac{R_3 m_{in}}{R_2 - R_1 \bar{s}_L} e^{-Gt} \tag{13}$$

It shows that the steam condensation front velocity decreases exponentially with increasing of time.

3. Lateral heat flux is caused by thermal conduction to the semi—infinite media surrounded

In this case the lateral heat flux can be obtained by solving the one—dimensional heat conduction equation. The resulting average lateral heat flux is

$$\bar{q}_f(\tau) = \frac{1}{X_o} \int_o^{X_o} \frac{K(T_r - T_o)}{\sqrt{\pi a (\tau - t)}} dx = \frac{1}{X_o} \int_o^\tau \frac{K(T_r - T_o) V_f(t)}{\sqrt{a\pi} \sqrt{\tau - t}} dt \tag{14}$$

where T_o is the surface temperature of the surrounding media. t is a time when heat loss is counted at distance x. Substituting equation (14) into equation (8), one obtains

$$R_3 m_{in} = \frac{P}{A} \frac{K(T_r - T_o)}{\sqrt{\pi a}} \int_o^\tau \frac{V_f(t)}{\sqrt{\tau - t}} dt + (R_2 - R_1 \bar{s}_L) V_f(\tau) \tag{15}$$

The solusion of equation (15) is

$$X_o(\tau) = \frac{A^2 a R_3 m_{in} (R_2 - R_1 \bar{s}_L)}{P^2 K^2 (T_r - T_o)^2} [e^{z^2} erfc(z) + \frac{2z}{\pi} - 1] \tag{16}$$

where

$$z = [\frac{PK(T_r - T_o)}{(R_2 - R_1 \bar{s}_L) A \sqrt{a}}] \tau^{1/2} \qquad \text{is nondimensional time.}$$

Thus the steam condensation front velocity is

$$V_f = \frac{R_3 m_{in}}{R_2 - R_1 \bar{s}_L} [e^{z^2} erfc(z)] \tag{17}$$

It is shown from equation (17) that V_f decreases approximately with inverse propotion to squar root of time. The calculated results for these three cases of lateral heat loss at given inlet mass flow rate are shown in Fig. 2. It is clear that even heat loss is small comparing with the condensation, it has significantly influence to the condensation front velocity V_f.

EXPERIMENTAL STUDIES

Experiments were conducted in a stainless steel tube of 0.21m long with an inside diameter of 35mm. A rubber sleeve of 5 mm thickness was inserted into the tube as shown schematically in Fig. 3. Glass beads of 0.3 mm in diameter were used as porous madia. The heat loss from the test section was by the natural thermal convection. Nine alumel — chromel thermocouples of 0.2 mm diameter were spaced along the centerline of the test tube to measure the temperature distribution inside the porous media in flow direction. The average velocity of moving steam condensation front was obtained from the measured temperature distribution.

A constant steam injection rate was set by fixed the valve openning in line from a steam generator which was controlled by a input power supple. The inlet steam flow rate was calibrated by the weighing method. Before the experiments, distilled water was used to saturate the porous medium. At the beginning of each experimental run, the steam from the generator was vented until a steady state flow was reached. The steam was then injected into the test tube and the measurements of temperature and steam flow rate were conducted. The effluent flow rate was calculated by falling liquid droplet number measured by a piezoelectric ceramic sensor developed in this work. Two sets of experiments were performed for the inlet mass flow rates of 0.0169 and 0.021 $kg/m^2 \cdot s$. The injection flow quality is one.

Fig. 4 shows the temperature—time curves measured at two different heat loss conditions. It is obvious that the larger the heat loss the slower the condensation front propagation. The experimental results show that the front propagation velocity increases with increasing inlet mass flow rate, for instance, when inlet mass flow rate m_{in} is 0.0169 $kg/m^2 \cdot s$, the measured average front velocity in 20 minutes is about $1.83 \times 10^{-4} m/s$, and when m_{in} is 0.021 $kg/m^2 \cdot s$, the average velocity is $2.22 \times 10^{-4} m/s$. Fig. 5 shows the comparison between the calculated front velocities by equation (13) and experimental values, the agreements are satisfactory.

CONCLUSIONS

1. Steam condensation front velocity depends mainly on the inlet injection rate, the lateral heat loss and liquid saturation in the two—phase region.

2. At lateral adiabatical case the condensation front velocity is independent of the time and has the maximum value, increase of the lateral heat loss will cause the front velocity to decrease.

3. Equation (11) is a generalized expression to predict the condensation front velocity of condensing flow in one — dimensional porous media. Satisfactory agreements were obtained between the predicted and experimental front velocity.

REFERENCES

1. Menegus, D. K., and Udsell, K. S., 1985, ASME HTD, Vol. 46.
2. Stewart, L. D., Basel, m. D., and Udell, K. S., 1987, ASME HTD Vol. 91, PP. 31—42.
3. Neuman, C. H., 1985, J. Pet. Tech., No. 1, PP. 163—169.
4. Nilson, R. H., 1980, Int. J. Heat Mass Transfer, Vol. 23, PP. 1461—1470.
5. Ni, S. S., 1989, J. Pet. Exp., (in Chinese), No. 2.
6. Bear, J., 1972, Dynamics of Fluids in Porous Media, American Elsevier Pub. Co., New York.

NOMENCLATURE

A	Cross—sectional area	s_L	Liquid satruation
a	Thermal diffusivity	T	Temperature
F,G,R	Parameter	q_f	heat flux
x	Distance	h	Enthalpy
X_o	Distance to condensation front	K	Thermal conductivity
V_f	Condensation front velocity	m	Mass flow rate
P	Wet periphery	u	Fluid velocity
ρ	Density	ε	Porosity
τ, t	Time	α	Quality

Subscripts

f	front	s	Solid
in	Inlet	s	saturation
L	Liquid	v	Vapor

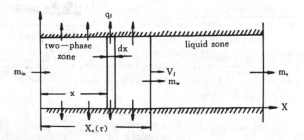

Fig. 1 one—dimensional condensing flow

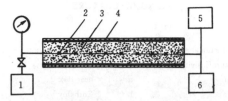

Fig. 2 Experimental apparatus

1. Steam generator   2. Thermocouple
3. Porous media   4. Rubber sleeve
5. Datalogger   6. Effluent flowmeter

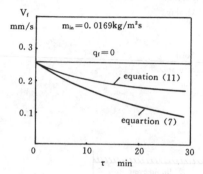

Fig. 3 Calculated condensation
front velocity

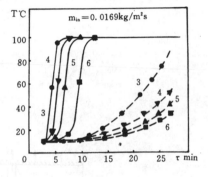

Fig. 4 Measured temperarture
curves for different q_f

$- - q_f = 524 \ w/m^2 - - - q_f = 753 w/m^2$

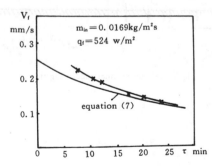

Fig. 5 Comparison between predicted
and measured front velocity

MAXIMUM DENSITY EFFECTS ON HEAT TRANSFER FROM A HEATED CYLINDER BURIED IN A POROUS MEDIUM FILLED ENCLOSURE WITH A COOLED TOP

P. H. Oosthuizen* and T. Hung Nguyen†
*Queen's University, Kingston, Ontario, Canada K7L 3N6
†Ecole Polytechnique de Montreal, Montreal, Quebec, Canada H3C 3A7

A numerical study of the steady heat transfer rate from a horizontal circular cylinder buried in a porous medium in a square enclosure has been undertaken. The cylinder lies on the vertical centre-line of the enclosure. The porous medium is saturated with water and the top wall of the enclosure is cooled to a uniform temperature that is lower than the temperature of the cylinder surface which is also uniform. The temperature of the cylinder surface is near the temperature at which the density maximum occurs with water. The flow has been assumed to be steady, two-dimensional and laminar and the usual Darcy type assumptions have been adopted. The properties of the unfrozen liquid have been assumed constant except for the density change with temperature which gives rise to the buoyancy forces, this being treated using a quadratic form for the relation between density and temperature difference. The governing equations, expressed in terms of stream function and temperature, have been written in dimensionless form and solved using the finite element approach. Solutions have been obtained for a wide range of the governing parameters and the effects of these parameters on the heat transfer rate from the cylinder and the flow pattern in the enclosure have been examined.

INTRODUCTION

The heat transfer from a horizontal circular cylinder buried in a porous medium in a square enclosure has been numerically studied. The centre of the cylinder lies on the vertical centre-line of the enclosure. The porous medium is saturated with water and the top wall of the enclosure is cooled to a uniform temperature that is lower than the temperature of the cylinder surface which is also uniform. The remaining walls of the cavity are adiabatic. The situation considered is thus as shown in Figure 1. The temperature of the cylinder surface is near the temperature at which the density maximum occurs with water, i.e. approximately 4C, and this can have a significant influence on the flow and heat transfer rate.

The situation considered has application, for example, in the prediction of the temperature history of foodstuffs being transported in boxes of wet frozen sawdust and in the prediction of heat transfer rates from pipes buried in backfilled trenches when the soil is saturated with water and the ambient temperature is near the freezing temperature of water. The values of the governing parameters considered in the present study differ from the values often encountered in such situations but the present results should give an indication of the heat transfer rates that exist in such situations and of the effects of variations in the governing parameters on the heat transfer rate.

There have been a number of studies of forced and free convective flow over bodies buried in a porous medium, see for example Frivik and Comini (1), Goldstein and Reid (2), Hashemi and Sliepcevich (3), Oosthuizen and Paul (4), (5),(6) and (7), Oosthuizen and Henderson (8) and Takashi (9). Some studies of freezing in porous media under such conditions that the induced free convective motion is important have paid particular attention to the effects of the density maximum, see for example Blake et al (10), Philip (11), Sugawara et al (12) and Zhang and Nguyen (13). There do not, however, appear to be any available studies of the flow situation here being considered nor do the available studies appear to allow an estimation to be made of the heat transfer rate that would exist with the situation here being considered.

GOVERNING EQUATIONS AND SOLUTION PROCEDURE

It has been assumed that the flow is two-dimensional and steady and that the Darcy type assumptions apply. It has also been assumed that fluid properties are constant except for the density change with temperature that gives rise to the buoyancy force, this being treated by assuming a quadratic type relationship, i.e. by assuming a relation of the form:

$$(\rho_M - \rho)/\rho = a(T' - T_M')^2 \tag{1}$$

The adequacy of this form of relation for describing the behaviour of water near the temperature of maximum density is discussed, for example, in (10) and (11). The flow has been assumed to be symmetrical about the vertical centre-line of the enclosure on which the cylinder lies. The surface of the cylinder is assumed to be at a uniform temperature and the top wall of the enclosure is assumed to be cooled to a uniform temperature that is lower than the temperature of the cylinder surface. The other walls of the cavity are assumed to be adiabatic.

Using the above assumptions, the governing equations for the unfrozen portion of the flow are, in terms of the coordinate system shown in Figure 2:

$$\frac{\partial u'}{\partial x'} + \frac{\partial v'}{\partial y'} = 0 \tag{2}$$

$$\frac{\partial p'}{\partial x'} = -\left(\frac{\eta_f u'}{K}\right) \tag{3}$$

$$\frac{\partial p'}{\partial y'} = -\frac{\eta_f v'}{K} + \rho g \tag{4}$$

$$u'\frac{\partial T'}{\partial x'} + v'\frac{\partial T'}{\partial y'} = \left(\frac{\lambda}{\rho_f c_f}\right)\left(\frac{\partial^2 T'}{\partial x'^2} + \frac{\partial^2 T'}{\partial y'^2}\right) \tag{5}$$

In these equations the superficial velocity has been used. The prime (') is used to denote dimensional quantities. The subscript, f, denotes fluid properties.

Because the flow is symmetrical about the vertical center-line, the solution has only been obtained for half of the flow domain i.e. for abcdef in Figure 2. The boundary conditions on the solution are then:

- On the walls of the cavity (afed in Figure 2): normal velocity component = 0,

- On the top wall of the cavity (de in Figure 2): $T' = T'_C$

- On the remaining walls of the cavity (ea in Figure 2): $\frac{\partial T'}{\partial n'} = 0$
  n' being the coordinate normal to the wall considered.

- On the cylinder walls (bc in Figure 2): normal velocity component = 0, $T' = T'_H$

- On the centre-line (ab and cd in Figure 2): $u' = 0$, $\frac{\partial v'}{\partial x'} = 0$, $\frac{\partial T'}{\partial x'} = 0$

The solution has been obtained by introducing the stream function defined as usual by:

$$u' = \frac{\partial \psi'}{\partial y'}, \qquad v' = -\frac{\partial \psi'}{\partial x'} \tag{6}$$

The following dimensionless variables have then been defined:

$$\theta = (T' - T'_M)/(T'_H - T'_C)$$
$$x = x'/d', \qquad y = y'/d', \qquad \psi = \psi'/\kappa_f \tag{7}$$

Because of the way in which the dimensionless temperature is defined, θ will be zero when the material is at the temperature of maximum density.

In terms of these variables, the governing equations become:

$$\frac{\partial^2 \psi}{\partial x^2} + \frac{\partial^2 \psi}{\partial y^2} = -2 Ra^* \theta \frac{\partial \theta}{\partial x} \tag{8}$$

$$\frac{\partial \psi}{\partial y} \frac{\partial \theta}{\partial x} - \frac{\partial \psi}{\partial x} \frac{\partial \theta}{\partial y} = \left(\frac{\partial^2 \theta}{\partial x^2} + \frac{\partial^2 \theta}{\partial y^2} \right) \tag{9}$$

where Ra^* is a modified Rayleigh number based on the cylinder diameter, d', i.e.:

$$Ra^* = a g K (T_H' - T_C')^2 d' / \nu_f \kappa_f \tag{10}$$

In terms of the dimensionless variables, the boundary conditions on the solution are:

- On the walls of the cavity (afed in Figure 2): $\psi = 0$
- On the top wall of the cavity (de in Figure 2): $\theta = \theta_C \; (= \theta_H - 1)$
- On the remaining walls of the cavity (ea in Figure 2): $\frac{\partial \theta}{\partial n} = 0$

  n being the dimensionless coordinate normal to the wall considered.
- On the cylinder walls (bc in Figure 2): $\psi = 0$, $\theta = \theta_H$
- On the centre-line (ab and cd in Figure 2): $\psi = 0$, $\frac{\partial \theta}{\partial x} = 0$

It has been noted that because of the way in which the dimensionless temperature, θ, is defined:

$$\theta_C = \theta_H - 1 \tag{11}$$

The governing equations, subject to the boundary conditions discussed above, have been solved using the finite element method. The solution directly gives the local Nusselt number distribution along the surface of the cylinder. The mean Nusselt number for the cylinder, defined here as:

$$\overline{Nu} = \bar{q} d' / (T'_H - T'_C) \lambda$$

can then be determined by integration of the local Nusselt number variation.

A highly non-linear nodal distribution was used in the present work, there being a concentration of nodal points near the surface. Calculations were undertaken with many different numbers and distributions of nodes and these calculations indicate that the present heat transfer results are grid independent to significantly better than 2%.

RESULTS AND DISCUSSION

The solution has as parameters:

- the dimensionless temperature of the cylinder wall temperature (this determines the dimensionless temperature of the enclosure walls)
- the modified Rayleigh number
- the size of the enclosure relative to the cylinder diameter
- the dimensionless position of the cylinder along the vertical centre-line of the enclosure

Solutions have been obtained for Rayleigh numbers of between 10 and 1000 for dimensionless enclosure sizes of between 4 and 10 for various cylinder positions and temperatures. It will be noted that a dimensionless temperature of zero exists when the temperature is equal to that at which the maximum density occurs and that negative dimensionless temperatures imply that the temperature is below that at which the maximum density occurs.

Although extensive studies of the streamline, isotherm and local heat transfer rate distributions have been undertaken, attention will here only be given to the mean heat transfer rate. Figure 3 shows the effect of dimensionless cylinder temperature on the mean Nusselt number for various fixed Rayleigh numbers. It will be seen that as the cylinder wall temperature rises above the maximum density temperature, a sharp rise in the mean Nusselt number occurs. When the cylinder temperature is below the maximum density temperature, the water near the cylinder will have a higher density than that of the water near the upper surface. Consequently, under these conditions, strong stratification results and the heat transfer rate is low. When the cylinder temperature is increased above that at which the density maximum occurs, the density of the water near the cylinder starts to decrease and when the temperature has reached a high enough value, a plume of heated water rising from the cylinder starts to impinge on the cooled upper surface bringing about the observed increase in the mean heat transfer rate. The effect of cylinder position on the mean heat transfer rate is illustrated by the results given in Figure 4. This figure shows the variation for three values of the dimensionless cylinder temperature for a fixed value of the modified Rayleigh number. When the cylinder temperature is close to the maximum density temperature the heat transfer rate remains low until the cylinder is close to upper surface when it starts to rise rapidly. When the dimensionless cylinder temperature is significantly above the maximum density temperature, both the water near the cylinder and near the upper surface are at a lower density than the maximum density in the enclosure. There are as a result, in general, two vortices in the enclosure, one associated with the flow about the cylinder and an oppositely rotating one associated with flow over the upper surface. When the cylinder is far below the upper surface, the first of these is dominant and the heat transfer rate increases slightly as the distance between the cylinder and the upper surface decreases. However, when the cylinder gets close to the upper surface, the strength of the first vortex decreases significantly while that of the second increases very significantly. This tends to result in a decrease in the mean heat transfer rate. Of course, when the gap between the cylinder and the upper surface becomes very small, the heat transfer rate again increases. The effect of modified Rayleigh number of the mean Nusselt number is illustrated by the results given in Figure 5. At low Rayleigh numbers, the heat transfer rate is essentially constant and equal to the conduction value. At modified Rayleigh numbers above approximately 20, however, the convective motion becomes significant and the Nusselt number rises sharply with Rayleigh number, being approximately proportional to the square root of the modified Rayleigh number at higher Rayleigh numbers. The effect of cavity size on the mean heat transfer rate is illustrated by the results given in Figure 6. As will be seen from this figure, this effect is relatively weak.

CONCLUSIONS

The results obtained in the present study indicate that the mean Nusselt number is dominantly determined by the value of the dimensionless cylinder temperature relative to the temperature of maximum density. The cylinder position can have a strong effect on the Nusselt number if the cylinder temperature is greater than the maximum density temperature and the temperature of the upper surface below this value. The cavity size has relatively weak effects on the mean Nusselt number.

ACKNOWLEDGEMENTS

This work was supported by the Natural Sciences and Engineering Research Council of Canada.

REFERENCES

1. Frivik, P.E., and Comini, G., 1982, ASME J. of Heat Trans. , 104 , 323-328.
2. Goldstein, M.E., and Reid, R.L., 1978, Proc., Royal Society of London , Ser. A, No. 364 , 45-73.
3. Hashemi, H.T., and Sliepcevich, C.M., 1973, ASCE Mech. and Found. Div. , 99 , (1), 267-289.
4. Oosthuizen, P. H., and Paul, J. T., 1989, Proc. 7th Int. Conf. on Finite Element Methods in Flow Problems , 726-731.
5. Oosthuizen, P. H., and Paul, J. T., 1989, Paper AIAA-89-1682 , AIAA 24th Thermophysics Conf.
6. Oosthuizen, P. H., and Paul, J. T., 1989, Proc. 6th Int. Conf. Num. Meth. in Thermal Problems , 545-554.

7. Oosthuizen, P. H., and Paul, J. T., 1990, Proc. 10th Symp. on Engg. Applications of Mech. , 279-284.
8. Oosthuizen, P. H., and Henderson, C., 1989, Proc. ASME Winter Annual Meeting , HTD-143 , 65-73.
9. Takashi, T., 1969, Special Report No. 103 , Highway Research Board, 273-286.
10. Blake, K.R., Bejan, A., and Poulikakos, D., 1984, Int. J. of Heat and Mass Trans. , 27 , 2355-2364.
11. Philip, J. R., 1988, PhysicoChemical Hydrodynamics , 10 , 283-294.
12. Sugawara, M, Inaba, H., and Seki, N., 1988, ASME J. Heat Transfer , 110 , 155-159.
13. Zhang, X., and Nguyen, T. H., 1990, 1990 ASME Winter Annual Meeting , HTD-156 , 1-6.

NOMENCLATURE

c_f = specific heat of fluid
d' = cylinder diameter
K = permeability
Nu = local Nusselt number based on d'
$\overline{Nu}$ = mean Nusselt number based on d'
p' = pressure
q' = local heat transfer rate at any point on the surface
$\overline{q}'$ = mean heat transfer rate over the surface
$Ra*$ = modified Rayleigh number based on d'
s' = side length of square enclosure
s = s' / d'
T' = temperature
T'_C = temperature of top wall of enclosure
T'_M = temperature at which the density maximum occurs
T'_H = cylinder temperature
u' = velocity in x-direction
v' = velocity in y-direction
V' = position of cylinder up center-line of enclosure
V = V' / d'
x' = coordinate in horizontal direction
x = x' / d'
y' = coordinate in vertical direction
y = y' / d'
κ_f = $\lambda / \rho_f c_f$
λ = effective thermal conductivity
ν = kinematic viscosity
θ = dimensionless temperature
θ_C = dimensionless top wall temperature
θ_H = dimensionless cylinder temperature
η_f = viscosity of liquid
ρ = density
ρ_M = maximum density
ψ = dimensionless stream function
ψ' = stream function

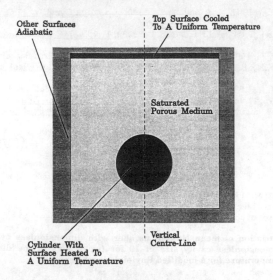

Other Surfaces
Adiabatic

Top Surface Cooled
To A Uniform Temperature

Saturated
Porous Medium

Cylinder With
Surface Heated To
A Uniform Temperature

Vertical
Centre-Line

Fig. 1 Flow situation considered

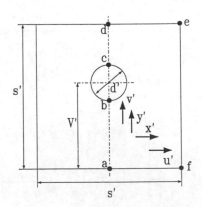

Fig. 2 Coordinate system adopted

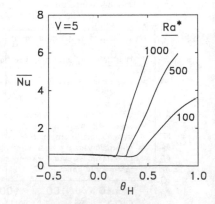

Fig. 3 Variation of mean Nusselt number with dimensionless cylinder temperature for various values of the modified Rayleigh number for a dimensionless cavity size of 10 and a centrally mounted cylinder.

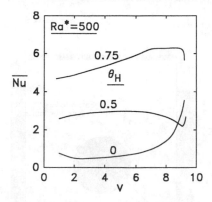

Fig. 4   Variation of mean Nusselt number with dimensionless cylinder position for a dimensionless cavity size of 10 for two values of the dimensionless cylinder temperature for a modified Rayleigh number of 500.

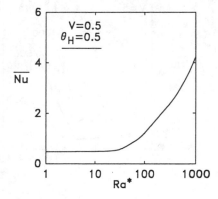

Fig. 5 Variation of mean Nusselt number with modified Rayleigh number for a dimensionless cavity size of 10, a dimensionless cylinder temperature of 0.5 and a centrally mounted cylinder.

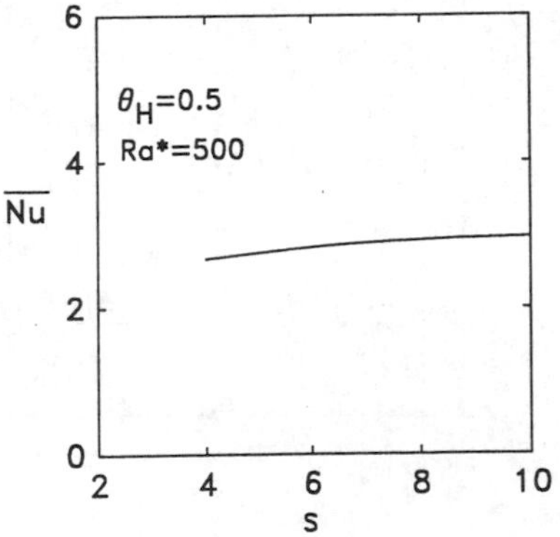

Fig. 6 Variation of mean Nusselt number with dimensionless cavity size for a modified Rayleigh number of 500, a dimensionless cylinder temperature of 0.5 and a centrally mounted cylinder.

UNSTABLE BOILING STATE IN A POROUS MEDIUM

Didier STEMMELEN, Christian MOYNE and Alain DEGIOVANNI
L.E.M.T.A. - 2, avenue de la forêt de Haye - 54516 Vandoeuvre les Nancy Cedex (France)

Time-periodic oscillations of temperature field with large amplitude are observed in a porous medium subjected at the bottom to a great heat flux and cooled at the top. A simplified linear stability study shows the possibility to obtain an unstable boiling state.

INTRODUCTION

Problems of boiling in porous media arise in many industrial processes and natural phenomena. Examples of such applications are : drying of porous materials, heat pipes, geothermal energy production, disposal of high-level nuclear waste in geological media or geothermal phenomena like geysers.

Many studies, involving boiling in porous media, deal with the steady state (1), (2), (3). Relatively little work has been carried out on the stability of boiling in porous media. Schubert and Straus (4) have studied the gravitational stability of a liquid region overlying a dry vapor region in geothermal systems. Bau and Torrance (2) have reported low oscillations of the temperature field when a water-saturated porous medium is heated from below and cooled from above. More recently, Ramesh and Torrance (5) have carried out, for the same configuration, a linear stability analysis of boiling in a porous medium. From a two-dimensional analysis, they showed that for liquid-dominated systems, the instability is driven by buoyancy in the liquid region while, for vapor-dominated systems, it is driven by gravitational instability of the overlying layers.

In this article, we are interested in the description of the boiling of a liquid in a horizontal porous layer. The configuration studied consists in heating the porous layer at constant heat flux and in cooling it above at constant temperature. The lower surface is impervious and the upper surface remains saturated by water at constant pressure. The temperature difference is such that the temperature at the heated end is above the saturation temperature of the liquid when the cooled end stays below this level.

Theoretical and experimental descriptions of boiling in these conditions generally concern the steady state, which is stable in most cases. Then, it is possible to distinguish two or three distinct regions in equilibrium in the porous layer :
- at the top, a zone with a low thermal gradient which is characteristic of a water saturated porous medium (liquid zone),
- under the liquid zone, an almost isothermal region dominated by countercurrent flows of liquid and vapor (two-phase or heat pipe zone),

- possibly, for great heat fluxes, a high thermal gradient zone at the bottom which is characteristic of a dry steam saturated porous medium (vapor zone).
Our measurements confirm this description for low heat fluxes. However, we find that very large oscillations of the field of temperature can develop in such a system when the vapor zone appears.

EXPERIMENTAL APPARATUS

The device is designed to study heat and mass transfer during boiling in a porous medium. A sketch of the apparatus is shown in figure 1. Experiments were carried out using a non-consolidated bed of glass beads of small diameter (200µm i.d.). The intrinsic permeability of such material is relatively large : $K \approx 10^{-11} m^2$. The porous bed is contained in a fiberglass tube of inner diameter $D = 50$ mm. The total height of the bed is 230 mm. The porous layer lies on a brass plate heated by an electric resistance heater. An O-ring seal ensures the tightness between the fiberglass tube and the brass plate.The top end of the porous layer is maintained at constant temperature by a cooling device regulated by a water flow. The temperature of the water flow is fixed at 20°C. A small pipe goes through this exchanger and is used to maintain water saturation on the cooled end. A screen is inserted between the porous layer and the exchanger to avoid fluidization of the bed. Finally, the lateral fiberglass tube surface is thermally isolated with insulating foam. In this way, the configuration can be considered one-dimensional .
The temperature measurements are made by ten chromel-alumel thermocouples (0.5 mm i.d.). They are placed along the porous layer centerline at about 20 mm intervals. A thermocouple is brazed on the brass plate and indicates its temperature. A power of 25.5 Watt is dissipated in the wire coil ; but, previous experiments allow us to estimate at about 50% the part of this heat flux which really goes through the porous layer. In principle, it will be possible to measure water content in the porous medium by a device using the attenuation of Am^{241} gamma rays. But, counting requires too much time and does not allow a good accuracy of the measurements because time evolutions of the mass transfer are rapid. So, measurements by attenuation of gamma rays will not be used in this study.

EXPERIMENTAL RESULTS

The porous bed at room temperature is initially saturated with water and then subjected to the specified conditions. We first observe a progressive rise of temperatures in the porous layer ; then, it is possible to see, above the brass plate, the formation and the development of an almost isothermal two-phase region. This zone is at saturation temperature (about 100°C). At the same time, we can notice a draining of liquid out of the porous medium through the upper surface. But later, instead of reaching a steady state, the entire temperature field oscillates with a period of about one hour. These oscillations of the temperature concern both the liquid zone and the two-phase zone. Figure 2 shows these oscillations in the liquid region. Their amplitude may be very large (up to 50°C). On the contrary, the range of temperature in the two-phase zone is low (less than 1°C). These oscillations of the temperature field are all in phase. On the other hand, the temperature of the brass plate always stays at about 188°C. This indicates that, adjacent to the plate, there exists a thin vapor layer. Concurrently, we observe a draining of liquid out of the layer with the rise of the temperatures followed by its reabsorption when the porous bed gets colder. The liquid exchanges through the upper surface are important : the amount of water contained in the porous medium varies between 87 ml and 132 ml when the value allowing the initial water saturation of the bed is 172 ml. The measurements show an up-and-down movement of the front which separates the liquid region from the two-phase region. Water saturation of the two-phase zone decreases when the front rises and increases when it goes back. Centerline temperature profiles illustrate the influence of convective heat transfer. The curvature of these profiles shows the convection following up-movement of hot liquid (figure 3-a), then downward movement of cold liquid (figure 3-b).

THEORETICAL STUDY

This study turns to a simplified linear analysis because first, it only concerns the heat transfer in liquid zone and secondly, it is a one-dimensional analysis. We are interested in the stability of the energy equation in the liquid zone :

$$(\rho \ C_p)_{ls} \frac{\partial T}{\partial t} = \lambda_{ls} \ \frac{\partial^2 T}{\partial x^2} - \rho_l \ C_{pl} \ v_l \ \frac{\partial T}{\partial x} \tag{1}$$

In this expression $(\rho \ C_p)_{ls}$ and λ_{ls} represent respectively the heat capacity and the thermal conductivity of the saturated porous medium, $(\rho_l \ C_{pl})$ is the heat capacity of the liquid and v_l is the superficial velocity of the liquid in the liquid region. To simplify the problem, we make the following assumption : near the point of equilibrium, v_l is proportional to the front velocity $\dot{X}$:

$$v_l = \alpha \ \dot{X} \tag{2}$$

This assumption can be validated by the mass balance at the interface between the liquid and two-phase regions. Mass-flux of liquid flowing in the liquid zone is given by :

$$n_l = \rho_l \ \varepsilon \ (\frac{v_l}{\varepsilon} - \dot{X}) \tag{3}$$

This amount is equal to the amount of water per unit area flowing out of the two-phase zone, i.e. :

$$- \frac{d \ m_{2\phi}}{dt} \quad \text{with} \quad : \quad m_{2\phi} \ = \ \varepsilon \ \int_0^X [\rho_l \ S + \rho_v \ (1-S)] \ dx \ \approx \ \varepsilon \ \rho_l \ \int_0^X S \ dx \tag{4}$$

So, the front velocity can be written as :

$$\dot{X} \ \approx \ \frac{v_l}{\varepsilon} + \ \frac{d}{dt} (\int_0^X S \ dx) \tag{5}$$

In assuming a static behaviour of the two-phase zone, $\frac{d}{dt} (\int_0^X S \ dx)$ can be approximated by :

$$\dot{X} \ S_0 \qquad \text{when the vapor zone does not exist} \tag{6}$$
$$0 \qquad \text{when the vapor zone exists.}$$

(S_0 is the water saturation just on the brass plate at x=0)

Finally, we obtain the following estimation for α :

$$\alpha = \varepsilon \ (1 - S_0) \qquad \text{without a vapor zone} \tag{7}$$
$$\alpha = \varepsilon \qquad \text{with a vapor zone}$$

A coordinate transformation is used to immobilize the interface. This coordinate transformation is defined as follows :

$$z = \frac{H_0 - x}{H_0 - X} \tag{8}$$

where H_0 is the height of the porous layer and X the position from the brass plate of the front which separates the two-phase and liquid zones.

So, the equation satisfied by the field of temperature can be written in the following form :

$$(\rho\, C_p)_{ls}\, (H_0 - X)^2\, \frac{\partial T}{\partial t} + \dot{X}\, ((\rho\, C_p)_{ls}\, z - \alpha\, \rho_l\, C_{pl})\, (H_0 - X)\, \frac{\partial T}{\partial z} = \lambda_{ls}\, \frac{\partial^2 T}{\partial z^2} \tag{9}$$

with the boundary conditions :

$$\begin{cases} T = T_0 & (z=0) \\[2mm] T = T_{sat} & (z=1) \\[2mm] \dfrac{\lambda_{ls}}{H_0 - X}\, \dfrac{\partial T}{\partial z} = q_0 & (z=1) \end{cases} \tag{10}$$

(T_{sat} is the saturation temperature and q_0 is the heat flux through the porous bed).

We consider small perturbations $\tilde{X}$ and $\tilde{T}$ about the equilibrium position X_{eq} and its temperature solution T_{eq} :

$$\begin{aligned} T &= T_{eq} + \tilde{T} \\ X &= X_{eq} + \tilde{X} \\ \dot{X} &= \dot{\tilde{X}} = \dot{\tilde{X}} \end{aligned} \tag{11}$$

After linearization and rewritting in a non-dimensional form, the governing equation and boundary conditions are obtained as follows :

$$\begin{cases} \dfrac{\partial \tilde{T}^*}{\partial t^*} + \dot{X}^*\, (z-C) = \dfrac{\partial^2 \tilde{T}^*}{\partial z^2} \\[3mm] \tilde{T}^* = 0 & (z=0) \\[2mm] \tilde{T}^* = 0 & (z=1) \\[2mm] \dfrac{\partial \tilde{T}^*}{\partial z} = -\tilde{X}^* & (z=1) \end{cases} \qquad \text{with} \qquad C = \frac{\rho_l\, C_{pl}}{(\rho\, C_p)_{ls}}\, \alpha \tag{12}$$

with : $t^* = \dfrac{\lambda_{ls}}{(\rho\, C_p)_{ls}}\, \dfrac{t}{(H_0 - X_{eq})^2}$; $\tilde{X}^* = \dfrac{\tilde{X}}{(H_0 - X_{eq})}$; $\tilde{T}^* = \dfrac{\tilde{T}}{(T_{sat} - T_0)}$

We look for solutions having the form :

$$\begin{cases} \tilde{T}^* = f(z) \exp(\omega t) \\ \tilde{X}^* = K \exp(\omega t) \end{cases} \qquad (\omega \text{ is complex number}) \qquad (13)$$

These solutions are unstable if the real part of ω is positive. Then, we come to the conclusion about the unstability of the front for :

$$C > \frac{\text{ch } \pi}{1 + \text{ch } \pi} \qquad (14)$$

CONCLUSION

This calculation shows that unstability of the interface between the two-phase and liquid regions can be considered. Still, an assessement of the value of α reaches in our configuration a top value of C=0.6, if we suppose that the two-phase zone (and possibly vapor) evolves in quasi-steady state. This value does not tally with the criterion put forward. However, we may suppose that, in the vicinity of the critical heat flux for which a vapor zone appears, the ratio α is much greater than the simplified analysis of operating in a quasi-steady state of the two-phase zone may have suggested at first sight. Moreover, if we take into account the temperature and flow disturbances, it seems unlikely that the linear assumptions may yield a precise result in that case.

Let us consider now the physical explanation of the time-periodic oscillations of temperature field : in the case the velocity at which the front moves may not be nil when the front reaches its position of balance (based on the theoretical steady state), there is convection in liquid zone and the temperature field is no longer linear in it. This makes delay in mechanisms of condensation possible. During that phase, the two-phase zone undergoes a global decrease in water saturation. So, the front can move up towards the exchanger whose temperature is T_0. Throughout that rise, the liquid velocity in liquid zone (and also the front velocity) slows down due to the increase of the mechanism of condensation as we move towards the exchanger. That velocity moves down to zero. The system, then, finds itself off balance. Return to equilibrium is visible in the re-humidification of the porous layer from the water overlying it. This time, the convection of the cold liquid through the liquid zone will involve a non-linear temperature field in the liquid zone and will reduce the mechanism of vaporization so that the front will be able to move down below its theoretical static position. During that phase, the front velocity decreases to the point to cancel for an off-balance position of the front. The phenomenon may then again go through the cycle described previously. Furthermore, the stability analysis tends to prove the unstability of the equilibrium position and, consequently, that the system may evolve towards that configuration.

An analysis taking into account both the liquid and two-phase zones is actually in progress. It must be completed by experiments carried out in order to determine more precisely when unstability appears and how it appears.

REFERENCES

(1) C. H. Sondergeld and D. L. Torrance, 1977, J. Geophys. Res., 82 , n°14, 2045-2053.
(2) H. H. Bau and K. E. Torrance, 1982, Int. J. Heat Mass Transfer, 25, n°1, 45-55.
(3) K. S. Udell, 1985, Int. J. Heat Mass Transfer, 28, n°2, 485-495.
(4) G. Schubert and J. M. Straus, 1980, J. Geophys. Res., 85, 6505-6512.
(5) P. S. Ramesh and K.E. Torrance, 1990, Int. J. Heat Mass Transfer, 33, n°9, 1895-1908.

NOMENCLATURE

C — dimensionless constant
C_p — specific heat (J/kg K)
H_0 — height of porous bed (m)
$m_{2\phi}$ — water amount in the two-phase region (kg/m^2)
n_l — mass flux of liquid in the liquid phase (kg/m^2 s)
q_0 — heat flux through the porous layer (W/m^2)
S — liquid saturation
S_0 — liquid saturation at x=0
t — time (s)
T — temperature (°C)
T_0 — temperature at $x=H_0$ (°C)
v_l — superficial velocity in the liquid phase (m/s)
x — spatial coordinate (m)
X — front position (m)
$\dot{X}$ — front velocity (m/s)
z — transformed spatial coordinate

Greek symbols :

α — constant
ε — porosity
λ — thermal conductivity (W/m K)
ρ — density (kg/m^3)

Subscripts and Superscripts :

eq — static value
l — liquid phase
ls — liquid saturated
sat — saturation (temperature)
v — vapor phase
2ϕ — two-phase zone
* — dimensionless quantity
~ — deviation from static value

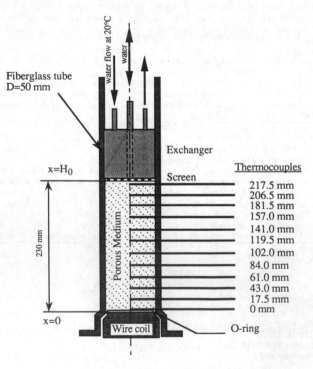

Fig. 1 - Experimental apparatus

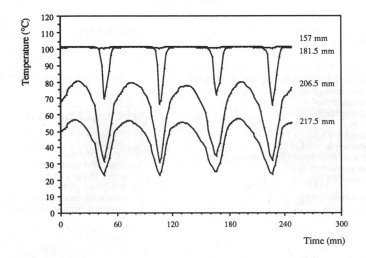

Fig. 2 - Time evolution of the temperatures during oscillatory boiling

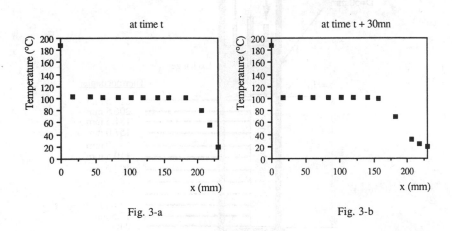

Fig. 3 - Temperature profiles at different instants - Convective effect in oscillatory boiling

ABSORPTIVE CHARACTERISTICS OF SPECULAR SPHERICAL CAVITIES

F. Kowsary and J. R. Mahan
Department of Mechanical Engineering, Virginia Polytechnic Institute and State University, Blacksburg, Virginia 24061 USA

An exact analytical method is described for calculating the directional absorptivity of a specular spherical cavity when collimated radiation enters through its mouth from a specified direction. Results show that each cavity opening angle has its own unique directional characteristics and that the absorptivity becomes increasingly more directional as the opening angle decreases. The apparent hemispherical absorptivity was calculated by numerically integrating the directional absorptivities over all directions. Results show that for spherical cavities, in contrast to the behavior of cylindrical and conical cavities, the diffuse cavity can be a more efficient absorber than the specular cavity.

INTRODUCTION

When thermal radiation enters a cavity it has a better chance of being absorbed than when it strikes a flat surface having the same area as the cavity opening. This is because of the possibility of multiple reflections within the cavity. Of course, it is this behavior of cavities which permits the radiant heat transfer between a surface and its surroundings to be enhanced by drilling, stamping, etching, or otherwise deforming the surface. Cavities are also widely used as blackbody calibration sources in radiometry.

The performance of a cavity may be described by its *apparent* radiative properties. For example, the apparent emissivity is defined as the ratio of the energy emitted through the opening of an isothermal cavity to the energy emitted by an imaginary black surface stretched across the cavity opening which is at the same temperature as the cavity wall. In the current work a new exact analytical method is used to predict the apparent radiative behavior of spherical cavities whose walls are specular reflectors. In their popular textbook, Sparrow and Cess (1979) make it clear that the solution of this problem would be a welcome contribution to the heat transfer literature. Our own interest in the problem, however, was stimulated by the possibility of using spherical cavities as field-of-view limiting baffles in earth-observing radiometric instrumentation applications.

ANALYSIS

Figure 1(a) shows a collimated beam of radiation of flux q_o (power per unit normal area) entering a spherical cavity of unit radius. The y,z-plane of the x,y,z coordinate system shown in figure 1(a) is arbitrarily oriented parallel to the direction of the incoming radiation. The direction of incoming radiation can then be identified by the angle γ it makes with respect to the z-axis.

Sparrow and Johnson (1962) show that, when the walls of the cavity are diffuse, the apparent absorptivity α_a is independent of the direction in which radiation enters the cavity and is given by

$$\alpha_a = \frac{\alpha}{1 - 0.5(1 - \alpha)(1 + \cos\psi)} \, , \tag{1}$$

where α is the absorptivity and ψ is the opening angle. According to Sparrow (1962) the only other cavity whose apparent absorptivity is directionally independent is the circular-groove cavity.

Now consider the case in which the cavity walls are specularly reflecting. To perform the analysis in this case it is much more convenient to use the x',y',z' coordinate system shown in figure 1(b) in which the z'-axis is parallel to the incoming radiation. The rays entering the cavity in the plane defined by the azimuthal angle ϕ' (measured from the positive x'-axis) undergo specular reflections as a group within that same plane, as can be easily demonstrated by applying the laws of specular reflection and recognizing that the normal to the surface of a sphere always points toward the center of the sphere. Hence, by referring to figure 1(b), the total energy absorbed by the cavity can be obtained by finding the portion of the energy that enters the cavity through a longitudinal band $d\phi'$ at ϕ' that is absorbed, and then integrating over the appropriate limits of the cavity opening in the ϕ'-direction.

One problem that needs to be solved is that of obtaining an equation for the boundary of the cavity opening with respect to the spherical coordinates based on the x',y',z' coordinate system of figure 1(b). Note that with respect to the coordinate system shown in figure 1(a) the boundary of the cavity opening is given simply by $\theta = \psi$, where θ is the zenith angle. Note also that x',y',z' in figure 1(b) can be obtained by rotating the y,z-plane in figure 1(a) by an amount γ about the x-axis. The relation between the spherical coordinates of the x,y,z and x',y',z' coordinate systems may be obtained by using the coordinate transformation matrix for the inverse of the rotation described above; that is,

$$\begin{bmatrix} 1 & 0 & 0 \\ 0 & \cos\gamma & -\sin\gamma \\ 0 & \sin\gamma & \cos\gamma \end{bmatrix} \begin{bmatrix} \sin\theta' \cos\phi' \\ \sin\theta' \sin\phi' \\ \cos\theta' \end{bmatrix} = \begin{bmatrix} \sin\theta \cos\phi \\ \sin\theta \sin\phi \\ \cos\theta \end{bmatrix} . \tag{2}$$

The boundary of the cavity opening in terms of the primed spherical coordinates can be obtained by first substituting the angle ψ, which defines the boundary of the cavity in the x,y,z coordinate system, for θ in equation (2). Then after multiplying out the third row of the transformation matrix and performing some simple algebraic manipulations, the boundary of the cavity opening in terms of the x',y',z' coordinate system is expressed implicitly by

$$\cos\theta' + (\tan\gamma \, \sin\theta') \sin\theta' = \frac{\cos\psi}{\cos\gamma} . \tag{3}$$

The intersection of the plane defined by the azimuthal angle ϕ' with the sphere is a great circle. For convenience this great circle is referred to as the *circle of reflection*. The circle of reflection is actually a spherical sector having differential thickness $d\phi'$. Each point on the circle of reflection is identified by the polar angle measured from the positive z'-axis. The rays, which in this arrangement are parallel to the z'-axis, enter the circle of reflection through the arc $[\psi^a, \psi^b]$ and undergo multiple reflections within the circle of reflection. The angles ψ^a and ψ^b, which delimit the arc of the opening for a given ϕ', can be obtained from equation (3). As a result of specular reflections some of the rays will be reflected into the arc $[\psi^a, \psi^b]$ and escape through it.

Consider an amount of energy $dQ(\psi^a \rightarrow \psi^b)$ which enters the cavity through a differential slit of angular width $d\phi'$ and circular arc $[\psi^a, \psi^b]$. Now suppose that the cavity walls are perfectly reflecting ($\rho^s = 1$). After a number n of specular reflections, some of this energy will have escaped through the opening. In general, there will be some number m of circular arcs $[\psi^L_1, \psi^U_1], [\psi^L_2, \psi^U_2], ..., [\psi^L_i, \psi^U_i], ..., [\psi^L_m, \psi^U_m]$ within $[\psi^a, \psi^b]$ for which incident radiation has not yet escaped after n reflections. This collection of circular arcs is identified by the symbol C_n. The procedure for finding C_n can be implemented on a digital computer in a straightforward manner provided that the formula for the reflection points of a given ray entering the circle of reflection is known. The procedure involves finding the illuminated region of the circle of reflection after each reflection and subtracting from it the circular arc which defines the opening.

To obtain the illuminated region, the reflection points of a given ray entering the circle of reflection must be determined. It is elementary to show that the polar angle between two consecutive reflection points for a ray entering the circle of reflection at a polar angle β_0 is given by $\pi - 2\beta_0$. Hence, the polar angle of the nth reflection point is given by

$$\beta_n = n\pi - (2n - 1)\beta_0 . \tag{4}$$

Through each circular arc $[\psi^L_i, \psi^U_i]$ in collection C_n enters an amount of energy which can be identified by the symbol $dQ_{n,i}$. The portion of this energy which is absorbed during the nth reflection is given by $(1 - \rho^s)(\rho^s)^{n-1}dQ_{n,i}$, where ρ^s is the specular reflectivity. Hence, the total energy which is absorbed due to that which has entered the cavity through the longitudinal band $d\phi'$ at ϕ' is given by

$$dQ_n = (1 - \rho^s)(\rho^s)^{n-1} \sum_{i=1}^{m} dQ_{n,i} . \tag{5}$$

The amount that is absorbed during all reflections is thus given by

$$dQ(\phi) = \sum_{n=1}^{\infty} dQ_n = (1-\rho^s) \sum_{n=1}^{\infty} [(\rho^s)^{n-1} \sum_{i=1}^{m} dQ_{n,i}] . \tag{6}$$

The quantity yet to be determined is $dQ_{n,i}$. The radiative flux q_0 enters the circle of reflection through a differential polar angle $d\beta_0$ at β_0. The differential area normal to q_0 where the first reflection occurs is given by $\cos\gamma_0 \sin\gamma_0 d\gamma_0 d\phi'$ for the unit spherical cavity. Therefore,

$$dQ_{n,i} = q_0 \int_{\psi^L_i}^{\psi^U_i} \sin\alpha_0 \cos\alpha_0 \, d\alpha_0 \, d\phi' = \frac{q_0}{2}(\sin^2\psi^U_i - \sin^2\psi^L_i) \, d\phi' . \tag{7}$$

By substituting equation (7) into equation (6), an expression for the amount of energy which is absorbed due to that which enters the cavity through the longitudinal band $d\phi'$ at ϕ' can be obtained. The amount of the energy which is absorbed by the cavity can be obtained by integrating equation (6) numerically over the appropriate limits of ϕ' which define the cavity opening. The amount of energy which enters the cavity is given simply by

$$Q_e(\gamma) = \pi q_0 \sin^2\psi\cos\gamma . \tag{8}$$

Dividing the energy absorbed by the total energy which enters the cavity gives the directional apparent absorptivity $\alpha_a(\gamma)$. The apparent (hemispherical) absorptivity α_a (which by Kirchhoff's law is the same as the apparent emissivity for an isothermal cavity) can be obtained by numerical integration of the directional absorptivity over the incident angle γ,

$$\alpha_a = 2 \int_0^{\pi/2} \alpha_a(\gamma) \cos\gamma \, \sin\gamma \, d\gamma \ . \tag{9}$$

RESULTS AND DISCUSSION

The results for the directional absorptivity are shown in figure 2. Each plot in figure 2 shows the apparent directional absorptivity $\alpha_a(\gamma)$ for a different opening angle ψ as a function of the angle of incidence γ for various surface absorptivities α. Apparent absorptivity results for the corresponding diffuse cavities, obtained using equation (1), are also plotted for comparison.

As expected, for a given opening angle and incidence angle, as the absorptivity α of the wall increases so does the apparent directional absorptivity. Moreover, for a fixed wall absorptivity α, as the cavity opening angle increases, the apparent directional absorptivity decreases for all angles of incidence γ. The apparent absorptivity of any cavity is usually greater than the absorptivity of the cavity surface. However, for large opening angles it is possible that for some angles of incidence the radiation entering the cavity will be reflected out after only one reflection. In these cases $\alpha_a(\gamma) = \alpha$; see for example the results for an opening angle of 150 deg.

The apparent hemispherical absorptivity results for both specular and diffuse cavities are shown in figure 3. Once again, the diffuse results were calculated directly from equation (1), and the specular results were obtained from equation (9). It should be mentioned that the Monte-Carlo technique was used to obtain an independent measure of the results presented in this paper, and the agreement between the two methods was excellent.

There is an important difference between the apparent absorptivity results of spherical cavities and those of unbaffled cylindrical and conical cavities. In spherical cavities, for a given wall absorptivity, the apparent absorptivity does not always increase as the cavity wall changes from diffuse to specular. Based on the results of figure 3, specular spherical cavities have a higher apparent absorptivity when the surface absorptivity is high. However, as the absorptivity of the surface decreases, curves representing the purely specular cavities begin to shift downward relative to the curves for the purely diffuse cavities. Eventually, as the absorptivity of the cavity wall becomes sufficiently small, the apparent absorptivity of the diffuse cavity becomes greater than that of the specular cavity. This type of behavior in spherical cavities may perhaps be explained by the fact that, unlike the case of conical and cylindrical cavities, some portions of the cavity surface face downward into the cavity.

CONCLUSIONS

The apparent directional and hemispherical absorptivity for specular spherical cavities has been obtained for the first time using an exact analytical approach. The apparent absorptivity has been shown to be highly directional for relatively small opening angles. Surprisingly, for low values of wall absorptivity diffuse spherical cavities are more efficient absorbers and emitters than specular cavities.

ACKNOWLEDGMENTS

The authors wish to thank the Radiation Sciences Branch of the Atmospheric Sciences Division of NASA's Langley Research Center for its partial support of this work under contracts NAS1-18471, Tasks 2 and 13, and NAS1-18106, Task 12.

REFERENCES

1. Sparrow, E. M., 1962, ASME Transactions, Journal of Heat Transfer, 84, pp. 283-293.
2. Sparrow, E. M., and Cess, R. D., 1979, Radiation Heat Transfer, p. 169, McGraw-Hill Book Company, New York.
3. Sparrow, E. M., and Jonsson, V. K., NASA Technical Note D-1289.

NOMENCLATURE

Q	Heat transfer rate (W)	γ	Angle of incidence (deg)
q	Heat flux (W/m^2)	θ	Zenith angle (deg)
Greek		ρ	Reflectivity (-)
α	Absorptivity (-)	ϕ	Azimuthal angle (deg)
β	Polar angle (deg)	ψ	Cavity opening angle (deg)

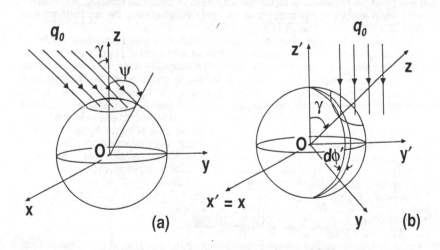

Figure 1. (a) A Collimated Beam of Flux q_0 Entering a Spherical Cavity at Angle γ, and (b) the Same Cavity Rotated Through Angle γ about the z-Axis.

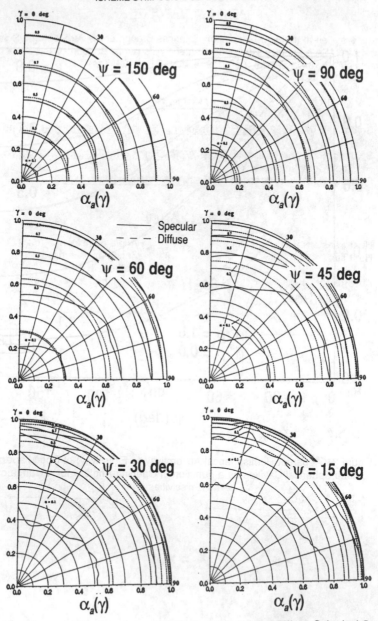

Figure 2. Apparent Directional Absorptivity for Specular and Diffuse Spherical Cavities for a Range of Wall Absorptivities α and Opening Angles, ψ.

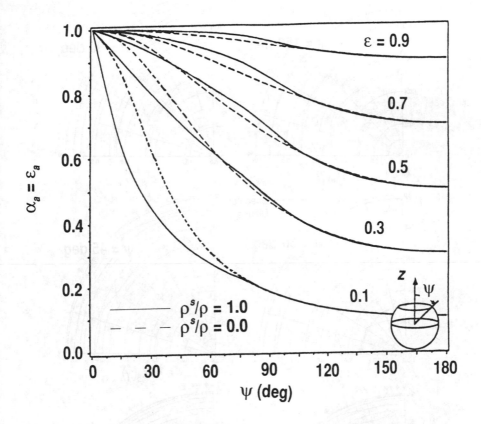

Figure 3. Apparent Hemispherical Absorptivity (= Emissivity) of Specular and Diffuse Spherical Cavities as a Function of Opening Angle ψ for a Range of Wall Emissivities ε.

EFFECT OF THERMAL GRADIENTS ON THE ADSORPTIVE DRYING OF SOLVENTS IN PACKED BEDS OF ZEOLITE

B D Crittenden and S Ben-Shebal
School of Chemical Engineering, University of Bath, Bath BA2 7AY

A packed bed of 3A zeolite has been used to dry ethanol solutions initially at 24°C and containing up to 6.6 wt% water. In view of the highly exothermic nature of the process, progress of the thermal wave through the bed has been followed, and its effect on product dryness has been identified. The thermal wave is found to leave the bed around the time at which breakthrough of water begins to occur. For a given flowrate, the water concentration in the initial effluent and the peak temperature rise are found to be directly proportional to the water concentration in the feed. For a given feed concentration, both the initial effluent water concentration and the peak temperature rise can be reduced by reducing the feed velocity. However, very low water concentrations should be achievable by removing the exotherm.

INTRODUCTION

Molecular sieve zeolites are suitable adsorbents for the removal of low levels of water from organic solvents. Water, because of its small molecular diameter, 0.28 nm, can easily enter zeolite channel structures, whilst most organic molecules, such as ethanol, 0.44 nm, are essentially excluded. For water in zeolites the electrostatic interaction forces can be very high, and can give rise to heats of adsorption comparable with heats of chemical reaction[1,2]. Hence the exotherm which occurs when water is adsorbed can become appreciable for a high water concentration in the feedstock[2,4]. For example, in the vapour phase adsorptive process to break the ethanol-water azeotrope (4.4 wt% water at 1 bar) temperature rises as high as 120°C can occur and have deleterious effects on the adsorption isotherm and product dryness[5]. Depending on the process conditions and the system under consideration, the thermal wave created by the exotherm can travel through a packed bed of adsorbent at a speed greater than, less than, or equal to that of the mass transfer wave of water. Garg and Ausikaitis[3] have indicated how control over the thermal wave can be used to advantage in the vapour phase drying of the ethanol-water azeotrope. However, little attention has been paid to the effect of the thermal wave on the dryness in liquid phase applications.

APPARATUS AND PROCEDURE

The apparatus consisted of a 0.0176m ID stainless steel column mounted vertically inside a thermostatically controlled oven, and packed to a height of 40cm with commercially available spherical beads of 3A zeolite (1.6 to 2.5mm diameter) manufactured by the Davison Chemical Division of W R Grace and Company. A 10cm long bed of similarly sized glass beads was placed at each end of the zeolite to provide good flow distribution.

Five thermocouples, protruding through the column wall into the adsorbent bed were used to monitor progress of the thermal wave. A constant flowrate in the range of 2.15 to

5.78 cm^{3}min^{-1} of feed of known composition in the range 2.46 to 6.55 wt% water was provided by a positive displacement metering pump. Flow was upwards and the liquid effluent passed into a fraction collector for water analysis by Karl Fischer titration. All adsorption experiments were carried out at feed and oven temperatures of 24°C. Experimental conditions are summarised in Table 1. Each experiment was preceded by thorough desorption of water from the zeolite.

Separate batch experiments were carried out to determine the equilibrium isotherms. An ethanol-water solution of known concentration was shaken with a known weight of freshly regenerated adsorbent for at least 8 hours in a flask placed in a thermostatted environment. The final concentration of water in the solution was obtained by Karl Fischer analysis. By assuming that no ethanol was adsorbed by the zeolite, for reasons given above, the uptake of water on the adsorbent was obtained by simple mass balance:

$$q^* = v(c_0 - c^*)/w \qquad (1)$$

RESULTS

Equilibrium and heat of adsorption

The equilibrium loading as a function of solution concentration is shown in Figure 1 for temperatures of 24, 29 and 34°C. Solid lines indicate fits obtained with the Langmuir isotherm equation:

$$q^* = K_L Q c^*/(1 + K_L c^*) \qquad (2)$$

the constants for which are given in Table 2. The Freundlich equation:

$$q^* = K_F(c^*)^{1/n} \qquad (3)$$

also fitted the data, although somewhat more poorly at low concentrations. The Freundlich constants are given in Table 2. As expected, both K_L and K_F decrease steadily with increasing temperature, whilst the monolayer loading Q is largely unaffected. The coefficient $(1/n)$ is directly proportional to temperature. Figure 2 shows the dependency of K_L on temperature written in the form of the van't Hoff equation:

$$K_L = K_L' \exp(\Delta H/RT) \qquad (4)$$

The average heat of adsorption at equilibrium conditions is calculated to be 96 kJ mol^{-1}, *ie* within the region normally expected for exothermic chemical reactions.

Breakthrough curves and thermal waves

An example breakthrough curve and the progress of its associated thermal wave through the bed are shown in Figure 3 for a feed flowrate of 3.3 cm^{3}min^{-1} and a feed water concentration of 2.67 wt%. The thermal wave passed out from the bed (indicated roughly by temperature T1) at about the same time as the concentration of water began to increase in the breakthrough curve. A temperature rise coincident with breakthrough is not desirable from an equilibrium viewpoint, Figure 1, although some marginal advantage could be gained from improved kinetics[4]. For this run, the maximum temperature rise was about 8°C, and the radial temperature gradient was found to be small.

The breakthrough curve and its associated thermal wave sharpen, and hence the maximum temperature rise is increased, as the feed concentration is increased. The effect of feed concentration on initial product water concentration and maximum temperature rise in the end bed temperature is shown in Figure 4. For the range of conditions studied both parameters increase linearly with the feed concentration. The origin is a valid data point for both lines on this Figure. The average bed loading is defined as follows:

$$ABL = (m_f - m_p - m_e)/w \tag{5}$$

Since ABL is calculated for conditions immediately after breakthrough has been completed, its value would be expected to be around that given by the equilibrium isotherms. As seen in Figure 1, ABL compares well with q^* obtained at the feed temperature. Thus despite the thermal transient passing through the bed, adsorptive loadings in equilibrium with the feed composition are re-established at a temperature very close to that of the feed.

In addition, the breakthrough curve and its associated thermal wave sharpen, and hence the maximum temperature rise is increased, as the flowrate is increased for a given feed concentration. The effect of flowrate on initial product water concentration, maximum temperature rise and ABL is shown in Figure 5. For the range of conditions studied, linear relationships between the three parameters and feed water concentration are observed. Figure 1 shows that the ABL becomes closer to q^* at the feed temperature of 24°C, when the feed velocity is the highest. This is probably the result of improved interparticle heat transfer at higher velocities, allowing a faster return of the bed to the feed temperature.

DISCUSSION

Teo and Ruthven[6] studied the adsorption of water from aqueous ethanol using 3A zeolite (1.6mm cylindrical pellets supplied by the Sigma Chemical Company). Breakthrough experiments were carried out in a 2.5cm diameter x 76cm jacketted glass column maintained at 24°C. The feedstock contained 4.7 wt% water and flow velocities were in the range 0.789 to 1.654 cm min^{-1}. Experimental breakthrough curves were compared with an analytical expression for constant pattern conditions derived by Weber and Chakravorti[7] who assumed isothermal conditions and a rectangular isotherm. Both external film and internal pore diffusional resistances were included. The comparisons between predicted and experimental curves were found to be poorest at the commencement of breakthrough. It was demonstrated that inclusion of the external film resistance in the model was essential, as Sowerby and Crittenden found in their vapour phase study[2].

Product drynesses prior to breakthrough were not reported by Teo and Ruthven[6], and in-column temperatures were not measured. The Weber and Chakravorti model appeared to predict zero water concentrations in the effluent up to breakpoint, but inspection of the experimental breakthrough data reveals that the initial values of dimensionless effluent concentration were in the range 0.01 to 0.02, a range which compares favourably with the value of 0.02 found in this study. The feed concentration of 4.7 wt% used by Teo and Ruthven[6] would have created a peak temperature rise of around 18°C, which would have been difficult to dissipate radially in a bed of zeolite contained in a 2.54cm diameter glass column.

For a feed of given water content, the product dryness can be improved somewhat by keeping the feed velocity low. The adsorbent vendor[8] recommends that the liquid velocity of ethanol should not exceed 7.62 cm min^{-1}, and should not be below 3.05 cm min^{-1} in order to avoid extremely large beds. Velocities used in this study and that of Teo and Ruthven[6] were mainly within this range. Cartón et al[9] studied the use of 3A zeolite in a 1.5cm diameter x 48cm column to dry ethanol at 22°C from a feed concentration of 6 wt% water. Purities were not reported but the treatment capacity (cm^3 ethanol per 100g zeolite) was found to be inversely proportional to velocity in the range 0.5 to 2 cm min^{-1}. Reducing the velocity allows a greater contact time, but unfortunately increases the external film resistance, which has a major impact on the shape of the early portion of the break-through curve[2,6].

Since the product dryness, which is determined by the shape of the leading edge of the breakthrough curve, determines whether the specification can be met, it is clearly necessary to adopt a more rigorous non-isothermal model for design and analysis, similar in concept to that reported by Garg and Ruthven for gas phase adsorption[10].

However, a simple expedient to produce very dry ethanol from a feed of high water concentration would be to use a series network of adsorption beds with interstage cooling. For the conditions studied in this work, the ratio of final water concentration to feed water concentration for j such beds, is given by:

$$c_{eff}/c_o \quad = \quad (0.02)^j \tag{6}$$

Thus a product containing less than 0.01% water should be obtainable from the azeotrope in two beds with interstage cooling back to 24°C.

CONCLUSIONS

The product dryness which is obtained when aqueous ethanol solutions are dried in a packed column of 3A zeolite is found to be directly proportional to the water concentration in the feedstock. The strongly exothermic nature of the adsorption process creates a thermal wave which begins to leave the column at a time around that at which breakthrough of water begins to occur. The exotherm at the end of the bed is directly proportional to feed water concentration, and thus creates unfavourable equilibrium conditions. The dryness obtained from a feed of given concentration can however be improved by reducing the linear velocity through the column. Accurate prediction of product dryness requires the development of a non-isothermal model.

References

1   Ruthven, D M, 1984, Principles of Adsorption and Adsorption Processes, John Wiley, New York
2   Sowerby, B and Crittenden, B D, 1991, *Trans IChemE*, 69(A), 3-13
3   Garg, D R and Ausikaitis, J P, 1983, *Chem Eng Prog*, 79 (4), 60-65
4   Sowerby, B and Crittenden, B D, 1988, *Gas Separation and Purification* 2, 77-83
5   Crittenden, B D and Sowerby, B, 1990, Advances in Separation Processes, IChemE Symposium Series 118, 55-69
6   Teo, W K and Ruthven, D M, 1986, *Ind Eng Chem Process Des Dev*, 25, 17-21
7   Weber, T W and Chakrovorti, R K, 1974, *AIChEJ*, 20, 228-238
8   W R Grace and Co, Ethanol Drying using Davison Molecular Sieves
9   Cartón, A, González, G, Iñiguez de la Torre, A and Cabezas, J L, 1987, *J Chem Tech Biotechnol*, 39, 125-132
10   Ruthven, D M, Garg, D R and Crawford, R M, 1975, *Chem Eng Sci*, 30, 803-810

Nomenclature

ABL	average bed loading defined by equation (5)	g water (g adsorbent)$^{-1}$
c^*	water concentration in solution at equilibrium	wt% or g cm^{-3}
c_{eff}	water concentration in effluent before breakthrough	wt% or g cm^{-3}
c_o	water concentration in initial or feed solution	wt% or g cm^{-3}
ΔH	heat of adsorption	kJ mol^{-1}

j	exponent in equation (6)	-
K_F	Freundlich constant	g soln (g water)$^{-1}$
K_L	Langmuir constant	g soln (g water)$^{-1}$
m_e	mass of water in bed voidage	g
m_f	mass of water fed into bed	g
m_p	mass of water collected in product	g
n	exponent in Freundlich equation	-
Q	monolayer loading - Langmuir equation	g water (g adsorbent)$^{-1}$
q^*	water loading on adsorbent at equilibrium	g water (g adsorbent)$^{-1}$
R	universal gas constant	kJ kmol^{-1}K^{-1}
T	temperature	K
v	volume of solution in batch equilibrium experiment	cm^3
w	mass of adsorbent	g

Table 1   Experimental conditions

Interstitial velocity (cm min^{-1})	Modified bed Re	Feed conc (wt% water)	Initial effluent conc (wt% water)	Average bed loading (g water/g bed)
3.57	0.082	2.68	0.068	0.109
3.57	0.082	3.45	0.069	0.121
3.57	0.082	4.38	0.097	0.134
3.57	0.082	6.55	0.129	0.151
2.32	0.053	2.46	0.049	0.096
3.57	0.082	2.68	0.068	0.109
4.71	0.108	2.67	0.101	0.111
6.25	0.144	2.67	0.141	0.131

Table 2   Isotherm parameters

| Temperature °C | Langmuir | | Freundlich | |
	K_L (g soln/g water)	Q (g solid/g water)	K_F (g soln/g water)	1/n (-)
24	0.325	0.255	0.065	0.548
29	0.185	0.245	0.040	0.682
34	0.092	0.254	0.023	0.772

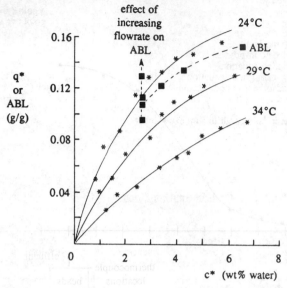

Figure 1   Equilibrium isotherms at 24, 29 and 34°C (asterisks indicate experimental data, solid lines represent Langmuir fits) and the effects of concentration and flowrate on ABL

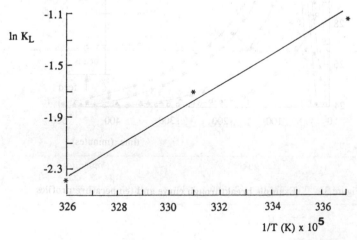

Figure 2   Effect of temperature on Langmuir constant K_L

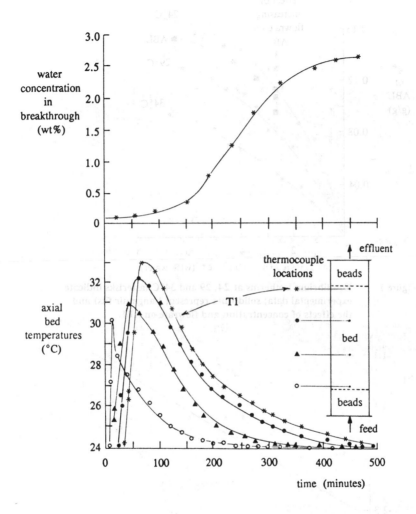

Figure 3   Example breakthrough curve and temperature profiles

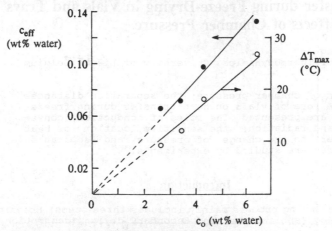

Figure 4   Effect of feed concentration on initial product water concentration and maximum rise in end bed temperature (constant feed flowrate)

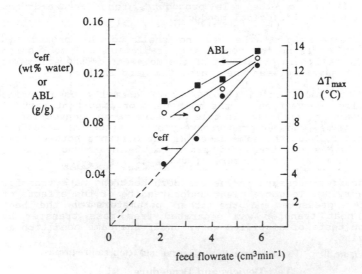

Figure 5   Effect of feed flowrate on initial product water concentration, average bed loading and maximum rise in end bed temperature (constant feed concentration)

Heat Transfer during Freeze-Drying in Vials and Trays and The Effects of Chamber Pressure

Shengwei Wang
Laboratory of Thermodynamics, University of Liège, Belgium

Effects of chamber pressure, the separation distances and the form of vials on heat transfer during freeze-drying are presented. The roles of conduction, convection and radiation; the sensitive location of heat transfer to the change of pressure and separation distances are studied by experiments.

Introduction

A freeze-drying process mainly includes three steps: Freezing; Primary drying (sublimation) and Secondary drying (desorption).

In the primary drying process, the ice is removed by means of sublimation. Physically, the sublimation process of freeze-drying is a combined heat and mass transfer process under vacuum. This paper focuses on the heat transfer of freeze-drying in vials and trays. It is one of the main procedures for freeze-drying pharmaceutical and biological products.

Heat transfer from the heating shelf to the product is performed by three means: conduction, radiation and convection /1/. The convection is the result of the movement of the molecules of vapour or gases between the vials, tray and shelf.

In the pressure range (0.01-2 mbar) normally used during freeze-drying, convection is the main part or important part of the heat transfer /2//3//4/. In this pressure range, convection is influenced by the chamber pressure very much, since the mean free path of vapour is at the same level as the distance between tray and shelf, vials and tray (for example, the mean free path of vapour is 0.4 mm at the pressure of 0.1 mbar /5/).

The chamber pressure influences both heat and mass transfer in freeze-drying. To get a clear understanding of the effects of the chamber pressure and the other parameters on the heat transfer, heat transfer was separated from mass transfer by sublimation tests of distilled water under the same condition as freeze-drying.

Test Devices and Procedure

The tests were performed on a table freeze-dryer. Fig.1 shows the arrangement of the test device. The pressure was regulated by controlling the automatic shut-off valve between the condenser and the pump. The temperature of the heat exchange liquid (alcohol) was regulated by the temperature controller of the thermostat.

The chamber pressure, ice temperature, shelf temperature and the temperature of the radiation protector were recorded.

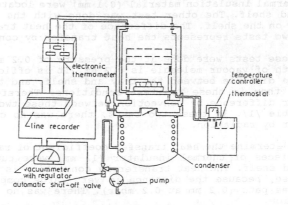

Figure.1 Schematic of The Test Device

A radiation protector (two layers of aluminium film) was used during the tests to limit the effect of the radiation from outside. The shelf was polished stainless steel.

The tray used was of aluminium. Two kinds of vials of different separation distance were employed for comparison: A. Moulded-glass vial (dm*: 0.9 mm); B. Tube-glass vial (dm: 0.4 mm).

The water in the vials or in the tray was frozen to -40°C for 2.5 hours on the shelf. Afterwards, the chamber was evacuated. When the pressure reached 1 mbar, heating of the shelf began. Since the period for evacuating and heating was very short, the effect of this period was ignored during the treatment of the test data. The ice temperature was measured during the freezing and sublimation.

When the pure ice sublimes in the vials, the thermal conduction begins to decrease after 35-50% of ice has sublimated /2/. In the tests presented in this paper, the total ice sublimed was less than 30% in all the tests, to be sure that there was no decrease in thermal conduction.

The sublimation time of 2.5 hours was chosen in the tests. The heat transfer rate was calculated by the weight loss. It was determined by weighing the vials or tray containing the ice before and after sublimation tests.

---

* Maximum separation distance between vial bottom and a flat surface.

Tests were performed in different situations: ice filled in vials standing in tray, ice filled in vials standing on shelf, and ice filled directly in tray. All the tests were repeated once.

To obtain the thermal conductivity between tray and shelf, two tests were done. One test was done when four small pieces of very thin thermal insulation material (0.1 mm) were located between the tray and shelf. The other test was done with the tray directly located on the shelf. The difference of the heat transfer between these two tests represents the heat transfer by conduction.

These tests were done at the pressure of 0.2 mbar. The mean free path of vapour molecule is 0.2 mm. It is efficiently larger than the distance between the tray and shelf, and the difference between them in these two test conditions. Therefore, the heat transfer difference by convection between these two tests can be negligible /1/. In these two tests, there was no change of heat transfer by radiation.

To determine the heat transfer coefficient of radiation, four small pieces of thermal insulator (12 mm) were put between the tray and shelf. The heat transfer by convection is negligible in this case, because the distance is very large compared with the mean free path (0.2 mm at 0.2 mbar). There was no difference of conduction.

Results and Discussion

Fig.2 shows the influence of the chamber pressure on heat transfer between tray and shelf. The chamber pressure influences convection very much. Therefore, the total heat transfer is influenced by the chamber pressure very much.

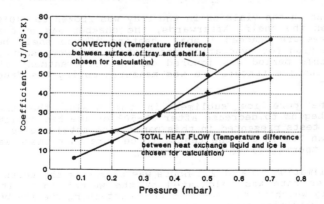

Figure.2 Influence of Chamber Pressure on Heat Transfer between Tray and Shelf

Table.1 shows the percentages of conduction, convection and radiation between tray and shelf when changing the chamber pressure in the tests. In the pressure range tested, the heat transfer by convection is always the important part. As pressure increases, the percentage of convection increases, and becomes the main part. This explains why the total heat transfer depends on the chamber pressure so much (see Fig.2).

Table.1 Percentage of Different Ways of Heat Transfer between Tray and Shelf (including radiation from above)

Pressure(mbar)	0.08	0.2	0.35	0.5	0.7
Conduction	6	5	3	2	1.4
Radiation	60	32	23	17	15.6
Convection	34	63	74	81	83

Fig.3 shows the influence of chamber pressure on heat transfer between the shelf and vials when the tray was used. Table.2 presents the resistance of heat transfer in the same tests. As the pressure increases, both of the resistances decrease. But comparing with the heat transfer resistance between tray and shelf, the heat transfer resistance between tray and vials decreases more slowly. Therefore, the percentage of it increases.

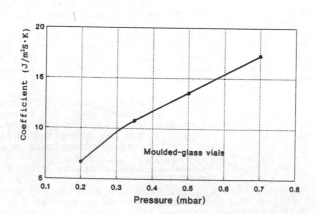

Figure.3 Influence of Chamber Pressure on Heat Transfer Coefficient between Vials and Shelf Surface except Radiation from Above (Vials Standing in Tray)

Table.2 Heat Transfer Resistance during Sublimation in Vials on Tray (unit of Res.(Resistance): m²s·K/J)

pressure (mbar)	0.2		0.35		0.5		0.7	
	Res.	%	Res.	%	Res.	%	Res.	%
vials <> tray	0.094	61	0.060	65	0.054	74	0.045	76
tray <> shelf	0.059	39	0.032	35	0.019	26	0.014	24

- moulded-glass vials were used in the tests

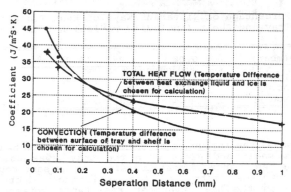

Figure.4 Influence of Separation Distance between Tray and Shelf on Heat Transfer Coefficient (under pressure of 0.5 mbar)

Fig.4 and Fig.5 show the influence of the separation distance on the heat transfer by convection. At the pressure of 0.5 mbar, the location where the convection decreases most rapidly is between 0.05 mm and 0.3 mm. At this pressure, the mean free path of vapour is 0.08 mm. At the pressure of 0.2 mbar, this location is between 0.15 mm and 0.4 mm. At this pressure, the mean free path of vapour is 0.2 mm.

Those show that the heat transfer by convection changes very much when the separation distance changes near the range of the mean free path of vapour under the corresponding pressure. In the second case, it is less obvious, since the convection plays less important role in the whole heat transfer under lower pressure.

Fig.6 Shows the difference in heat transfer with different kind of vials. In the pressure range from 0.2 mbar to 0.7 mbar, heat transfer of Vial B changes much quicker than that of Vial A. The mean free path of vapour changes from 0.2 to 0.06 mm under the pressure range. It is closer to 0.4 mm (dm for Vial B) than to 0.9 mm (dm for Vial A). It shows again that the pressure change has a bigger effect on heat transfer when the mean free path of vapour under this pressure is closer to the range of separation distance.

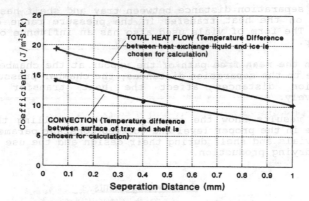

Figure.5 Influence of Separation Distance between Tray and Shelf on Heat Transfer Coefficient (under Pressure of 0.2 mbar)

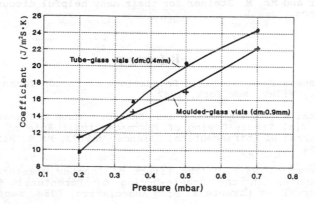

Figure.6 Influence of Pressure on Heat Transfer Coefficient between Vials and Shelf Surface except Radiation from Above (Vials Directly Standing on Shelf)

Conclusions

Convection is affected by the variation of chamber pressure very much in the pressure range of the freeze-drying. The total heat transfer during freeze-drying is sensitive to the change of pressure, since the main part of the heat is supplied by the convection.

The separation distance between tray and shelf has important effects on the heat transfer in the pressure range of freeze-drying. The form of vials used also has an influence on the heat transfer.

When the mean free path of the vapour at the chamber pressure is close to the separation distances, the chamber pressure and the separation distances affect the heat transfer the most sensitively.

The results show the importance to controlling the chamber pressure in the proper level and to keep constant parameter of the trays, vials and shelf during their design and the use of them in freeze-drying production.

Acknowledgements

The experiment results presented in this paper were carried out in the Laboratory of Freeze-Drying, Leybold AG, Cologne, F.R. Germany by the author. The author would like to thank Ms. H. Willemer and Mr. M. Steiner for their many helpful discussions and assistance.

References

1. Dushman S. <Scientific Foundations of Vacuum Technique> 2nd ed., Wiley, New York, 1962

2. Pikal M. J., Roy M. L., and Shah S. 'Mass and Heat Transfer in Vial Freeze-Drying of Pharmaceutical: Role of the vial' Journal of Pharmaceutical Science, Vol.73, No.9, P1224, Sept. 1984

3. Steven L. Nail 'The Effect of Chamber Pressure on Heat Transfer in the Freeze-Drying of Parenteral Solutions' Journal of parenteral Drug Association, P358, Sept/Oct 1980

4. Livesey R. G., Row T. W. G. 'Effect of System Pressure on Primary Freeze-Drying in Vials' 17th International Conference of Refrigeration, Volume C, P113 Viana, 1987

5. Leybold <Vacuum Technology; Its Foundations; Formulae and Tables> Cologne, F. R. Germany, 2, 1982

PREDICTING THE VARIATION OF TEMPERATURE AND WEIGHT LOSS OF PRODUCTS IN ARRAYS COOLED BY AIR FLOW

D. Burfoot
AFRC Silsoe Research Institute, Wrest Park, Silsoe, Bedford MK45 4HS

Cooling and drying of products using convective air are common to many industries, including agriculture, food and forestry. Numerical models which predict the air flow distribution and temperature and moisture changes of the air and product have been developed to assist in process design. Simulated flow profiles are in good agreement with measurements but predicted temperature and mass changes are less reliable. The models need extending to allow for dimensional changes of the products and the condensation of moisture.

INTRODUCTION

In many industries products are cooled or dried by blowing air through piles or arrays. For example, in the food industry, products such as ready prepared meals or meat joints are placed on the shelves of rack trolleys, and then cooled in batch systems by blowing refrigerated air towards the foods. Rapid cooling of the products to temperatures below 10°C is essential to restrict the growth of vegetative pathogenic organisms (1). A parallel problem exists for the forestry industry when seeking to use the residues which remain after the better quality wood has been removed from felled trees for the subsequent manufacture of paper and board. The use of the residue as a 'green' fuel source is being investigated but a primary problem is the need to dry and store the material so deterioration and heating due to mould growth are restricted. This requires the moisture content to be reduced from about 50% to less than 30% (2). In addition, this must be done with minimal use of energy because of the low unit value of the product.

We are developing mathematical models and experimental programmes to assist in the design of safe and cost effective processes using both forced and natural convection. This work builds on past developments of similar models for the agricultural sector, where grain is dried and many vegetables cooled and stored by blowing air through piles of the product (3). The aim of this paper is to describe the mathematical models which have been used and are still being developed. Results from some of the models are presented and existing problems and potential applications highlighted.

MATHEMATICAL DEVELOPMENT

There are two components to the mathematical simulations of these problems; air flow distribution and the simultaneous transfer of heat and moisture. Many of the piles of particles, such as beds of grain, seeds or nuts, have the structure typical of porous media. The flow of air through such media can be described by Darcy's law (equation 1a) which can be derived from a momentum balance on the air and relates the velocity to the pressure gradient such that in cartesian coordinates:

$$v_x = -R_x \frac{\partial p}{\partial x} \; ; \; v_y = -R_y \left(\frac{\partial p}{\partial y} - \rho g \right) \; ; \; v_z = -R_z \frac{\partial p}{\partial z} \qquad [1a]$$

where R_x, R_y, R_z, are the flow coefficients which may be estimated from equations given in basic texts (4). In the agricultural sector, empirical equations of the following form (5) are used for these coefficients:

$$R = \frac{\ln(1 + b_1 v)}{a_1 v} \qquad [1b]$$

or

$$R = a_2 v^{b_2} \qquad [1c]$$

Equations 1b and 1c apply to air flows, but other correlations, incorporating fluid viscosity, have been determined for a wider range of fluids.

The similarity with porous media becomes less evident as the particle, or package, size and spacing increase. In addition, Darcy's law is limited to conditions of laminar flow. However, the use of empirical equations such as those given above enable the application of the law to be extended to turbulent flows with higher fluid velocities and larger particle sizes, provided that wall effects do not become dominant.

The second equation is a mass balance on the air which gives:

$$\frac{\partial(\rho v_x)}{\partial x} + \frac{\partial(\rho v_y)}{\partial y} + \frac{\partial(\rho v_z)}{\partial z} = -\frac{\partial \rho}{\partial t} \qquad [2]$$

In most common applications, the flow is steady so that the accumulation term on the right hand side of equation [2] can be ignored. Solutions of equations [1] and [2] provide the air pressure and velocity distribution throughout the pile.

In applications where the materials forming the pile are not sealed then both mass and heat transfer occur as water evaporates from the surface of each particle or product in the batch. A mass balance on the moisture shows that

$$\frac{\partial(\rho v_x A H)}{\partial x} + \frac{\partial(\rho v_y A H)}{\partial y} + \frac{\partial(\rho v_z A H)}{\partial z} = -\frac{\rho_b}{\rho} \frac{\partial M}{\partial t} \qquad [3]$$

The rate of evaporation ($\partial M / \partial t$) from the particles could be determined by solving a diffusion equation but, for small products, it is more common to use an empirical equation derived from experiments with a thin layer drier (6). The simplest form is:

$$\frac{\partial M}{\partial t} = -K(M - M_e) \qquad [4]$$

An energy balance on the air shows that

$$\frac{\partial(\rho v_x AcT)}{\partial x} + \frac{\partial(\rho v_y AcT)}{\partial y} + \frac{\partial(\rho v_z AcT)}{\partial z} = -h_v(T-\theta) + \rho_b c_v \frac{\partial M}{\partial t}(T - \theta) \qquad [5]$$

and on the solids gives

$$\rho_b c_{gd} \frac{\partial \theta}{\partial t} = h_v(T-\theta) - \rho_b(\lambda - c_w \theta)\frac{\partial M}{\partial t} \qquad [6]$$

These equations can be solved using various numerical methods, including finite difference, finite element and control volume techniques. Discussion of the relative merits of each method is beyond the scope of this paper. All three methods should lead to the same solutions given adequate discretisation into imaginary segments and the appropriate choice of time step. Existing models do, however, differ in the complexity of the problem that they have been designed to handle. For example, some simplify problems into the 1- or 2-dimensional cases, while others consider the full 3-dimensional solution. The following section shows examples of the use of these approaches.

EXAMPLES OF MODELS

1-Dimensional

Many problems can be simulated by assuming that air flows essentially in one direction only and such 1-dimensional models have been widely applied to the drying of grain (7,8,9,10,11). The early work of Bakker-Arkema et al. (7) did not provide reliable predictions of product moisture content. Better predictions were achieved by Spencer (8) using an improved equation, based on a simple diffusion model, to describe the drying rate. These models concentrated on the drying of static beds. Further improvements were achieved by O'Callaghan et al. (9) and Nellist (10) using the empirical drying rate model (equation [4]). Their work, whilst still employing the 1-dimensional approach, also provided simulations of the more complex and practical drier applications, including cross-, contra- and co-flow grain driers. These models allowed predictions of the grain and air outlet moisture and temperature, based on the inlet conditions, but provided no reliable information on the changes at specific points within the bed.

2-Dimensional

Several researchers have modelled the air flows, but not the simultaneous heat and mass transfer, in a 2-dimensional bed, using finite difference (11) and finite element methods (12). These models have also ignored the density changes of the air and the gravitational effects included in equation (2). Several papers have reported on measured pressure distributions through 2-dimensional beds, but few of them give sufficient detail for comparison with numerical models. Figure 1 shows some unpublished pressure distributions measured in a bin of grain and the results of a finite difference model developed by the author to solve equations [1] and [2]. Measured and predicted pressures, during this and other tests, agree to within 30 Pa (probably the measurement accuracy achieved using an open ended static pressure tube). However, there are few data in the region of high pressure gradients where the air velocities are high, and are beyond the range covered by the empirical equations used to calculate the flow coefficients.

The work by Williamson (13), while still not providing all the necessary data for model validation, does allow some comparisons. Figure 2 compares the lines of equal moisture content obtained by measurement by Williamson and those predicted using a model developed at Silsoe by Mao and Nellist (14). The model predicts the expected trend of increasing moisture content away from the air supply duct, but the predicted moisture profiles provide a poor comparison with the measured data. Confidence in the model is restricted by several factors. The equations used to predict the properties of the grain were not entirely appropriate for the grain used by Williamson, and insufficient experimental details were provided by Williamson to enable a more reliable estimation. Also, the methodology of the model may lead to some errors. The model uses a finite element scheme to predict pressure and velocity distributions and a finite difference method to predict the temperature and moisture changes of the air and grain. Initially, the air pressure and velocity distribution are predicted throughout the bed of grain and the positions of streamlines are then calculated starting from an air inlet and moving in the direction of the air velocity. This requires interpolation between the finite element nodes. The change of moisture content and temperature of the grain and air along these lines are computed using a 1-dimensional finite difference scheme. Interpolation between the streamlines is then used to calculate the conditions (temperature and moisture content) at the finite element nodes. The changes in properties of the air lead to a new air pressure and velocity distribution which is again calculated using the finite element scheme. The potential inaccuracies in interpolation, and the use of an insufficiently fine numerical grid and number of streamlines, may be the cause of the poor predictions of moisture profiles. Further work is needed to refine the model and test it against more complete experimental data.

3-dimensional

Not surprisingly, given the problems that are still encountered with 2-dimensional studies there are few reports of work in the more practically realistic case of 3-dimensions. Smith (15) presents a model for the prediction of 3-dimensional air flows through granular beds but could not verify the model due to a lack of experimental data. We have also developed a 3-dimensional model and will be investigating its performance, particularly in the regions of high velocities in packed beds. These are the regions in which the models are most likely to fail, and where the empirical factors such as flow and heat transfer coefficients have not been measured. Figure 3 shows some of the predictions which have been made.

CONCLUSIONS

Much of the work at Silsoe has concentrated on the cooling and drying of agricultural products such as grain and hay, where ducts are located within a bed of product. This situation is ideally suited to simulation using solutions of Darcy's law with simultaneous heat and mass balances. Work has progressed to the development of 3-dimensional models although there is still a need for further work to include detailed effects, such as shrinkage of the bed, heating due to respiration by moulds, and to allow the simulation of moisture condensation which can occur in practice.

Work is now proceeding on new applications, such as the drying of forest residues and coppice wood, where the problems are very similar if air supply ducts, located within the pile, are used. Further work will be needed to simulate the interaction of the wood pile with the surroundings, due to factors such as rainfall or direct sunlight. Natural convection could also be important during any periods when the pile is not being force ventilated.

Application of the models to processes in the food industry, such as the cooling of arrays of processed and packaged products is a realistic possibility. It might appear to be simple, since mass transfer can be ignored in some cases, and the product geometry is fixed. However, in many conventional food processing operations, the fan and heating elements or evaporator are located outside the array and no ducting is used to direct the air flow through the batch. In such cases, the use of more complex flow models would be required, although the heat and mass transfer analyses should remain unchanged. Changes in the attitudes of the food industry towards cooling systems could also see further use of the models described in this paper. For example, the use of air socks as flexible ducts within the product batch could lead to improved air flow distribution and more effective cooling which could be simulated and optimised using the models described.

Despite the simplicity of the models, they have potential for applications in many industries. Their use is expected to increase as the need to restrict process and product variability, which are often caused by temperature and moisture variations, become more important.

ACKNOWLEDGEMENTS

I wish to thank Drs Bill Day and Martin Nellist (Silsoe Research Institute) for their comments on this manuscript and for permission to use the experimental data in Figure 2.

REFERENCES

(1) Department of Health (1989) Chilled and Frozen. Guidelines on Cook-Chill and Cook-Freeze Catering Systems. HMSO, London.

(2) Burfoot, D. (1991) Storage and drying of comminuted wood fuels - Mathematical modelling of heat and moisture transfer. Proc. Joint Meeting IEA/BEA Activity 4 and 5, June 1991.

(3) Nellist, M.E. (1988) Near-ambient grain drying. Agricultural Engineer, 93-101.

(4) Coulson, J.M., Richardson, J.F. (1968) Chemical Engineering, Volume 2, 2nd Ed, Pergamon Press, Oxford.

(5) Hukill, W.V., Ives, N.C. (1955) Radial air flow resistance of grain. Agricultural Engineering, 36(5), 332-335.

(6) Sokhansanj, S., Cenkowski, S. (1988) Equipment and methods of thin-layer drying - A review. Proc. 6th Int. Drying Symp., Versailles, France, 5-8 Sept, Vol. 1, 159-170.

(7) Bakker-Arkema, F.W., Bickert, W.G., Morey, R.V. (1967) Gekoppelter Warme-und Stoffaustausch Wohrend des Trockungsvorgangs in einem Behalter mit Getreide. Lantech. Forsch., 17(7)(4), 175-180.

(8) Spencer, H.B. (1969) A mathematical simulation of grain drying. J. Agric. Engng Res., 14(3), 226-235.

(9) O'Callaghan, J.R., Menzies, D.J., Bailey, P.H. (1971) Digital simulation of agricultural drier performance. J. Agric. Engng Res., 16(3), 223-244.

(10) Nellist, M.E. (1987) Modelling the performance of a cross-flow grain drier. J. Agric. Engng Res., 37(1), 43-57.

(11) Brooker, D.B. (1961) Pressure patterns in grain drying systems established by numerical methods. Trans. Am. Soc. Agric. Engrs., 4(1), 72-75.

(12) Marchant, J.A. (1976) Prediction of fan pressure requirements in the drying of large hay bales, J.Agric. Engng Res., 21, 333-346.

(13) Williamson, W.F. (1965) Pressure losses and rying rates of grain ventilated with various on-floor duct systems, J. Agric. Engng Res., 10(4), 271-276.

(14) Mao, Z., Nellist, M.E. (1991) Two dimensional heat and mass transfer simulation

program. Div. Note DN1600, AFRC Institute of Engineering Research, Silsoe, February 1991 (CONFIDENTIAL).
(15) Smith, E.A. (1982) 3-dimensional analysis of air velocity and pressure in beds of grain and hay, J. Agric. Engng Res., 27, 101-117.

NOMENCLATURE

a_1, b_1	Constants in equation [1b]	M_e	Equilibrium moisture content
a_2, b_2	Constants in equation [1c]	p	Pressure
A	Flow area	R, R_x, R_y, R_z	Flow coefficient, in x-, y-,
c	Specific heat capacity of air		and z-directions
c_{gd}	Specific heat capacity	t	Time
	of dry material	T	Air temperature
c_v	Specific heat capacity	v_x, v_y, v_z	Velocity components in x-,
	of water vapour		y-, and z-directions
c_w	Specific heat capacity	x, y, z	Cartesian coodinates
	of water	θ	Solids temperature
h_v	Volumetric heat transfer	λ	Latent heat of water in the
	coefficient		material
H	Absolute humidity	ρ	Density of air
K	Drying rate constant	ρ_b	Bulk density of dry solids
M	Moisture content, dry basis		

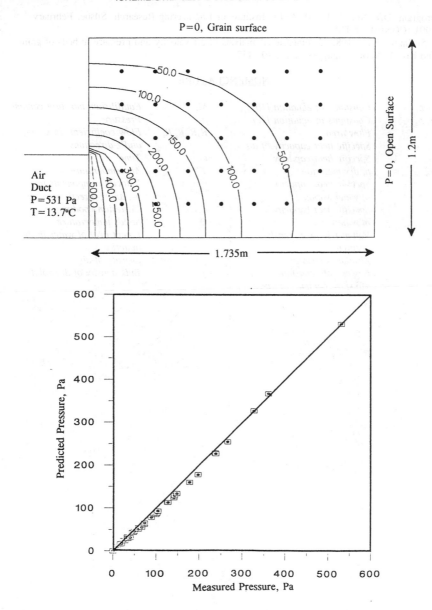

Figure 1.   Measured and predicted pressure distributions at the mid-plane of a 0.375m wide bin of grain. Upper diagram shows the computed isobars (——) and positions where pressures were measured (●).

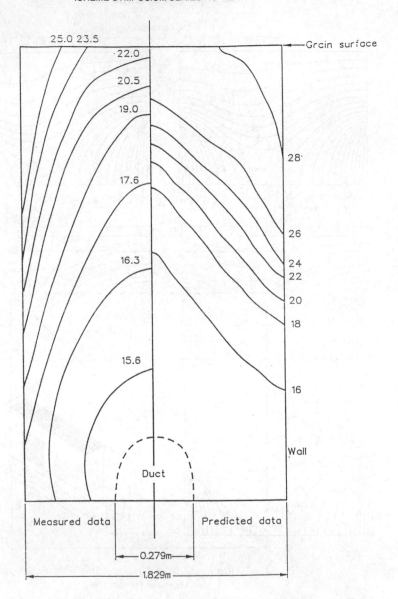

Figure 2.   Measured (12) and predicted (14) moisture profiles in a 1.64 m deep bed
of Attle wheat, initially at 28.2% moisture (dry basis) and dried for 192
hours. Air supplied at 15°C and 420 Pa.

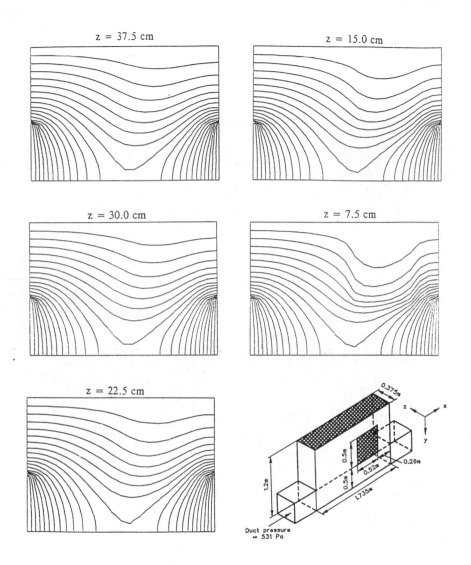

Figure 3.   Example of predicted isobars in an open-topped bin of grain (with a side opening) ventilated with air from 2 ducts. Isobars vary from 0 to 500 Pa at 25 Pa intervals. Pressure distributions shown at planes 7.5, 15.0, 22.5, 30.0 and 37.5cm from the side with the opening.

HEAT TRANSFER IN AN IN-LINE TUBE BUNDLE WITH A GAS-PARTICLE CROSS FLOW

D.C. Stewitt, D.B. Murray and J.A. Fitzpatrick.

Department of Mechanical Engineering, Trinity College,
Dublin 2, IRELAND.

The effect of solid particles in suspension on heat transfer tubes located at different positions within a square, in-line array has been investigated. Tests conducted with dilute particle concentrations of two mean diameters at a flow Reynolds number of 8000 indicate that heat transfer is enhanced at all locations by the presence of the particles. A preliminary assessment of the mechanisms of heat transfer modification is given and the significance of the findings for fluidised bed heat exchangers is discussed.

INTRODUCTION

Heat exchangers located high in the freeboard of fluidised bed combustors operate with a shell-side environment of gas with a low concentration of less than 1kg/m3 of fine dust particles. These operating conditions can lead to changes in the heat transfer characteristics with a consequent decrease or increase in overall system performance. For example, Biyikli et al. [1] investigated the local and overall heat transfer coefficients for a single horizontal tube located in the freeboard region of an air fluidised bed and reported increased heat transfer for the gas-particle environment. For this bundle, the performance of a staggered array of tubes in the gas-particle concentrating splash zone of a fluidised bed was investigated by Wood et al. [2] and Xu et al. [3] who reported substantial increases in heat transfer. However, it was shown by Murray & Fitzpatrick [4] that particles at low mass loadings can cause either an increase or a reduction in heat transfer for a triangular tube array. For in-line configurations, the effect of graphite particle-gas suspension on heat transfer has been investigated by Woodcock & Worley [3] and their results indicate a substantial enhancement of heat transfer with increasing mass loading ratio. However, such closely pitched arrays (P/D=1.16) are rarely if ever used in fluidised bed applications. The performance of in-line tube banks with more widely spaced tubes is of specific interest for fluidised bed combustors as the problems associated with axial passage of the tubes in these heat exchangers may be less severe. Again, a disadvantage is that, currently, there are no relevant heat transfer data reported for in-line tube arrays operating in dilute solid suspension flows.

This paper reports on an investigation of the effect of solid particles in suspension on heat transfer in a square, in-line tube array with two diameter ratios. Heat transfer results from different positions in the array are given and enhanced pressure distributions are reported to explain the observed asymmetries in the local heat transfer data.

HEAT TRANSFER IN AN IN-LINE TUBE BUNDLE WITH A GAS-PARTICLE CROSS FLOW

D.C. Sterritt, D.B. Murray and J.A. Fitzpatrick

Department of Mechanical Engineering, Trinity College, Dublin 2, IRELAND.

The effect of solid particles in suspension on heat transfer for tubes located at different positions within a square, in-line array has been investigated. Tests conducted with dilute particle concentrations of two mean diameters at a flow Reynolds number of 6000 indicate that heat transfer is enhanced at all locations by the presence of the particles. A preliminary assessment of the mechanisms of heat transfer modification is given and the significance of the findings for fluidised bed heat exchangers is discussed.

INTRODUCTION

Heat exchangers located high in the freeboard of fluidised bed combustors operate with a shell side environment of gas with a low concentration (less than 1kg/kg) of fine dust particles. These operating conditions can lead to changes in the heat transfer characteristics with a consequent decrease or increase in overall system performance. For example, Biyikli et al. (1) investigated the local and overall heat transfer coefficients for a single horizontal tube located in the freeboard region of an air fluidised bed and reported increased heat transfer for the gas-particle environment. For tube banks, the performance of a staggered array of tubes in the high particle concentration splash zone of a fluidised bed was investigated by Wood et al. (2) and Ku et al. (3) who reported substantial increases in heat transfer. However, it was shown by Murray & Fitzpatrick (4) that particles at lower mass loadings can cause either an increase or a reduction in heat transfer for a triangular tube array. For in-line configurations, the effect of graphite particles in suspension on heat transfer has been investigated by Woodcock & Worley (5) and their results indicate a substantial enhancement of heat transfer with increasing mass loading ratio. However, such closely pitched arrays (P/D=1.16) are rarely if ever used in fluidised bed applications. The performance of in-line tube banks with more widely spaced tubes is of specific interest for fluidised bed combustors as the problems associated with metal wastage of the tubes in these heat exchangers may be less severe (Brain & Minchener (6)). Currently, there are no relevant heat transfer data reported for in-line tube arrays operating in dilute solid suspension flows.

This paper reports on an investigation of the effect of solid particles in suspension on heat transfer in a square, in-line tube array with pitch to diameter ratio of 1.75. Heat transfer results from different positions in the array are given and surface pressure distributions are reported to explain observed asymmetries in the local heat transfer data.

EXPERIMENTAL FACILITIES AND PROCEDURES

The heat transfer measurements were carried out in a closed loop circulating rig with a removable test section such that different heat exchanger configurations can be tested. Solid particles are introduced to the flow upstream of the test section and a cyclone is used to separate out the solids for re-introduction to the flow. An individual instrumented and electrically heated tube is placed at different locations within the tube array and local heat transfer coefficients are determined using a measurement system based on a single surface mounted microfoil heat flux sensor. Rotation of the tube with respect to the flow permits local heat flux to be measured at 10° intervals on the tube surface. The uncertainty in local Nusselt numbers is estimated at ±7%. A more detailed description of the experimental facilities and heat transfer instrumentation has been given in (4). For the present study, a square, in-line array with a pitch to diameter ratio of 1.75 was used and tubes of 25mm diameter and 400mm length were arranged in a 10 row by 9 column tube bank. This is shown in Figure 1.

Tests were carried out at a Reynolds number of approximately 6000, based on the tube diameter and maximum gap velocity within the array. Spherical glass beads with mean particle diameters of 58μm and 127μm were used at solids mass loading ratios of around 0.5kg/kg.

RESULTS

A series of tests was conducted in air and a comparison of the mean Nusselt numbers for the inner tubes of the array with values obtained from the correlation of Zukauskas (7) for an array of the same geometry showed good agreement. Thus, the measurements provide suitable base-line data for identifying the influence of particles on heat transfer in in-line arrays.

Figure 2 shows the local Nusselt numbers at the first row of the array for the single phase flow and for the suspension flow with 127μm particles. The local variation in heat transfer at this position is similar to that observed for a single tube in cross flow or for a tube in the first row of a staggered array. Although some asymmetries in the data exist, it is evident that the main effect of the particles is to increase the Nusselt numbers over the front of the tube. Figure 3 shows the Nusselt numbers at the second row for the 127μm particles and for the single phase flow at the same Reynolds number. For both flows, the Nusselt number rises sharply from the front stagnation point to a maximum at about 60°, corresponding to flow reattachment, and then falls off to a minimum at around 110° -120°. Following this, there is a small increase in heat transfer in the wake region. From the suspension results, it is clear that an appreciable enhancement of heat transfer again occurs over the front of the tube, up to the reattachment point at 60°. In the wake close to 180°, there is a modest but consistent increase in heat transfer. Figure 4 shows the local heat transfer for a tube in the third row with the 127μm particles and Figure 5 shows the Nusselt numbers at the tenth row for both particle sizes. In each case, the main effect of the particles is an increase in Nusselt number around the heat transfer maxima at reattachment. This enhancement continues up to the point at which the boundary layer separates. Broadly similar trends were observed at other positions within the array.

Row	Re	S_L kg/kg	d_p μm	Nu_a	Nu_{su}	% Change
1	6000	0.50	127	40.8	44.1	+8.2%
2	6000	0.50	127	52.1	54.2	+4.1%
3	6000	0.50	127	50.6	51.5	+1.7%
5	6000	0.50	127	51.0	52.4	+2.8%
8	6000	0.50	127	50.2	52.9	+5.4%
10	6000	0.50	127	51.9	55.3	+6.5%
1	6000	0.48	58	39.0	45.8	+17.3%
2	6000	0.48	58	50.4	52.0	+3.2%
3	6000	0.48	58	49.3	51.3	+4.1%
5	6000	0.48	58	49.9	51.3	+2.8%
10	6000	0.48	58	48.8	50.7	+4.1%

TABLE 1   MEAN NUSSELT NUMBERS FOR IN-LINE ARRAY

Table 1 summarises the effect of the particles on the mean Nusselt numbers for both particle sizes and a range of tube locations. From these data, it is evident that there is a net increase in the mean Nusselt number for all tests carried out at the Reynolds number of 6000. The greatest increase in Nusselt number is observed for the tube located in the first row. Except at the first row, where the enhancement is greater for the smaller particles, the effect of varying particle size is not significant.

DISCUSSION

As evident from Figures 3 and 4 for the single phase flow, there is a marked asymmetry in the magnitude of the peak Nusselt numbers on either side of the tube. Furthermore, the side of the tube which has the higher peak Nusselt number at the second row has the lower peak at the third row. It has been observed by Bressler (8) that, for in-line arrays of small pitch to diameter ratio, the pressure at one of the flow reattachment points may be higher than at the other. In addition, the respective positions of the higher and lower surface pressure levels may change from row to row. Figure 6 shows the variations in mean surface pressure coefficients around the circumference of tubes at the first, second, third and tenth rows for the single phase flow at the Reynolds number of 6000. Some asymmetry in the surface pressure measurements is evident at the second and third rows and the highest pressure coefficient does occur at different sides for the two tube locations. Thus, the flow characteristics at the second and third rows are not uniform and asymmetry in the heat transfer results would be expected as a consequence.

The localised enhancement of heat transfer over the front of the tube in the first row (Figure 2) is slightly greater than that observed previously (4) for a tube in the first row of a staggered array with tube pitch to diameter ratio of 2. This enhancement has been attributed to an increase in the thermal capacity of the suspension and is more significant for smaller particles, as found here. The difference in enhancement between in-line and staggered arrays may be due to the smaller transverse pitch and greater flow blockage for the in-line array. The enhancement evident over the front of the tube in the second row (Figure 3) is less readily understood as it was not originally anticipated that significant numbers of particles would be convected into the wake zone behind tubes. It may be that the particles, once entrained in the wake region, perhaps as a result of rebounds from the surrounding tubes or as a consequence of the flow asymmetry, have relatively long residence times between the first and second rows.

From Figures 4 and 5, it is evident that the greatest changes in heat transfer generally occur around the reattachment points. Clearly, the influence of the particles on heat transfer is strong over the sides of the tube where local solids concentrations will be high and where boundary layer thinning as a result of momentum transfer from the particles may take place. The suspension heat transfer results exhibit asymmetries for all of the tests carried out. The pressure distribution at the tenth row shown in Figure 6 also indicates significant non-uniformity. It is apparent that, as a consequence of the non-uniform characteristics of the flow through the array, the local particle concentration is likely to vary.

CONCLUSIONS

The local heat transfer characteristics for tubes located at different positions within a square array with a tube pitch to diameter ratio of 1.75 have been investigated for suspension and single phase flows. 0.5kg/kg of glass beads of mean diameter 58μm and 127μm were used at a Reynolds number of 6000.

From the results, it is clear that at the Reynolds number of 6000, a net increase in heat transfer for the complete array is recorded for both particle sizes. At similar upstream conditions, a significant overall reduction in heat transfer performance was previously observed for a staggered tube array. This suggests that judicious selection of appropriate heat exchanger geometry may lead to enhanced performance for heat exchangers located in the freeboard of fluidised bed combustors.

ACKNOWLEDGEMENTS

Financial support for this work was provided by DGXII of the E.C. under research grant JOUF/0031/C (EDB).

REFERENCES

1. Biyikli,S., Tuzla,K. and Chen,J.C., 1983 AIChE Journal 29 712-716.

2. Wood,R.T., Kuwata,M. and Staub,F.W., 1980 in Fluidization, eds.J.R.Grace and J.M.Matsen, pp.235-242, Plenum, New York.

3. Ku,A.C., Kuwata,M. and Staub,F.W., 1981 AIChE Series 77 no.208, 359-367.

4. Murray,D.B. and Fitzpatrick,J.A., 1991 ASME J. Heat Transfer 113 865-873.

5. Woodcock,M.T. and Worley,N.G., 1967 Proc.I.Mech.E. 181 17-33.

6. Brain,S.A. and Minchener,A.J., 1991 Proc. 1991 Int.Conf. on Fluidised Bed Combustion, Montreal, vol.2, pp.619-630.

7. Zukauskas,A., 1972 Adv. in Heat Transfer 8 93-154.

8. Bressler,R., 1958 Kaltetechnik 11 365.

Figure 1   In-Line Tube Bundle
Pitch/Diameter=1.75

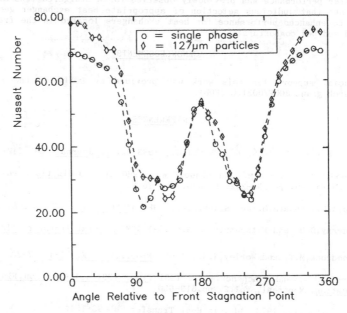

Figure 2   Local Nusselt Numbers at First Row
Re=6000; Loading Ratio=0.5kg/kg

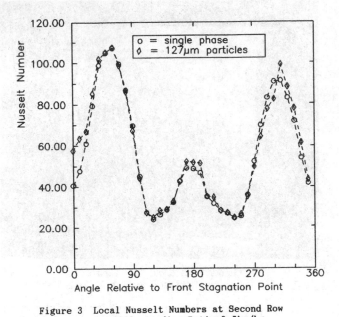

Figure 3   Local Nusselt Numbers at Second Row
Re=6000; Loading Ratio=0.5kg/kg

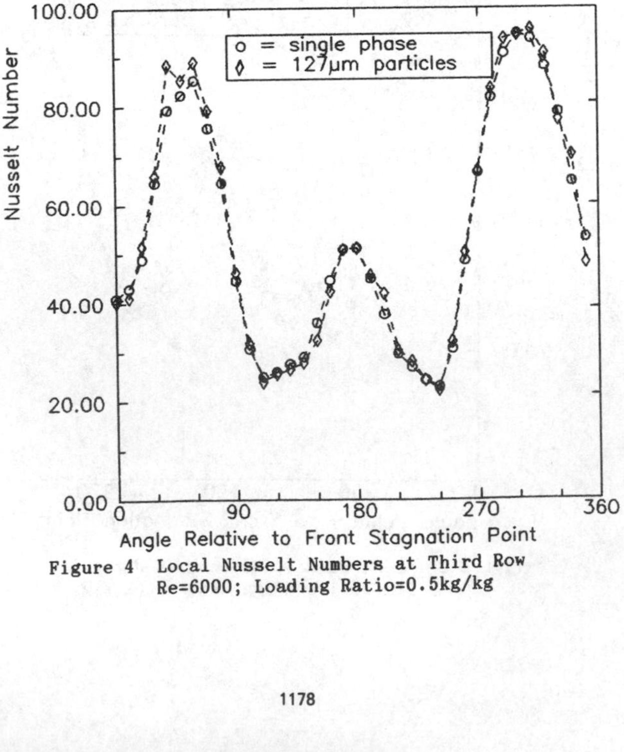

Figure 4   Local Nusselt Numbers at Third Row
Re=6000; Loading Ratio=0.5kg/kg

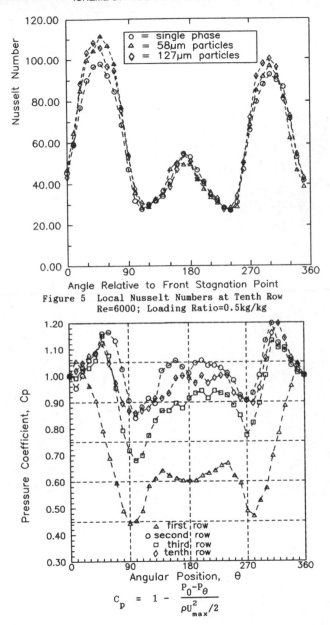

Figure 5   Local Nusselt Numbers at Tenth Row
Re=6000; Loading Ratio=0.5kg/kg

Figure 6   Surface Pressure Coefficients for Single Phase Flow

$$C_p = 1 - \frac{P_0 - P_\theta}{\rho U_{max}^2 / 2}$$

STUDY OF THE HEAT TRANFERS DURING THE CRYSTALLIZATIONS AND THE MELTING
OF THE DISPERSED DROPLETS OF AN EMULSION. EXPERIMENTS AND MODEL.

J.P. Dumas, Y. Zeraouli, M. Strub, M. Krichi
Laboratoire de Thermodynamique et Energétique, Université de Pau et des
Pays de l'Adour, Avenue de l'Université, 64000 PAU (France)

Two models to describe the heat transfers inside an
emulsion when the droplets crystallize or melt are
presented. Upon cooling, the undercooling
phenomenon and its erratic character is taken into
account. On the contrary, upon heating, the melting
always occurs, for each droplet, at a fixed
temperature. Experiments on dispersed hexadecane
are in good accordance with the models.
Keywords:heat transfer, emulsion, undercooling,
crystallization, melting

Introduction

For the droplets inside an emulsion, the two processes of crystallization and
melting are not symmetrical because of the large undercooling induced by the
smallness of their sizes (with liquids dispersed within emulsions,
undercoolings from 10 to 200 K are possible [1][2]). At the crystallization of
the droplet, the release of energy is practically instantaneous because it
occurs far from the thermodynamic equilibrium but, at the melting, the
absorption of energy is at the fixed melting temperature T_{sl} and its kinetics
depends on the exchanges with the surrounding medium.

In this paper, we will present the models for the determination of the heat
transfers and the kinetics of the transformations either upon cooling during
the crystallizations of the droplets or during their melting upon heating.
Comparison will be made with experimental results[3][4].

The main feature of crystallization is its stochastic character, i.e. samples
which are apparently identical will not transform at the same temperature
during the cooling process. Nucleation theories(i.e. see [5]) explain that
inside the liquid the fluctuations create small aggregates which initiate the
crystallization if they have a size greater than a critical value depending on
the temperature. The erratic character due to the fluctuations imposes that it
is only possible to determine $\mathcal{J}(T)$, the probability of crystallization of one
droplet per unit time. Generaly, this function is practically zero down to a
given temperature and increases very sharply afterwards (see figure 1).

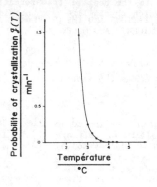

Figure 1   Experimental curve of the
crystallization probability per unit
time $\mathcal{J}(T)$ for hexadecane

Experimental

The investigated substance is an Analytic Grade hexadecane (T_{sl} = 18 °C)
dispersed within an emulsifying medium made of a mixture of water, glycerol
and Tween 80 as surfactant[6].

The experimental cell has been presented in details elsewhere [3][4]. It is a
vertical metallic tube about 500 cm^3 in volume   closed by two isolated caps.
Its height is sufficiently larger than its diameter to consider that only the
radius r and the time t are the variables of the study. This cylinder contains
a "cage" which supports 12 thermocouples tied up on streched nylon threads.
The solders are located regularly in a median horizontal plane at different
radii. The cell is immersed in a bath at temperature $T_\infty(t)$ heated or cooled at
a constant rate β by a thermostat.

Models

The droplets of the emulsion are so small (≤ 1 μm^3 in volume), and the
dispersed system is so visquous, that only the conduction must be taken into
account. Hence, we shall use the energy equation:

$$\rho\,c\,\frac{\partial\,T(r,t)}{\partial\,t} = \nabla\,(\,\lambda\,\nabla\,T(r,t)\,)\,+\,\dot{q} \qquad (1)$$

where λ is the thermal conductivity of the emulsion, ρ its mass density and c
its specific heat. The heat source $\dot{q}$ is different from zero when the phases
transformations occur and is different at the crystallization or at the
melting.

Model for the crystallization of the dispersed undercooled liquids

This model has been already presented in details[4], and only the results are
given here. The two variables to calculate are the local temperature $T(r,t)$
and $\varphi(r,t) = n(r,t)\,/\,n_t$ the local proportion of the crystallized droplets.

$n(r,t,)$ is the number of droplets crystallized at the time t per emulsion unit
volume and n_t the total number of droplets per unit emulsion volume.

The volumetric heat source $\dot{q}$, is proportionnal to the heat of transformation of one droplet and to the number of droplets which crystallize per unit time and unit emulsion volume $n_t \, d\varphi/ \, dt$. We have:

$$\dot{q} = \rho_0 \, V \, h_{sl} \, n_t \, d\varphi \, / \, dt \qquad (2)$$

where ρ_0 is the mass density of the dispersed phase (hexadecane), h_{sl} the latent heat of fusion ($h_{sl} > 0$) and V the droplet volume.

The two equations to solve are[4]:

$$\rho \, c \, \frac{\partial T(r,t)}{\partial t} = \nabla \, (\, \lambda \, \nabla \, T(r,t) \,) + \rho \, P \, h_{sl} \, \mathcal{J}(T(r,t)) \, (1 - \varphi(r,t)) \qquad (3)$$

$$\frac{d\varphi}{dt} = \mathcal{J}(T(r,t)) \, (1 - \varphi(r,t)) \qquad (4)$$

The first equation is the energy equation (1) where P is the mass fraction of the emulsion (mass of hexadecane/mass of emulsion). The second equation expresses that the number of droplets crystallizing per unit time is proportionnal to the probability of crystallization of a droplet and to the number of droplets which remain unfrozen. These equations are non linear because of the function $\mathcal{J}(T(r,t))$[7] (see figure 1).

The boundaries conditions are classical:

for $r = 0$, $\left(\dfrac{\partial T}{\partial r} \right)_0 = 0$

for $r = R_0$, $-\lambda \left(\dfrac{\partial T}{\partial r} \right)_{R_0} = h^{ext} \, (\, T(R_0,t) - T_\infty(t))$ $\qquad (5)$

h^{ext} is an exchange coefficient which expresses a thermal resistance between the emulsion and the bath (and thus which globaly takes into account the exchanges through the metallic tube and a part of the bath) and $T_\infty(t)$ is the imposed temperature of the bath. We have chosen for the bath, either a full cooling with a linear variation such as:

$$T_\infty(t) = \beta \, t + cte \qquad (\, \beta < 0 \,) \qquad (6)$$

or the same cooling but limited at a temperature T_i followed by a stabilization at this temperature afterwards.

For the initial conditions, we have chosen, as for the experiments, a stabilized temperature for the cell.

In fact, to have an accurate comparison between the model and the experimental results we must take into account, for λ and c, the notable difference for these values due to the fact that the disperse phase is liquid or solid. So, λ or c depend mainly on φ the fraction of droplets already crystallized and linear functions can be chosen[8][9]:

$$\lambda = \lambda_L + (\, \lambda_S - \lambda_L \,) \, \varphi(r,t) \qquad (7)$$

and $\qquad c = c_L + (\, c_S - c_L \,) \, \varphi(r,t)$ $\qquad\qquad (8)$

where λ_L and c_L are the heat conductivity and the specific heat of the emulsion when the dispersed phase is liquid and λ_S and c_S the corresponding values when the dispersed phase is solid.

The equations (3) and (4) taking into account the boundaries conditions (5), the initial conditions and the equations (7) and (8) are solved by an explicit finite differences method.

Examples of the experimental and calculated curves of the temperature of the solders versus time is given on figures 2 and 3 for the emulsion of hexadecane. The curves are in good accordance and this fact confirms the validity of our model.

As expected, the droplets near the inner side of the tube first crystallize; but a part of the released energy heats up the axis region of the cylinder delaying the crystallizations of the droplets. Generally, the result is that the temperature near the axis is practically constant over a lapse of time τ. Let us notice that, the temperature beeing much lower than the melting temperature T_{sl}, no thermodynamic reason can explain this fact.

This quasi steady temperature named T^{**} and taken for $\tau/2$ is about 4°C and varies very little with the different emulsions and the different parameters we have studied. On the contrary, τ depends on these parameters[8]:
- τ increases when the cooling rate β decreases
- τ decreases when the mass fraction P decreases
- τ largely increases when the final temperature T_1 increases approaching T^{**}.

More informations are given by the model if we plot the values of $\varphi(r,t)$ the fraction of crystallized droplets. As indicated on figure 4, giving for each value of r, the function $\varphi(r,t)$ versus time, and comparing with the corresponding curves on figure 3, we see that the proportion of crystallized droplets is very small when the temperature is stabilized around T. φ increases sharply only at the end of the temperature plateau and the temperature decreases only just after the crystallizations of all droplets. We conclude that, for a full cooling, the region where the droplets are crystallizing corresponds to a narrow radius range which slowly moves from r = R_0 to the axis.

Model for the melting of the crystallized droplets

At the melting there is no delay for the transformation and as soon as the temperature of the droplet is T_{sl}, the droplet acts as a heat source (in fact a heat sink) during the lapse of time where the melted fraction of the droplet $X(r,t)$ is such as $0 < X(r,t) < 1$. In this case, the heat source will be proportionnal to the number of droplets whose temperature is T_{sl} and to the fraction of the droplet which melts per unit time dX/dt.

During the melting the energy balance of one droplet is given by:

$$\rho_0 \, V \, h_{sl} \, \frac{dX}{dt} = - \, h \, (\, T_{sl} - T(r,t) \,) \, S \qquad (9)$$

S is the external area of the droplet and h an exchange coefficient between the droplet and the surrounding emulsifying medium expressing a thermal resistance due to the surfactant adsorbed on the droplet. If R is the mean radius of the droplets, the energy source (sink) is:

$$\dot{q} = - \rho \, h_{sl} \, P \, \frac{dX}{dt} = \frac{3 \, h \, P}{R} \, \frac{\rho}{\rho_0} \, (\, T_{sl} - T(r,t) \,) \qquad (10)$$

The system to be solved is when $0 < X(r,t) < 1$:

$$\rho \, c \, \frac{\partial T}{\partial t} = \nabla \, (\lambda \, \nabla T) + \frac{3 \, h \, P}{R} \, \frac{\rho}{\rho_0} \, (\, T_{sl} - T(r,t) \,) \qquad (11)$$

$$\frac{dX}{dt} = - \frac{3 \, h}{\rho_0 R \, h_{sl}} \, (\, T_{sl} - T(r,t) \,) \qquad (12)$$

When $T(r,t) < T_{sl}$ ($X = 0$) or when the melting is complete ($X(r,t) = 1$) the only conduction equation have to be solved:

$$\rho \, c \, \frac{\partial T}{\partial t} = \nabla \, (\lambda \, \nabla T) \qquad (13)$$

The numerical calculation begins for $t = 0$ at a temperature lower than T_{sl} solving first the equation (13) as long as $T < T_{sl}$. But as soon as T is equal to T_{sl}, the system of the equations (11) and (12) is solved as long as X is such as $0 < X(r,t) < 1$. As soon as the calculated value of $X(r,t)$ is 1, we solve again the equation (13).

The initial and boundaries conditions (equations (5)) are the same as for the cooling except that $\beta > 0$ and that the heating can be limited to a higher temperature T_m. We have the same problem as for the cooling because λ and c are different when the dispersed phase is whether liquid or solid. A good approximation for the values of λ and c is given by the linear equations:

$$\lambda = \lambda_L + (\, \lambda_S - \lambda_L \,) \, X(r,t) \qquad (14)$$

and
$$c = c_L + (\, c_S - c_L \,) \, X(r,t) \qquad (15)$$

On figure 5 we present the experimental curves of the temperatures versus time for different values of the radius r. The heating is limited at $T_m = 29.8°C$ because a full heating would necessitate a very high temperature where the emulsion would be destroyed. We observe the curves present a quasi plateau whose temperature of its end is T_{sl} the melting temperature. The length of this plateau is all the greater as r is nearer 0. For the axis this length is τ' and will be characteristic of the melting of the dispersed phase in the cylinder.

On figure 6 we have the corresponding calculated curves, which they are in good accordance with the experimental curves of the figure 5.

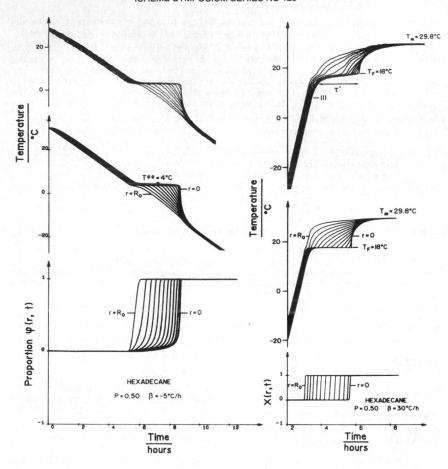

Figure 2   Experimental temperatures T(r,t) for different radii versus time for an hexadecane emulsion (p=.50) cooled at $\beta = -5°C/h$

Figure 3   Calculated temperatures T(r,t) at different radii versus time for an hexadecane emulsion (P=.50) cooled at $\beta = -5°C/h$

Figure 4   Calculated proportions of crystallized droplets at different radii versus time for an hexadecane emulsion (P=.50) cooled at $\beta = -5°C/h$

Figure 5   Experimental temperatures T(r,t) for different radii versus time for an hexadecane emulsion (P=.50) heated at $\beta = 30°C/h$

Figure 6   Calculated temperatures T(r,t) at different radii versus time for an hexadecane emulsion (P=.50) heated at $\beta = 30°C/h$

Figure 7   Calculated proportions of crystallized droplets at different radii versus time for an hexadecane emulsion (P=.50) heated at $\beta = 30°C/h$

For the melting it seems more logical to have a plateau of temperature because the transformations of the dispersed phase occurs at a fixed temperature T_{sl} which corresponds to the thermodynamical equilibrium if we assume that the droplets are all melting during the lapse of time τ'. But on figure   giving $X(r,t)$ the fraction of the droplet which is melted, we see that, in fact, the melting of the droplets actually occurs at the end of the plateau. Analysing more accuratly the curves, we observe that before the end of the plateau, the temperature is close to T_{sl} but always slightly lower except at the end. Like for the cooling, we have a front of fusions which moves from the cylinder to the axis and not the progressive melting of all droplets together.

For the axis the duration of the plateau τ' [9][10]:
- decreases when the heating rate β increases
- decreases when the mass fraction P decreases
- largely increases when the final temperature T_m is lowered approaching

T_{sl}.

Conclusion

We have presented two models to explain our experimental results on the crystallization or the melting of dispersed droplets inside an emulsion. In both cases, we observe a stabilization of the temperature in the axis region of the cell. For the crystallization, the stabilization is 14°C below the melting temperature and is explained by a self-regulation of the erratic nucleation phenomenon due to the release of heat. For the melting, the stabilization is at the melting temperature and is normally due to the phase equilibrium inside the droplets. More details on these experimental results and models will be given on the Conference.

References

1   Clausse, D.,Dumas, J.P., Meijer, P.H.E. & Broto F., 1987 Journal of Dispersion Science and Technology, vol 8(1),pp 1-28(Part I)
    Dumas, J.P., Tounsi, F., Babin, L, 1987 Journal of Dispersion Science and Technology, vol 8(1), pp 29-54 (Part II)
2   Perepezko, J.H. & Rasmussen, D.H. 1979 Proc. 17th Aerospace Sci. Meet., New Orleans
3   Dumas, J.P., Strub, M., Krichi, M., Broto, F. & Zeraouli, Y. 1988 Eurotherm Seminar n° 5 Compiègne4   Dumas, J.P., Strub,M., Broto,F. 1990 9th International Heat transfer Conference Jerusalem
5   Turnbull, D. 1956 Solid State Physics Academic Press Inc Pub.,New York Vol 2, pp 226-306
6   Zeraouli, Y. 1991 Thesis Pau
7   Vallet, G. 1991 XXIII Congrés National d'Analyse Numérique Royan (Fr)
8   Kenneth, R., Foster, E.C., Jonathan, B.L. 1984 Biophys. J. Biophysical Society Vol 45, 975-984
9   Krichi, M. 1992 Thesis Pau
10   Dumas, J.P., Zeraouli, Y., Strub, M., Krichi, M. 1992 J. Heat Mass Transfer (summited)

EXPERIMENTAL METHODS FOR THERMAL CHARACTERIZATION OF INTERFACES THIN

FILM - SUBSTRATE AND SEMICONDUCTOR - SUBSTRATE

Y. Scudeller, N. Hmina

Heat transfer is playing an essantial role on semiconductor power components. Due to the high local heat densities, defects can bring about temperature rises of several tens of degrees. We present two original techniques for thermal interface resistance measurement in small scale. One is suitable for contact of deposit layers a few μm thick. Defects are mainly brought by impurities and vacancies. The other is for characterization of chip bonds. The contact layer is an organic glue or a solder with some non-adhesion regions, gas bubbles an cracks. A sensisity analysis gives the limits of the two methods.

NOMENCLATURE

T : Temperature
R : Theoretical impulse response
R_c : Thermal contact resistance
τ, τ' : Time constant
C_1 : Thermal capacity of circuit
C_2 : Thermal capacity of spreader
$C\rho$: Heat capacity per unit volume of the film
$C_0\rho_0$: Heat capacity per unit volume of the substrate
ℓ : Film thickness
λ_0 : Thermal conductivity of the substrate
b : Thermal effusivity of the substrate

INTRODUCTION

We report two experimental techniques that we are setting up to investigate contact layers used in microelectronic and electronic devices. The structures are represented on figure 1. The question is to quantify on one hand, contact between a semiconductor circuit and its substrate (1.ab) and, on the other hand, the one between a thin deposit layer and its substrate (1c). These techniques identify a thermal interface resistance which characterizes at once adhesion and defects. Defects in interfacial structure induce an interface resistance variation.

Fig. 1a shows a contact between a semiconductor circuit and an heat spreader. The joint is formed of an organic glue or a solder. Thickness is

Laboratoire de Thermocinétique URA CNRS 869, ISITEM - La Chantrerie
CP 3023, 44087 NANTES cx 03

about 100 μm. Analysis of the interfacial region shows clearly some porosities, cracks, unadhesive areas and detachments.

Fig. 1b shows the same type of contact but joint is an eutectic. Substrate is a dielectric material. The two contact layers are about 5 μm thick. The interface is also heterogeneous. In it is found some non contact regions and cracks which may break the two layers.

Fig 1c represents a very thin film such as amorphous material that has been deposited on a substrate. Here, the interfacial boundary is characterized by strains, contaminations, impurities and vacancies. Moreover, due to the diffusion, chemical interactions can modify local thermal properties. It follows that the boundary is not abrupt. Detailed knowledge of the interfacial properties and structure is difficult to obtain. It is the reason why experiment is fundamental.

This report describes the methods and their sensitivity analysis.

CHARACTERIZATION OF A CONTACT SEMICONDUCTOR - SUBSTRATE

The method is based on analysis of transient heating produced by internal heat generation of the circuit. If T denotes the temperature rise of the spreader, experiment identifies a contact layer resistance R_c by means of the following relation :

$$T = \int_0^t q(t - \tau) R(\tau) d\tau \tag{1}$$

where R is the theoretical impulse response which depends on R_c. A thermal model produces R and its dependence with R_c. Identification is performed from data recording $T(t)$ and $q(t)$ during a period where R is very sensitive to R_c. The setting up has been done on a DIL component. It consists of a silicon transistor of 6 mm^2 bonded on a Copper spreader of 12 mm^2 in connexion with a grid. All is moulded in a resin package. R is obtained from assumptions that q is uniform and the circuit and spreader are isothermals. The capacity effect of the contact layer is considered. On the other hand, to avoid an important bias, 3D effects are also taken into account. As a matter of fact, heat flow is expanded through package and grid, and circuit and spreader constitute very small sources. The effect appears at very early times. It is essential and is the reason why, we have set up an original calculation technique for R. It uses a network of 3D transfer functions associated with integral transforms /1/.

Experimental results are on figure 2. $R_c = 7.3 \ 10^{-5} \ \text{Km}^2/\text{w}$ has been obtained from 50 to 250 ms. Temperature detection is achieved by means of a semi intrinsic thermoelectric method. The spreader is one element of a thermoelectric junction on which is bonded a 80μm Cu Ni Wire. It may be shown that time response for this sensor is about a few μs /2/. Local implementation does not disturb significantly the transient temperature field. Bias does not exceed 0.5%. Consequently, it is an accurate temperature detection.

We have calculated that precision is better than 15%. It may be demonstrated that if τ denotes a time constant :

$$\tau = R_c \ C_1 \ C_2 / (C_1 + C_2)$$

the maximum of sensitivity is for $t = 2\tau$ to 3τ. C_1, C_2 are thermal capacities of the circuit and spreader. Accuracy is indicated by the parameter $\theta = q\,R_c$ and resolution is given by $\theta \simeq 0.5$ (fig. 3).

Experiments use $q = 1$ to 4.10^5 W/m^2. It follows that resolution is $R_c \simeq 10^{-6}$ Km^2/w.

CHARACTERIZATION OF A CONTACT THIN FILM SUBSTRATE

The experimental arrangment is in fig. 4. A transient temperature rise is induced on the surface film with the absorption of a short optical heating pulse. This is performed by means of a YAG laser which applies a pulse of about 15 ns. Absorbed energy q causes at first an ultrafast heating of the layer. After that, the thin film falls progressively in temperature through the substrate. During this phase, temperature rise analysis allows to identify the thermal interface resistance. Detections have a duration between 0.01 and 100 μs. Experiments are set up on 1 μm of copper. Calculation shows that isothermicity on the film is brought about after 4.10^{-9} s, (fig. 5). This concerns thermal contact resistance of 10^{-9} to 10^{-5} Km^2/w and substrate such as Si O_2 or Si. The cooling phase is very dependent on contact conditions. It may be shown that it is fixed by the following parameter :

$$\alpha = (\frac{\ell}{\lambda_0\,R_c})\,(\frac{c\rho}{c_0\,\rho_0})$$

ℓ denotes the film thickness $c\rho$ its heat capacity per unit volume. λ_0 is thermal conductivity and c_0 ρ_0 heat capacity per unit volume of the substrate. Time scale is given by the following fundamental constant :

$$\tau' = R_c\,C\rho\,\ell$$

The different sensitivity coefficients are indicated on figure 6. Figure 7, which represents incertitudes in R_c, shows that identification has to be achieved at $t = \tau'$. It appears that optimum experiments are approximately set up from $\alpha \lesssim 0.01$. This means that a conductor substrate presents a better sensitivity. The assumption that the film is isothermal remains true. Also maximum temperature rise is different to $q/C\rho\,\ell$. It is because there is heat diffusion in the substrate during the heating of the film. It is the reason why we do identification on a ratio of two temperatures corresponding to two times. Like this, we exclude an important bias. Sensitivity study gives the detection resolution corresponding to a value $\alpha \simeq 25$. This corresponds to a resistance of about 5.10^{-8} Km^2/w for a glass substrate and 5.10^{-10} Km^2/w for a silicon substrate.

Two temperature measurement techniques are used. The first uses an electrical resistivity measurement of the film itself. The second uses a potential measurement of two separated thermoelectric wires (12 μm copper 12 μm constantan) which are bonded on the surface. Unavoïdable implantation disturbs the transient temperature field of the film. To avoid a bias it is convenient to take into account this effect in modelling.

CONCLUSION
We are setting up two experimental techniques to measure contacts semiconductor device - substrate and thin film - substrate. A thermal contact resistance is identified. It holds information on interfacial structures. In the first, a transient heating is performed and substrate temperature is analysed. A semiintrinsic technique is used for detection. Sensivity is mainly linked with heating. With devices studied give a resolution of $10^{-6} Km^2/w$.

Fast transient heating is used to characterize a contact thin film - substrate. Cooling phase analysis gives thermal contact resistance. Optimum experimental conditions have been defined. It appears that resolution, depending on substrates, is about $10^{-9} Km^2/w$ for 1 μm of copper.

REFERENCES
/1/ Y. SCUDELLER, N. HMINA Unpublished, 1992

/2/ A. BOUVIER, PHD Thesis, University of Nantes, 1987

/3/ H. STEFEST, Gaver, Comm. ACM 13, 47, 49, 1970

/4/ C.A. PADDOCK, G.L. Eesley, J. Appl. Phy. 60 (1), 1986

/5/ Y. SCUDELLER, J.P. BARDON, I.J.H.M.T. 591, 24, 1990

FIGURE CAPTIONS

Fig. 1 : Interfaces

Fig. 2 : Experimental detection
a) DIL microcomponent
b) Temperature rise
c) Heat flux

Fig. 3 : Accuracy of the detection $\theta = q R_c$ vs t/τ

Fig. 4 : Schematic of experimental arrangment

Fig. 5 : Transient temperature on the surface film (1 μm of copper, silicon substrate)

□ : approximate model (isothermal film)

Fig. 6 : Sensitivity coefficients vs t/τ'

$$S_{R_c} = |R_c \frac{\partial T}{\partial R_c}| \quad S_c = |C \frac{\partial T}{\partial C}| \quad S_b \quad |\sqrt{b} \frac{\partial T}{\partial b}|$$

$$C = c\rho \ell \quad b = \sqrt{\lambda_0 C_0 \rho_0}$$

Fig. 7 : a) dimensionless temperature of the film
b) accuracy vs t/τ and $\alpha = (\frac{\ell}{\lambda_0 R_c}) (\frac{c\rho}{C_0 \rho_0})$

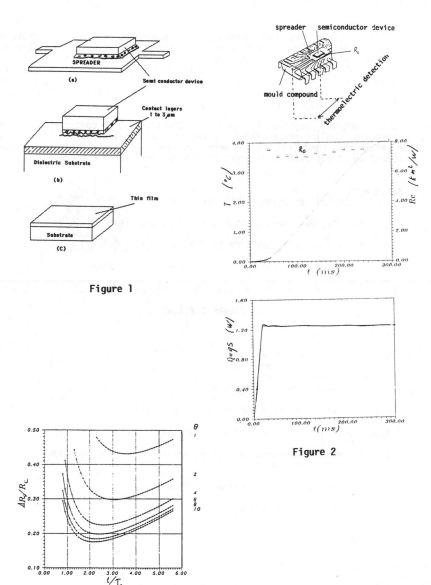

Figure 1

Figure 2

Figure 3

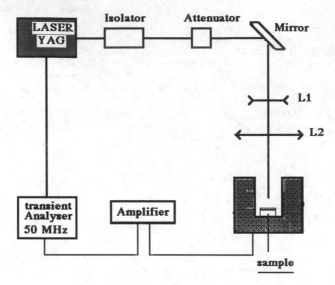

Figure 4

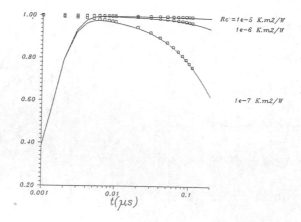

Figure 5

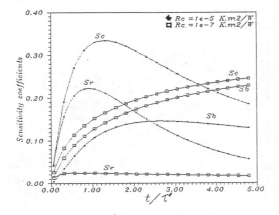

Figure 6

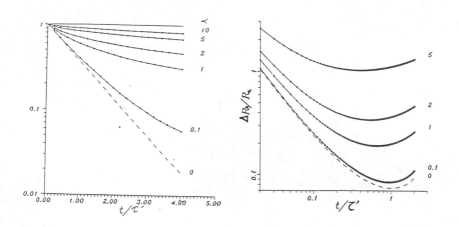

Figure 7

AN EXTENSION OF THE FLASH TECHNIQUE IN THE 300K TO 800K TEMPERATURE RANGE: APPLICATION TO THERMAL DIFFUSIVITY MEASUREMENT OF SEMI-TRANSPARENT MATERIALS

Stéphane ANDRE and Alain DEGIOVANNI
L.E.M.T.A. - 2, avenue de la forêt de Haye - 54516 Vandoeuvre les Nancy Cedex (France)

The flash method is extended for the measurement of thermal diffusivities of big solid samples in the 300K to 800K temperature range. It allows an experimental approach of the effect of temperature on the thermal characteristics of a material, especially the study of the combined radiative-conductive heat transfer in semi-transparent materials (STM). A theoretical model confirms the experimental observations and is used to predict if an "apparent" diffusivity can be associated to the STM sample, and if it is the case, what does it represent.The effects of both thickness and radiative boundary conditions on the identified diffusivity are clearly shown.

INTRODUCTION

For purely economical and practical considerations - such as the low speed of earliest computers - the transient study of both conductive and radiative modes of heat transfer has received few attention until very recent times. The current fields of application are of great importance: the glass technology for thermal stresses in moulding operations, the space engineering for the design of spacecraft's thermal shields, the metrological aspects of the thermal characterization of STM, the storage of nuclear waste for calculating the ideal dimensions of the vitreous coatings,... But experimental works are quite scarce.

We present first the experimental set-up that has been developped for measuring the diffusivity of glass until 800K. Some results will then be discussed, to show how the measured diffusivity varies with temperature and radiative boundary conditions. Finally, theoretical results will precise under which conditions of thickness and radiative limits, our metrological system leads, on the one hand, to a phonic diffusivity, that is the diffusivity derived from the phonic conductivity - the true conductivity of Fourier's law ,and on the other hand, to an apparent diffusivity which takes into account both radiative and conductive mechanisms.

EXPERIMENTAL APPARATUS

The experimental device, designed from the original apparatus built by Degiovanni (1) is composed of different parts providing the following functions:

- Heating system: This point is one of the originalities of the measurement system, because it has an important repercussion on the final geometry of the apparatus (accessibility, cost) and on the temperature measurement techniques as well.(surface thermocouples versus I.R pyrometry. The important advantage of working with big samples is preserved - this allows the study of heterogeneous materials as well - and the classical constraints associated with both a furnace and a laser to generate the heat pulse have been avoided. As a result, the temperature level cannot exceed 800K. The heating system consists of an hot-air stream circulating around the sample.

- Flash energy: It is released by the discharge of condensers in flash-tubes. Once the pulse is triggered, the lamp is removed by rotation around an axis, to prevent any parasitic radiation from the heated tubes, that would change the kind of energy excitation (the theoretical model assumes the heat flux condition imposed on the front face to be a dirac). At the same time, the acquisition of the signal begins, that also includes the storage of an adjustable number of data points preceding the instant of the flash.

- The temperature measurement: It is carried out by semi-conductors surface thermocouples. The electric circuit built by two $FeSi_2$ tablets (p- and n- doped) is closed by the rear face of the sample, which is locally covered by a thin layer of silver paint for electrically insulating samples. The tablets are brazed at the bottom of two cylindrical fingers made of molybden, located in cylindrical holes bored in an isothermal ceramic block that has to be in thermal equilibrium with the sample before the experiment. Springs push each finger against the rear side of the sample. This type of thermocouple - which measures the temperature difference between a region on the rear side of the sample and the ceramic block - presents the advantage of high thermoelectric power (of the order of $400\ \mu V/K$) and of correct surface temperature measurement.

- Measurement cell: The warm-air stream circulate inside the inner-space of a quartz double-wall bell. The sample is located in the inner enclosure of the bell and kept in primary vacuum conditions in order to avoid convective heat-losses and oxydation of the constitutive materials. The sample, which is cylindrical in shape has a thickness e ranging from 1 to 10 mm. Fig 1 shows the lay-out of the different components.

- Signal treatment: The tension from the thermocouple enters an amplifier with a variable gain, set in our case at 3000 and a permutable frequency filter set to 100 Hz (glass sample). This value prevents from any signal cut in frequency of the response (of characteristic frequency $\frac{a}{e^2}$) even for thin samples.

The amplified signal is then digitized into 4000 points by a 12 bits digital oscilloscope (Nicolet 310). The resolution is $1\mu V/point$ at a maximum rate of $1\mu s/point$. In these conditions, the electronic noise on the signal is of the order of $0,3\mu V/K$.

A statistical analysis carried out on 50 measurements (Glass sample with gold coated surfaces at room temperature) has proved the reproductibility of our metrological set-up and assessed the confidence bounds on the measured thermophysical parameter to less than 1%. The comparison with a predictive theoretical model of the errors made when identifying the diffusivity, is excellent. This has lead Maillet et al (4) to prove the superiority of the temporal partial moments method to other identification techniques.

The accuracy of the method has been tested by comparing our diffusivity measurements on pure iron and stainless steel at different temperatures with data found in the litterature. The agreement is globally within 2%.

EXPERIMENTAL RESULTS

Identification algorithm:

The signal described above is the temperature response of the rear face of the sample. This information can be analysed in order to estimate the value of the physical parameters, namely the thermal diffusivity and the heat-losses through a Biot number. A numerical algorithm has been developped to that end. It is based on the partial temporal moments method of Degiovanni and Laurent (2) and uses all the points of the rising part of the experimental thermogram.

Experimental thermograms:

We present two experimental thermograms (fig.2), which point out the two extreme behaviors of heat transfer in a S.T.M in the transient case.

They show, that in the case of reflecting boundaries, we obtain a "classical" thermogram: the rear face temperature grows continuously from time $t_0 = 0$ (time of the heat pulse) to a maximum, function of the input energy and the heat losses.

The shape of the experimental thermogram of the same sample with black frontiers is strikingly different. Our theoretical approach (3) confirms this result. Two peaks can be observed on this thermogram. One appears at very early times. The second one occurs at a characteristic time next to the one that would have been obtained in a purely conductive case of given diffusivity or that has been obtained on the same sample in the case of reflecting walls.

The first peak can be explained as follows: the energy that is released on the front face, due to its high emissivity (the black paint is given to have a gray emissivity of 0,93) will be immediately dissipated in the medium - if it is not a too strong absorbent - by the propagation of electromagnetic radiations and will contribute to increase instantaneously the temperature of the rear face. Very soon, its emissive contribution will therefore become quite important. Like the front face and for the same reasons, its temperature will decrease until it is submitted to the gradual coming of a heat flux connected to the coupled mechanism of both conductive and radiative transfer instead of radiative transfer only.

The last conclusion, which is more perceptible physically, is that the temperature and the condition of black walls accelerates the heat transfer. This will be seen by analysing the measured diffusivities in the following chapter.

It is obvious, that no conclusion can be made on the intensity of the maxima finally reached in the two cases, and this because the heat losses are clearly different at these two different temperatures and because we did not get under control the heat input from one experience to the other.

Thermal diffusivity measurements:

Effect of the radiative boundary conditions: Fig.3 shows results obtained on a green float glass TSA3+ (ref 7206) from temperature 300K up to 800K. The curves in both cases of reflecting and black walls are concave with a minimum value of the diffusivity reached around 500K. In the case of reflecting walls, the thermal diffusivity at 800K is quite identical to the value at 300K.

In the case of black walls, a strong increase of the diffusivity above 500K is observed. The radiative mechanisms of emission at the boundaries accelerates the heat transfer. These results confirmed the apparent thermal conductivity measurements that have been carried out at St

GOBAIN Company. The polynomial expression of $\lambda(T)$ derived from their measurements is used

together with the expression of $\rho C_p(T)$ given by Scholze (5) for float glasses, and reported in terms of diffusivity on fig.3.On a quantitative point of vue, one can notice that their measurements are systematically greater than ours, even at room temperature. This may prove that the hot wire technique can not separate both conductive and radiative modes of transfer with a sufficient efficiency. For the flash method, the influence of the radiative boundary conditions (sample having black painted or gold coated limits) is clear.

The first conclusion is, that reflecting boundaries seem to considerably reduce the radiative part of the transfer. In which proportion? Theoretical results will show under which conditions the heat transfer can be completely free from the radiative contribution. We then could identify the phonic conductivity of the glass.

If our results in the case of reflecting walls are analysed in terms of conductivity, fig. 4 shows that the function $\lambda(T)$ so obtained is linear in temperature. This result agrees qualitatively with the model given by Blazek (6).

Effect of the thickness: It is studied from results yielded by our theoretical model. The heat transfer level is characterized by an equivalent conductivity (in dimensionless form), that depends on the sample's thickness. The curves presented in figs. (5,6) stemed from two different models based on two different assumptions on the radiative transfer. The hypothesis of the optically thick medium leads to a radiative conductivity expressed in terms of the optical thickness τ_0 and the emissivities of the limits and is known as the Poltz/Jugel approximation (P/J). The other model, the radiative equilibrium (R.E) can also lead to an expression of equivalent conductivity but will not be discussed here.

Our results, presented in terms of an apparent dimensionless diffusivity from theoretical rear face thermograms, are plotted on the same figures. They corroborate the tendencies observed with the other models, that are surely more restrictive than our. The most important conclusion is that these results prove that the flash method can lead to the identification of a phonic diffusivity if both the thickness of the sample and the radiative boundary conditions are selected in an adequate manner. For exemple fig. 6 shows that the determination of the phonic diffusivity of a glass whose absorption coefficient is in the range of 15 m^{-1}, impose to consider a thickness below 1cm and to coat the sample's surfaces with a gold substrate. Large thicknesses and black walls rather lead to the identification of an apparent diffusivity which takes into account both radiative and conductive transfer with the radiative part of the transfer being maximum (the radiative conductivity of Rosseland can be used in that case).

Consequently, the same experimental device allows the measure of apparent conductivities actually performed by techniques such as the hot wire. Furthermore, one can put forward that for all intermediate situations, the Poltz/Jugel approximation may be accurate enough to estimate the radiative part of heat transfer.

CONCLUSION

A simple experimental device has been built, that measure thermal diffusivities of solids in the 300K to 800K temperature range, by a transient method. Applied to the thermal characterization of semi-transparent materials, this experimental system:
- has allowed the validation of our theoretical model and a best knowing of the combined radiative-conductive heat transfer mechanics in the transient case, with regards to various physical parameters;
- has brought reliable data of the phonic diffusivity of various glasses (600 experiments);
- seems to demonstrate, that a very simple model like the Poltz/Jugel approximation can be used to calculate the radiative part of an apparent conductivity.

REFERENCES

(1) A. Degiovanni, 1977, Revue Générale de Thermique, n°185, p 417-442.
(2) A. Degiovanni, M. Laurent, 1986, Revue Phys. Appl., n°21, p 229 237.
(3) S. Andre, A. Degiovanni, 1992, Proc. of Eurotherm Seminar n°21, p 249-258.
(4) D. Maillet, S. Andre, A. Degiovanni, (submitted for acceptation), J. of Phys.III.
(5) H. Scholze, 1980, Le Verre. Institut du verre, Paris.
(6) A. Blazek, 1983, Review of thermal conductivity data of glass. Institut National du Verre, Charleroi

ACKNOWLEDGEMENTS

The authors are grateful to Mr S.KLARSFELD and St GOBAIN Recherche company for supporting this study.

NOMENCLATURE

a	thermal diffusivity (m^2/s)	
C_P	specific heat (J/kg K)	
e	thickness of the slab (m)	

Greek symbols

λ	thermal conductivity (W/m K)
ρ	density (kg/m^3)
τ_0	optical thickness

Fig. 1 - Experimental apparatus

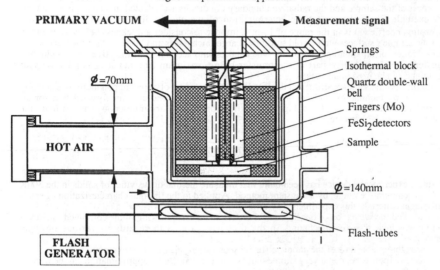

PRIMARY VACUUM ← → **Measurement signal**

ϕ =70mm

HOT AIR

Springs
Isothermal block
Quartz double-wall bell
Fingers (Mo)
$FeSi_2$ detectors
Sample

ϕ =140mm

Flash-tubes

FLASH GENERATOR

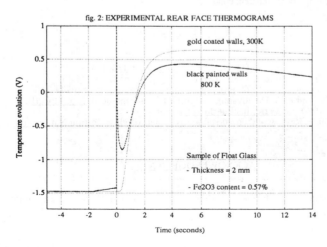

fig. 2: EXPERIMENTAL REAR FACE THERMOGRAMS

gold coated walls, 300K

black painted walls
800 K

Sample of Float Glass

- Thickness = 2 mm

- Fe2O3 content = 0.57%

Temperature evolution (V)

Time (seconds)

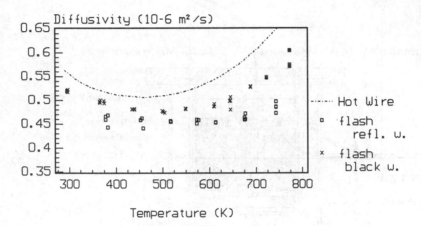

fig.3: Diffusivity of float glass (TSA3+)
vs temperature
(ref 7206, e=4mm, %Fe O =0.57)

fig.4 : Conductivity of
float glass (TSA3+) vs temperature
(ref 7206, e=4mm, %Fe O =0.57)

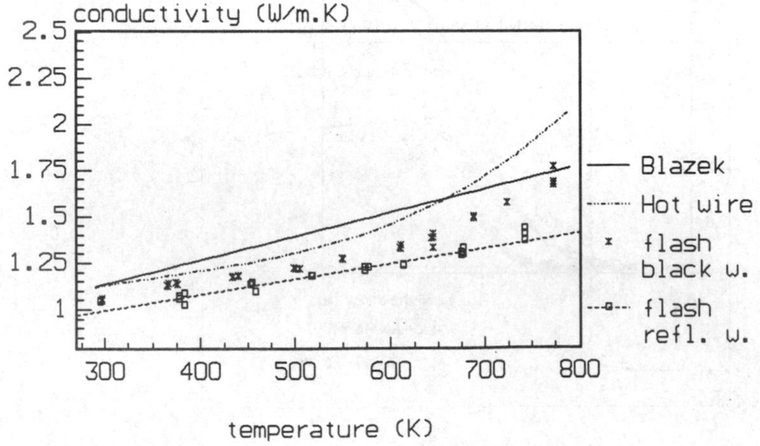

fig. 5: Heat Transfer in a S.T.M vs thickness
(T=800K, N=8 , χ=250 m-1)

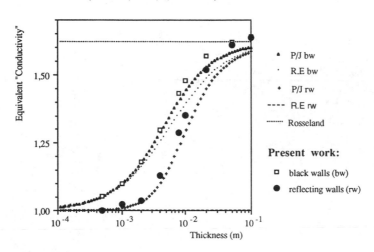

Present work:

□ black walls (bw)

● reflecting walls (rw)

fig. 6 : Heat Transfer in a S.T.M vs Thickness
(T=800K, N=0.5, χ=14.5 m-1)

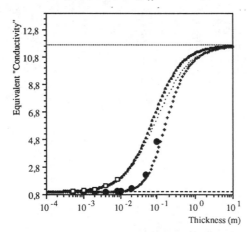

S-T Jurkowski, Y. Jarny, D. Delaunay

Summary

A method enabling the identification of thermal conductivity without using
interior sensors is described. The presence of thermal contact resistance is
taken into account. They are identified simultaneously with the
conductivity. The method is tested by using simulated data. The method
gives confidence intervals for the identified parameters.

I. Introduction

The problem consists in determining the thermal properties (conductivity or heat capacity) of materials for which there is no possibility of independent of sensors. It is assumed that the thermal properties are varying with the temperature.

The main idea is to construct a three-layered material, the sample is put in the centre, and these thermocouples are implanted in the external layers. The sensors measure temperature evolution in there layers, in particular the temperature on the interfaces with the sample. Because the thermal properties of the sample layers are known, it is possible to model, in each state channel flux passing through the interfaces. The set is thermally excited from one side, and the problem is studied in the transient state.

The essence of the problem is related to the presence of the thermal contact resistance (TCR) in two interfaces, which induces a temperature drop. This phenomenon does not exist for the interior sensor methods et al. The weakness of the method without the use sensors comes from the difficulty in identifying the TCR. Neglecting them is considered to introduce errors, hence the case to identify correctly with the thermal parameters. Here we consider the thermal centre of the sample configuration, assuming that the heat capacity is known.

We then solve an inverse problem, where the measured data are $T_1(t), T_2(t), T_3(t)$ and we identify $\lambda(t), R_1(t)$ and $R_2(t)$ simultaneously.

II. Description of the method

We consider a thermal system which is described by a set of non-linear partial differential equations in the temporal $\theta(t)$ and a spatial space $(0,1)$ domains. The system limits the "MODEL" and the solution Equations θ of this model system, problem is named "MODEL RESPONSE". The response depends on unknown parameters of the equation. These parameters formulate vector θ. A solution is to be identified, is called the vector of admissible parameters, and in practice the set defined by their maximal and minimal values, to the problem which we deal here, is defined by:

$$\theta = N\{\lambda\} + N\{R_1\} + \ldots$$

This paper is a contribution to URA CNRS xxx, IUT, Mechanics de France, CP 3133, xxxx, xxxxxxxxxxx xxxx.

SIMULTANEOUS IDENTIFICATION OF THERMAL CONDUCTIVITY AND THERMAL CONTACT RESISTANCES WITHOUT INTERNAL TEMPERATURE MEASUREMENTS

T. Jurkowski* , Y. Jarny* , D. Delaunay*

Summary
A method enabling the identification of thermal conductivity without using interior sensors is described. The presence of thermal contact resistance is taken into account. They are identified simultaneously with the conductivity. The method is tested by using simulated data. The method gives confidence intervals for the identified parameters.

I Introduction

The problem consists in determining the thermal properties (conductivity or heat capacity) of materials for which there is no possibility of implantation of sensors. It is assumed that the thermal properties are varying with the temperature.

The main idea is to construct a three-layered test-set : the sample is put in the centre and the thermocouples are implanted in the external layers. The sensors measure temperature evolution in these layers, in particular the temperature on the interfaces with the sample. Because the thermal properties of the outer layers are known, it is possible to calculate the heat flux passing through the interfaces. The set is thermally excited from outside and the problem is studied in the unsteady state. The essence of the problem is related to the presence of the thermal contact resistance (TCR) on two interfaces, which causes a temperature step. This problem does not exist for the interior sensor methods [1,2]. The weakness of the method without interior sensors comes from that disturbing effect introduced by the TCR-s. Neglecting them, it could lead to important errors, hence the idea to identify them together with the thermal parameters. Here we consider the identification of the sample conductivity, assuming that the heat capacity is known.

We treat here an <u>inverse problem</u>, where the measured data are: $T_1(t), T_2(t), \phi_1(t)$ and we <u>identify</u> $\lambda(\theta)$, $R1(t)$ and $R2(t)$ <u>simultaneously</u>.

II Description of the method

We consider a thermal system which is described by a set of non-linear partial differential equations in the temporal $(0, t_f)$ and in the space $(0, L)$ domains. The system forms the "MODEL" and the solution calculated for the measurement points is named "MODEL RESPONSE". This response depends on p unknown parameters of the equation. These parameters form the vector $\beta \in \Lambda$, which is to be identified. Λ is called the "set of admissible parameters" and in practice, the set is defined by their maximal and minimal values. In the problem which we deal here, β is defined by :

$$\beta = (\lambda_1, ..., \lambda_i, ..., \lambda_{NI}, R1_1, ..., R1_j, ..., R1_{NJ}, ..., R2_1, ..., R2_l, ..., R2_{NL}) \qquad (1)$$
$$p = NI + NJ + NL \qquad (1a)$$

*Laboratoire de Thermocnétique, URA CNRS 869, ISITEM-Université de Nantes, CP3023, 44087 NANTES, France

where : λ_i is the conductivity for the temperatures θ_i, and $R1_j$, $R2_l$ the thermal contact resistances for the instants j and l, respectively.

The thermal system is submitted to a known excitation, then an evolution of the temperature in time results for each measurement point. This evolution is called the OBJECT RESPONSE. The identification consists in determination of β^* - a value of β - in order that the MODEL RESPONSE be as close as possible - the criterion remains to define - to the OBJECT RESPONSE. To find β^* a DESCEND METHOD is used. It is an iterative method which consists in calculating a sequence $\beta^0, \beta^1, ..., \beta^i$ (in Λ), which satisfies : $\qquad J(\beta^0) > J(\beta^1) > ... > J(\beta^i)$ (2)

The temperature of the sample $\theta(x,t)$ in the spatial domain $(0,L)$ limited by the two interfaces and the temperature θ_1 of the first layer at the interface $x=0$, are solution of :

$$C(\theta) \frac{\partial \theta}{\partial t}(x,t) - \frac{\partial}{\partial x}\left[\lambda(\theta)\frac{\partial \theta}{\partial x}(x,t)\right] = 0 \qquad 0 < x < L \qquad (3)$$

$$\theta_1(t) = \theta(0,t) - R1(t)\,\phi_1(t) \qquad x = 0 \qquad (3a)$$

$$-\lambda(\theta)\frac{\partial \theta}{\partial x}(0,t) = \phi_1(t) \qquad x = 0 \qquad (3b)$$

$$-R2(t)\left[\lambda(\theta)\frac{\partial \theta}{\partial x}(L,t)\right] + \theta(L,t) = T_2(t) \qquad x = L \qquad (3c)$$

The data are :
- measurements : $\qquad \phi_1(t)$, $T_1(t)$ and $T_2(t)$, $\qquad$ for $0 < t < t_f$
- parameters : $\qquad C(\theta) = \rho(\theta)c_p(\theta)$, and L
- initial conditions : $\qquad \theta_1(0)$ and $\theta(x,0)$ $\qquad$ for $0 < x < L$

The general expression of the criterion J is :

$$J(\beta) = \sum_{k=1}^{n}(\theta_1(t_k, \beta) - T_1(t_k))^2 w_k + \sum_{i=1}^{p}(\mu_i - \beta_i)^2 u_i \qquad (4)$$

where : w_k - is a weighting positive coefficient for instant k , which may include some information regarding the experimentation errors [3, chapt.6],

$\qquad u_i$ - is a weighting non-negative coefficient for i-th component of β,

$\qquad \mu_i$ - is an initial estimation of β.

Using the nomenclature of Beck and Arnold [3] :
- the observation vector Y, with the components $Y_k = T_1(t_k)$,

- the model vector $\eta(\beta) = \left[\theta_1(t_k, \beta)\right]_{k=1}^{n}$ and,

- the weighting matrices W and U, which correspond to coefficients w_k and u_i ,
the criterion J is written in the form :

$$J(\beta) = \left[Y - \eta(\beta)\right]^T W \left[Y - \eta(\beta)\right] + (\mu - \beta)^T U (\mu - \beta) \qquad (4a)$$

To minimize the criterion, the principle of the Gauss-Newton method is applied. In the minimum ($\beta = \beta^*$), the gradient of the criterion is null :

$$\nabla_\beta J(\beta^*) = 0 \qquad , \text{i.e. :} \qquad (5)$$

$$\nabla_\beta J(\beta^*) = 2\left[\nabla_\beta \eta^T(\beta^*)\right] W \left[Y - \eta(\beta^*)\right] - 2 U (\mu - \beta^*) = 0 \qquad (6)$$

The sensitivity matrix is defined as : $\quad X(\beta) \equiv \left[\nabla_\beta \eta^T(\beta)\right]^T \qquad (7)$

its elements are : $\qquad X_{ji}(\beta) = \nabla_{\beta i}\,\eta_j(\beta) \qquad i = 1 \text{ to } p , j = 1 \text{ to } n \qquad (7a)$

Let $\hat{\beta}$ be an estimation of β^*, then the equation (6) and (7) give :

$$X^T(\hat{\beta}) \, W \left[Y - \eta(\hat{\beta}) \right] + U \, (\mu - \hat{\beta}) \approx 0 \qquad (8)$$

To resolve the equation (8), $\eta(\hat{\beta})$ is developed about b in a Taylor series :

$$\eta(\hat{\beta}) \approx \eta(b) + X(b) \left[b - \hat{\beta} \right] \qquad (9)$$

then $X(\hat{\beta})$ is replaced by $X(b)$ in the equation (8) and b is taken as μ,

The matrix P is defined : $\qquad P^{-1} \equiv X^T W X + U \qquad (10)$
then after transformations we get :

$$\hat{\beta} = b + P(b) \left[X^T(b) \, W \, (Y - \eta(b)) \right] \qquad (11)$$

hence the sequence of iterates $b^{(m)}$ is calculated by :

$$b^{(m+1)} = b^{(m)} + \varepsilon \, P(b^{(m)}) \left[X^T(b^{(m)}) \, W \, (Y - \eta(b^{(m)})) \right] \qquad (12)$$

where : ε is a coefficient close to 1.

III Results

The identified functions are piece-wise approximated : let $(\theta_1,...,\theta_i,...,\theta_{NI})$ - be the set of values of temperature for which λ is searched. The functions $R1(t)$ and $R2(t)$ are approximated similarly in time. The solution $\eta(\beta)$ is obtained by a finite-difference method and Crank-Nicolson scheme. The weighting matrices W and U are taken, according to :

$$W = I \,, \qquad U = c \times \mathrm{diag}(X^T W X) \quad , \text{ with } \; 10^{-6} < c < 10^{-4} \qquad (13)$$

The method was tested in simulation. A three-layered set was taken (conductor - insulator - conductor) and the conductivity of the insulator was studied. The thickness of the layers are : 0,04m - 0,01m - 0,04m , respectively. The thermal properties of the conductor (i.e. outer layers of the set) are : $\lambda_c = 40$ W/m/K for thermal conductivity, $C_c = 3,6 \times 10^6$ J/m3/K for volumic heat capacity, and the volumic heat capacity of the insulator (i.e. the sample) is : $C = 2,0 \times 10^6$ J/m3/K.

The influence of the measurement noise on the results of parameter identification was studied by adding a uniform noise ($\sigma = 0,05$ K) to the exact data. The identified values of the parameters are given with their confidence intervals, deduced from the relation [3] :

$$\sigma_j = \sigma \, (P_{jj})^{0,5} \qquad (14)$$

where : σ_j - is the standard deviation of parameter β_j.

The initial temperature of the set is uniform : $100°$ C. At the initial instant ($t = 0$), the temperature $0°$ C is applied on the external layers. Four cases are considered. The values of λ of the sample and the values of RTCs are given below :

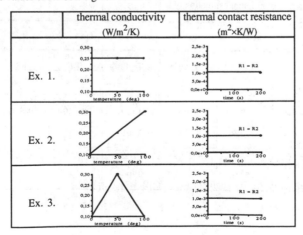

	thermal conductivity (W/m^2/K)	thermal contact resistance (m^2×K/W)
Ex. 1.		
Ex. 2.		
Ex. 3.		

Ex. 4.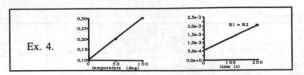

Figure 1 presents heat fluxes and figure 2 shows temperatures at x=0 for the four cases above (the temperatures are very closed one to the other).

The thermal conductivity was identified for three temperatures : 0, 50 and 100° C. In the case 1 to 3, the thermal contact resistances R1 and R2 were identified as constants parameters, and in the case 4 as a time varying parameter. The initial values of the parameters were : $\lambda = 0,5$ W/m/K , R1 = R2 = 0,0001 m²×K/W. In each case, both exact data and noised data were treated. For the identification results from the noised data, we give the confidence intervals (with a probability of 95% , i.e. $\pm 2\sigma_j$ for j-th parameter).

We compare in the table 1 to 4 the TCRs disturbing influence on the identification of the parameter λ. One can observe that the identification errors of λ can attain 50%, if TCRs are not taken into account.

Figures 3 to 8 show several examples of plots (for noised data every time) of criterion evolution, sensitivity coefficients and residuals. It can be noticed that residuals have a form of a random noise. The data and the results are put together in tables 1 to 4. We should precise that, as the criterion, we have taken $\sqrt{J/n}$,where J is given by relation (4), and the sensitivity coefficients are defined for each parameter β_i by : $\beta_i \times \partial \eta / \partial \beta_i$.

Example 1   Table 1

Identified parameter :	$\lambda(T = 0\ C)$	$\lambda(T=50C)$	$\lambda(T=100C)$	R1	R2	Criterion
unit :	W/m/K	W/m/K	W/m/K	K×m²/W	K×m²/W	K
initial value	0,5	0,5	0,5	0,1E-3	0,1E-3	12,0
exact value	0,25	0,25	0,25	1,0E-3	1,0E-3	
identified value (exact data)	0,2500	0,2503	0,2504	1,0119E-3	1,0202E-3	0,141E-2
identified value (noised data)	0,2482	0,2499	0,2488	0,9891E-3	1,0587E-3	4,943E-2
confidence interval	±0,0040	±0,0005	±0,0030	±0,0276E-3	±0,1042E-3	
erroneous identified value (if TCR neglected)	0,2122	0,2323	0,1573			

Example 2   Table 2

Identified parameter :	$\lambda(T = 0\ C)$	$\lambda(T=50C)$	$\lambda(T=100C)$	R1	R2	Criterion
unit :	W/m/K	W/m/K	W/m/K	K×m²/W	K×m²/W	K
initial value	0,5	0,5	0,5	0,1E-3	0,1E-3	15,0
exact value	0,10	0,20	0,30	1,0E-3	1,0E-3	
identified value (exact data)	0,0996	0,1999	0,3000	1,0075E-3	1,0489E-3	0,140E-2
identified value (noised data)	0,1059	0,2005	0,3016	1,2079E-3	0,4909E-3	5,123E-2
confidence interval	±0,0028	±0,0002	±0,0022	±0,0181E-3	±0,3549E-3	
erroneous identified value (if TCR neglected)	0,0827	0,1993	0,1968			

Example 3   Table 3

Identified parameter :	$\lambda(T = 0\,C)$	$\lambda(T=50C)$	$\lambda(T=100C)$	R1	R2	Criterion
unit :	W/m/K	W/m/K	W/m/K	$K\times m^2/W$	$K\times m^2/W$	K
initial value	0,5	0,5	0,5	0,1E-3	0,1E-3	13,5
exact value	0,10	0,30	0,10	1,0E-3	1,0E-3	
identified value (exact data)	0,0988	0,2995	0,0986	0,9857E-3	1,0541E-3	1,580E-2
identified value (noised data)	0,0988	0,2993	0,0987	0,9832E-3	1,0685E-3	5,085E-2
confidence interval	±0,0019	±0,0017	±0,0021	±0,0415E-3	±0,2552E-3	
erroneous identified value (if TCR neglected)	0,1043	0,2766	0,0417			

Example 4   Table 4

Identified parameter :	$\lambda(T = 0\,C)$	$\lambda(T=50C)$	$\lambda(T=100C)$	R1 and R2 (t = 0s)	R1 and R2 (t =200s)	Criterion
unit :	W/m/K	W/m/K	W/m/K	$K\times m^2/W$	$K\times m^2/W$	K
initial value	0,5	0,5	0,5	0,1E-3	0,1E-3	15,4
exact value	0,10	0,20	0,30	0,5E-3	2,0E-3	
identified value (exact data)	0,0995	0,1994	0,2992	0,5078E-3	1,9576E-3	0,116E-2
identified value (noised data)	0,0992	0,1985	0,2977	0,5007E-3	1,8770E-3	4,874E-2
confidence interval	±0,0064	±0,0052	±0,0082	±0,0183E-3	±0,4386E-3	
erroneous identified value (if TCR neglected)	0,0791	0,1866	0,2252			

IV Comments

Example 1 : For the constant λ and R1 and R2 also constant, the results are very satisfactory, i.e. the identified values are sufficiently close to the exact values - even for noised data, and the confidence intervals are small (of order of ±1%). R1 and R2 are identified correctly as well. The minimum value of the criterion corresponds to mean square deviation of the data noise and the residuals have the form of that noise (see : figure 3 and 4).

Example 2 : For a linearly varying $\lambda(\theta)$, the identified values are found precisely, but those of R1 and R2 (particularly for the noised data) are less correct. This may be explained by the low values of the sensitivity coefficients of R1 and particularly of R2 - compared to those of λ (it is shown on figure 5).

Example 3 : In this case of $\lambda(\theta)$ variation, the criterion shows a tendency to converge to a local minimum which corresponds to erroneous identified values of the parameters. However, a "good" choice of the initial values (i.e. relatively close to the exact values) ensures a good convergence to the minimum. Looking at figure 6, one remarks that the λ sensitivity coefficients are fairly larger than those of R1 and R2. Thus, we can define a "strategy" of identification : fixing first the less "influential" parameters (R1 and R2), and then identifying λ. In this way, a first approximation of λ is obtained. Then we use these values as the initial values to restart the identification, of all searched parameters together. Subsequently we find the results precisely. Figure 7 - of the criterion evolution - shows quite well the two-stage identification procedure.

Example 4 : Here, we also look for the R1 and R2 values for different instants : t = 0 and t = 200 s. Unfortunately, the sensitivity coefficient values of R2 are so small (figure 8), that it is impossible to identify all the parameters simultaneously i.e. the procedure is unstable. So we have to add a hypothesis about the character of R2. Due to the symmetry of the set, it is assumed that R2 = R1 for all instants. With this hypothesis the procedure becomes stable and the results are correct (see : table 5 - the results, figure 9 - the criterion evolution [two-stage identification] , and figure 10 - residuals).

V Conclusions

The tests we have carried out, confirm the possibility of thermal conductivity identification in the presence of the unknown thermal contact resistances. We can perform the simultaneous identification of all parameters.

The optimization procedure can either appear unstable or to converge to the local minima. This can be caused on the one hand by the very small sensitivity coefficients for the TCR or on the other hand by very poor X^TWX matrix conditioning (unbalanced). To assure a good convergence of the procedure, it was necessary to :

- introduce a reducing coefficient ε (eq. 12),
- adjust coefficient c for the matrix U (eq. 13),
- carry out a two-stage identification (examples 3 and 4)
- eliminate the parameters having low sensitivity coefficient values (example 4).

Measurement noise influence on the identification results is relatively weak. The residuals diagrams and criterion values in the minima, confirm the correctness of the results.

VI References

1. JURKOWSKI, T., *Mise en oeuvre d'une méthode et réalisation d'un appareillage de mesure de la conductivité thermique d'un polymère.* Internal report, Laboratoire de Thermocinétique Nantes, March 1991.

2. JARNY, Y., DELAUNAY, D., BRANSIER, J. - *Identification on nonlinear thermal properties by an Output Least Square Method.* Proc. of the 8-th International Heat Transfer Conference, San Francisco 1986.

3. BECK, J.V. and ARNOLD, K.J. *Parameter Estimation in Engineering and Science*, Wiley, New York, 1977.

Nomenclature

C - volumic heat capacity , $[J/m^3/K]$
J - least square criterion , $[K^2]$
L - thickness of the sample , $[m]$
n - number of observations
p - number of identified parameters
R1 - thermal contact resistance of the interface 1 (vector : NJ×1) , $[m^2K/W]$
R2 - thermal contact resistance of the interface 2 (vector : NL×1) , $[m^2K/W]$
t - time , $[s]$
T - temperature (measurement) , $[°C]$
u - weighting coefficient
U - weighting matrix (p×p)
w - weighting coefficient
W - weighting matrix (n×n)
x - distance , $[m]$
X - sensitivity matrix (n×p)
Y - observation vector ; measurements (n×1) , $[°C]$
β - identified parameter vector (p×1)
$\beta*$ - parameter vector (minimizes the J)
ϕ - heat flux (measurement) , $[W/m^2]$
λ - thermal conductivity (vector : NI×1) , $[W/m/K]$
Λ - set of allowed parameters
μ - initial estimation of β
η - model vector (n×1) , $[°C]$
σ - mean square deviation, standard deviation
θ - model temperature (calculated) , $[°C]$
∇_β - matrix derivative operator

1 - subscript, for the interface at x = 0
2 - subscript, for the interface at x = L

Figure 1 :

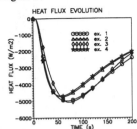

Figure 2 :

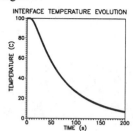

Figure 3 :

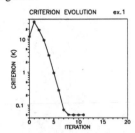

Figure 4 :

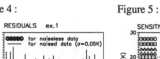
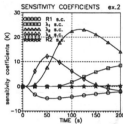

Figure 5 :

Figure 6 :

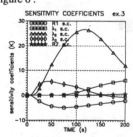

Figure 7 :

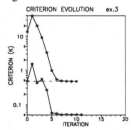

Figure 8 :

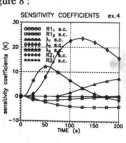

Figure 9 :

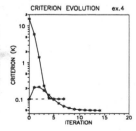

Figure 10 :

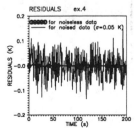

THE SIMULATION OF DC COOLING DURING THE CASTING OF ALUMINIUM ALLOYS

S.C.Flood[†], D.T.Hughes[†], L.Katgerman[†], H.Read[‡]

† Alcan International Ltd, Banbury Laboratories, Banbury, Oxon OX16 7SP

‡ University of Cambridge

The convective heat transfer due to the DC (direct chill) water during the casting of aluminium alloys has been investigated as a function of the flowrate and temperature of the cooling water. The DC water impinges directly on to the surface of an ingot and then travels down the surface as a turbulent falling film. The results have been reduced to a simple correlation and this has been used in a mathematical model to investigate the effect of the flowrate and temperature of the cooling water on the temperature and solidification profiles of ingots during DC casting.

INTRODUCTION

A casting model which can accurately predict the temperature distribution inside an ingot is essential for the development of optimised casting practices. For good predictions, it is important that the external heat transfer boundary conditions are specified accurately. This work was performed to obtain reliable values for the heat transfer coefficients due to the cooling water in the casting process.

Heat is lost by the ingot to the mould, to the DC cooling water and to the starter block (Fig 1). The heat transfer to the cooling water is due to both nucleate boiling and convection. However, nucleate boiling, because of its high cooling efficiency, only exists close to the mould; the cooling over most of the ingot is due to convective heat transfer. A correlation for the convective component of the DC water cooling has been put forward by Weckman[1] :the convective heat transfer coefficient is expressed as a function of the water flowrate and the temperatures of the ingot surface and cooling water.

$$h = (\text{-}1.67 \times 10^5 + 704\, T_{av})Q^{1/3} \qquad (1)$$

where

h = convection coefficient ($Wm^{-2}K^{-1}$)
T_{av} = arithmetic mean of the local ingot and water temperatures (K)
Q = secondary cooling water flowrate per unit
width of ingot (m^2s^{-1})

In this expression, the heat transfer coefficient is independent of the distance from the mould; it treats the water as a fully developed falling film and therefore underestimates the coefficient in the vicinity of the mould.

In order to provide a better description of the <u>convective</u> component of the secondary cooling, local heat transfer coefficients were measured experimentally. It was possible to use a steady state technique for the measurements because the constant heat fluxes that needed to be supplied were readily obtained with electrical heaters. The advantage of a steady state technique is that the conditions during measurement can be established and reproduced with accuracy more easily. In contrast, the heat fluxes required to investigate the boiling coefficients (on the scale of this process) are so large that they cannot be sustained easily in the laboratory and so an investigation of boiling in this system would necessitate quenching experiments.

The convective heat transfer correlations from the steady state experiments have been used, together with existing boiling correlations, in a mathematical model of DC casting[2] to predict the temperature distributions in a 2D aluminium ingot

EXPERIMENTAL METHOD

An aluminium block (250x201x50 mm) heated by 10x1.5 kW cartridge heaters was used to simulate the ingot surface. One of the large faces was cooled by a falling film of water from a mould section positioned at the top of the block. The cooling water was recirculated in a circuit consisting of a pump, heat exchanger, control valve and flowmeter (Fig 2). The other faces of the block were protected from water splashing by a thin aluminium shield and were insulated with ceramic fibre (Fig 3). The temperatures in the block were recorded by six K type thermocouples positioned 2 mm from the water-cooled surface. A further K type thermocouple was used to monitor the temperature of the water in the mould.

The water flowrate to the block and the power supplied to the cartridge heaters could be adjusted and measured. Measurements were made when the system had attained thermal equilibrium: the water flowrate, the power supplied to the heaters and the thermocouple temperatures were recorded.

The temperatures in the block were determined by the coolant flowrate and the power to the heaters. The flow of water to the heat exchanger controlled the temperature of the water in the mould.

CALCULATION OF THE HEAT TRANSFER COEFFICIENTS

The temperature distribution 2mm below the water-cooled surface of the block was obtained from interpolation between the readings of the embedded thermocouples. Taking the impingement point of the water at the top of the block to be the origin of the 2D coordinate system (x,z) (see Fig.3), the temperature distribution in the aluminium block for x > 2mm was calculated from each set of experimental results by solving the equation

$$\frac{\partial^2 T}{\partial x^2} + \frac{\partial^2 T}{\partial z^2} = 0 \qquad\qquad (2)$$

subject to the boundary conditions

$$T = T(z, x=2mm) \quad \text{interpolated from the} \qquad (3a)$$
$$\text{experimental data}$$

$$\frac{\partial T}{\partial x} = 0 \quad \text{at } x = 50 \text{ mm} \qquad (3b)$$

$$\frac{\partial T}{\partial z} = 0 \quad \text{at } z = 0 \text{ and } 250 \text{ mm} \qquad (3c)$$

$$q = q_i \quad \text{at the surface of each heater} \qquad (3d)$$

Equation (2) was solved numerically by the finite volume method. The domain was discretised in 1x1 mm squares. Each heater was modelled with a square cross section of equivalent cross-sectional area. The appropriate value of q_i was calculated from the heater power output.

An example of the calculated temperature distribution is shown in Fig 4. The block surface temperatures and heat fluxes were evaluated numerically from the temperature solution. The water flowrate and water temperature at $z = 0$ are known, and so the increase in the temperature of the water down the block could be obtained.

The heat transfer coefficient, h(z), was calculated as a function of distance from the mould from the local values of the surface temperature, the bulk water temperature and the surface heat flux ($T_s(z)$, $T_w(z)$ and q(z) respectively):

$$h(z) = \frac{q(z)}{T_s(z) - T_w(z)} \qquad (4)$$

Fig 5 shows a typical graph of heat transfer coefficient as a function of distance from the mould.

RESULTS AND CORRELATION

A graph of the type shown in Fig 5 was obtained from each experiment. A total of 53 experiments were performed in which the water flowrate per unit width was varied between 8.3×10^{-4} and 5.0×10^{-3} $m^2 s^{-1}$. Each graph showed a high impingement heat transfer coefficient at $z = 0$ and a decay which levelled off at approximately $z = 35mm$. The thermocouple measurements were interpolated using a cubic spline and this is the cause of the undulation in the coefficients at $z > 35mm$. A cubic spline was used in preference to linear interpolation to define the boundary condition (3a) after it was found that a linear interpolation generated "spikes" in the heat transfer graphs caused by the sudden change in the gradient $\partial T/\partial z$ at each thermocouple location.

The data from the experiments were correlated by expressing the heat transfer coefficients obtained at $z = 0$ and $z = 125$ mm as a function of water flowrate per unit width and the arithmetic mean of the surface ingot temperature and water temperature. The data were fitted to an equation of the form

$$h = (A + B T_{av}) Q^C \tag{5}$$

using least squares multiple non linear regression, where A,B,C were the parameters to be determined.

The heat transfer coefficients (h_o) obtained in the tests at the impingement point $(z=0)$ were best described by the equation

$$h_o = (-3.71x10^5 + 2020 T_{av})Q^{0.352} \tag{6}$$

and those halfway down the block at $z=125mm$ (h_{125}) were correlated by the equation

$$h_{125} = (-1.27x10^5 + 741 T_{av})Q^{0.401} \tag{7}$$

The correlation for h_o was obtained from experimental data in which T_{av} varied from 298 to 344 K. The range of T_{av} in the h_{125} correlation was 303 to 359 K. The range of water flowrates per unit width used in both correlations was $8.3x10^{-4}$ to $4.95x10^{-3}$ m^2s^{-1}.

APPLICATION OF THE CORRELATIONS

The correlations were used to investigate the change in temperature distribution down the centre line of an ingot as the water cooling flowrate is changed from a low value of 1 litre/min per cm $(Q=1.66x10^{-3}$ $m^2s^{-1})$ to a high value of 3 litres/min per cm $(Q=4.98x10^{-3}$ $m^2s^{-1})$.

The correlations (6) and (7) were used and it was assumed, that over the range $0<z<35mm$, the convection coefficient (h_c) falls linearly from h_o to h_{125} , i.e

$$h_c = h_o - (h_o - h_{125})\left[\frac{z(mm)}{35}\right] \quad 0<z<35mm \tag{8}$$

This approximation to h_c probably overestimates the coefficient in the impingement zone, but a better description requires further experiments to examine more closely the nature of the heat transfer decay near the mould.

For distances greater than 35 mm, the convection coefficients are given by

$$h_c = h_{125} \qquad z>35mm \tag{9}$$

After Weckman [1], it was assumed that the total heat flux leaving the ingot surface due to the DC water, $q_t(z)$, consists of separate components for convective cooling, $q_c(z)$, and nucleate boiling, q_n:

$$q_t(z) = q_c(z) + q_n \qquad (10)$$

The local convective component is given by

$$q_c(z) = h_c(T_s(z)-T_w(z)) \qquad (11)$$

where

$T_s(z)$ = local ingot surface temperature (K)
$T_w(z)$ = local bulk water temperature in the film (K)

The nucleate boiling component was assumed to be active when the convective component $q_c(z)$ is below the boiling incipience heat flux q_b. The boiling correlations were taken from Weckman:

$$q_n = 20.8 (T_s(z) - T_{sat})^3 \quad \text{if } q_c(z) < q_b$$

and $q_n = 0 \qquad \text{if } q_c(z) \geq q_b$

where q_b is given by

$$q_b = 3910 (T_s(z)-T_{sat})^{2.16} \qquad (12)$$

and T_{sat} = boiling point of water (K)

The casting parameters used in this example were

Ingot geometry = 2D rectangular, 0.5m wide, 2m long
Metal   = Aluminium
Pouring temp.   = 700C
Casting speed   = 40 mm/min
Initial water temp = 25C
Water flowrate   = 1 l/min per cm and 3 l/min per cm

The temperature distribution in the centre of the ingot from the top $(z=0)$ to the base $(z=2m)$ for the two water flowrates is shown in Fig 6. It can be seen that the temperature distribution can be significantly altered by changing the water flowrate.

Ingot temperature distributions may be used as input data for casting stress models, from which the tendency to butt curl and crack may be evaluated as a function of the casting parameters.

DISCUSSION

A new method has been developed to study the forced convection heat transfer from a vertical heated surface to a falling film of water issuing from a casting mould. The heat transfer coefficients decay rapidly on leaving the mould over a distance of approximately 35 mm. Coefficients from the experiments have been correlated as a function of water flowrate per unit width and the average value of the water and ingot surface temperatures.

The resulting correlations are

$$h_o = (-3.71 \times 10^5 + 2020\ T_{av})Q^{0.352} \qquad (6)$$

and

$$h_{125} = (-1.27 \times 10^5 + 741\ T_{av})Q^{0.401} \qquad (7)$$

The experimental technique that has been developed will be useful for examining the effect of different coolants and the mode of application. Furthermore, metal from cast ingots can be used to investigate the effect of the irregular as-cast surface topology on the heat transfer.

Further work is required to develop the technique. In particular, the number of thermocouples in the block should be increased to resolve more closely the rapid decay of heat transfer coefficient in the impingement zone.

Incorporation of the coefficients in the casting model show that the temperature distribution can be significantly affected by changes in water flowrate, and this will, in turn, affect the stress distribution in the ingot.

REFERENCES

1.   Weckman D.C., Niessen P.,(Dec. 1982), Metallurgical Transactions B, 13B, pp539-602

2.   Flood S.C., Kasai K., Katgerman L.,(1989), 6th Int. Conf. Numerical Methods in Thermal Problems, 3-6 July, Swansea,Wales UK, pp1591-1599.

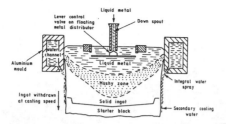

FIGURE 1. Schematic diagram of the DC semi–continuous casting process

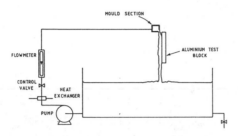

FIGURE 2. Schematic diagram of DC–Casting heat transfer rig

FIGURE 3

CROSS SECTION THROUGH TEST BLOCK

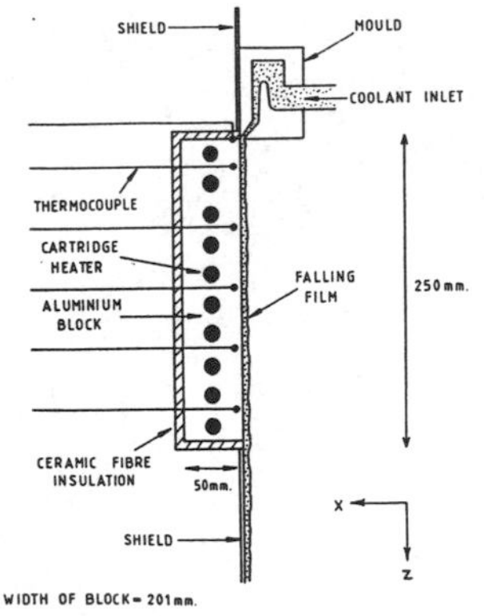

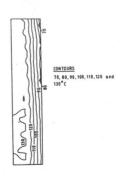

CONTOURS
70, 80, 90, 100, 110, 120 and 130°C

FIGURE 4. Calculated temperature distribution inside the block

WIDTH OF BLOCK= 201mm.

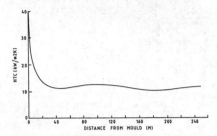

FIGURE 5. Heat transfer coefficient as a function of
distance down surface

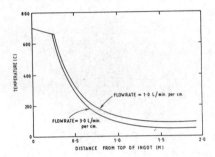

FIGURE 6. Centreline temperature distributions in the 2D ingot
at two flowrates

NODAL MODELLING AND EXPERIMENTS OF CIRCULAR GROOVED HEAT PIPE FOR SPACE APPLICATIONS ON 0-G ENVIRONMENT

SULEIMAN A*, ALEXANDRE A*

A new thermal control system based on the integration of a heat pipe is being studied to support the development of a lithium/thionyl chloride (Li/SOCL2) battery designed for future space applications. Therefore, a good understanding of the behavior of heat pipe in zero gravity (0-g) environment is necessary to make an accurate nodal model to predict its thermal performances integrated in space systems. This paper starts with a theoretical study of the heat pipes. We describe then an experimental rotating assembly constructed to simulate the 0-g operating conditions of the heat pipe in laboratory and to correlate the heat pipe model in 0-g environment (model based on the nodal concept). We compare then the thermal behavior of the heat pipe with the mathematical model results to validate the nodal model. Results show that this new mathematical model can be used effectively to predict the thermal behavior of lithium cell integrating circular grooved heat pipe in 0-g environment.

INTRODUCTION

Heat pipes are widely used in the space, especially in satellites (Ref. 1,2). This is due to the fact that the electronic equipment used in telecommunication satellites have increasing heat generation rate. Heat pipes became an element of the general thermal control network. At the present time, military satellites and shuttles require fuel cells or lithium cells to provide an important electrical power source. The main advantage of lithium is the large ratio of its electrical capacity to weight. The most recently developed version can deliver an energy density of 300 Ah/kg, which is comparable to that of complex fuel cells (Ref. 3). However, it can be noted that lithium cell performance depends strongly on the thermal conditions. Experiments carried out by mean of calorimetry method have shown that the performances of lithium cells are optimum at 40 °C, and that the lower the temperature gradient across the cell, the higher the performance of the cell. Furthermore, experiments have also shown an increase of battery heating with increased temperature due to the rising thermoneutral potential (Ref. 4). So, a good cooling system is required to optimize Lithium cells. We think that heat pipes offer an excellent approach for minimizing cell temperature and temperature gradient.

In this paper we study particularly the performance of heat pipe during the 0-g phase. A global study including integration of heat pipe in the battery has already been performed (Ref. 5,6).

THERMODYNAMIC OF HEAT PIPE

Since there are no body forces to help liquid return from the condenser to the evaporator in 0-g environment, the heat pipe includes a capillary structure to maintain the circulation of fluid against the liquid and vapor flow losses. As the energy conversion process occurs in the phase change

---

* Laboratoire d'Etudes Thermiques URA CNRS 1403 - ENSMA - F 86034 Poitiers

across the liquid-vapor interface, the heat pipe generally has a high heat transport capability. The system operating is limited by the ability of the capillary structure to maintain the fluid circulation. Otherwise, the circulation of the working fluid inside the heat pipe involves pressure drops in both liquid and vapor regions. For a steady state operation the sum of the pressure drops must be less than the maximum capillary pressure ΔP_{cmax}, in the heat pipe.

$$\Delta P_v + \Delta P_l \leq \Delta P_{c\,max} \tag{1}$$

By substituting the appropriate expressions (Ref. 7) for the pressure drops and the maximum capillary pumping pressure in the following equation:

$$\Delta P_v + \Delta P_l = \Delta P_{c\,max} \tag{2}$$

we obtain the maximum heat transfer rate of the heat pipe, called capillary limit:

$$Q_{max} = \cfrac{2\,\sigma/r_p}{\cfrac{\mu_l\,L_{eff}}{\rho_l\,A_l\,K\,h_{lv}}(1+0.3\,D\,\phi^2) + \cfrac{8\,\mu_v\,L_{eff}}{\pi\,\rho_v\,r_v^4\,h_{lv}}} \tag{3}$$

This equation assumes (Fig. 1)
  - Uniform heat density applied to the evaporator and removed from the condenser.
  - Uniform wick properties

Additional thermodynamic phenomena (called operation limits) could restrict the heat transport capability of heat pipe. They can be encountered at low temperature (near triple point of the working fluid). These are the viscous limit, sonic limit, entrainment limit and boiling limit. The latter rises because of the occurrence of boiling inside the wick. The first limits encountered with a Nickel/freon heat pipe are the boiling and capillary pumping limit.

It is difficult to predict exactly the boiling limit because of the complexity of many interfacing parameters. It depends to a considerable extent on the working fluid, the capillary structure, the internal heat pipe surface finish and the operating conditions. The maximum axial heat transfer rate for the boiling limit could be calculated by the following expression (Ref. 7):

$$Q_{cb} = \cfrac{2\pi\,L_c\,\lambda_e\,T_v}{h_{lv}\,\rho_v\,Ln\left(\cfrac{r_i}{r_v}\right)}\left(\cfrac{2\sigma}{r_n} - \cfrac{2\sigma}{r_p}\right) \tag{4}$$

HEAT PIPE DESIGN

The details of the heat pipe used for cooling cell are given in figure 2. The advantage of this annular form of groove is a decrease of the pressure losses in liquid. This increases the heat transfer capability but the high transport capability of this structure depends on a good filling of grooves at the operating temperature.

The heat pipe was designed to be used in space application, integrated to a reactive chemical element. These considerations led us to be careful in choicing heat pipe material and working fluid. We have choice Nickel as container because it is compatible with lithium battery components. The choice of working fluid is dictated by a good chemical compatibility with the container (nickel) and with the electrolyte (thionyl chloride). Tests have permit to choice freon 11 as a suitable work fluid. A complete study (Ref. 8) shows that, for a heat pipe with the following characteristics, the first limit encountered is the capillary one as shown Fig. 3.

ZERO-GRAVITY SIMULATION TEST APPARATUS

The apparatus used for zero-G thermal simulation is shown in figure 4. The heat pipe is placed in a horizontal position and rotates about its axis using a variable speed motor. The sense of rotation changes every 2π, which eliminate the need for contactless interfaces for electrical supply, water supply and thermal sensors.
The heat pipe is heated by means of an electrical heater wrapped around the evaporator, and cooled by a water exchanger.
When the heat pipe is rotating about its axis, the fluid is uniformly condensing and evaporating over all the internal wall, which will prevent puddle formation in the bottom. The presence of a puddle affects thermal performance of heat pipe, especially the wall temperatures: the upper grooves may partially drain and a local burnout can occur. Furthermore, a difference of liquid film thickness between the upper and bottom wall leads to difference of the wall temperatures at the same axial location.

Heat pipe instrumentation

The heat pipe wall temperatures are measured by 31 Ni-Cr thermocouples distributed over 3 generating lines (Fig. 5). Good homogenization of the liquid film is demonstrated when the temperatures are nearly equal at every axial location.

Test results and discussion

Experiments show that when the heat pipe was operated in various motionless positions (Fig. 6 - zones A, B, C), great difference occured in measured temperatures for the same axial location. This could be explained by the gravity field effect that causes puddle formation in the bottom of the heat pipe. In the region of the puddle film increases the wall-vapor thermal resistance and this induces temperature differences.
As the heat pipe is operated at rotational speed of 10 rpm (zone D), the temperature difference at the same axis location decreases to about 0.5-1 °C, nevertheless, small fluctuations are observed.

We have observed that the amplitude of the temperature fluctuations varies with speed (Fig. 7). It is minimal for a rotation speed of about 12.5 rpm. In all heat pipe test this speed was used to measure heat pipe performances and temperature distribution over the pipe.

MODELLING

A thermal model of heat pipe of about 50 nodes (Fig. 8) is described in ESACAP software (Ref. 8). The model is based on the nodal concept (Ref. 9,10) and takes account of the different parts of heat pipe:

* Pipe wall, capillary structure, liquid, vapor.
* Heat conductances through the heat pipe wall.
* Heat conductances through capillary structure.
* Heat conductances for evaporation and condensation.
* Uniform heat source input and heat sink temperature.

Modelling results

Predicted temperature distribution for a steady state is calculated with a heat input rate of 80 W and the use of 20°C cooling water. The value of 80 W has been chosen to correspond a special safety case heat dissipation in a lithium cell.
Figure 9 shows the predicted and measured temperatures for a steady state test at the same conditions. The results are the wall temperature at the interface with heater and cooling exchanger regions. The curves indicate a good agreement between simulation and test results. A good correlation between the model and test's results have been observed in other runs.
With the validated.heat pipe model we can write a global nodal model of the heat pipe integrated within the lithium cell. It includes the heat generation of the lithium couples versus temperature and the thermal conductances toward heat pipe. Figure 10 shows the computational thermal behavior of the cell for a realistic current discharge profile (no more than 30 Amp.).
Furthermore, we observe that, not only does the temperature level decrease strongly, but also the temperature gradient becomes insignificant by comparison with a cooling system based on pure conduction and implemented using aluminum corset surrounding the cell and put on a coldplate.
On the other hand, the Nickel Pipe acts as a cathode and reinforces the cell structure. Another advantage is the weight reduction of the whole system that lead to an important gain on electrical capacity versus weight (more than 350 Ah/kg).

CONCLUSIONS

Theoretical and experimental studies allowed to the design and construction a nickel circular grooved heat pipe filled with freon 11. An experimental apparatus was set up in the out in laboratory to simulate zero-G heat pipe thermal operating conditions, and results outlined in this paper permit a good determination of heat pipe wall temperatures during ground tests.
The heat pipe nodal model realized allows the calculation of the temperature distribution in the heat pipe. Thus, a global model of the whole system (heat pipe plus electrical battery) could be used to predict battery thermal behavior under different discharge current profile.

ACKNOWLEDGMENT

The authors wish to express their gratitude to J.C. Pret and H. Doreau for their contributions in realising the test apparatus and the data acquisition system. Thanks are also to SAFT company for their financial contribution.

REFERENCES

1- EDELSTEIN F., FLIEGER H., Satellite battery temperature control.
2- CHALMERS D.R. et al Design trends in communications satellite thermal subsystems, ESA SP-288 1988.
3- SAFT - Internal rapport.
4- SULEIMAN A., Doctorate thesis, LET-ENSMA,Poitiers (France),1992.
5- ALEXANDRE A, SULEIMAN A., FIRMIN J.L., SAE 911483 - San Francisco 1991
6- SULEIMAN A., ALEXANDRE A., FIRMIN J.L, ESA SP-286, Florence 1991
7- DUNN P. D & REAY D.A., Heat pipes, pergamon press 1978.
8- STANGERUP P., ESA SP-288 1988.
9- SAULNIER J. B. & ALEXANDRE A., congrès modélisation et simulation en thermique , ENSMA - Poitiers -France,1984.
10- SAULNIER J.B., ESA Spacecad, Bologne 1978.

NOMENCLATURE

D	Liquid/vapor shear parameter	
h_{lv}	Latent heat of vaporization	J/kg
$\dot{O}_l$	Surface tension	N/m
A_l	Wick cross section area	m^2
r_p	Capillary radius	m
r_v	Rayon vapor	m
ρ_l	Density of liquid.	kg/m^3
ρ_v	Density of vapor	Kg/m^3
μ_l	Dynamic viscosity of liquid	$N\,s/m^2$
μ_v	Dynamic viscosity of vapor	$N\,s/m^2$
ΔP_{Cmax}	Maximum capillary head	N/m^2
ΔP_l	Pressure drop in liquid	N/m^2
ΔP_v	Pressure drop in vapor	N/m^2
ϕ	Groove aspect ratio b/2a.	
Q_{cap}	Capillary limit $= Q_{max}$	W
Q_{eb}	Boiling limit	W
K	Wick permeability	
L_{eff}	Effective length of heat pipe	m
M	Coefficient M#0.3 for F< 0.5	
m	Mass flux	kg/s
N	Number of grooves	

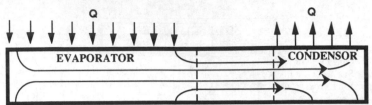

Fig. 1- Heat Pipe

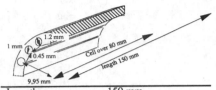

Length	150 mm
In Diameter	17.5 mm
Working fluid	Freon 11
Capillary structure:	cylindrical grooves
number	50
diameter	1 mm
interface width	0.4 mm

Fig 2 details of the heat pipe

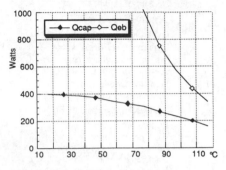

Fig 3 Main heat rate limits

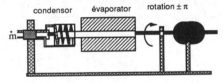

Fig. 4 Schema of experimental equipment

Zone A : Heat pipe is motionless , thermocouples locations as shown

Zone B : Heat pipe is motionless , thermocouples locations as shown

zone C : intermediate

zone D : alternate rotation 10tr/min

Fig. 5 Thermocouples position

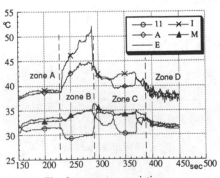

Fig. 6 temperature variations

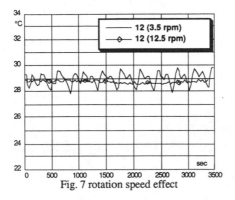

Fig. 7 rotation speed effect

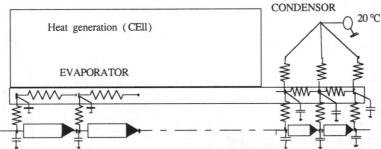

Fig. 8 nodal breakdown of heat pipe.

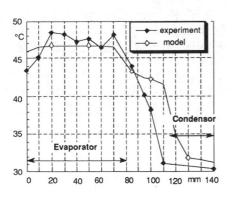

Fig. 9 - Wall temperatures

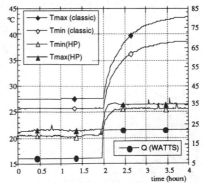

Fig 10 - Battery thermal behavior with two cooling methods

EHD ENHANCED HEAT TRANSFER: EFFECT ON THE PERFORMANCE OF A REFRIGERATION SYSTEM.

R.K. Al-Dadah[*], T.G. Karayiannis[*], R.W. James[*], P.H.G. Allen[**]

The performance of a practical refrigeration system utilizing an innovative flooded evaporator is predicted using a mathematical model. At the evaporator, the boiling heat transfer coefficient is enhanced by the application of a high intensity electric field. A review of the recent experimental implementation of electrohydrodynamic (EHD) heat transfer enhancement in multitube shell/tube evaporators and condensers of refrigeration/heat pump systems is presented. Subsequently, a brief description of the model and its results is described.

Introduction

Many techniques of heat transfer enhancement have been investigated in the past in order to improve the heat transfer performance and thus reduce the size of heat exchangers. Enhancement methods can be classified broadly as passive ones, which include special heat transfer surfaces and induced mixing devices, and active ones, which include surface vibration, fluid vibration, injection and electrostatic fields (1, 2). In the past, two-phase heat transfer enhancement was mostly effected by using passive techniques. However, the high controllability of active heat transfer techniques is now recognised as being essential for extremely precise thermal control. This makes the active method of applying a high electric field (termed electrohydrodynamic and abbreviated EHD) to heat transfer fluids one of the most promising methods for enhancing and controlling heat transfer.

The effect of electric fields on heat transfer rates has been known for seventy years (3). Since then, many studies of EHD enhanced two-phase and single-phase heat transfer have been carried out and these have been reviewed by Jones (4) and recently by Karayiannis and Allen(5). However,it was only during the last decade that research in this field was directed towards practical applications. An experimental study of EHD enhanced boiling and condensation in a single-tube heat exchanger was·performed at Imperial College (London) (6), followed by a study of a nine-tube heat exchanger at City University (London) (7). While the research in the U.K. was mostly concerned with horizontal evaporators and condensers, the research in Japan was concerned with vertical condensers (8, 9).

In this paper an EHD evaporator (i.e. an evaporator with electrically enhanced heat transfer on the shell side) was integrated with a refrigeration system and the overall performance of the system is assessed through a mathematical model.

[*] : Institute of Environmental Eng., School of Eng. Systems and Design, South Bank University, London.SE1 0AA
[**] : Thermofluids Eng. Research Centre, City University, London. EC1V 0HB

Recent Developments In Practical EHD Research

EHD enhancement mechanism lies in the electrohydrodynamic force F_E per unit volume generated by an electric field strength E, in a fluid of dielectric permitivity ε, density ϱ and at temperature T. This was derived by Landau and Lifshitz (10) and given in SI units by Parmar and Labroo (11) as:

$$F_E = qE - \frac{1}{2}E^2\nabla\varepsilon + \frac{1}{2}\nabla[E^2(\frac{\partial\varepsilon}{\partial\varrho})_T \varrho] \tag{1}$$

where q is the electric charge density in the fluid. The first term of the right hand side of equation (1) is called the electrophoretic (Coulomb) force and it is due to the action of the electric field on the free charges within the fluid. The second and third terms are the dielectrophoretic and electrostrictive forces respectively and they relate to the nature of the dielectric permitivity and the electric field and their spatial distribution.

To the best of the authors' knowledge, there are two separate and distinctive attempts to design practical EHD systems. Firstly, in Japan, the EHD force is employed effectively in a vertical condenser using an electrode system capable of stripping off the condensate film, see figure 1 (a). This results in the reduction of the thermal resistance and thus enhances the heat transfer coefficient (8, 9). A 50 kW EHD condenser was designed and integrated in a Super Heat Pump. Enhancement ratios of up to 5 times for both R113 (8), and Fluorohexane (9) were reported. However, there were no data supplied for the overall performance of the heat pump. Secondly, in the U.K., the EHD effect was studied in a single-tube evaporator/condenser unit (6). A design of simple and practical electrode system suitable for multitube evaporators and condensers was developed (12) which consisted of mesh or perforated plates and rods, see figure 1 (b). The electrode system was optimised to give a strong and radially uniform electric field distribution around the tubes. Tests using the electrode in 9-tube evaporator and condenser were performed using smooth and low fin surface tubes. The EHD enhancement of condensation heat transfer coefficient on the smooth tube bundle was moderate and in the range of 10% (7). The small degree of enhancement can possibly be explained by the fact that the increase in condensation on the upper rows causes over flooding of the lower rows thus counteracting the effect of EHD in disturbing the condensate film. In boiling, EHD enhancement of the heat transfer on low fin tubes was substantial (7). Figure 2 shows the ratio of EHD heat transfer enhancement as a function of the applied voltage. As can be seen in the figure, ratios of heat transfer coefficient with EHD to that without (α_{mE}/α_{m0}, where α_{mE} and α_{m0} are the mean heat transfer coefficients with and without the electric field respectively) of up to 2.5 can be obtained. This figure of 2.5 is by no means the maximum enhancement obtained by Damianidis (7) or in the earlier work of Cooper (6), but is limited by the fact that it was calculated at available heat flux rates. In fact, much greater enhancement is possible (up to 60 see table in Karayiannis and Allen (5)). The pronounced effect of EHD on boiling over low fin tubes was explained in (6). In addition to the heat transfer enhancement, the application of an electric field has proved to be beneficial in eliminating the boiling hysteresis. Also, the EHD effect on R114 was shown to overcome the heat transfer suppressing effect of the compressor lubricant (6, 9).

The Mathematical Model

Mathematical modelling has proved to be a helpful tool in investigating the performance of practical systems. A mathematical model predicting the dynamic and steady state responses of the refrigeration system (figure 3) was derived. The model divides each component into a number of zones representing the thermal behaviour of the system. Figure 4 shows the complete conceptual model of the plant depicted in figure 3. Each zone was analyzed assuming complete mixing of the zone content. Taking zone Ie as an example, neglecting the effect of kinetic and potential energy changes, the energy and mass balance equations can be written as follow:

$$\dot{M}_{4g} + \dot{M}_5 - \dot{M}_1 = \frac{d(M_{Ie})}{dt} = \frac{d}{dt}(\varrho_1 V_{Ie}) \tag{2}$$

$$\dot{M}_{4g}(h_{4g}) + \dot{M}_5(h_5) - \dot{M}_1(h_1) = \frac{d}{dt}(\varrho_1 h_1 V_{le}) \qquad (3)$$

where $\dot{M}$ and h are the mass flow rate and the specific enthalpy of the refrigerant at specified points, V and M are the volume and the mass of the refrigerant in the specified zone (see figure 4). Similar equations were written for the different zones of each component. These equations and the equations describing the thermodynamic properties of R22, gave a set of algebraic and differential equations characterizing the thermal behaviour of the plant and were solved using a personal computer.A brief description of the system components and their zones is given below. A detailed explanation of the mathematical model can be found in Al-Dadah (13).

The compressor was rotary vane with swept volume flow rate $(\dot{V}_{sw})$ of 12.75 m³/h at a speed of 1500 rpm. It was represented by one zone where the mass flow rates at inlet and outlet are equal. The compressor was driven by a variable speed motor and therefore the compressor speed (N) was programmed as input giving the mass flow rate as:

$$\dot{M}_1 = \dot{M}_2 = N \dot{V}_{sw} \eta_v \varrho_1 \qquad (4)$$

where η_v is the volumetric efficiency of the compressor and ϱ_1 is the refrigerant density at the compressor inlet. The specific work done by the compressor (w) was calculated using equation 5 assuming no internal compression and a polytropic compression from the evaporating pressure (p_e) to the condensing pressure(p_c). The compression index (n) for R22 was assumed to be 1.17.

$$w = \frac{p_e}{\varrho_1} * \frac{n}{n-1} * [(\frac{p_c}{p_e})^{\frac{n-1}{n}} - 1] \qquad (5)$$

The condenser was a shell and tube water cooled one. It consisted of 18 tubes arranged in 18 passes. Each tube is 23.8 mm outside diameter, and 20.6mm inside diameter. Superheated vapour entered the vessel at connection 2 (see figure 3) which was then represented by zone Ic (see figure 4). A proportion of the vapour in this section entered the thermal boundary layer represented by IIc, IIIc and IVc where it was cooled. Once cooled it was less buoyant and, depending on its position, it either moved away from the tube wall via return links 22, 23, 24 or condensed on it. The ratio of the vapour escaping from the boundary layer to that condensing was assumed constant and estimated to be 4 (14). Three zones for representing the tube metal and the water were time efficient and accurate enough for the purpose of this study. The familiar Dittus-Boelter formula was used for determining the water side heat transfer coefficient and the Nusselt formula was used for the condensation heat transfer coefficient over the tube bundle.

The evaporator was of the flooded shell and tube type where the water passes inside the low fin tubes and the refrigerant boils on the outside surface. The tube specification (outside diameter, inside diameter, length, surface area ratio) and the number of the tubes per pass were programmed as inputs. The number of passes was fixed by the number of zones assumed for metal and water representation as each zone represents one pass. The refrigerant entered the evaporator as a mixture of liquid and vapour, the vapour, represented by 4g, entered the vapour phase zone Ie and the liquid, represented by 4l, entered the liquid phase zone IIe. Part of the liquid in the liquid zone evaporated by absorbing heat from the circulating water and entered the vapour phase zone via point 5. The tube spacing and the shell size were determined taking into consideration the dielectric strength of the refrigerant and providing space over the tube/electrode bundle for vapour-liquid separation.

A key factor in the design of the evaporator was the boiling heat transfer coefficient over a bundle of low fin tubes. To date, no general correlation exists which takes into account the surface effect, fluid properties, bundle effect and the distribution of vapour-liquid mixture entering the evaporator. Danilova and Dyundin (15) performed experiments and produced a correlation for the mean boiling heat transfer coefficient over a bundle (α_{mo}) for R12 and R22 which can be used in the design of flooded evaporators. Their bundle consisted of 19 tubes arranged in six rows and two types of low fin tubes were tested over a wide range of evaporating temperatures (-20 to +30 °C). A correlation of the following form was proposed:

$$\alpha_{mo} = k_1 * \dot{a}^{k_2} * p_e^{k_3} \tag{6}$$

where $\dot{a}$ is the heat flux in W/m^2 and p_e is the evaporating pressure in bar. k_1, k_2 and k_3 are empirical constants and for R22 they are equal to 33, 0.45, 0.25 respectively. For the water side, the heat transfer coefficient was calculated using the Dittus-Boelter formula.

The Model Results

The model was used to optimise the specifications of the flooded evaporator taking into consideration several basic requirements such as water temperature drop, the evaporating temperature, the evaporator approach temperature and the heat flux. All these requirements were studied before and after the introduction of fouling resistance. Prediction of the system performance after the application of the electric field at the evaporator was then studied using a heat transfer enhancement ratio, see figure 2. This ratio is used as a multiplication factor for the heat transfer coefficient without EHD.

Performance criteria produced by Webb (16) were used in assessing the benefits of heat transfer enhancement. It was found that the use of an enhancement ratio of 1.8 resulted in a reduction of 15% in the heat transfer area of the evaporator. Figure 5 shows how the refrigeration load increases with the application of the electric field for constant compressor lift (p_c - p_e). The gradual increase in the evaporator load with EHD provides an effective and reliable method of controlling the refrigeration capacity of the plant. Upon the application of EHD, the evaporating temperature increases, figure 6. This could then result in smaller log mean temperature difference thus increasing the suction pressure (temperature) to the compressor and consequently enabling the compressor to work at lower speeds (lower power consumption) while achieving the same evaporator load. The effect on applying EHD at the evaporator on the coefficient of performance of the plant is seen in figure 7.

Conclusion

The results of this investigation clearly demonstrate that EHD can have a significant role in future heat exchanger design. 15% reduction in the size of the heat exchanger was obtained for a very conservative enhancement ratio. The application of the electric field also resulted in increasing the refrigeration capacity for a constant compressor lift or the reduction of power consumption for a constant load. In addition, the gradual increase of the refrigeration load due to the gradual increase of the applied voltage can be a very reliable capacity control method.

An experimental facility has been constructed comprising an EHD shell and tube evaporator integrated in a refrigeration plant. This will be used to test both EHD enhancement and overall plant performance for various refrigerants.

References
1. BERGLES, A.E., 1985, Handbook of Heat Transfer Applications, 2nd edition, McGraw Hill.
2. REAY, D.A., 1991, Heat Recover Systems & CHP, Vol. 11, No. 1, pp. 1-40, Pergamon Press plc.
3. CHUBB, L.W., 1916, U.K. Patent No. 100796.
4. JONES, T.B., 1978, Advances in Heat Transfer, 14, pp. 107-148, Academic press.
5. KARAYIANNIS, T.G. and ALLEN, P.H.G., 1991, Eurotech Direct 91, C413-036, pp. 165-181, Birmingham, U.K.
6. COOPER, P., 1986, PhD Thesis, University of London, U.K.
7. DAMIANIDIS, C., COLLINS, M.W., KARAYIANNIS, T.G. and ALLEN, P.H.G., 1991, Interklima 91, pp.10-26, Int. Symp. of Heating, Refrigerating and Air Conditioning, Zagreb
8. The development of Super Heat Pump Energy Accumulation System, 1988, Progress Report, New Energy and Industrial Technology Development Organisation, pp. 108-124, Tokyo.
9. The development of Super Heat Pump Energy Accumulation System, 1990, Progress Report, New Energy and Industrial Technology Development Organisation, pp. 105-117, Tokyo.

10. LANDAU, L.D. and LIFSHITZ, E.M., 1963, Electrohydrodynamics of Continuous Media, pp. 68, Pergamon, New York.

11. PARMAR, D.S. and LABROO, B., 1985, Phys. Letters, Vol. 108A, pp. 115-118.

12. ALLEN, P.H.G. and COOPER, P., 1985, U.K. Patent No. 8522680.

13. AI-DADAH, R.K. 1991, Research Progress Report, Institute of Environmental Eng., South Bank University, UK.

14. JAMES, R.W., 1991, Direct Communication. Institute of Environmental Eng., South Bank University, U.K.

15. DANILOVA, G.N. and DYUNDIN, V.A., 1972, Heat Transfer-Sov. Res., 4(4), pp. 48-54.

16. WEBB, R.L., 1988, Heat Transfer Equipment Design,Hemisphere Publication, New York.

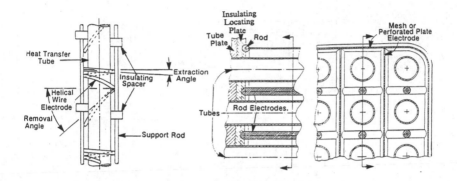

Figure 1. Practical Electrode Systems (6, 8).

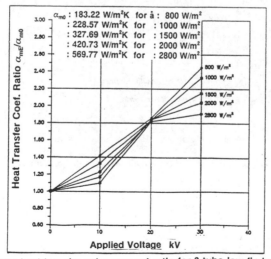

Figure 2. Mean heat transfer enhancement ratio for 9-tube low fin bundle (7).

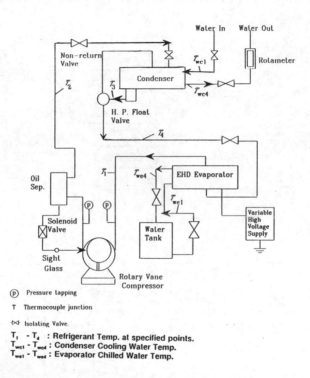

P Pressure tapping

T Thermocouple junction

⋈ Isolating Valve.

T_1 - T_4 : Refrigerant Temp. at specified points.
T_{wc1} - T_{wc4} : Condenser Cooling Water Temp.
T_{we1} - T_{we4} : Evaporator Chilled Water Temp.

Figure 3. Schematic of the experimental refrigeration rig.

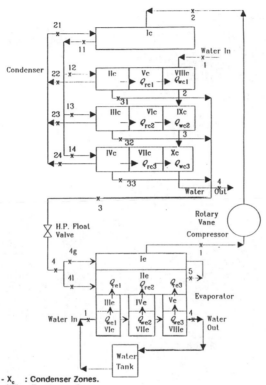

I_c - X_c : Condenser Zones.
I_e - VIII_e : Evaporator Zones.
Q_{rc} - Q_{wc} : Heat Transfer Rate For Condenser Refrigerant and Water Sides.
Q_{re} - Q_{we} : Heat Transfer Rate For Evaporator Refrigerant and Water Sides.

Figure 4. Conceptual model of the refrigeration system.

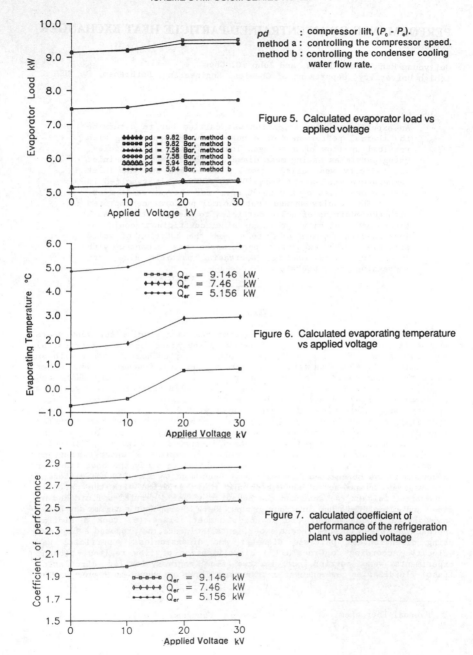

pd : compressor lift, $(P_c - P_e)$.
method a : controlling the compressor speed.
method b : controlling the condenser cooling water flow rate.

Figure 5. Calculated evaporator load vs applied voltage

Figure 6. Calculated evaporating temperature vs applied voltage

Figure 7. calculated coefficient of performance of the refrigeration plant vs applied voltage

PERFORMANCE OF AN ENTRAINED-PARTICLE HEAT EXCHANGER

Gui-young Han[1], Kemal Tuzla, and John C. Chen
Lehigh University, Department of Chemical Engineering, Bethlehem, PA, USA

An experimental investigation was carried out to determine the thermal performance of a heat exchanger operating with vertical up flow of a hot gas loaded with solid particles. Using particles having mean diameters (88-157 μm), the inlet gas velocity was varied from 1.5 to 13 m/s and the inlet temperature was varied from 100 to 600°C. This resulted in mixture densities of 2.0 to 30 kg/m³ for the gas-solid flow.

The results showed that thermal performance improved with the addition of solid particles to the gas stream. At a given gas velocity, the gas-side coefficient could be increased by factors of 2 to 4 with the addition of solid particles. The exchanger performance was enhanced with increasing solid loading, decreasing particle size, and increasing gas temperature.

INTRODUCTION

Upward flow of gas with solid particulates has been studied for many years because of extensive use for heat recovery equipment in the petrochemical industry. Farbar and Morley (1) experimentally studied heat transfer with an air-alumina-silica catalyst mixture system in a vertical isothermal tube. Other systematic experiments were conducted by Wen and Miller (2), and Depew and Farbar (3). They reported that the addition of solid particles to a gas flow increased the heat transfer by virtue of increasing the volumetric heat capacity and disturbing the gas boundary layer. In spite of their investigation, a detailed analysis of effect of parameters on the heat transfer coefficient has not been achieved, because of the complexity of flow patterns in the two-phase (gas-solid) system. As illustrated in Figure 1, gas-solid systems show a broad range of flow regimes. Since the understanding of hydrodynamics of gas-solid suspension is essential to study the heat transfer mechanism of two-phase flow, attempts have been made to classify the different flow regimes in gas-solid suspension and their relation with the particle properties. Geldart (4) analyzed the behavior of solid particles fluidized by gases and classified them into four groups. Reh (5) proposed a regime map based on modified Froude number and Reynolds number. Grace (6) took a similar approach and based on more recent experimental evidence, he proposed a flow map using dimensionless particle diameter and dimensionless superficial gas velocity. According to Grace's (6) classification of flow regimes, present experiments were carried out in two flow regimes, called circulating (fast)-fluidization and pneumatic-transport flows as shown in Figure 1. The

---

[1]Hyundai Petrochemical Co., Ltd. SeoSam, Chungnam, Seoul, Korea

objective was to obtain experimental information on effect of various operating parameters on heat transfer.

EXPERIMENTAL SET-UP

The heat transfer test facility used in these experiments is shown in Figure 2. The facility consisted of a combustion chamber, a cylindrical heat transfer test section, and solid recycle and feeding system. The test section was 0.9 m long shell-and-tube type heat exchanger having 5 cm ID tube and 7.5 cm ID shell size. The suspension flowed upward through the tube and cooling water flowed downward through the shell side. The heat exchanger section was instrumented with thermocouples for measurement of suspension and water temperatures at the top and bottom of the test section. Pressure taps were also mounted in order to measure the average suspension density. Solids recycle rates were determined by temporarily closing the valve located above the particle reservoir and timing the accumulation rates of a packed bed of solids through the three sight windows which are mounted along the return column. To obtain uniform flow of water through the annulus, 40 μm porous metal disks were placed at the inlet and outlet of cooling water jacket. Because of inherent fluctuating characteristics of two-phase flow, a computer-based data acquisition system was employed to measure the time average pressure drop gradient and temperatures of water and gas-particle suspension. Inlet and outlet temperatures of water and suspension and corresponding tube side heat transfer coefficient were recorded every three seconds and pressure drops were recorded simultaneously. Heat transfer coefficient and pressure drop were averaged over 5 minute recording time. In this present work, overall heat transfer coefficient (U_i) of the heat exchanger was determined from the logarithmic mean temperature difference (LMTD) of the heat exchanger and the temperature rise of cooling water. Suspension-to-wall heat transfer coefficient (h_i) was determined by subtracting the cooling water side thermal resistance.

RESULTS AND DISCUSSIONS

Suspension density effect

Time and cross-sectional average suspension-to-wall heat transfer coefficients (h_i) are plotted against the suspension densities in Figure 3 for FCC particles of mean diameter of 88 μm. As seen in Figure 3 the heat transfer coefficient increased with suspension density. Obviously a higher solid concentration will result in a higher frequency of bombardments of particles against the wall and a higher heat capacity of mixture, thus a higher rate of heat transfer. Therefore, time and cross-sectional average suspension density gives a way to represent the heat transfer characteristic of gas-solid suspension flow system. In Figure 3, it is seen from the log-log plot that the heat transfer coefficient increased linearly with the cross-sectional average suspension density.

Superficial gas velocity effect

A wide range of superficial gas velocity was employed to investigate the gas velocity effect on the heat transfer coefficient. In order to compare the superficial gas velocity effect for different size of particles, a normalized superficial gas velocity U_n was introduced. U_n was defined as the ratio of superficial gas velocity to particle terminal velocity, where particle terminal velocity was calculated from information in Kunii and Levenspiel (7). Superficial gas velocity and particle terminal velocity were calculated at the operating temperature and pressure. Figure 4 shows the gas velocity effect on the heat transfer coefficient at the same suspension density of 10-12 kg/m^3. As the normalized gas velocity increased up to U_n of 8, heat transfer coefficient increased. A decreasing trend of heat transfer coefficients at the higher superficial gas velocity was observed in spite of increasing gas convection. The explanation for this unexpected effect may lie in hydrodynamics of circulating-fluidization and pneumatic-transport flows. According to Grace's (6) classification of flow regime, in certain operating conditions circulating-fluidization and pneumatic-transport flows are not easily distinguished. But as the gas velocity increase further for a specific particle, flow regime changes from circulating-fluidization to pneumatic-transport flow. It is believed that in a pneumatic flow radial solid concentration profile is more evenly distributed than that of fast fluidization-flow at the same suspension density. Because of this decreased wall solid concentration which is predominant parameter of particle convective heat transfer in the suspension-to-wall heat transfer, particle convective heat flux was reduced. Such hydrodynamics characteristic of radial solid concentration profile change was experimentally reported by Herb et al. (8). These experimental data on superficial gas velocity effect show that heat transfer mechanism is closely tied with hydrodynamics of gas-solid suspension flow and emphasizes the importance of flow regime with respect to operating conditions.

Suspension temperature effect

The influence of suspension temperature on the heat transfer coefficients is shown in Figure 5 for the 157 μm sand particle. Result showed that heat transfer coefficient increases with temperature. This increase is predominantly due to radiation. The contribution of radiation to the heat transfer coefficient was about 10-20% in this experimental temperature range (100-500°C) and suspension densities (5-20 kg/m^3). Therefore two-phase flow enhanced not only particle convection but also radiation due to increased suspension emissivity. Similar results were observed for the FCC particles of two different sizes.

Comparison with enhanced single-phase heat exchangers

It is worthwhile to compare the performance of two-phase flow heat exchanger with enhanced single-phase heat exchanger. While there is no general method to compare the performance of different types of heat exchangers, it is possible to compare on basis of total pressure drop losses. Two types of conventional single-phase enhanced heat transfer techniques were selected for the comparison. Evans (9) used axial promoter and Thomas (10) used turbulence promoter in order to increase the single-phase heat transfer coefficient. Figure 6 shows the comparison of performance of different types of enhanced heat transfer technique based on the pressure drop losses. As seen in Fig. 6 at a pressure drop gradient of 20 mm water/m in 2 inch pipe, axial promoter type heat exchanger enhanced heat transfer coefficient up to 50 W/m^2°C and turbulence promoter enhanced heat transfer coefficient up to 60 W/m^2°C. For the dilute two-phase flow type heat exchanger, heat transfer coefficients can be maintained about 85 to 110 W/m^2°C for two different size of FCC particles. Therefore, the dilute two-phase flow heat exchanger gives 50-100% higher heat transfer coefficient over enhanced single-phase heat exchangers, at the same power consumption. Furthermore, radiative heat flux of particles to heat transfer surface will be additional enhancement over single-phase enhancement techniques at the higher operating temperature because of higher emissivity of particle suspension. Another advantage of two-phase flow heat exchanger over enhanced single-phase flow heat exchanger is the operation flexibility.

Unlike the single-phase enhancement, dilute two-phase flow heat exchanger have several operation variables (such as particle size and solid recycle rates) in addition to superficial gas velocity which is the major operation variable in single-phase enhancement heat exchanger. For example, even at low gas velocity, by choosing smaller particles and higher solid recycle rate, two-phase flow heat exchanger can be maintained at higher suspension density and it can obtain higher heat transfer rates. As a conclusion, the dilute two-phase flow heat exchanger can provide advantages of increased heat transfer coefficient and flexible operation.

CONCLUSIONS

Suspension-to-wall heat transfer coefficient has strong dependence on the suspension density. Changing flow regime (from circulating-fluidization to pneumatic-transport flow) due to increased superficial gas velocity could result in lower heat transfer coefficients. Radiation contribution to total heat flux was determined to be about 10-20% in the operating condition of this study. Comparison of experimentally determined heat transfer coefficient for two-phase flow heat exchanger with that of enhanced single-phase heat exchanger showed that dilute gas-solid suspension heat exchanger has advantages over enhanced single-phase heat exchangers.

REFERENCE

1.   Farbar, L. and M.J. Morley, 1957, Ind. Eng. Chem. 49, 1143
2.   Wen, C.Y. and E.N. Miller, 1961, Ind. Eng. Chem. 53, 51
3.   Depew, C.A., and L. Farbar, 1963, Trans. ASME. Series C. J. Heat Transfer 85, 164
4.   Geldart, D., 1973, Powder Technol. 7, 285
5.   Reh, L., 1971, Chem. Eng. Progr., 67, 58
6.   Grace, J.R., Can. J., 1986, Chem. Eng. 64, 353
7.   Kunii, D. and O. Levenspiel, 1977, Fluidization Engineering, 77
8.   Herb, B.E., K. Tuzla, and J.C. Chen, 1989, in Fluidization VI, Eds. J.R. Grace, L.W. Shemilt and M.A. Bergougnou, 65.
9.   Evans, L.B. and S.W. Churchill, 1963, Chem. Eng. Progress, 41, 36
10.   Thomas, D., 1967, I&EC Process Design and Development, 3, 385

NOMENCLATURE

h_i suspension-to-wall heat transfer coefficient (W/m^2°C)
h_o water side heat transfer coefficient (W/m^2°C)
LMTD   logarithmic mean temperature difference (°C)
T_b bed temperature (°C)
U_g superficial gas velocity (m/s)
U_n normalized gas velocity (U_g/U_t)
U_t particle terminal velocity (m/s)
U_o overall heat transfer coefficient (W/m^2°C)

   Greek letter

ρ_{sus} suspension density (kg/m^3))

ACKNOWLEDGEMENT

This work is supported in part by the Gas Research Institute. Contract No. 5087-232-1545

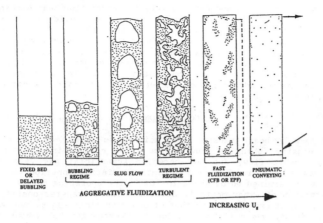

Figure 1.   Principal flow regimes for upward flow of
gas through solid particulate materials

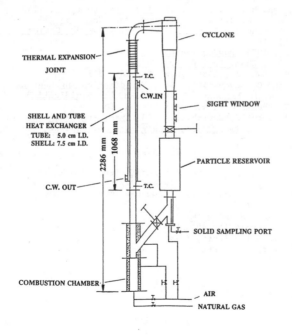

Figure 2.   Schematic of Heat Transfer Test Loop

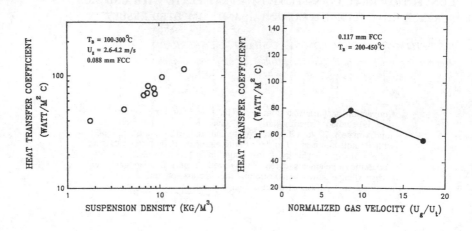

Figure 3. Heat transfer coefficient as a function of suspension density

Figure 4. Superficial gas velocity effect on heat transfer coefficient

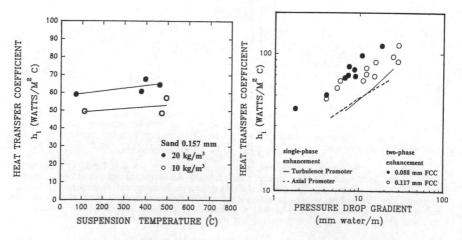

Figure 5. Suspension temperature effect on heat transfer coefficient

Figure 6. Comparison of heat transfer coefficients for single and two-phase enhancements

CONVECTION HEAT TRANSFER FROM A FLAT PLATE WITH UNIFORM SURFACE HEAT FLUX; EXPERIMENTAL AND COMPUTED RESULTS

J.F.T.PITTMAN, J.F.RICHARDSON, A.O.SHARIF, C.P.SHERRARD [*]

Department of Chemical Engineering, University of Wales, Swansea,U.K.

Convection heat transfer from a flat plate with uniform surface heat flux is studied experimentally at plate Reynolds numbers from approximately 20 to 1500, for Newtonian and shear thinning liquids with Prandtl numbers 50 to 580. Finite element computer simulations using a Streamline Upwind Petrov Galerkin formulation on biquadratic isoparametric elements agree closely with experimental local plate surface temperatures demonstrating the success of this formulation as a tool for analysis of process heat transfer.

INTRODUCTION

The present paper forms the introduction to a project on heat transfer from surfaces immersed in fluids in laminar flow. The emphasis is on liquids of high viscosity, some with non-Newtonian flow properties, many of which exhibit variations of consistency and density with temperature, - conditions which are relevant in many industrially important processes involving polymeric liquids and solutions, suspensions and emulsions. The non-linearity of the coupled momentum and energy conservation equations which apply in this situation renders an analytical treatment impossible, except in certain special cases. Consistencies are often high, and at the low Reynolds numbers involved the boundary layer thickens too rapidly for the classical boundary layer theory to be valid; in any case, uncertainties remain in boundary layer theory for variable property problems such as those with which we are concerned here. Additionally, the geometry of the heat transfer surfaces may not be simple, again rendering approximate analytic approaches difficult.

There is thus a well recognized need for numerical methods which are sufficiently powerful and flexible to encompass the phenomena outlined above. The literature, however, shows very few instances where a numerical solution of the present class of problems has been tested against careful experimental results.

The twin objectives of the present work are therefore, on the one hand, to establish an experimental technique for the precise and detailed measurement of local heat transfer coefficients, and, on the other, to compare measurements with numerical predictions for a range of conditions. Numerical results are obtained using the finite element method, chosen for the convenience with which it can be applied to problems of irregular and various geometries, and the ease which variable physical properties and rheology can be taken into account.

By validating experimental and numerical results against each other, we establish confidence in an experimental technique that can be used to investigate heat transfer to a range of materials in a number of fairly simple geometries, and a numerical method that is applicable in a wide variety of geometries.

* Present Address: Lloyds Register of Shipping, 29 Wellersley Rd. Croydon. Surrey, U.K.

HEAT TRANSFER EXPERIMENTS

Experiments were carried out [1] using the flat plate heat transfer element illustrated in Figure 1. This provides on almost exactly uniform surface heat flux. The element was mounted vertically across the diameter (104mm) of a vertical glass tube, and liquid was circulated upwards through the tube via a flow straightener and thence over the element.

In use, the element was supplied with AC power from a variable transformer. Power dissipation was calculated from the calibrated resistance of the element and the measured potential drop. A check was made for current by-passing through the liquid by measuring the element apparent resistance in distilled water with specific conductivity $\varepsilon = 1.3$ $\mu\Omega^{-1}$ and in salt solution, $\varepsilon = 4.7$ $m\Omega^{-1}$. No significant difference was detected. Plate surface temperatures were obtained from thermocouple outputs, measured with a precision of 1 μV using a Solartron 7043 multimeter. The staggered arrangement of the thermocouples near the plate centreline serves to prevent any flow disturbance resulting from slight surface imperfections at the weld points from influencing measurements downstream. Two thermocouples welded to the foil 20 mm from the element centreline were used to provide a direct check that the surface heat transfer was two-dimensional, and a further thermocouple remote from the plate surface recorded the bulk fluid temperature.

Velocity profiles across the pipe, 15mm upstream of the heat transfer elements, were obtained using laser anemometry.

Experiments were carried out using aqueous Newtonian and shear-thinning liquids, with Reynolds numbers ranging from 20 to 1520, based on plate length in the flow direction, and Prandtl numbers from 50 to 580. Heating rates ranged from 2 to 10 kW/m^2, giving temperature differences between the plate surface and the bulk fluid ranging from a fraction of a degree to approximately 35 °C. Results are reported here in terms of local plate surface temperatures, since these are directly measured quantities. Local heat transfer coefficients can easily be derived from these values and the rates of electrical heating.

MATHEMATICAL MODELLING AND FINITE ELEMENT ANALYSIS

The steady two-dimensional nature of the heat transfer near the plate centreline was demonstrated experimentally, and the appropriate equations of momentum, energy, and mass conservation are accordingly, (with i,j=1,2 and the summation convention in force):

$$\rho \, u_j \, u_{i,j} = - \, p_{,i} + \tau_{ji,j} + f_i \qquad (1)$$

$$\rho \, C \, u_i \, T_{,i} = k \, T_{,ii} + \dot{S} \qquad (2)$$

$$u_{i,i} = 0 \qquad (3)$$

In the momentum equation, (1), viscous stress is given by

$$\tau_{ij} = \mu \, \dot{\gamma}_{ij} \qquad ; \qquad \mu = \mu \, (\, \dot{\gamma}, T) \qquad (4)$$

The power law model characterized the liquids well.

$$\mu = \mu_o \dot{\gamma}^{\,n-1} \, e^{-b(T-T_o)} \qquad (5)$$

The body force (buoyancy) term is taken as

$$f_i = [\, \rho(T) - \rho(T_{in}) \,] \, g_i \qquad (6)$$

with density elsewhere taken as constant, $\rho(T_{in})$, (the usual Boussinesq assumption).

In the energy equation, specific heat, C, and thermal conductivity, k, may be taken as functions of temperature. The source term, $\dot{S}$, represents viscous energy dissipation in the flow and, although this is very small here, the term is retained for completeness.

Heat transfer from the plate is modelled including simultaneous heat conduction in the steel foil, which is also treated using equation (2), with velocities set to zero, and $\dot{S}$ then being the rate of electrical energy dissipation in the steel foil.

The system of equations (1) to (6) was solved using finite elements, with the Reduced Integration Penalty method for continuity enforcement (see .e.g. [2]). Convection terms are significant , particularly in the energy equation, and the Streamline Upwind Petrov-Galerkin (SUPG) formulation is used to obtain stable, convergent solutions. The form used here is an extension to isoparametric biquadratic elements of the formulation introduced originally [3] for bilinear elements, desirable because of the greater geometrical flexibility and efficiency of the higher order elements. The essence of the method is the use of a weight function:

$$\bar{N}_I = N_I + \tau P_I \,, \qquad P_I = u_j N_{I,j}$$

with the field variables discretised in the usual way

$$u_i = N_I u_{iI} \,, \qquad T = N_I T_I$$

As a consequence of the discontinuous form of P_I, terms weighted by this are evaluated only on elements interiors, and second derivatives of the shape functions then arise. Though these were omitted as negligible in the original form of SUPG, they must be taken into account here, and are evaluated exactly using an algorithm described elsewhere [4]. The multiplier, τ, is calculated at integration points, as a function of local mesh Peclet or Reynolds numbers, using a form similar to that proposed by Brooks and Hughes [3], but incorporating the optimal upwinding parameters of Donea et al [5]. Details of this will be given in another publication, where the formulation is tested successfully against analytic results for a severe scalar convection- diffusion problem [4].

The resulting finite element equations are,

momentum:

$$u_{1J} \int_{\Omega} \rho \, (N_I + \tau P_I) \, (N_{J,1} N_K \, u_{1K} + N_{J,2} N_K \, u_{2K}) \, d\Omega$$

$$+ u_{1J} \left[\int_{\Omega} \mu(2 \, N_{I,1} N_{J,1} + N_{I,2} N_{J,2}) \, d\Omega - \sum_e \int_{\Omega e} \left[\tau P_I \mu \, (2 N_{J,11} + N_{J,22}) \right. \right.$$

$$\left. \left. + \tau \, P_I (2 \, \mu_{,1} N_{J,1} + \mu_{,2} N_{J,2}) \right] d\Omega \right]$$

$$+ u_{2J} \left[\int_\Omega \mu \, N_{I,2} N_{J,1} d\Omega - \sum_e \int_{\Omega e} \left(\tau P_I \mu \, N_{J,12} + \tau \, P_I \mu_{,2} N_{J,1} \right) d\Omega \right]$$

$$+ u_{1J} \left[\int_\Omega \lambda \, N_{I,1} N_{J,1} d\Omega - \sum_e \int_{\Omega e} \tau P_I \lambda \, N_{J,11} d\Omega \right]$$

$$+ u_{2J} \left[\int_\Omega \lambda \, N_{I,1} N_{J,2} d\Omega - \sum_e \int_{\Omega e} \tau P_I \lambda \, N_{J,21} d\Omega \right] = \int_\Omega (N_I + \tau P_I) \, f_1 d\Omega + \int_\Gamma N_I t_1 d\Gamma \qquad (7a)$$

$$u_{2J} \int_\Omega \rho \, (N_I + \tau P_I) \, (N_{J,1} N_K u_{1K} + N_{J,2} N_K u_{2K}) \, d\Omega$$

$$+ u_{2J} \left[\int_\Omega \mu (N_{I,1} N_{J,1} + 2 N_{I,2} N_{J,2}) \, d\Omega - \sum_e \int_{\Omega e} \left(\tau P_I \mu \, (N_{J,11} + 2 N_{J,22}) \right. \right.$$

$$\left. \left. + \tau \, P_I (2 \, \mu_{,1} N_{J,1} + \mu_{,2} 2 N_{J,2}) \right) d\Omega \right]$$

$$+ u_{1J} \left[\int_\Omega \mu \, N_{I,1} N_{J,2} d\Omega - \sum_e \int_{\Omega e} \left(\tau P_I \mu \, N_{J,21} + \tau \, P_I \mu_{,1} N_{J,2} \right) d\Omega \right]$$

$$+ u_{2J} \left[\int_\Omega \lambda \, N_{I,2} N_{J,2} d\Omega - \sum_e \int_{\Omega e} \tau P_I \lambda \, N_{J,22} d\Omega \right]$$

$$+ u_{1J} \left[\int_\Omega \lambda \, N_{I,2} N_{J,1} d\Omega - \sum_e \int_{\Omega e} \tau P_I \lambda \, N_{J,12} d\Omega \right] = \int_\Omega (N_I + \tau P_I) \, f_2 d\Omega + \int_\Gamma N_I t_2 d\Gamma \qquad (7b)$$

Energy:

$$T_J \int_\Omega \rho \, C \, (N_I + \tau P_I) \, (N_K u_{1K} N_{J,1} + N_K u_{2K} N_{J,2}) \, d\Omega$$

$$+ T_J \left[\int_\Omega k (N_{I,1} N_{J,1} + N_{I,2} N_{J,2}) \, d\Omega - \sum_e \int_{\Omega e} \tau P_I \left[k \, (N_{J,11} + N_{J,22}) + (k_{,1} T_{,1} + k_{,2} T_{,2}) \right] d\Omega \right]$$

$$= \int_\Omega (N_I + \tau P_I) \, \dot{S} \, d\Omega - \int_\Gamma N_I q_n d\Gamma \qquad (8)$$

RESULTS AND DISCUSSION

We can   show here only a small selection from the results that have been obtained. We compare   computed and experimental results for four cases; two for shear thinning aqueous Carbopol solutions at low low Reynolds number; and two for Newtonian aqueous glycerol at higher Reynolds number. The numerical results were obtained after careful mesh refinement experiments, and we are confident that they are adequately converged, the final meshes being somewhat more refined than that shown in Figure 2.

Figures 3 and 4 compare computed and measured plate surface temperatures for the shear thinning solutions.(Experimental conditions and fluid properties are summarized in the figure captions) Agreement is close, within $0.4 \, ^\circ C$ or 3.6% of the maximum plate surface temperature rise. The results confirm the success of the computations in handling the non-linearities in the problem, due primarily to the significant shear rate and temperature dependence of viscosity. The importance of the temperature dependence of viscosity (and, to a lesser extent, the thermal properties) is also illustrated Figure 4, by showing results for the case where physical properties are fixed at values corresponding to the fluid inlet temperature.

Figures 5 and 6 refer to results for Newtonian liquids at higher plate Reynolds numbers, up to 1520. Since the diameter of the tube in which the heat transfer plate is mounted is approximately equal to the flow direction plate length, tube and plate Reynolds numbers are approximately equal. A Reynolds number of around 1500 is thus close to the maximum for which laminar flow conditions can be guaranteed. Again, computed and experimental plate surface temperatures are in close agreement. Discrepancies are generally less than 5% of the maximum surface temperature rise, though differences up to $2 \, ^\circ C$, 8%, occur at upstream thermocouple positions in Figure 5. The pattern of isotherms near the plate surface, for the case of Figure 5 is shown in Figure 7.

CONCLUSIONS

The good agreement between computed and experimental plate surface temperatures, generally within ± 5% of the maximum surface temperature rise, gives considerable confidence in both the experimental and numerical procedures. The capabilities of the numerical method in handling the non-linearities arising from non-Newtonian flow behaviour and temperature dependent physical properties have been demonstrated. Additionally, the effectiveness of the extended from of SUPG on isoparametric biquadratic elements in providing stable solutions to convection-diffusion problems with significant Peclet and Reynolds numbers in laminar flow has been demonstrated. The finite element procedure, of course, yields comprehensive information on velocity and temperature fields, heat fluxes, point-wise surface heat transfer coefficients, etc, and provides a powerful technique for heat transfer calculations, including those on geometrically complex domains.

REFERENCES

1. Sherrard, C.P. 1987. Ph.D Thesis, University College, Swansea.
2. Pittman, J.F.T. 1989, in 'Fundamentals of Computer Modelling for Polymer
   Processing' Ed. C.C. Tucker III, Pub. Hanser.
3. Brooks, A.N., and Hughes, T.J.R.,1982, Comp.Methods in Appl. Mech. and Eng.,32,199-259,
4. Petera, J, Nassehi, V. and Pittman J.F.T. To appear.
5. Donea, J., Belytschko, T. and Smolinski, P. 1985, Comp. Methods in Appl.
   Mech. and Engng,48, 25-43

SYMBOLS

b	temperature coefficient of viscosity	θ^{-1}
C	specific heat at constant pressure	$L^2 T^{-2} \theta^{-1}$
d_i^e	element characteristic dimension in i-direction	L
f_i	body force/volume	$M L^{-2} T^{-2}$
g_i	acceleration of gravity	$L T^{-2}$
k	thermal conductivity	$M L T^{-3} \theta^{-1}$
L	flow direction plate length	L
n	power law model exponent	-
n_{el}	number of elements	-
N_I	shape function at node I	-
p	pressure	$M L^{-1} T^{-2}$
P_I^d	discontinuous Petrov term in shape function at node I	T^{-1}
P_e^d	mesh Peclet number $ud\rho C/k$	-
Pr	Prandtl number $\mu C/k$	-
Re_L	plate Reynolds number $\rho u L / \mu$	-
q_n	normal heat flux	$M T^{-3}$
q_s	rate of electrical energy dissipation per unit area of plate surface	$M T^{-3}$
$\dot{S}$	rate of energy dissipation per unit volume	$M L^{-1} T^{-3}$
t_i	resultant force /area	$M L^{-1} T^{-2}$
T	temperature	θ
T_{in}	fluid inlet temperature	θ
T_o	reference temperature for fluid consistency.	θ
u_i	fluid velocity	$L T^{-1}$
x_i	Cartesian co-ordinates	L
$\tilde{x}_i$	mapping of local to global co-ordinates	L
y	downstream co-ordinate measured from plate leading edge	L
$\dot{\gamma}_{ij}$	rate of deformation tensor	T^{-1}
$\dot{\gamma}$	second invariant of the above	T^{-1}
Γ	boundary of analysis domain	L
ΔT_s	surface temperature rise of plate	θ

η	finite element local co-ordinate	-
λ	penalty parameter in R.I.P formulation	$M L^{-1} T^{-1}$
μ	viscosity	$M L^{-1} T^{-1}$
μ_o	consistency coefficient in power-law model	$M L^{-1} T^{n-2}$
ξ	finite element local co-ordinate	-
ρ	density	$M L^{-3}$
τ_{ij}	viscous stress tensor	$M L^{-1} T^{-2}$
τ	multiplier on Petrov term in *SUPG*	T
Ω	domain of analysis	L^2
Ω^e	domain of element e	L^2
—	overbar denotes the approximate, numerical solution.	

Subscripts:

I	indicates nodal value
,i	denotes differentiation with respect to x_i
jj	Einstein summation convection applies j=1,2

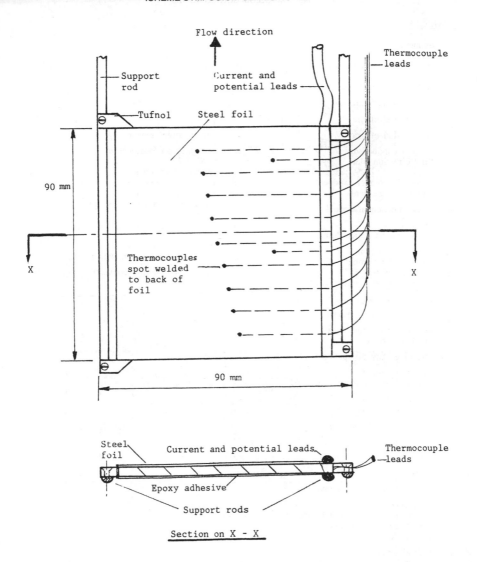

Figure 1. The flat plate heat transfer element.

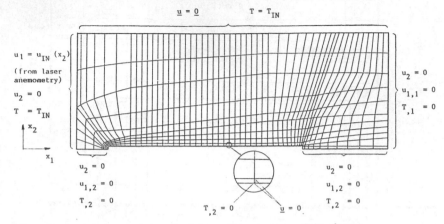

Figure 2. The 2-dimensional domain of the analysis, showing a finite element mesh and boundary conditions.

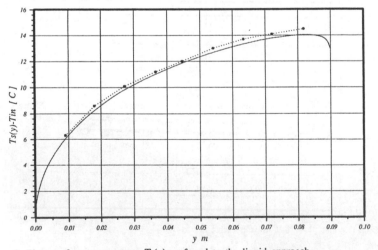

Figure 3. Plate surface temperature, $T_s(y)$ referred to the liquid approach temperature, T_{in}, as a function of downstream distance, y, from the leading edge. Aqueous Carbopol: n=0.53, b=0.024 K^{-1}, q_s=6.66 kW/m^2, Re_L=22, Pr=813. Computed ——Experimental •

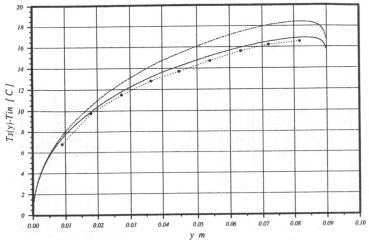

Figure 4. $T_s(y)-T_{in}$ vs. y.
Aqueous Carbopol: n=0.54, b=0.020 K^{-1}, q_s=8.57 kW/m^2, Re_L=36, Pr=582.
Computed, including temperature dependence of physical properties
─── Computed, omitting temperature dependence, and taking all
properties at T_{in} ─·─·─ Experimental •

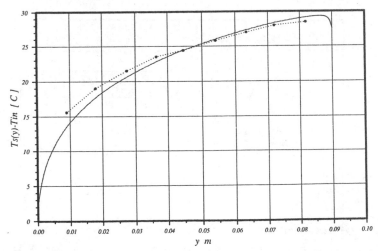

Figure 5. $T_s(y)-T_{in}$ vs. y.
Aqueous glycerol: Newtonian, b=0.043 K^{-1}, q_s=9.38 kW/m^2, Re_L=964, Pr=49.

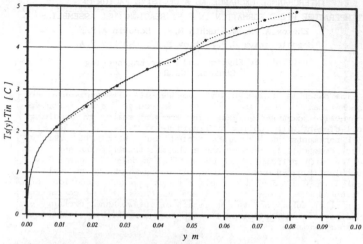

Figure 6. $T_s(y)$-T_{in} vs. y.
Aqueous glycerol: Newtonian, b=0.042 K^{-1}, q_s=1.99 kW/m^2, Re_L=1520, Pr=75.

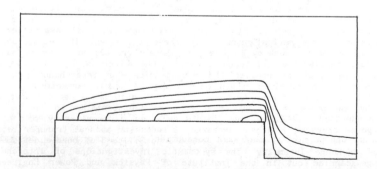

Figure 7. Temperature contour plots for case shown in Figure 5.
Contour interval 5 C°

Scale in direction perpendicular to plate surface expanded x 15.

INTERCHANNEL EXCHANGE AS A DECISIVE FACTOR OF
TEMPERATURE FIELD FORMATION IN FAST REACTOR FUEL ASSEMBLIES

Zhukov A.V., Matjukhin N.M., Sorokin A.P.,
Bogoslovskaya G.P., Titov P.A., Ushakov P.A.

Institute of Physics and Power Engineering
Obninsk, USSR

The paper is concerned with some results of experimental and theoretical investigations on molecular and turbulent convective exchange of mass, impulse and energy in bundle of fuel element with cooling by liquid metal.

Experiments have been carried out with the use of electromagnetic transducer and a thermal tracer technique.

The main principles of theoretical models for convective, turbulent interchannel impulse and energy exchange as well as heat transfer due to fuel element heat conduction are presented. Relations obtained are analyzed in the comparison with other author's data. An impact of interchannel exchange on a temperature distribution through fuel assembly is illustrated by some examples

Introduction

The requirement of high efficiency and at the same time high reliability of fast breeder reactors have rendered an urgent solution of a wide class of problems concerned with the thermophysical validation of reactor fuel assemblies operability under nominal and non-nominal conditions. The development of subchannel codes for fuel assembly thermohydraulics calculation [1] has required both an experimental investigation of interchannel exchange structure in fuel assemblies and development of theoretical models for these processes. The early studies surveyed e.g. in the work by Todreas [2], the authors [3] have revealed individual regularities of these processes, the subsequent hydrodynamic and thermal investigations enabled the dependencies for interchannel exchange coefficients to be obtained within a wide range of bundle parameters.

Diffusive interchannel heat transfer in liquid metal as compared with water or gas contains a significant exchange component due to fuel element heat conduction. Essential are points such as: 1)peculiarity of impulse exchange as compared with heat exchange, 2)intensity of heat transfer between the cells due to fuel element heat conduction, 3)impact of bundle deformation on the exchange intensity. The systematic investigations of interchannel exchange carried out in the Institute of Physics and Power Engineering (Obninsk) allowed these questions to be answered and a set of completing coefficients for the solution of macro transport equations in the channel-by-channel approximation [8] to be obtained.

The paper presents some results of these investigations.

I. Interchannel convective exchange.

Central area of fuel assembly. The experimental investigations carried out

on the basis of electromagnetic method of flow rate measuring through bundles cooled with liquid metal[5] have shown that the behavior of transverse transfer of mass through the pin-to-pin gaps follows the periodic sine (Fig.1), that defines the periodic variation of local interchannel mixing coefficient. **The local coefficient of interchannel mixing representing a mass flow rate running from i-th to j-th channel through unit of gap length referred to longitudinal coolant flow rate in a channel is described with a sufficient accuracy by one harmonic:**

$$\mu_{ij}^{m}(\varphi_{ij}) = (G_{ij} / G_{in}) \; \mu_{ij}^{m \; max} \; \sin \varphi_{ij} \, , \tag{1}$$

where

$\varphi_{ij} = 2\pi z / h - \alpha_{ij}$ (α_{ij}- is the phase of wire coiling entering the j-th cell from the i-th cell).

The amplitude $\mu_{ij}^{m \; max}$, $[\; m^{-1}]$, in (1) is calculated according to the formula from [7,8]:

$$\mu_{ij}^{m \; max} = (1.047/h) \; \phi^{m}(s/d) \; \psi \; (Re), \tag{2}$$

where $\phi^{m}(s/d) = 2.57(s/d) - 3.57e^{-119(s/d-1)^{2.12}} + 1,$ $\tag{3}$

$$(1.01 \le s/d \le 1.4 \; ; \qquad 2 \le h/d \le 50)$$

$$\psi(Re) = 1.087 - 0.754e^{-0.132 \; 10^{-3}Re}, \tag{4}$$

$$(2 \cdot 10^{3} \le Re \le 2 \cdot 10^{5})$$

The average mixing coefficient through gap length calculated with (1)-(4) is [7,8]:

$$\mu_{con}^{m} = \frac{1.047}{\pi \; h} \; \phi^{m}(s/d) \; \psi \; (Re). \tag{5}$$

The formula error is set at about ± 10%.

The experimental data [6] have shown that due to flow deceleration and downstream wave effect the low pressure takes place under the wire wrap. The interchannel convective mass transfer intensity derived from the pressure difference between the channels is in a good agreement with experimental data in the range $s/d < 1.25$.

The experimental studies of interchannel heat transfer [10] carried out with the thermal track technique have demonstrated that the local coefficient of interchannel heat transfer is defined by the relation(fig.2):

$$\mu_{conij}^{t} = \frac{Q_{ij}}{\rho \; \frac{W_{i} + W_{j}}{2} H^{*} \bar{\omega}} \propto \beta^{t} \mu_{conij}^{m}(\varphi_{ij} - \pi/3), \tag{6}$$

where $\beta^{t} \approx 0.7$; H^{*} - is enthalpy in a donor channel.

Having averaged the transverse heat flow in the interchannel gap over the initial angle of wire coiling orientation, the integral values of interchannel heat transfer coefficients are obtained:

$$\mu_{con}^{t} = \frac{\int_{0}^{2\pi} Q_{ij} \; d\alpha_{ij}}{\rho[(W_{i} + W_{j})/2](H_{i} - H_{j})\bar{\omega}}, \tag{7}$$

where $\quad \mu_{con}^{t} = (\mu_{conij}^{t})^{max} /\pi = \beta^{t}\mu_{con}^{m} = \dfrac{\beta^{t}}{3h}\phi^{m}(s/d)\ \psi\ (Re).$ (8)

Dependence (8) is in a fair agreement with the experimental data on convective interchannel heat transfer (effective value disregarding both heat conduction and turbulent diffusion) obtained by different authors.

The coefficient of non-equivalence between heat and mass transport (β) have been calculated based on the model of mole convective exchange with fuel elements spaced by wire wrap (as well as impulse and mass). Assuming the inter-cell substance exchange to be accomplished by moles (vortices), whose interaction with environment is described by the equation

$$\psi(M) = \overline{\psi}(M_0) + [\overline{\psi}(M) - \overline{\psi}(M_0)]\dfrac{1-\exp(-\alpha_\psi)}{\alpha_\psi}$$ (9)

where $\quad \alpha_\psi = (10.5 + \dfrac{2.7}{Pr_\psi})\dfrac{\nu}{R^2}\dfrac{r}{W},$

(R - is the mole radius, r - mole path distance, W - velocity of mole transverse motion in a flow), ψ - substance (either impulse or energy) and also assuming the mole defining the effect of inter-channel substance convective exchange in the cross-section z is formed in the cross-section z - h/6 (according to the experimental data: $\Delta\varphi= -\pi/3$), follows the wire coiling path and has a radius equal to the inter-fuel element gap width, we shall have:

$$\beta^{\psi} = 1- 0.4\ \dfrac{1-\exp(-\alpha_\psi)}{\alpha_\psi},$$ (10)

$$\beta^{W}= \dfrac{4,4}{2\pi Re}\ \gamma(h/d,s/d),\qquad \beta^{t}= \dfrac{1}{2\pi Re}(\dfrac{0.9}{Pr}+3.5)\ \gamma(h/d,s/d)),$$

where $\quad \gamma(h/d,s/d) = \dfrac{(h/d)^2[\dfrac{2\sqrt{3}}{\pi}(s/d)^2-1]}{(s/d-1)^2 s/d}$

Results obtained with the formula (10) is in a satisfactory agreement with the experimental data (fig.3)

2. Interchannel exchange due to fuel element heat conduction.

In the case of non-uniform coolant heat-up in channels surrounding a fuel element non-uniformity of temperature distribution over the fuel element perimeter is observed, which defines the heat transfer process between channels due to fuel element heat conduction.

Using the connection between temperature and heat flow distribution over a fuel element perimeter obtained by P. A. Ushakov we shall have a relationship [8]:

$$\mu_\lambda^{t} = \dfrac{16}{3}\dfrac{\varepsilon_1}{1+(2d_h/d)(\varepsilon_1/Nu)\,Pe}\dfrac{1}{d}$$ (11)

This formula shows that the coefficient of interchannel heat transfer due to fuel element heat conduction reduces as the Pe number and fuel element diameter grows, it increases as the parameter of equivalent fuel element heat conduction calculated by the first harmonic grows, and is insignificantly dependent on a relative pitch of fuel element arrangement.

Conclusions

In fuel subassembly thermohydraulics calculation it should be born in mind that the accurate calculation of temperature behavior under different conditions within fuel assemblies is specified by the correct consideration of exchange mechanisms, reliability of data for each of the interchannel exchange mechanisms. In case of significant deformation of fuel assembly geometry associated with the effects of radiative deformation of fuel element cladding or fuel assembly wrapper tubes, the effects of convective and turbulent exchange in close lattices and due to fuel element heat conduction are of a great importance(fig.4). Neglect of exchange leads to substantial errors in design values of fuel element temperature behavior in fuel assemblies.

Nomenclature

d - fuel element diameter,
d_h - hydraulic diameter,
G - coolant flow rate,
H - enthalpy,
M - coordinate of point,
Q - thermal flux,
s - lattice spacing,
t - temperature,
W - coolant velocity,
z - axial coordinate,
α_{ij} - initial phase of wire wrap,
α - coefficient,
β - transport non-equivalence coefficient
ε - effective thermal conductivity of fuel element,
ξ - hydraulic resistance coefficient,
ρ - density,
φ - wrapping phase,
ω - surface area,
ψ, γ, ϕ, δ - functions

Subscripts

clad - cladding
con - convective
i,j - number of cells
ij   - number of pin-to-pin gap
in   - central cell
m   - mass
λ - exchange due to thermal conduction
t   - thermal
tur - turbulent
──   - average value
ψ - relating to substance (impulse or energy)

References

1. Zhukov A.V.,et.al. 1981, Atomic energy, v.51, N5, p.307-311.
2. Rogers J.T., Todreas M.E. 1968, Heat Transfer in Rod Bundle. ASME.
3. Zhukov A.V.,et.al. 1973. Preprint IPPE-417, Obninsk (in Russian).
4. Ingesson L., Hebderg S. Heat Transfer between Subchannels in a Rod Bundles. 1970, Heat Transfer, Amsterdam, v.3, p.FC7.11,1-11.
5. Zhukov A.V.,et.al. 1976, Preprint IPPE-665, Obninsk(in Russian).
6. Patch L.,et.al. 1979, Report on the International Meeting on Reactor Heat Transfer, Karlsruhe.
7. Zhukov A.V.,et.al. 1985, Thermohydraulic Calculation of Liquid Metal Coolant Fast Reactor Fuel Assemblies. M. Energoatomizdat.
8. Zhukov A.V.,et.al. 1989, Interchannel Exchange in Fast Reactor Fuel Assemblies. M. Energoatomizdat.
9. Kazachkovsky O.D.,et.al. 1988, Atomic energy, v.65, N2, p.89-97.
10. Zhukov A.V.,et.al. 1975, Preprint IPPE-556, Obninsk (in Russian).
11. Baumann W.,et.al. 1971, Report on International Heat Transfer Seminar, Trogir, Yugoslavia.
12. Gryaznov V.M.,et.al. 1978, Thermophysics and Hydrodynamics of Core and Steam Generator for Fast Reactor, Prague, Edit.ChSKAE, v.1, p.182-209.

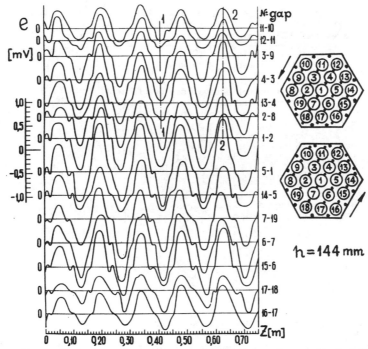

Fig.1. Variation of transverse mass flow-over through the height of gaps in diverse cells of fast reactor fuel assembly mock-up (the arrows indicate transverse flow directions in cross-sections 1-1, 2-2).

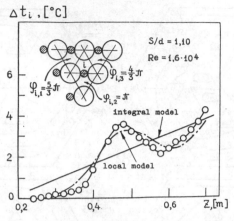

Fig. 2.

Variation of coolant heating over the cell length with central rod energy release in a mock-up fast reactor assembly: -O- - experimental data, -·-·- - local model of exchange, ——————— - integral model.

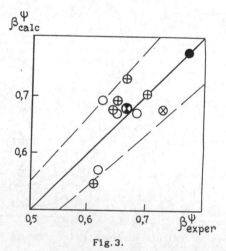

Fig. 3.

Comparison of calculated and experimental values of substance and mass heat transfer non-equivalence

⊗ - [10]; ● - [11]; O,⊕ - [5, 7, 8]; ⊗ - [12].

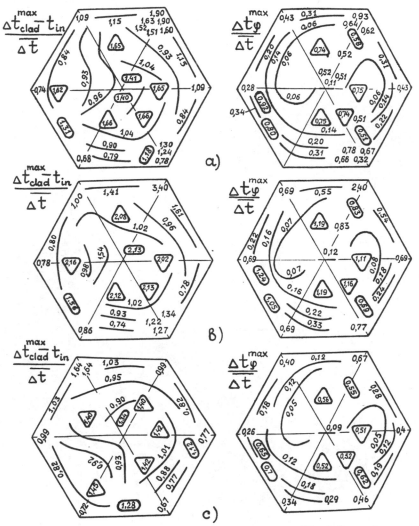

Fig. 4. Comparison of calculated results for "Phenix" reactor fuel deformed assembly taking into account centrifugal effect and heat transfer due to fuel element heat conduction (a), without taking into account these effects (b), taking into account effects at λ_{fuel}=30 W/m grad(c).

LOCAL THERMOHYDRAULICAL CHARACTERISTICS
OF LIQUID METAL HEAT EXCHANGER

Bogoslovskaya G.P., Zhukov A.V., Sviridenko E.Ya.,
Sorokin A.P., Ushakov P.A.

Institute of Physics and Power Engineering, Obninsk, USSR

This paper presents some results of experimental studies on hydrodynamics and heat transfer in liquid metal heat exchanger with transverse inlet and exit of flow. Electromagnetic measurement technique of axial and transverse velocity components in the space between pipes as well as procedure of local thermal modeling of pipe bundle have been applied. Subchannel thermal-hydraulic calculating approach in the context of the pipe heat exchanger has been developed.

Introduction

Engineering development of effective and reliable heat exchanger is one of the most important problem of nuclear power plant design for reactor of various type Especially this is concerned with coolant such as liquid metal (sodium or alloys).

It should be noted that electro-magnetic method is the poweful technique in studying of flow in heat exchanger model . Experience has shown that sometimes to ensure the modelling with liquid metal is easier than to make the cumbersome and expensive model for experiments with air or water.

The subchannel calculation approach used now in prediction of reactor core operating conditions is found to be effective in heat exchanger thermal-hydraulics calculation.

1. Hydrodynamic experiments

Electromagnetic technique for the measurement of local characteristics in complex liquid metal flow in pipe system has been developed in the Institute of Physics and Power Engineering, Obninsk. Detailed studies of the longitudinal-transverse flow in the model of BN intermediate heat exchanger have been fulfilled with the use of this technique [1].

Experimental model is a rectangular box, 7 sizing pipes ⌀ 19 × 0,5 mm and 16 displacers placed inside the box. Pipes and displacers simulate the triangular arrangement with relative pitch s/d=1,32. The box is inserted into cylinder shell, which is provided by the apparatus for side coolant supply and removal.

The electromagnetic local operation transducer, sensible element of which is consist of the permanent magnet, screen and shunt rings and two perpendicular each to another electrodes (principal scheme and graduation are described in [2,3]) is used for simultaneous measurements of longitudinal and transvere components of velocity.

Turning of the pipe is carried out discretely in a 10^0, the axial movement of pipe is carried out by power-drive. Na-K alloy is used as a working medium.

Velocity distribution is greatly nonunion at the inlet and exit of pipe bundle around the pipes 1 and 2. Maximal velocity of coolant takes place at $\varphi=180^0$ and $\varphi=0^0$ for entrance and exit sections, respectively.

Axial component of velocity behavior through model length is given in the Fig.1. We can see that it varies according to exponential law, approximately. At the distance from the upper edge of inlet window in 350-450 mm the short decreasing of transverse flow rate is observed, but the closer to exit window (distance more than 800 mm from lower pipe plate) the more transverse velocity value. In the entrance section velocity smoothly increase with model high and depth. For cross-section appropriated to inlet window edge the flow rate distribution becomes uniform . In further moving the flow near first row of pipes falls due to inertia. Equalization takes place at 800 mm length, but influence of inlet window is just felt here. It should be marked that Re changing by a factor of 10^2 from 10^4 to 10^6 does not cause essential deformation of velocity field. In this case we can say about quasi-self-similarity of flow with Re.

The main effect in inlet window size decreasing to 150 mm is the noticeable decreasing of nonuniformity of flow distribution with model depth and high: at the distance 150 mm from upper edge of inlet window flow is close to stabilized flow, that is why there is section were small changing of flow rates is observed with channel length (250 mm < 1 < 600 mm). In model with window size 150 mm flow rate nonuniformity becomes $v^{max}/v^{min}= 1.7$ instead of $v^{max}/ v^{min}= 3$ for window size 350 mm.

It should be noted that if window is wide ($H_{in}/H_{Hx} = 2$) inertial effects are small and flow is close to "potential" one with maximal velocity $w^{max}/w=1.19$ at the channel periphery. For s/d=1.3 the optimal ratio $(H_{in}/H_{Hx})_{op} = 1$. For other pitch, as it is shown in [4], p.211 we can adopt:

$$(H_{in}/H_{Hx})_{op} \approx 3.3 \ (s/d - 1) \tag{1}$$

where

H_{in} - inlet window high, H_{Hx} - heat exchanger radius.

2. Thermal experiments

Experimental studies of temperature behavior as well as heat transfer in Na-Na heat exchanger have been carried out on the model, which is a box, 19 sizing pipes ø 19 × 16,4 mm with triangular arrangement, 18 displacers attached to the box (relative pitch s/d = 1,315).

Heating coolant goes from the upper plenum through the window into space between pipes, that simulates a "big volume", then it goes downwards and exits through the window for bottom collector. Cooling liquid goes to lower section of model, then it runs inside pipes and flows out the circulation loop through the slots in pipes.

Every subchannel and every pipe are provided with tension and plugged capillaries with thermocouples enclosed. Thermocouples are placed also at inlet and exit flow of heating and cooling coolant. Thermocouples 1-67 are moved simultaneously by special device.

Method of experimental data processing is based on the local thermal modelling principle: heat transfer coefficients are calculated in pipe section where mainly longitudinal flow takes place and influence of model windows is insignificant.

Nusselt number for space between pipes (Nu_s) is taken to be

$$Nu_s = \frac{d_h/\beta}{(\lambda_f/k)_{exp} - d_1/(\beta \cdot Nu_p) - \lambda_f/\lambda_w \cdot d/2 \ln(d/d_1)} \tag{2}$$

where k - local heat transfer coefficient calculated with the use of maximal temperature difference ($t_s^{max} - t_p^{min}$), which has been measured directly in experiments (t_s^{max}- maximal temperature in cross-section of inter-pipe space, t_p^{min}- minimal temperature in cross-section). Nu_p number for round pipes is taken to be [6]:

$$Nu_p = 5 + 0.025 \, Pe^{0,8} \tag{3}$$

Coefficient β is designated as 0,65 on the basis of experiments and predictions of temperature behavior in liquid metal flow in round pipe and pipe bundle [6,7].

Experimental data are approximately the same for different pipe rows, this fact can be explained over the equalized temperature field of heating coolant flow front.

Average points with small spread (which does not exceed ± 10%) are disposed near dependence (fig.2):

$$Nu_S = 7.2 + 0.0235 \, Pe^{0.81} \tag{4}$$

$$(10 \leq Pe \leq 350)$$

This formula gives lower data that correlation for heating coolant flow from bottom to top (fig.2):

$$Nu_S = 9.36 + 0.0235 \, Pe^{0.81} \tag{5}$$

$$(20 \leq Pe \leq 1200)$$

However, distinction is not so large to say about directional flow effect.

It should be noted that large range of Peclet number covers regimes with natural convection.

Combination of (4) and (5) gives simple averaging formula:

$$Nu_S = 8.4 + 0.0235 \, Pe^{0.81} \tag{6}$$

$$(10 \leq Pe \leq 1200)$$

which can be used for heat transfer prediction within space between pipes in heating coolant flow above or below at Peclet number range $10 \leq Pe \leq 1200$, and which differs from (4) and (5) as not more than ± 10%.

The main result of experiments is the integral heat transfer coefficients calculated with volume average temperature in collectors, as it should be waited, are just lower than heat transfer coefficients for "infinite" pipe bundle defined by local thermal modelling method.

Conclusions

Experiments and predictions carried out have shown:
- geometry of heat exchanger inlet region has an effect on non-uniformity of flow rate distribution through the heat exchanger and in the end on its efficiency;
- heat transfer integral coefficients calculated with volume average temperature in collectors are just lower than heat transfer coefficients for

"infinite'pipe bundle defined with local thermal modelling method

$$Nu_s = 8.4 + 0.0235 \ Pe^{0.81} \ ;$$

This work has shown also the advantage of combine hydrodynamics study on the base of powerfull electromagnetic technique for experiments and subchannel methodics for predictions.

References (in Russian)

1. Ushakov P.A.,et.al. Local hydrodynamics characteristics of longitudinal-transverse flow in fast reactor heat exchanger model // Thermophysical Investigations-77. Moscow, VINITI, 1977, p.5-16.
2. Subbotin V.I., et.al. Hydrodunamics and heat transfer in nuclear power instalations (base of calculations). Moscow, Atomizdat, 1075, p.140.
3. Subbotin V.I., et.al. Experimental and theoretical statements of electro-magnetic method for liquid metal velocity measurement//Report on the 2-seminar "Heat Transfer and Hydrodynamics in Core and Steam Generators of LMFBR". Nove Mesto, Chekoslovakia,1973.
4. Mitenkov F.M., et. al. Nuclear Power Plant Heat Exchanger Design. Moscow, Energoatomizdat, 1988.
5. Subbotin V.I., et. al. Heat exchange in round tube liquid metal flow // Journal of Physical Engineeing, 1963, v.4, N 4, p.16-21.
6. Kirillov P.L., et.al. Heat transfer in round tibe to Na-k alloy and to Hg //Atomic Energy, 1959, v.6, N 4, p.382-390.
7. Ushakov P.A., et. al. Azimuthal nonuniformity of fuel pin temperature in turbulent flow of liquid metal//Thermophysics of High Temperatures, 1977, v.15, N 1, p.76-83.

Nomenclature

d   - external pipe diameter, (m);

d_1 - inner pipe diameter, (m);

d_h - hydraulic diameter of subchannel, (m);

s   - pipe pitch, (m);

Nu   - Nusselt number;

Pe·   - Peclet number;

λ   - heat conductivity ,(Wt/m·K)

Subscripts

exp   - experimental;

f   - fluid;

p   - pipe;

s   - space between pipes;

w   - wall

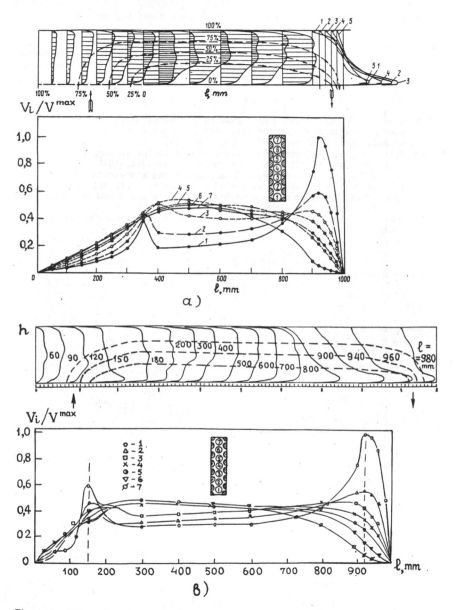

Fig. 1. Profiles of velocity axial components and flow lines for inlet window size 350 mm (a) and 150 mm (b)

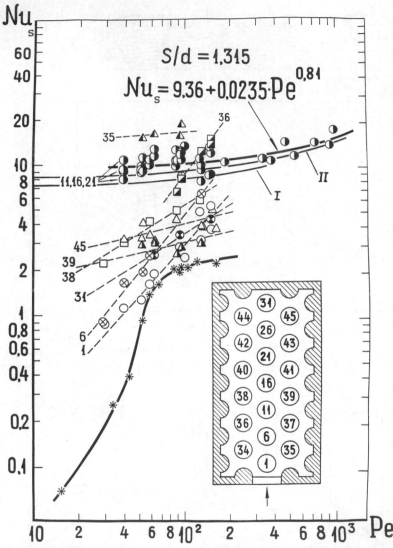

Fig. 2. Heat transfer in space between pipes *vs* Peclet number
-*- - calculation with mean-log temperature difference based on
the volume average temperature in collectors
— - according formula (5) for central pipe
I, II - according formulas (4) and (6), respectively
--- - peripheral pipes

TRANSITION TO CHAOS IN CAVITIES HEATED FROM THE SIDE

P. Le Quéré

LIMSI-CNRS, BP133, 91403-ORSAY CEDEX FRANCE

We numerically investigate several mechanisms responsible for onset of unsteadiness and subsequent transition to chaos in cavities heated from the side. The numerical algorithm uses Chebyshev spatial expansions and a second order finite difference time-stepping scheme. Although the bulk of the results are obtained for the 2D Boussinesq equations and a fluid of Prandtl number corresponding to air, some Prandtl number and non-Boussinesq effects are investigated. We also examine the stability of the 2D solutions with respect to 3D perturbations. Comparisons are presented with experimental results obtained elsewhere. We finally present some direct simulations of fully chaotic solutions for Rayleigh number values up to 10^{10}.

INTRODUCTION

Over the last ten years, due to improvements in algorithms and in computing resources, substantial progress have been achieved in the computation of natural convection flows in differentially heated cavities. Let us recall that, the largest Rayleigh value that was computed for the numerical contest of 1980 was 10^6 [1] while it is now common practice to compute flows corresponding to Rayleigh values several orders of magnitude larger.

Over this range of Ra values, solutions have lost one important property, steadiness. For the highest values, fully chaotic solutions can even today be simulated with confidence, which correspond to real configurations which are fully turbulent. On the other hand, prediction of turbulent heat transfer is mainly achieved through resolution of Reynolds averaged equations closed with turbulence models, amongst which k-ϵ models with or without low Reynolds number correction are undoubtedly most popular.

Validation of these models requires comparisons with experimental data which can be provided both from laboratory experiments and from direct simulations of turbulent flows. In fact owing to the difficulty of imposing specified boundary conditions in laboratory experiments and the resulting uncertainty in these boundary conditions, direct simulations will certainly provide as reliable data as those from laboratory experiments. It is also clear that better modelling will be achieved if one knows what are the instability mechanisms responsible for turning a laminar flow into a turbulent one, what is the location of transition, what are the effects of stratification,...

The aim of this paper is to review the basic instability mechanisms responsible for the onset of unsteadiness in air-filled differentially heated cavities and some typical routes to chaos. We will also investigate some Prandtl number effects and non-Boussinesq effects. Since these results were obtained from ab initio assumption of two-dimensional flow, it is also therefore important to examine their stability with respect to three-dimensional perturbations. We will lastly present some recent results of fully chaotic flows for Rayleigh values up to 10^{10}.

PROBLEM DESCRIPTION and ALGORITHM

The basic configuration is a cavity of height H and width W filled with a newtonian viscous fluid of thermal diffusivity α and kinematic viscosity ν. It is submitted to a temperature difference ΔT at the vertical walls while the top and bottom walls are adiabatic. We generally assume that the flow is modelled with the two-dimensional unsteady Boussinesq equations (using these equations does not prevent from obtaining solutions which break centro-symmetry as will be shown later). These equations are made dimensionless with the following reference quantities: H for length, $\frac{\alpha}{H}Ra^{0.5}$ for velocity and $\frac{H^2}{\alpha}Ra^{-0.5}$ for time, where Ra is the Rayleigh number $Ra = \frac{g\beta\Delta T H^3}{\nu\alpha}$ and Pr the Prandtl number $Pr = \frac{\nu}{\alpha}$. These equations are integrated in primitive variable formulation, by a pseudo-spectral algorithm combining spatial expansion in series of Chebyshev polynomials with a semi-implicit second order finite-difference time-stepping scheme on the computational domain $(x, z) \in [0, 1/A] \times [0, 1]$, where A is the aspect ratio $A(= H/W)$. The incompressibility condition is treated by the use of an influence matrix technique. The details of the algorithm are given in [2].

ONSET of UNSTEADINESS

Air-filled cavities
Fig.1 summarizes the stability characteristics of solutions of the 2D Boussinesq equations for a fluid of Prandtl number 0.71 in a cavity of aspect ratio ranging from 1 to 10 for the two usual types of boundary conditions adiabatic and perfectly conducting top and bottom walls. It can be seen that, over the range of aspect ratios, perfectly conducting boundary conditions consistently have a destabilizing influence upon the adiabatic case. One however notices that for large enough aspect ratio, both curves follow the same trend while for small values they diverge and the difference between the critical values reaches 2 orders of magnitude for the square cavity which indicates that the instability mechanisms are very different.

It has indeed been shown that, depending on the aspect ratio and on the boundary conditions, this transition is due to three different instability mechanisms. For large values of the aspect ratio $(A > \simeq 3)$, the instability is an instability of the vertical boundary layers independently of the thermal boundary conditions on the top and bottom walls. For small values of the aspect ratio and conducting walls, the instability is a thermal instability which takes place in the bottom left corner [14,16,5] . The instability results from the horizontal flow driven by a horizontal pressure gradient being more and more heated from below as it approaches the hot wall. In the case of adiabatic walls, it is on the contrary the detached region close to the ceiling which becomes unstable first [3,4,5,6,7].

We have also shown that the transition to unsteadiness possesses the characteristic features of a Hopf bifurcation which can be used to determine with accuracy the critical value if needed. It is one characteristic of the Hopf bifurcations that the period of oscillations varies continuously in the vicinity of the bifurcation point. There thus exists a time scaling for which the period of oscillation, measured in that time unit, remains constant in the vicinity of the bifurcation point and determination of this scaling can give some hints of the associated instability mechanism.

We have shown that for the boundary layer instability, this time unit scales with the characteristic time $\frac{H^2}{\alpha}Ra^{-0.5}$ built with the characteristic boundary layer velocity $\frac{\alpha}{H}Ra^{0.5}$ and the height H of the cavity. It is also the characteristic time associated with the internal wave oscillation frequency. This is also the case for the thermal instability mechanism. This means that, in laboratory experiments where the Ra value is increased by increasing the temperature difference, the physical oscillation period should behave like $\Delta T^{-1/2}$.

On the other hand for small aspect ratio and adiabatic walls, the time scaling is $\frac{H^2}{\alpha}Ra^{-1/4}$ which indicates that the nature of the mechanism responsible for unsteadiness is very different. In this case the physical oscillation period should behave like $(\frac{H^5}{\Delta T})^{1/4}$. This result needs confirmation, but due to the difficulty of imposing adiabatic conditions in air filled cavities, this instability mechanism has not yet been observed in laboratory experiments.

Prandtl number effect

The other most common fluid, water, has a Prandtl number 10 times larger than that of air, at sufficiently low temperature. It is thus of some interest to investigate the stability characteristics of fluids with that Prandtl number value, which furthermore has been shown to well approximate infinite Prandtl number fluids. Calculations were carried out only in a cavity of aspect ratio 10 for both types of thermal boundary condition of the horizontal walls.

It was a great surprise to discover that the critical Ra value for adiabatic walls was more than 30 times larger than that for perfectly conducting walls, in sharp contrast with the case of air. The large value of Prandtl results in thin thermal boundary layers and the perfectly conducting boundary condition promotes the developement of a much earlier triggering of the thermal instability along the top and bottom walls.

The critical value in the case of adiabatic walls was found to be comprised between 8×10^9 and 9×10^9. Since the corresponding solution exhibits very thin boundary layers which comprise a large stratified core region (characterized by a dimensionless stratification of 0.55 in units of $\frac{\Delta T}{H}$, the solution can be considered to be in the asymptotic flow regime and the value of the critical Rayleigh number is probably quite independent of the aspect ratio. This is in agreement with the result obtained by Henkes who, with a finite volume algorithm, located the critical Rayleigh number in the square cavity with adiabatic walls between 4×10^9 and 5×10^9 [7,8].

Non-Boussinesq effect

Since the Boussinesq approximation is only valid in the limit of infinitely small temperature difference, it is of interest to understand how the conclusions drawn above are modified when the temperature difference becomes sufficiently large than one has to take into account variations of physical properties with temperature.

The situations is different for gases or liquids since for the latter compressibility is negligible and one only has to take into account variations of physical properties with temperature while keeping a linear or quadratic relationship between density and temperature.

For gases, allowance for density variation requires the specification of an equation of state which is usually taken as the perfect gas law. The numerical treatment of this problem is not straightforward since abrupt replacement of the pressure by the product $\rho\Theta$ would result in a system of equations capable of describing acoustic waves and in over-restrictive stability criteria. It is thus desirable to filter out these acoustic waves which can be done by a small Mach number expansion which results in a splitting of the pressure in two terms, one mean value in space and one small correction responsible for the satisfaction of the continuity equation. We have developed a Chebyshev collocation algorithm for integrating the corresponding governing equations, but due to lack of space, we do not give any further detail on the numerical scheme which is presented elsewhere [9].

We have used that algorithm to investigate the influence of the temperature difference on the loss of stability of the solution in a cavity of aspect ratio 8 with adiabatic top and bottom walls. The results are shown on fig. 2 in the form of a stability diagram in the plane (Ra, ϵ) where ϵ is the temperature difference $\frac{\Delta T}{2T}$. For these computations it was assumed that the viscosity follows the Sutherland law. Increasing the temperature difference to .6 thus significantly affects the critical Rayleigh value.

They furthermore have a large effect on the structure of the fluctuating temperature field as can be seen in fig. 3 which shows that non-Boussinesq effects result in lost of centro-symmetry. While in the Boussinesq case, the temperature fluctuations are symmetrically amplified and damped in the cavity, one even reaches at largest ϵ investigated a totally dis-symetrical situation in which fluctuations are constantly amplified in the upward boundary layer and constantly damped in the downward boundary layer.

ONE ROUTE TO CHAOS

In this section we will consider a cavity of aspect ratio 4 and investigate in some detail the transition to chaos. That particular aspect ratio was selected because it is the smaller value of the aspect ratio for which the boundary conditions imposed on the top and bottom walls has a minor influence on the instability mechanism.

The diagram shown on fig. 4 summarizes extensive computations that were made. It presents the dimensionless oscillation period (in units of $\frac{H^2}{\alpha} Ra^{-0.5}$) of stable unsteady solutions as a function of the Rayleigh number. According to figure 1, for this aspect ratio and adiabatic walls unsteadiness appears for a critical value of 1.61×10^6. In the vicinity of the critical number, the oscillations have a dimensionless period of 2.46. For increasing values of the Rayleigh number, the solutions remain periodic with increasing oscillation amplitudes, until for a Rayleigh number of 2×10^6, the period of the oscillation drops suddenly to 2.22. One can then increase or lower the Rayleigh value and the solution then resumes periodic behavior with the same characteristic period. For sufficiently large Rayleigh value ($\simeq 2.8 \times 10^6$) the oscillation period drops to a third value (1.82), and the diagram shows that in total 6 branches of solutions were found, each of them characterized by different dimensionless oscillation periods.

It is also apparent from the figure that for increasing Rayleigh number, one is more likely to be on a branch of smaller dimensionless period. The explanation is that, the size of the structures scales with the boundary layer thickness which decreases with increasing Ra. Since the phase velocity of the structure is directly linked to the fluid velocity, which remains constant in the convective time unit, one should therefore find branches of increasing number of structures with increasing Rayleigh number. This is exactly what happens, as can be seen from fig. 5, which presents typical instantaneous fluctuating temperature fields belonging to different branches of solutions. It is clear that each branch possesses one structure more than the branch corresponding to the immediately larger value of the dimensionless period. The number of structures for the branch that appears at criticality is 7 and the number of structures of the branch on which the first chaotic solution is found is 12 as indicated on the diagram. It is worth remarking that an odd number of circulating structures is still compatible with the centro-symmetric property while solutions with an even number of structures inevitably break that symmetry.

For increasing Ra values, one also finds solutions which present a quasi-periodic time behavior, and eventually for a Ra value of 3.6×10^6 the first appearance of chaotic solution. We have investigated in some detail the transition to chaos and, even though some uncertainty remains, it is very likely that chaos appears through intermittency.

Most of these findings have been qualitatively and to a very good extent quantitatively confirmed in laboratory experiments conducted in the group of F. Penot in Poitiers [12,13]. We will later come back to a major discrepancy between numerical predictions and the experimental observations.

A SECOND ROUTE TO CHAOS

Let us know consider a square cavity with conducting walls which has been investigated experimentally by Briggs and Jones [14,15] and numerically by Winters [16].

In this case, unsteadiness sets in in the solution of the 2D Boussinesq equation at a Rayleigh value close to 2.1×10^6, whereas Briggs and Jones report on appearance of unsteadiness around a value of 3×10^6 in a cubical enclosure.

We have made extensive computations of unsteady solutions for Ra values up to 2×10^7, that is to say one order of magnitude larger than the onset of unsteadiness. Computations were not pursued for larger Ra values because the assumption of two-dimensionality probably looses any physical meaning.

The conclusions we have obtained show that, like in the previous case, unsteady solutions are also organized in several branches which can be characterized by different dimensionless frequencies in units of $(\frac{g\beta\Delta T}{H})^{1/2}$. This result was found experimentally by Briggs and Jones who report on 3 different solutions branches. We could find numerically 4 solutions branches which are shown on fig. 6. The characteristic frequencies for these solutions are in very good quantitative agreement with the values found experimentally. They are also in good agreement with the imaginary parts of the eigenvalues of the Jacobian of the linearized evolution operator computed by Winters.

It is interesting to note that, in this configuration, for values of the Rayleigh number one order of magnitude after the onset of instability, the solution is still periodic in time in constrast to the aspect ratio 4 cavity where the first chaotic solution was reached for a Rayleigh value slightly larger than twice the critical value.

Like in the previous case, solutions are characterized by increasing number of structures, 5 for the solution which appears at criticality up to 8 for the branch with highest dimensionless frequency.

It is also interesting to note that on the solutions with an odd number of structures period doubling solutions can be found, whereas on even numbered branches quasi-periodicity is found. This is linked to symmetry properties of the solution. These results will be presented in more detail elsewhere [10].

THREE DIMENSIONAL INSTABILITY

As we have already said, the laboratory experiments performed in a cavity of aspect ratio 4 have confirmed to a large extend the numerical results obtained under the assumption that the flow is adequately modelled with the 2D Boussinesq equations. A major discrepancy though is that the laboratory experiments show the existence of a very low frequency which can appear after or before the high frequency travelling wave instability depending on the horizontal aspect ratio of the cavity.

Since all the other phenomena are well predicted by the 2D Boussinesq equations, the fact that this phenomena is not reproduced by the numerical model means that it is not described by the model which we consider and the discrepancy has therefore to be ascribed to 3D or non-Boussinesq effects. Since non-Boussinesq effects have not brought any new type of time-dependency, the origin of this low frequency modulation probably lies in 3D effects.

Rather than performing directly fully 3D calculations in cavities with a specified horizontal aspect ratio (calculations that will eventually have to be performed) it seems interesting to investigate the stability of 2D solutions with respect to 3D perturbations and thus assume periodicity in the direction orthogonal to the base flow solution. All the flow variables are then expanded in Fourier series in the y-direction and the 3D algorithm merely reduces in as many 2D problems as Fourier modes.

The methodology simply consists of integrating the 3D equations starting from an initial condition which is built by juxtaposing plane by plane the 2D solution, on which can be superimposed some noise. One has to specify the length of the fundamental mode in the y-direction and this methodology can then give access to the stability of the solution with respect to 3D modes of given wavelength. One can also cancel the non-linear terms and if an instability is found, determine the growth rates of modes of different wavelength, which gives an indication of the typical size of 3D structures one should find in an actual laboratory experiment.

This stability investigation was carried out for the 2D solution found in a cavity of aspect ratio 4 with adiabatic top and bottom walls for a Rayleigh value of 1.5×10^6, that is slightly below the onset of unsteadiness in the 2D equations (which happens as already said for a Ra value of 1.61×10^6). The 2D solution was found unstable to 3D perturbations of typical wavelength equal to the cavity width, but, still in discrepancy with the experimental results, the 3D solution resulting from that instability was found steady.

Visualizations of the regions of maximum amplitude of the transverse velocity component show that the instability mechanism for this 3D instability lies in the end regions. The recirculating structures in this region are weak weak enough so that they cannot sustain 3D perturbations. It is also very likely that the unsteadiness in the experiment is due to non-linear interactions between several modes and to end effects in the horizontal direction.

We have also investigated the stability of some 2D unsteady solutions with the same methodology and found that they also undergo a 3D instability. The resulting 3D solution is still periodic with the same frequency as the 2D solution which shows that the 3D instability is sufficiently weak not to perturb the triggering of the 2D insability. This is also because the two instability mechanisms concerne different regions of the flow, the boundary layers and the end regions respectively.

The (partial) conclusions up to now is that, even though the 2D solution is indeed unstable to a 3D instability, this 3D instability seems sufficiently weak not to perturb the 2D route to chaos established previously. Further investigations could invalidate this optimistic conclusion.

INVESTIGATION OF FULLY CHAOTIC FLOWS

By increasing the Ra value far above the range of values corresponding to the onset of unsteadiness, one can obtain fully chaotic solutions which mimic turbulent flows. In this process one forgets about detailed investigation of transition and one focuses on investigating directly chaotic solutions which exhibit temporal as well as spatial randomness. Let us here acknowledge the pioneering work of Fromm [17] more that 20 years ago and the more recent simulations of Paolucci [18].

We have performed two such series of computation, one in the cavity of aspect ratio 4, the other in a square cavity for values of the Rayleigh number based on cavity height of 10^{10}. The computations in a cavity of aspect ratio 4 are reported in more detail in a companion paper in this conference [19] while those in a square cavity are described in [20,21].

For each Rayleigh number value considered, statistically steady-state solutions were obtained by integrating long enough in time, generally using a solution corresponding to a smaller Rayleigh number as the initial condition. For these computations, it is thus obviously essential to integrate the equations long enough in time so that one feels confident that a true statistically steady state has indeed been reached.

It has been shown by Patterson and Imberger [24] than, for the convection dominated laminar regime, the time to achieve steady state is of order $0.1Ra^{1/2}$ in the time unit considered here. This very long time scale is the time needed to damp through diffusion effects the internal gravity waves which can be sustained by the stratified core region, and which are generated during the transients following the sudden change in Ra value.

It is clear that if the time needed to get to an statistically steady state were of the the same order as this time scale, because of the stability criteria, computations would not be feasible. Paolucci for instance, started from rest and stopped his computation at a dimensionless time of 100 approximately, which is 1/100th of this time scale [18]. However, because the solutions at these Ra values are very chaotic and show very large amplitude fluctuations which enhance long scale mixing, it is clear that one should expect a substantial decrease of the time needed to achieve asymptotic behavior.

The simulations that were performed in cavities of aspect ratio 1 and 4 for Ra values of 10^{10} were carried out over dimensionless times larger than 200 in units of convective time $\frac{H^2}{\alpha}Ra^{-0.5}$, which approximately corresponds to 0.002 time units of thermal diffusion time over cavity height.

These simulations give access to large quantities of information, firstly on the dynamics of the solutions but also and perhaps more importantly on time-averaged quantities such as the structure of the time-averaged solution, the Reynolds stress tensor, turbulent kinetic energy and temperature variance, viscous and thermal disspation, etc.

We will briefly summarize here some of the salient features of the solutions corresponding to Ra values of 10^{10}.

Time-averaged Flow Structure

The time-averaged solution in the square cavity for a Ra value of 10^{10} is displayed in fig. 7. Although this solution shows the expected general flow organization, thin boundary layers and a stratified core region, it is clear that this solution displays some new features which are also found in the cavity of aspect ratio 4 at the same Ra value: the recirculating structure which, at lower Ra was located in the top left corner, has now moved upstream in the boundary layers and it is located in the mean around $z = 0.7$. This corresponds to the location where the waves travelling downstream the boundary layer have grown to a point where they totally disrupt the boundary layers and large eddies are ejected from the boundary layer into the core.

The effect of that recirculating structure is that hot fluid is thrown into the cavity core at much lower altitude that for lower Ra value and the same is true for the downward boundary layer (cold fluid is thrown at higher height than at lower Rayleigh number). This has an important effect of the temperature distribution in the core region which now shows a region of increased stratification around mi-height (the average stratification is approximately $\simeq 1.15$ in units of $\frac{\Delta T}{H}$ for $0.4 \leq z \leq 0.6$). As a result, the top and bottom parts of the core region become much more isothermal and also because of enhanced mixing in these regions due to the large eddies which are expelled from the boundary layers.

The finding that the dimensionless stratification around mid-height remains larger than 1 for values of Ra up to 10^{10} disagrees with what was observed by Paolucci [18]. This increased stratification is however in good agreement with the fact that some solutions obtained with k-ϵ models do also predict such an increase of the stratification for Ra values in the range 10^9–10^{10} [22,7,23] before it eventually decreases to much smaller values [7]. This is also particularly intriguing since, one knows from laboratory experiments and from everyday's experience (convection in a room), that the stratification will eventually drop to a very small value around .3 or .4 as also found in solutions obtained with turbulence models [7]. One can then anticipate that the temperature field will undergo a **stratification crisis** for Ra values larger than 10^{10} with a corresponding dramatic change in the flow structure. How far above 10^{10}?

Instantaneous Flows Features

Time traces of temperature at three locations situated in the upward boundary layer, one at mid-height, the next two at z=0.6 and 0.72 respectively are shown in fig. 8 together with the corresponding density power spectra. At the intermediate position, the temperature history clearly exhibits high frequency oscillations modulated by a somewhat smaller frequency. The high frequency corresponds to the travelling waves in the boundary layers and the peak frequency obtained from a density power spectrum of the signal is equal to 2.0. At the position further downstream, the waves have grown and begin to interact, resulting in fluctuations of much larger amplitude which are also somewhat more irregular in time. On the contrary, the time trace at mid-height shows a long period large amplitude dependancy on top of which higher frequency modulations appear. This long period phenomenon is due to the internal waves in the core region which are permanently excited by the large eddies which are ejected from the boundary layers.

The present simulations suggest that the large amplitude fluctuations due to the sidewall boundary layer instability are capable of permanently exciting the internal waves which oscillate at their characteristic frequency, the Brunt-Väisälä frequency. In time unit $\frac{H^2}{\alpha}Ra^{-0.5}$, the dimensionless Brunt-Väisälä frequency reads: $f_{BV} = \frac{(CPr)^{0.5}}{2\pi}$ where C is the dimensionless stratification measured in units of $\frac{\Delta T}{H}$. Taking $C = 1.15$ gives $f_{BV} = 0.144$. This value compares well with the value of 0.122 of the low frequency found in the temperature trace at mid-height and is also present in the time oscillations of of the mean Nusselt number Nu_c through the vertical mid-plane, which is known to characterize well the internal waves. The agreement between the two values is good and supports the assertion that internal waves do get permanently excited by the waves propagating in the sidewall boundary layers, or perhaps more truely by the large structures which are ejected from the boundary layers. (This assertion is also supported by video-animations made from the results which show the phenomenon very clearly.)

Turbulent Quantities

These direct simulations can give access to average quantities which characterize turbulent motion. We only present in fig. 9 the turbulence kinetic energy of the mean motion, the turbulent kinetic energy, the temperature variance, the viscous dissipation rate and the thermal dissipation rate. These plots show that, as expected, turbulent fluctuations are only significant in the downstream part of the boundary layer. They also show that the temperature fluctuations are well spatially correlated to the thermal dissipation rate, but this does not seem to hold so well for the turbulent energy and its viscous dissipation.

Heat Transfer

The time and space average Nusselt numbers for $Ra = 10^{10}$ are equal to 100 and 102 for aspect ratio of 1 and 4, respectively. All the values still very well follow a laminar $Nu \propto Ra^{1/4}$ relationship which shows that the large amplitude fluctuations have still a minor influence on the mean heat transfer. Fig. 10 presents local Nusselt number along the heated wall, which confirm that the boundary layer is laminar up to mid-height and undergoes then an abrupt transition with very large local fluctuations of the heat transfer coefficient.

Conclusions

The main conclusions drawn from these direct numerical simulations of 2-D turbulent convection in cavities of aspect ratio 1 and 4 for Ra values of 10^{10} are the following:

(a) the vertical boundary layers remain laminar in the first half of the cavity, undergo rapid transition at mid-height. A large recirculating structure sets in the mean around $z \simeq 0.7$, throwing hot fluid into the cavity core

(b) the mean stratification in the core remains larger than 1 over the range of Ra values investigated; internal waves in the core region get permanently excited by the large eddies and oscillate at the Brunt-Väisälä frequency;

(c) turbulent fluctuations are only important in the downstream region of the boundary layers; the turbulent kinetic energy k and dissipation rate ϵ do not seem very well spatially correlated;

(d) over the range of Ra values investigated, the mean Nusselt number still follows a $Ra^{1/4}$ dependency, even though large local fluctuations are observed in the turbulent part of the boundary layers.

CONCLUSIONS

We have shown that direct simulations of the unsteady governing equations can give some insight into the basic mechanisms responsible for turning natural convection flows from laminar to turbulent. We have also shown that direct simulations of 2D chaotic flows can also be performed with a high degree of accuracy. They can give access to some turbulent quantities for the purpose of comparison with solutions of the Reynolds averaged equations. It is however clear, than 2D solutions, no matter how chaotic they might be, cannot simulate actual 3D turbulent flows, and full 3D direct simulations will have to be performed. This is the subject of future research.

Acknowledgements:

The computations were performed on the Cray-2 at CCVR and on the VP200 at CIRCE. This work is supported by DRET under contract 91-150.

References

[1] De Vahl Davis G. and Jones I.P., 1983, Int. J. Num. Meth. Fluids, vol. 3, 227-248

[2] Le Quéré P. and Alziary de Roquefort T., Computation of natural convection in two-dimensional cavities with Chebyshev polynomials, J. Comp. Phys., 1985, 57, 210-228

[3] Haldenwang P., 1984, Thèse d'Etat, Univ. de Provence

[4] Chenoweth D.R. and S. Paolucci S., Natural convection in an enclosed vertical air layer with large horizontal temperature differences, J. Fluid Mech., 1986, 169, 173-210

[5] Le Quéré P., 1987, Thèse d'Etat, Univ. de Poitiers

[6] Paolucci S.and Chenoweth D.R., Transition to chaos in a differentially heated vertical cavity, J. Fluid Mech., 1989, 201, 379-410

[7] Henkes R.A.M.W., 1990, PhD Thesis, Univ. of Delft

[8] Henkes R.A.W.M. and Hoogendorn C.J., Bifurcation to unsteady convection for air and water in a cavity heated from the side, 9th Int. Heat Transf. Conf., Jerusalem, August 1990, paper 2-NC-17, Hemisphere Pub., 257-262

[9] Le Quéré P., Masson R. and Perrot P., A Chebyshev collocation algorithm for non-Boussinesq convection, J. Comp. Phys, in press

[10] Le Quéré P., Unsteady convection in a square cavity with conducting horizontal walls, in preparation

[11] Le Quéré P. and Penot F., 1987, ASME HTD, vol. 94, 75-82

[12] Penot F., N'Dame A. and Le Quéré P., 1990, 9th IHTC, vol. 2, 417-422

[13] N'Dame A., Thesis, June 1992, University of Poitiers

[14] Briggs D.G. and Jones D.N., Two-dimensional periodic natural convection in a rectangular enclosure of aspect of aspect ratio one, J. Heat Transf., 1985, 107, 850-854

[15] Briggs D.G. and Jones D.N., Periodic two-dimensional cavity flow: effect of linear horizontal thermal boundary conditions, J. Heat Transf., 1989, 111, 86-91

[16] Winters K.H., Hopf Bifurcation in the double-glazing problem with conducting boundaries, J. Heat Transf., 1988, 109, 894-898

[17] Fromm J.E., 1971, IBM J. Research and Development, 186-196

[18] Paolucci S., 1990, J. Fluid Mech. 215, 229-262

[19] S. Xin and Le Quéré P., 1st Eur. Heat Trans. Conf., Birmingham, UK, 1992

[20] Behnia M. and Le Quéré P., 4th Int. Fluid Mech. Conf., Alexandria, April 1992

[21] Le Quéré P., A modified Chebyshev collocation algorithm for the simulation of turbulent convection at high Rayleigh number, ICOSAHOM 92, June 1992, Montpellier, France

[22] Nobile E., Sousa A.C.M., Barozzi G.S., 1990, 9th IHTC, vol. 2, 543-548

[23] Le Breton P., Thèse, Univ. de Bordeaux, 1991

[24] Patterson J. and Imberger J., 1980, J. Fluid Mech., 100, 65-86

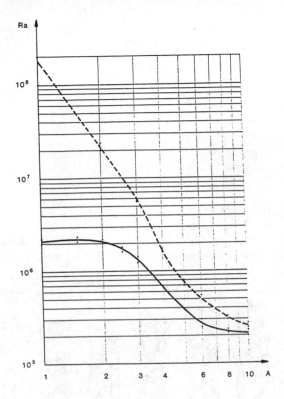

Fig. 1 Stability diagram in (A, Ra) plane: (Ra is based on cavity width): solid line, perfectly conducting top and bottom walls; dashed line, adiabatic top and bottom walls

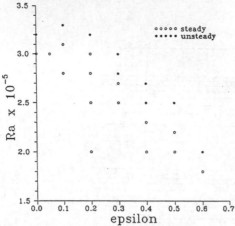

Fig. 2 Stability diagram in (ϵ, Ra) plane in a cavity of aspect ratio 8 with adiabatic top and bottom walls (Ra is based on cavity width): ($\bullet$) unsteady solutions ; ($\circ$) steady solutions

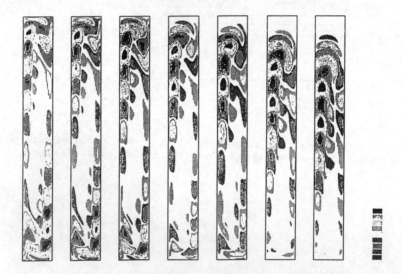

Fig. 3 Instantaneous fluctuating temperature fields. From left to right: $Ra = 3.2 \times 10^5$, $\epsilon = 0$; $Ra = 3.3 \times 10^5$, $\epsilon = 0.1$; $Ra = 3.2 \times 10^5$, $\epsilon = 0.2$; $Ra = 2.8 \times 10^5$, $\epsilon = 0.3$; $Ra = 2.5 \times 10^5$, $\epsilon = 0.4$; $Ra = 2.5 \times 10^5$, $\epsilon = 0.5$; $Ra = 2.0 \times 10^5$, $\epsilon = 0.6$. (Ra is based on cavity width) Isovalue ranges correspond to the greyscale shown; smallest positive and negative values around zero are in white; the other isovalue ranges are symmetrical with respect to zero

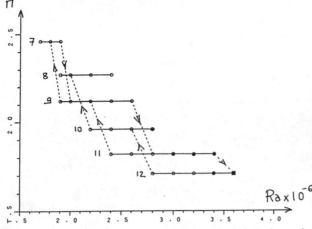

Fig. 4 Diagram   showing stable unsteady solutions which have been obtained in a cavity of aspect ratio 4 with top and bottom adiabatic walls; Solutions are classified according to their dimensionless period in units of $\frac{H^2}{\kappa} Ra_H^{-1/2}$: (o) periodic solutions (•) quasi-periodic solutions; (■) Chaotic solutions. The number of structures on each branch is indicated on its left.

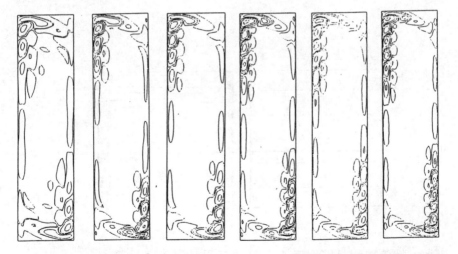

Fig. 5 Instantaneous fluctuating temperature fields for one solution on each solution branch. From left to right; $Ra = 1.8 \times 10^6$, $\Pi = 2.46$; $Ra = 2.0 \times 10^6$,$\Pi = 2.27$; $Ra = 2.0 \times 10^6$,$\Pi = 2.12$; $Ra = 2.6 \times 10^6$,$\Pi = 1.96$; $Ra = 2.8 \times 10^6$,$\Pi = 1.82$; $Ra = 3.2 \times 10^6$,$\Pi = 1.71$. Note the increasing number of structures.

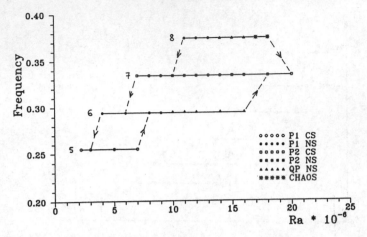

Fig. 6 Diagramm showing stable unsteady solutions which have been obtained in a square cavity with perfectly conducting top and bottom walls; Solutions are classified according to their dimensionless frequency in units of $\left(\frac{g\beta\Delta T}{H}\right)^{1/2}$.

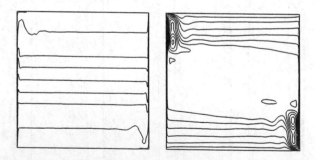

Fig. 7 Time averaged temperature and stream function fields: $Ra = 10^{10}$

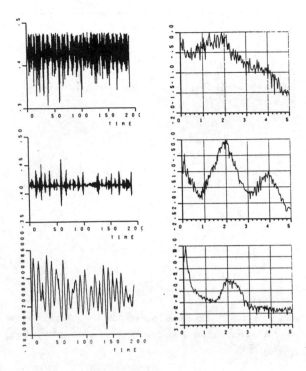

Fig. 8 Temperature traces (left) and corresponding power spectra (right) for 3 monitoring points located respectively at $x = 0.0013$, $z = 0.75$ (top); $z = 0.6$ (middle) ; $z = 0.5$ (bottom)

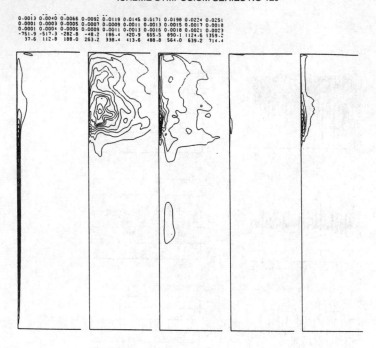

```
0.0013  0.0040  0.0066  0.0092  0.0119  0.0145  0.0171  0.0198  0.0224  0.0251
0.0001  0.0003  0.0005  0.0007  0.0009  0.0011  0.0013  0.0015  0.0017  0.0018
0.0001  0.0004  0.0006  0.0009  0.0011  0.0013  0.0016  0.0018  0.0021  0.0023
-751.9  -517.3  -282.8  -48.2  186.4  420.9  655.5  890.1  1124.6  1359.2
37.6  112.8  188.0  263.2  338.4  413.6  488.8  564.0  639.2  714.4
```

Fig. 9 Turbulent fields. $Ra = 10^{10}$. from left to right: kinetic energy of mean field, turbulent energy $\overline{u_i'^2}$, temperature variance $\overline{\theta'^2}$, viscous dissipation $\overline{\frac{\partial u_i'}{\partial x_j}\frac{\partial u_i'}{\partial x_j}}$ and thermal dissipation $\overline{\frac{\partial \theta'}{\partial x_j}\frac{\partial \theta'}{\partial x_j}}$

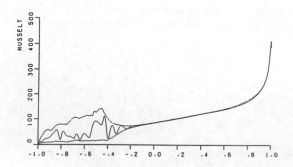

Fig. 10 Local Nusselt number distribution along heated wall, $Ra = 10^{10}$: maximum and minimum envelopes in time and a typical instantaneous distribution

EVAPORATION HEAT TRANSFER IN LIQUID FILMS FLOWING DOWN THE HORIZONTAL SMOOTH AND LONGITUDINALLY-PROFILED TUBES

V.G.RIFERT[*], Ju.V.PUTILIN[**], V.L.PODBEREZNY[**]

Heat transfer mechanism of liquid film flowing down the horizontal tubes has been studied by experimental determination of local and average heat transfer coefficients. Method of its enhancement has been developed based on the break-down of heat boundary layer by longitudinal fins and grooves. The calculated functions and recommendations on geometrical parameters of profiling, which allow to enhance the heat transfer from a surface to a liquid film by a factor of 1.4-1.9 and overall "vapour-liquid" heat transfer by a factor of 1.3-1.5 are given.

Introduction

Horizontal-tube film apparatus (HTFA) is widely used nowadays both as evaporators at desalination and evaporation plants and as a heat exchange equipment in different branches of industry. Raising the rate of heat transfer through the use of profiled heat exchange surface, which results in decrease of limiting thermal resistance is the most efficient method for reducing the specific quantity of metal and product costs. The development of this method for HTFA is hindered by insufficient study of heat transfer mechanism for liquid films flowing down even the smooth horizontal tubes, that caused the data deficiency on the choice of profiling type for heat transfer enhancement in literature. This results from the fact that most of investigations determine the integral mean-over-the tube surface heat transfer coefficients, which are insufficient for understanding heat exchange mechanism. A few investigations of local-along-the tube perimeter heat transfer have been performed only on smooth surfaces and the data obtained are contradictory and are of a qualitative character.

The present investigation through the use of specially developed original method and technical means allowed to obtain local heat transfer characteristics useful for quantitative analysis. As a result, the physical model of heat exchange and method of its enhancement are developed, and generalized functions for calculation of average heat transfer are suggested. Besides, the most rational shapes and parameters for profiling of external tube surfaces of commercial HTFA are determined.

[*]Kiev Polytechnical Institute, Kiev, Ukrain

[**]Scientific Research Institute of Chemical Machine Building, Ekaterinburg, the Russian Federation

Experimental technique

Experimental investigation of variation of local-along-the tube perimeter heat transfer was performed in a test plant described in detail in ref. /1/. Its main element was a 6-tube model of HTFA. The tubes are in one vertical row, their effective length is 205 mm, outside diameter - 38 mm. A tube space (S_T-d) could be changed from 10 to 20 mm.

Heat exchange surface of the test element was made of 0.1 mm thick metal (constantan) foil, attached to fabric-based laminate and heated by direct electric current that provided for $q = q_\varphi$=const. The surface temperature was measured by special flat, tape-like resistance microthermometers about 0.2 mm width, made of copper wire with 0.02 mm in diameter which were glued to inner surface of foil along the tube generatrices. Seven microthermometers, arranged along one half-perimeter at the distance of 30° from each other were placed into smooth-tube test element. The scheme of microthermometers arrangement in two profiled elements differs, it is shown in Fig.1. The measuring means, mentioned above, being applied instead of standard point thermocouples, allowed to increase the accuracy (±0.1°C) and stability of local-along-the perimeter temperature measurement of heat transfer surface and hence the reliability of α_φ determination as compared to the known works.

To investigate the average-over-the surface heat transfer for smooth and profiled tubes another experimental plant was used /2/. Its main difference was application of hot water as a heat carrier, which circulates inside the test tubes at high velocity (7.8-14.6 m/s). Shapes of external tube surface are shown in Fig.2. α_{sm} and α_{pr} from the tube surface to liquid film were determined by additivity of thermal resistances method. α_{pr}-values are related to total enhanced outside surface without effect of the latter.

The tests were performed under heating and surface evaporation conditions of water film without nucleate boiling.

Results and discussion

Fig.3 shows the experimental data obtained under evaporation conditions at atmospheric pressure, which characterize the change of heat transfer rate streamwise the liquid film along the tube perimeter (angular coordinate φ). The same relation $\alpha_\varphi = f(\varphi)$ was obtained for film heating at T_f = 96 and 44°C and two-fold increase (from 10 to 20 mm) of tube space. The decrease of heat transfer rate allows to conclude that at the most part of tube perimeter the heat exchange takes place in the initial region of thermal boundary layer development. Some increase of α_φ-values, associated with the film breaking-away from the tube surface, occurs at all Γ in the part of perimeter, which corresponds to $\varphi > 120°$.

Experimental data obtained (all α_φ-values at $\varphi \leqslant 120°$) were treated according to the system of dimensionless groups, accepted for heat exchange in the initial region, and the resulting equation for local heat transfer was

$$Nu_{dx} = 0,77 \left(Re_d/X\right)^{0,45} \cdot \left(Pr\right)^{0,33}, \tag{1}$$

where $X = x/d_x'$ - dimensionless length of film way. According to eq.(1), the rate of heat transfer is proportional to $(Re_\sigma/X)^{0,45}$ which practically corresponds to theoretical $\alpha_x \sim (Re/X)^{0,5}$-value, obtained by H.Schlichting, W.M.Kays, A.A.Zukauskas, B.S.Petukhov for heat exchange in initial thermal region with simultaneous development of velocity and temperature profiles in laminar boundary layer. Raising the rate of average heat transfer with increasing mass flow rate, noted by a number of investigators, is caused by increase of initial thermal region length and by decrease of developing boundary layer thickness at all angular locations (φ coordinates) along the tube perimeter, but not by transition to turbulent regime at stable flow and heat exchange.

Comparison of $\alpha_\varphi = f(\varphi)$, obtained for gravity falling film with that for cross liquid flow shows the heat exchange processes to be analogues for both cases (it was for the first time suggested in /2/). Hence, the experimentally determined values of average heat transfer coefficient in liquid film were treated according to the system of demensionless numbers, accepted for heat exchange in cross flow. The empirical dependence obtained

$$\alpha_{sm} = 0,295 \, \frac{\lambda}{d} \, (Re_d)^{0,63} \cdot (Pr)^{0,36}, \tag{2}$$

is almost the same as applied for calculation of heat exchange for cross flow over the in-line tube bundles and generalizes well not only our test data, but also that of the other investigators for film heating, cooling and evaporation. Re_d comprising the value of falling film velocity, U, was calculated with due account of liquid flow rate, tube space, character of tube surface spraying and according to method in ref./2/.

Proceeding from the revealed mechanism of heat exchange in film flow over the smooth horizontal tubes, a new method of its enhancement is proposed, in which the building-up boundary layer is broken by grooves or fins along the tube generatrix. Mechanism of enhancement was studied on the basis of local heat transfer investigation. Two elements (d = 38 mm) with 10 and 28 longitudinal triangular grooves of 3 mm wide, 1 mm deep and a pitch of 11.9 and 4.26 mm respectively were tested. The scheme of microthermometers arrangement 1-10 along the perimeter of these elements is shown in Fig.1. ($S_T - d$) was unchanged and was equal to 10 mm.

Fig.4 shows a sharp increase of heat transfer on each element of profiled surface and gradual decrease of α_φ-values on smooth areas between the neighbouring grooves (grooves arrangement along the tube perimeter is given under x-axis). In these areas the influence of film way length on heat transfer does not change after passing over the groove by liquid film and is close to theoretical $\alpha_x \sim x^{-0,5}$ (as well as for the smooth tube) for laminar boundary layer, i.e. profiling does not replace the flow pattern in boundary layer by a turbulent one for which $\alpha_x \sim x^{-0,2}$. The analyses of experimental data obtained allows to make a conclusion that enhancement in heat transfer on each element of profiled surface is caused by partial destruction of laminar boundary layer by grooves but not by turbulization of the last. This conclusion is very important for understanding the mechanism of heat exchange.

As shown in Fig.4, the influence of longitudinal profiling on heat transfer is ambiguous. The areas of low heat transfer, the so-called, stagnant pockets (expecially in.the grooves at upper half-perimeter of the tube) are formed due to building-up of film thickness and origin of flow along the grooves. Simultaneous action of two opposite effects, namely destruction of heat boundary layer and formation of stagnant pockets, influences the rate of heat transfer being obtained over longitudinally profiled surface, and is determined by parameters of profiling and process operating conditions. So, the enhancement of heat transfer increases up to definite limit with increasing mass flow rate (Fig.4) and amount of profiled surface elements (breaking-down the boundary layer) as is evident from the test of the sample with 28 longitudinal grooves (Fig.5).

To obtain the quantitative results for the influence of profiling geometry on the rate of heat transfer, the best profile parameters of external tube surface for HTFA and generalized dependence for calculation of enhanced heat exchange, the investigations of average heat transfer were performed. 20 test specimens of metallic tubes with 3 types of profile shape (Fig.2) and different combination of profile geometry were tested (groove width - 1.4-8.0 mm, groove depth or fin height - 0.5-2.0 mm, pitch - 3.0-11.8 mm). The external diameter of tubes was 30 and 38 mm, and brass, copper and aluminium alloy were used as a material.

The experiments reveal the most low rate of heat transfer to be on profile surfaces which have favourable conditions for stagnant pockets formation: narrow rectangular grooves or triangular grooves with low δ/Z ratio. The highest α_{pr} were obtained for tubes which have on their external surface: triangular grooves with the width of ≈ 3 mm, depth 0.5-1.0 mm and pitch 4-6 mm; wide rectangular grooves with $\delta = 6$-8 mm, $Z \approx 1$ mm in combination with narrow $(\delta' = S-\delta \approx 1.5$ mm) or wide $(\delta' \approx 7.0$ mm) fins as well as being finned in the form of wire $(d_w = 1.2$ mm, $S = 7.5$ mm - profile "3", Fig.2). However, the results obtained showed the most effective profiling according to "1"-type at $\Gamma < 0.25$ kg/m s and its advantage over the profiling according to "2"- and "3"-types increases with decreasing mass flow rate. That is why we used only "1"-profile in the form of triangular grooves with different combination of δ, z and S-values when developing the calculation procedure of heat exchange on longitudinally profiled tubes for commercial HTFA.

Experimental data (Fig.6) show the degree of heat transfer enhancement to be dependent on Re_δ and profile geometry but not practically dependent on Pr of liquid in the temperature range tested. The width-depth ratio for a groove (δ/z) may be used as a parameter, which displays the influence of profile geometry on enhancement of heat transfer. The rate of heat transfer is almost the same (with one and the same S-value) for the tubes with different $\delta = 1.4$-4.0 mm and $Z = 0.7$-2.0 mm but identical $\delta/z = 2$. The decrease of groove pitch results in growth of heat transfer with the largest α_{pr}/α_{sm} values at $S = 4$-6 mm and then it sharply decreases. The experimental data, given in Fig.6, for the whole range of $2 \leqslant \delta/z \leqslant 6$-values investigated are well approximated with regression dependence

$$\alpha_{pr_0} / \alpha_{Sm} = 0,53 (Re_\delta)^{0,11} \cdot (\beta/z) \cdot exp\left[-0,23 (\beta/z)\right]. \tag{3}$$

Here, α_{pr_0} is heat transfer coefficient at $S = S_0 = 4.25$ mm, i.e. maximum, obtained from experiments for triangular profile with all dimensions investigated. To take into account the influence of fin pitch in the range of 3.0-11.8 mm the following regression dependence is obtained:

$$\alpha_{pr} / \alpha_{pr_0} = 1,33 (S/S_0)^{0,45} \cdot exp\left[-0,31 (S/S_0)\right]. \tag{4}$$

Equations (3, 4) allow to calculate the rate of heat transfer from the heat exchange surface, profiled in the form of longitudinal triangular grooves, in a wide range of changes both of profile geometry: 1.4 mm $\leq \beta \leq$ 4.0 mm; 0.5 mm $\leq z \leq$ 2.0 mm; $2 \leq \beta/z \leq 6$; 3.0 mm $\leq S \leq$ 11.8 mm and operating variables.

The experiments resulted in the best profile geometry, which gives the largest enhancement of heat transfer: $\beta = 3$ mm, $z = 0.7$-0.75 mm, $\beta/z \approx 4$, $S = 4.0$-4.5 mm. From the experimental data obtained (Fig.6) it may be stated that profiled tubes with such geometrical characteristics will increase the rate of heat transfer by a factor of 1.3-1.7 as a function of Re_δ. Taking into account the increase of heat exchange surface, which is slight and equals to ≈ 1.08 for the parameters recommended, the enhancement is 1.4-1.9. The validity of the calculated data and profiling parameters recommended is confirmed by heat transfer tests of pilot model for HTFA with heat exchange surface of 1 m^2 and water vapour as a heat carrier. The range of operating variables covered operating conditions of appropriate equipment: $\Gamma = 0.04$-0.40 kg/m s; q = 15-75 kW/m^2; T_S = 42-100°C. Satisfactory coincidence of calculated and experimental values for heat transfer coefficients (discrepancy exceeds no 2-8%) is obtained and it is shown that the increase of heat transfer from the tube surface towards the liquid film allows to enhance the overall heat transfer from condensing vapour to evaporating liquid by a factor of 1.3-1.5. The results of investigation allow to recommend longitudinally profiled tubes for enhancement heat exchange in commercial HTFA.

Nomenclature

d, d_w- external diameter of tube, wire fin, respectively; x - distance along film way; β, z - width and depth (height) of groove or fin, respectively; S - space between the axis of neighbouring grooves or fins (fin pitch); S_T- space between the axis of horizontal tubes, vertically (tube bundle pitch); φ - angular coordinate; δ - liquid film thickness; U - average film velocity; α - average heat transfer coefficient from tube surface to liquid film: α_{sm}- for smooth tube, α_{pr} - for profile tube; α_φ, α_x- local along the tube perimeter heat transfer coefficient; q - heat flux; G - mass flow rate per unit length over one side of the horizontal tube; T_f - film temperature; T_s - liquid saturation temperature; ν - kinematic viscosity; λ - thermal conductivity; ρ - density.

Dimensionless groups: $Re_d = U \cdot d / \nu$ - Reynolds number; $Re_\delta = \Gamma / \rho \cdot \nu$ - Reynolds number of liquid film; $Nu_{\delta_x} = \alpha_x \delta_x / \lambda$ - local film Nusselt number; Pr - Prandtl number.

References

1. Putilin Ju.V., Rifert V.G. et al., 1986, Izvestija vusov SSSR, Energetika, N 8, pp. 99-101.

2. Putilin Ju.V., Podberezny V.L., Rifert V.G. et al., 1980, Proceedings of 7-th Intern. Symposium on Fresh Water from the Sea, vol.1, pp. 241-252.

Illustration

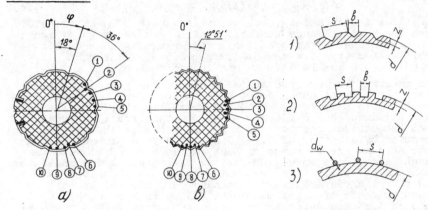

Figure 1. Arrangement of resistance micro-thermometers in profiled tubes with 10(a) and 28(b) longitudinal grooves.

Figure 2. Shape of profiling over the external tube surface for average heat transfer study.

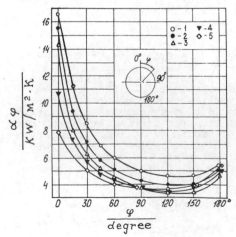

Figure 3. Local heat transfer in evaporation of water film on smooth horizontal tube.

d = 38 mm, $(S_T - d)$ = 10 mm, T_S = 100°C, q = 10 kW/m², at G, kg/m·s:
1 - 0.40; 2 - 0.25; 3 - 0.16; 4 - 0.12; 5 - 0.08.

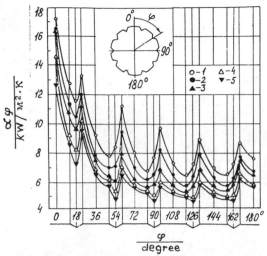

Figure 4. Local heat transfer in water film evaporation on profiled tube, Figure 1(a).
$T_5 = 100°C$, $q - 10$ kW/m^2, at G, kg/m·s:
1 - 0.40; 2 - 0.32; 3 - 0.25; 4 - 0.16;
5 - 0.12.

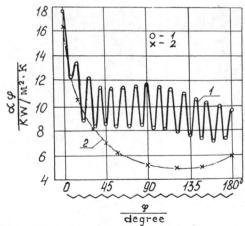

Figure 5. Comparison of local heat transfer for smooth and profiled tubes.
Fig.1(b), $G = 0.25$ kg/m·s, $T_5 = 100°C$,
$q = 10$ kW/m^2:
1 - profiled tube; 2 - smooth tube.

CONDENSATION OF MIXED VAPOURS OF R22 AND R114 REFRIGERANTS INSIDE HORIZONTAL TUBES

Ph. GAYET, A. BONTEMPS*, Ch. MARVILLET

Groupement pour la Recherche sur les Echangeurs Thermiques (GRETh)
Centre d'Etudes Nucléaires - 85 X - 38041 GRENOBLE CEDEX

Abstract
Condensation of non-azeotropic binary mixtures of R22 and R114 refrigerants inside a horizontal tube was experimentally studied. Tests have been conducted using two concentric tubes, with the vapour mixture flowing inside the inner tube and the cooling water in the annular space. The inner tube was fitted with thermocouples in order to provide wall temperature data for comparison to theoretical models. Experimental data were obtained for heat fluxes varying from 1 to 6 kW/m^2 and vapour mass fluxes varying from 50 to 250 kg/(m^{2}s) and were analysed with standard correlations.

Introduction

The use of non-azeotropic mixtures in refrigeration processes should grow up in the next years, because of direct advantages generated. These advantages may include improvement of the Coefficient Of Performance (COP), variable capacity of refrigeration, better oil miscibility, and substitution of the classical ChloroFluoroCarbons (CFC). However, because of a specific behaviour (non isothermal condensation, resistance to heat and mass transfer at liquid/gas interface) classical methods of design are not allowed for the prediction of heat and mass transfer during the condensation of non-azeotropic mixtures. Contrary to the vertical case, very few informations are available for the condensation inside horizontal tubes. This article describes an experimental study on condensation of R22/R114 mixtures inside a smooth horizontal tube. Moreover, a comparison of existing theories is carried out : equilibrium theory (based on a predetermined temperature/enthalpy relationship called the condensation curve) and the classical film theory (based on a mass and heat transfer analysis).

Experimental apparatus and procedure

Experimental loop
Since experimental data are rather scarce [1,2] experimental apparatus to measure local and global heat flux was built. The mixed vapours generated in a helicoidal double pipe evaporator flow through the test section constituted of a straight, horizontal smooth tube. This geometry is simple enough to allow the analysis with theoretical models and is commonly used in industrial applications as finned tube heat exchangers. The residual vapours and the formed condensate after being condensed and subcooled in a post-condenser are pumped toward the evaporator.

Test section
The test section is a straight, smooth, horizontal stainless steel tube having a 3 meter length and a 16 mm inner diameter. It is cooled by means of cold water flowing in an external annulus. This heat transfer test section is preceded by a 1 meter long calming length. Water and refrigerant temperatures are measured on both side of the test section.

* Université Joseph FOURIER, GRENOBLE, FRANCE

The test section is divided in 12 units. Wall and cold water temperatures are measured at the center of each unit. The wall temperature measuring thermocouples were inserted and soldered in holes and the water temperature measuring thermocouples were inserted in the middle of the water flow (fig. 1).

Experimental procedure

Global heat flow rate is calculated from the thermal balance on the cold water side by :

$$Q = \dot{M}_c \ c_p \left(T_{c,out} - T_{c,in} \right) \tag{1}$$

where $\dot{M}_c$ is the mass flow rate, c_p the specific heat of the cooling water and where $T_{c,in}$ and $T_{c,out}$ are the inlet and the outlet temperatures of the cooling water. Local heat flow rates Q_l are determined by means of the 12 thermocouples regularly embedded in the inner wall in front of 12 thermocouples inserted in the cooling fluid at the same longitudinal position. They are given by :

$$Q_l = \alpha_c A \left(T_w - T_c \right) \tag{2}$$

where T_w and T_c are the inner wall and the coolant temperatures respectively. The coolant side heat transfer coefficient α_c is calculated by the GNIELINSKI correlation [3]. corrected by the PETHUKHOV & ROIZEN factor for an annulus [4].

Wall temperature measurements

Because of contact resistances and fin effect on the thermocouples implanted in the wall (fig. 1) we developed a method to correct the measured wall temperatures [5].

Experiments were carried out by replacing the refrigerant fluid with hot water flow. Mass flow rates and inlet temperatures were chosen in order to obtain Reynolds numbers and heat fluxes in the same range of those expected. Then, the hypothesis was made that local heat fluxes with and without thermocouples are proportional. These experiments provided us correcting factors of the temperature difference between wall and cold water.

Flow regimes

To determine the nature of the flow regime, the test section is able to rotate around its longitudinal axis. The wall temperature being representative of the thickness of the condensate layer inside the tube, the angular distribution of temperatures allows us to estimate the role of the condensate film and what kind of ideal flow (annular or stratified) is encountered.

Test fluids and operating conditions

Because of lack of physical data for new refrigerants we began to investigate pure R22 and R114 and their mixtures. Other fluids will be later studied. During our experiment 5 compositions were tested (Table I). The experimental and operating conditions are summarized in table II .

Heat transfer coefficient calculations

Condensation curve methods

Basic hypotheses are :
- vapours and condensate are in equilibrium and it is assumed that the system follows a predetermined temperature-enthalpy relationship,
- mass transfer is neglected and the ordinary dry-gas coefficient is used for sensible heat transfer.

The total heat flux is given by :

$$\varphi_t = \alpha_{sc} \left(T_s - T_c \right) \tag{3}$$

where T_s is the gas/liquid interface temperature and T_c the coolant temperature ; α_{sc} is the heat-transfer coefficient between interface and coolant, and it is given by :

$$\alpha_{sc} = \left(\frac{1}{\alpha_c} + R_w + \frac{1}{\alpha_l}\right)^{-1}$$ (4)

where α_l is the heat-transfer coefficient across the condensate liquid film, α_c the coolant side heat-transfer coefficient and R_w the wall thermal resistance.

In the same way, the sensible heat flux from gas to condensate is :

$$\varphi_s = \alpha_s(T_b - T_s)$$ (5)

where α_s is evaluated from the vapour film parameters and where T_b is the bulk gas temperature.

The combination of the equation (3) and (5) leads to :

$$\varphi_t = \frac{T_b - T_c}{\dfrac{1}{\alpha_{sc}} + \dfrac{\varphi_s}{\varphi_t} \times \dfrac{1}{\alpha_s}}$$ (6)

allowing us to define the overall heat-transfer coefficient by :

$$U = \left(\frac{1}{\alpha_{sc}} + \frac{\varphi_s}{\varphi_t} \times \frac{1}{\alpha_s}\right)^{-1}$$ (7)

In the above equations, the $Z = \varphi_s/\varphi_t$ ratio is calculated following Bell and Ghaly [6].

Classical film theory
The classical film theory is based on the existence of a laminar film in the vapour phase near the gas liquid interface. All resistances against heat and mass transfer are concentrated in this film. By integration of mass and energy balance equations across the diffusion film the total molar rate of condensation of a binary mixture n_t is obtained [7].

Neglecting subcooling effects the heat balance at the gas-liquid interface leads to :

$$\alpha_{sc}(T_s - T_c) = \alpha_s^*(T_b - T_s) + n_t\Delta h$$ (8)

where Δh is a weighted average value of the mixture enthalpy difference between the vapour and the liquid state [7].

The gas-phase sensible heat transfer coefficient was corrected by Ackermann [8] and Colburn and Drew [9] to take into account the effect of mass transfer. It is given by :

$$\alpha_s^* = \frac{a}{\exp a - 1}\alpha_s$$ (9)

where $a = \dfrac{n_t \tilde{C}_{pm}}{\alpha_s}$, $\tilde{C}_{pm}$ being the molar heat capacity of the mixture.

Results and discussion

Behaviour of condensate

By observing wall temperatures, the nature of flow regime can be determined. For example, for two experiments with M2 composition (table I), it can be seen in figure 2, that two regimes occur, annular in figure 2b (inlet Reynolds number Re = 450980) and stratified in figure 2a (inlet Reynolds number Re = 112 420).

Heat transfer data reduction

Global and local heat flow rates were deduced from temperature and flow rate measurements. It can be remarked that the sum of the twelve local heat flow rates is equal to the global one within 10 %.

Measured heat flow rates are compared with those obtained from theoretical predictions using either condensation curve method or film theory. Comparisons between global heat flow rates are made through the deviation factor q_{Gj} defined by :

$$q_{Gj} = \frac{Q_{mj} - Q_{cal.j}}{Q_{mj}} \tag{10}$$

where Q_{mj} and $Q_{cal.j}$ are the measured and calculated heat flow rates respectively for the jth experiment (at a given composition and mass flow rate).

For all n_M mass flow rates (at a given composition) the mean value is defined as :

$$q_G = \frac{1}{n_M} \sum_{j=1}^{n_M} q_{Gj} \tag{11}$$

For local heat fluxes, comparisons are made by defining :

$$q_L = \frac{\Delta T_m - \Delta T_{cal.}}{\Delta T_m} \tag{12}$$

where $\quad \Delta T_m \quad = \quad T_{wm} - T_{cm}$
and $\quad \Delta T_{cal.} \quad = \quad T_{w\,cal.} - T_{c\,cal.}$

T_{wm} and T_{cm} being the measured local wall and cooling liquid temperatures, $T_{w\,cal.}$ et $T_{c\,cal.}$ being the calculated local wall and cooling liquid temperatures respectively.

Computer simulations

Calculations were carried out with the CETUC software developed by GRETh [10]. This software, using a Bell-Delaware method is able to predict global and local behaviour of fluids in a tube and shell or coaxial heat exchanger. At each step, the local heat transfer coefficients α_{sc} and α_s (or α_s^*) are calculated. For these calculations, several well-established correlations were introduced to determine the heat transfer coefficient α_s.

After experiments using pure fluids, a preliminary choice allows us to select three correlations :

-   the correlation of Shah   [11],
-   the correlations proposed by the VDI-Wärmeatlas   [12],
-   the correlation proposed by Cavallini and Zecchin   [13].

In all cases, the mixtures are considered to give rise to miscible condensates.

Results

Figure 3 illustrate the comparison between the heat power predicted by the 2 methods (condensation curve and film theory) and the experimentally observed values at the M2 concentration.

As shown in figure 3, the Cavallini-Zecchin correlation gives a better agreement with experimental data for the two methods. Moreover it can be observed that data are best fitted by the condensation curve calculations.

The local measurements and the corresponding calculated values are compared in figure 4. As for global data, the best agreement is obtained for the condensation curve method coupled with the Cavallini-Zecchin correlation.

To explain these results, the film theory was modified to take into account the effect of mass transfer on film thickness by introducing a very simple turbulence model. Preliminary results show that the agreement is better but the deviation factor is still important [5].

A tentative explanation of the better representation of data by condensation curve methods could be that, at high Reynolds number (>100 000), the classical film theory does not apply. New methods have to be developed to design condensers in such conditions.

Conclusion

An experimental investigation was carried out on condensation of R22/R114 mixtures inside a smooth tube. Global and local heat flow rates were experimentally determined and compared to values from two theoretical methods : condensation curve and film theory. Within the range of operating conditions, the classical film theory does not correctly account for experimental data.

References

1. Stoecker WF, Kornota E, 1985, ASHRAE Trans., 91, 1353-1367

2. Tandon TN, Varma HK, Gupta C.P, 1983, Proc. XVI Int. Cong. Refrig., Paris

3. Gnielinski V., 1988, VDI Wärmeatlas, 5 Auflage, Gd3

4. Pethukhov B.S., Roizen, 1988, VDI Wärmeatlas, 5 Auflage, Gd3

5. Gayet Ph., 1992, Thèse de Doctorat, Univ. Joseph Fourier, Grenoble

6. Bell K.S., Ghaly M.A., 1973, AIChE Symp., 69, 72-79

7. Webb D.R., McNaught J.M., 1980, Developments in Heat Transfer Technology, Applied Science Publishers London

8. Ackermann G., 1937, Forschungschaft, 382, 1-116

9. Colburn A.P., Drew T.B., 1937, AIChE Trans., 33, 197-215

10. Mercier P., Ratel G., 1990, Note GRETh 90/218 (Unpublished)

11. Shah M., 1979, Int. J. Heat Mass Trans., 22, 547-558

12 VDI Wärmeatlas, 1988, 5 Auflage, Ja13-Ja16

13 Cavallini A., Zecchin R., 1974, Proc. Int. Heat Trans. Congress, Tokyo

Nomenclature

a		Ackermann correction factor
C_p	J/kg.K	Specific heat capacity of the coolant
$\tilde{C}_{pm}$	J/mol. K	Molar heat capacity of the mixture
n_t	mol/s.m^2	molar condensation rate
q		deviation factor
A	m^2	heat transfer area
$\dot{M}$	kg/s	mass flow rate
Q	Watt	heat flow rate
R	m^2.K/W	thermal resistance
T	°C or K	temperature
U	W/m^2.K	overall heat transfer coefficient
α	W/m^2.K	heat transfer coefficient
$\alpha^{\cdot}$	W/m^2.K	corrected heat transfer coefficient
Δh	J/mol	enthalpy difference
ΔT	°C or K	temperature difference
φ	W/m^2	heat flux

Subscripts

b	bulk
c	coolant
cal	calculated
j	experiment number
l	liquid
m	measured
s	interface
sc	from interface to coolant
t	total
w	wall
G	global
L	local

TABLE 1
R22/R114 Mixture compositions

	R22/R114 % mol	R22/R114 % mas
R22	100%/0%	100%/0%
R114	0%/100%	0%/100%
M1	18%/82%	10%/90%
M2	45%/55%	30%/70%
M3	62%/38%	45%/55%

TABLE 2
Experimental conditions

Refrigerant mass velocity	50–250 kg/(m²s)
Inlet vapour velocity	1–30 m/s
Heat power	1–6 kW/m²
Test section length	3 m
Inner tube diameters	16–20 mm
Heat exchange area	0.19 m²
Material	stainless steel

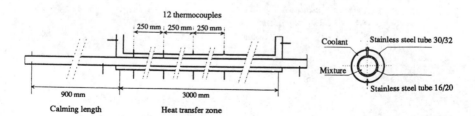

Figure 1 - Test section.

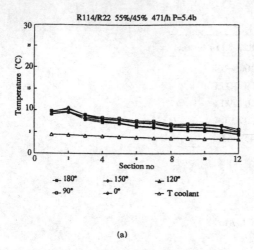

(a)

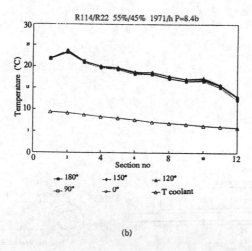

(b)

Figure 2 - Local wall temperature : (a) stratified flow regime , (b) annular flow regime . Angles define thermocouple positions (0° is for upper position).

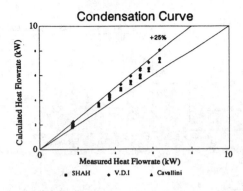

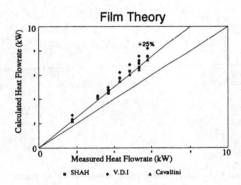

Figure 3 - Comparison of calculated and measured heat powers.

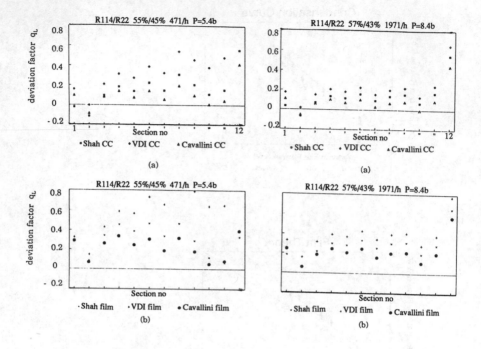

Figure 4 - Comparison between deviation factors for the three proposed correlations :
(a) condensation curve method , (b) film theory.

THE CONDENSATION HEAT TRANSFER CHARACTERISTICS OF LIQUID NITROGEN IN SERRATED, PLATE-FIN PASSAGES

R.H. Clarke
HTFS, AEA Technology, Harwell Laboratory, Didcot, Oxon. OX11 0RA

An experimental study has been carried out to investigate the characteristics of downflow condensation in a serrated-fin plate-fin test section under conditions similar to those used in industrial, cryogenic process plant. This work complements that already reported for a plain-fin test section (Robertson et al, 1986). Heat transfer coefficients are presented for a pressure of 6 bar with mass fluxes in the range 30 to 80 kg/m²s. Comparison is made between the results for the two fin types, and with various predictive methods.

INTRODUCTION

The boiling and condensing characteristics of nitrogen in plain plate-fin passages have been extensively investigated in the laboratory by HTFS (the Heat Transfer and Fluid Flow Service) using cryogenic test facilities. The Harwell Laboratory rig used for condensation studies was refurbished in 1988 and a serrated-fin test section was installed. Schematic construction is shown in Figure 1, which also illustrates the two forms of finning used in this study. The serrated finning has the following dimensions: fins 6.35 mm high, 710 fins/m, 0.2 mm fin thickness, 3.18 mm long serrations. The test section was made up of 12 pads of finning, each 283 mm long and 76 mm wide. A gap of 3 mm was left between each pad to permit thermocouple access for bulk temperature measurements.

The serrated finning has the same hydraulic diameter (2.0 mm) as the plain finning used in tests reported by Robertson et al. (1986), so the new data may be used to assess the effect of serrations on the condensation process. A series of tests was carried out at 6 bar with pure nitrogen as the test fluid. Mass fluxes were varied from 30 to 80 kg/m²s, with outlet qualities ranging from 0.7 to zero (complete condensation). Heat flux was not a controlled variable. Heat transfer results are compared with the predictions of the Boyko and Kruzhilin (1967) correlation and the models of Yung et al (1980), Nusselt (1916), and a new correlation developed for HTFS.

Operating Details and Single Phase Heat Transfer Tests

The test rig used for cryogenic condensation studies generates superheated nitrogen vapour which is fed to the top of a test section which consists of three plate-fin passages. Cooling is achieved by boiling nitrogen at low pressure in the outer layers. A calibrated thermal resistance separates the boiling and condensing streams so that local heat transfer coefficients may be measured at 12 locations along the test section with the aid of bulk and wall thermocouples.

As with the plain fin test section, described by Haseler (1980), the 100 test section thermocouples were calibrated in-situ to enable temperature differences to be determined to within +/- 0.1 K. The thermal resistances between the two boiling streams and the central condensing stream were calibrated by carrying out a heat balance on a known condensing flow and were found to be almost identical to those found for the plain fin test section ($1/680$ m^{2}K/W). No recalibration of the condensing stream Annubar gas flowmeter was found necessary as the change in test section did not affect its operation.

Single-phase gas tests were carried out with cryogenic nitrogen passing through the condensing channel and heated with the external heater mats placed on the outside surfaces of the boiling streams (see Figure 1). Colburn-j factors were determined at several Reynolds numbers ranging from 3000 to 10000 (Figure 2). This means that there is no data for the laminar region in which most of the liquid-portion heat transfer coefficients used in correlations normally lie. The manufacturers' curve lies about 20% above the measured data.

Predictions of Colburn-j factor using the regression correlation of Wieting (1975) are shown on Figure 2. In the turbulent region there is good agreement with the experimental data up to a Reynolds number of 10000. Lack of experimental Colburn-j data for the laminar region makes it difficult to rely on the Boyko and Kruzhilin (1967) correlation for condensation heat transfer. Wieting presented an equation for this region, but his data are for warm air which has a Prandtl number (Pr) of about 0.7 whereas liquid nitrogen has a Prandtl number of about 2. Since Nusselt numbers may be assumed constant at low Reynolds numbers, Wieting's equation for the laminar region has been pre-scaled by $(0.7/Pr)^{0.33}$.

The transition region (Reynolds numbers between 1000 and 2000) is not covered by either of Weiting's equations. For this region the highest value of Colburn-j from the turbulent and laminar equations has been used with the Boyko and Kruzhilin correlation to make predictions of the condensing heat transfer coefficient.

In the condensation tests (Figure 3), the gas phase heat transfer coefficients are predicted well using the experimental Colburn-j factors unlike the results for the all liquid coefficients which show that the predicted Colburn-j factors are too high. Unfortunately, the heat fluxes at the outlet of the test section were too low to be able to accurately determine any experimental Colburn-j factors for liquid nitrogen.

CONDENSATION HEAT TRANSFER RESULTS

All heat transfer results presented in this paper have been obtained using the standard fin efficiency assumption relating to extended surface heat exchangers. It is acknowledged that the heat transfer coefficient varies around the periphery of the sub-channels formed by the finning. This does not matter provided the correlations and data are compared on the same basis, as is the case in this paper.

The effect of serrations on the heat transfer coefficient

Figure 3 shows typical results for the plain and serrated finning. The plain fin results were reported by Robertson et al (1986). At a mass flux of 50 kg/m^{2}s the condensing coefficients for the serrated fin are, at maximum, only 40 percent above those measured for the plain fin. At a quality of 0.7 the coefficient increases from 2250 W/m^{2}K for the plain finning to 3100 W/m^{2}K for the serrated finning. The manifold enhancement which the serrations provide for single phase flows are not present during two phase heat transfer.

The effect of mass flux on the heat transfer coefficient

The effect of varying the mass flux on the condensing heat transfer coefficient at 6 bar pressure is shown in Figure 4. There is a very slight increase in coefficient with increasing mass flux at moderate to high qualities. Above a quality of 0.8 this effect was masked by the presence of wet-wall desuperheating. At the lower qualities the coefficients tend towards the coefficients observed in the plain fin tests.

Boyko and Kruzhilin (1967) correlation

Boyko and Kruzhilin predictions of the condensing heat transfer coefficient are shown on Figure 4. The trend of increasing coefficient with mass flux is followed, but the magnitude of the calculated increase is much too large. The method does not predict the correct effect of quality. It indicates a steady fall in heat transfer coefficient as the quality falls, whereas the experimental results show nearly constant coefficients at mid range qualities, then a gradual fall to 2000 W/m²K at near zero quality. In general, the Boyko and Kruzhilin method significantly overpredicts the condensing heat transfer coefficients at higher mass fluxes although this could be attributable to uncertainties in the Colburn-j factor data for the laminar/transition region.

Film models

Apart from the Boyko and Kruzhilin method, comparison has been made with the predictions of Nusselt (1916) and Yung et al (1980). The Nusselt analysis assumes a laminar falling film with no vapour shear and in Figure 5 the predictions lie well under the experimental data because, even at the lowest mass fluxes, vapour shear and film "waviness" effects are significant. A similar observation was reported for the plain fin condensation tests (Robertson et al (1986,1987). Except at the lowest qualities, Dukler's (1960) dimensionless shear stress (β) is usually greater than 10, which indicates that shear is always likely to be important in condensation with serrated finning.

Yung et al. present a film analysis in which the flow of condensate around the serrations is assumed to be analogous to flow past a vertical bank of horizontal tubes. The liquid drains, uninterrupted from one serration to the next-but-one serration. Use of this model requires knowledge of the total pressure gradient and interfacial friction factor appropriate to a laminar film flowing along the primary and secondary surfaces (fins). In this paper the model has been amended to make use of

(1) the measured total pressure gradients, and

(2) *plain fin* friction factors, which being only about 20% of the *serrated fin* friction factors are assumed to approximate the two phase *interfacial* friction factors (the remainder of the pressure drop is accounted for in drag terms).

The results of these predictions are shown in Figure 5 for a pressure of 6 bar. At high qualities, the Yung et al predictions are high, but at qualities in the range 0.5 to 0.8 the agreement with experimental heat transfer coefficients is better. The predicted, weak effect of mass flux is approximately correct, unlike the exaggerated effects found with the Boyko and Kruzhilin correlation (Figure 4).

HTFS CORRELATION

Illustrative predictions of the often-used Boyko and Kruzhilin correlation, and of the various film models, demonstrate the need for an improved correlation. With serrated finning it is believed that the interrupted flow path may lead to significant entrainment in both boiling and condensing flows. This may be similar to the turbulent, "homogeneous flow" condensation in pipes for which the Boyko and Kruzhilin method was developed. However, at low mass fluxes it is unlikely that the vapour shear is high enough to disturb the wavy-film flow which exists mostly on the primary surfaces.

Taking these two features into account, HTFS developed in 1991 a new correlation based on the Boyko and Kruzhilin method which has been suitably amended to account for wavy, falling film effects. Predictions made with this correlation give a good representation of the 6 bar serrated fin data shownn in Figure 6.

The correlation developed for HTFS is intended for general applicability over a range of fin types, fluids, mass fluxes and pressures. However, for the 6 bar serrated fin results shown here for condensing nitrogen the correlation used is given by Equation (1).

$$\alpha_{cond} = \frac{1}{2}(\alpha_{BK} + \alpha_{WF}) \quad [W/m^2K] \tag{1}$$

where

α_{cond} = predicted condensation coefficient
α_{BK} = Boyko and Khruzhilin (1967) coefficient
α_{WF} = "wavy-film" coefficient, Butterworth (1983), given in Equation (2).

$$\alpha_{WF} = 0.756k_L \left\{ \frac{\rho_L(\rho_L - \rho_v)g}{\mu_L^2} \right\}^{1/3} Re_f^{-0.22} \quad [W/m^2K] \tag{2}$$

where

k_L = liquid thermal conductivity [W/mK]
ρ_L = liquid density [kg/m^3]
ρ_v = vapour density [kg/m^3]
g = gravitational acceleration [m^2/s]
μ_L = liquid dynamic viscosity [kg/ms]
Re_f = $(1-x) G d_e \mu_L$ (liquid film Reynolds number)
x = vapour mass fraction [-]
G = total flow mass flux [kg/m^{2}s]
d_e = plate-fin sub-channel hydraulic diameter [m]

DISCUSSION AND CONCLUSIONS

A series of condensation tests at 6 bar has been carried out using the HTFS Cryogenic Condensation Rig. A limited number of Colburn-j factor tests were carried out using heated cryogenic gas. The results obtained were found satisfactory at high Reynolds numbers, but less so at the low Reynolds numbers important for use with the Boyko and Kruzhilin correlation.

The serrated-fin tests were carried out under identical conditions of mass flux and pressure to those used for plain finning of the same dimensions. The heat transfer coefficients exhibited the same features already found in the plain fin results, namely a complex trend with quality and a slight effect of mass flux. For the serrated finning there was at most a 40% increase in heat transfer coefficient. This result suggests that the use of serrated finning for condensing duties would be worthwhile provided the considerable two phase pressure drop penalty incurred is acceptable.

The commonly used Boyko and Kruzhilin method failed to predict major trends in the data and significantly overpredicted the higher mass flux results. In the absence of anything better, use of this correlation may be adopted for mass fluxes up to 50 kg/m^{2}s. An adaptation of Yung et al's method better predicted trends in the data but is still inadequate for design purposes, except inasmuch as it does not require any single phase coefficients to be calculated. To overcome these difficulties, a new correlation has been developed by HTFS. Although based on the Boyko and Kruzhilin method, it has been suitably amended to account for film flow effects which are a principal feature of condensation in both plain and serrated-fin passages. This correlation has been found to be much more satisfactory than the Boyko and Kruzhilin correlation which it is intended to replace.

ACKNOWLEDGEMENTS

The work was carried out under the Research Programme of the Heat Transfer and Fluid Flow Service (HTFS) and was supported by the Department of Trade and Industry.

The author wishes to acknowledge the valuable contributions of N. Blundell, who carried out the experimental work, and L.E. Haseler and J.M. Robertson for their guidance during this study.

REFERENCES

Boyko, L.D. and Kruzhilin, G.N. (1967). Heat Transfer and Hydraulics Resistance During Condensation of Steam in a Horizontal Tube and in a Bundle of Tubes. Int. J. Mass and Heat Transfer, V.10, pp.361-373.

Butterworth, D. (1983). Film condensation of pure vapour. Heat Exchanger Design Handbook, Section 2.6.2, pp.2.6.2-1 to 2.6.2-6, Hemisphere Pub. Corp. New York.

Dukler, A.E. (1960). Fluid Mechanics and Heat Transfer in Falling Film Systems. Chem. Eng. Progress. Symposium Series, Vol.56, No.30, pp.1-10.

Haseler, L.E. (1980). Condensation of Nitrogen in Brazed Aluminium Plate-Fin Heat Exchangers. ASME/AIChE Transfer Conf., Orlando, Florida, USA, 27-30 July. 9pp.

Nusselt, W. (1916). Die Oberflächenkondensation des Wasserdampfes. Zeit. Ver. Deutsch. Ing., 60, pp.541-569.

Robertson, J.M., Blundell, N. and Clarke, R.H. (1986). The Condensing Characteristics of Nitrogen in Plain, Brazed Aluminium, Plate-Fin Heat-Exchanger Passages. Heat Transfer 1986 (Proc. 8th Int. Heat Transfer Conf., San Francisco, CA, USA, 17-22 Aug. 1986), C.L. Tien et al (Ed.), Hemisphere Pub. Corp., Vol.4, pp.1719-1724.

Robertson, J.M., Blundell, N. and Clarke, R.H. (1987). Characteristics of Pressure Gradients in Downflow Condensing of Nitrogen in Plain, Brazed Aluminium, Plate-Fin Heat Exchanger Passages. AIChE Symp. Ser., Vol.83, No.257, pp.59-64.

Yung, D., Lorenz, J.J. and Panchal, C. (1980). Convection Vaporization and Condensation in Serrated Fin Channels. Heat Transfer in OTEC Systems, V. HTD12, pp.29-37.

Wieting, A.R. (1975). Empirical Correlations for Heat Transfer and Flow Friction Characteristics of Rectangular Offset-Fin Plate-Fin Heat Exchangers. J. Heat Transfer, Vol.97, Series C, No.3, pp.488-492.

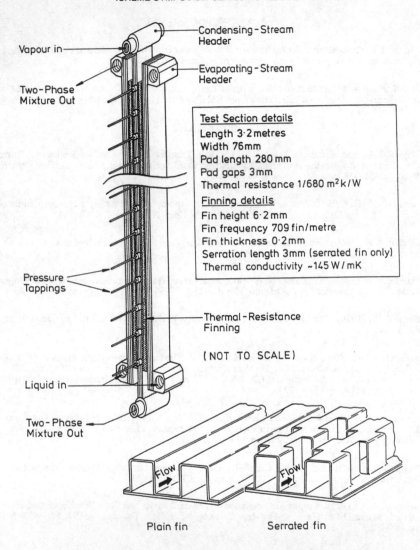

Condensing-Stream Header

Vapour in

Evaporating-Stream Header

Two-Phase Mixture Out

Test Section details
Length 3·2 metres
Width 76mm
Pad length 280 mm
Pad gaps 3mm
Thermal resistance 1/680 m²k/W

Finning details
Fin height 6·2 mm
Fin frequency 709 fin/metre
Fin thickness 0·2mm
Serration length 3mm (serrated fin only)
Thermal conductivity ~145 W/mK

Pressure Tappings

Thermal-Resistance Finning

(NOT TO SCALE)

Liquid in

Two-Phase Mixture Out

Plain fin   Serrated fin

Figure 1: The test section and two forms of finning used in tests.

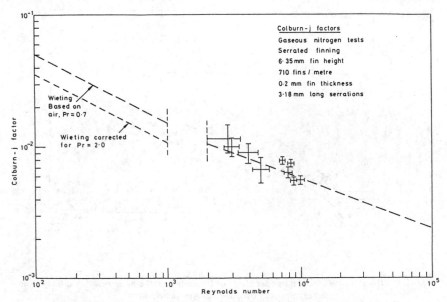

Figure 2: Colburn-j factor - Reynolds relationship for the serrated finning.

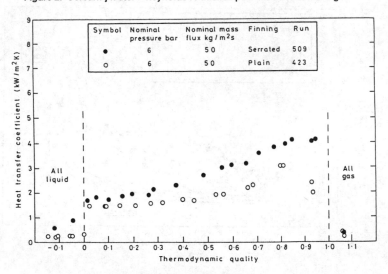

Figure 3: Comparison of the condensing heat transfer coefficients in plain and serrated finning.

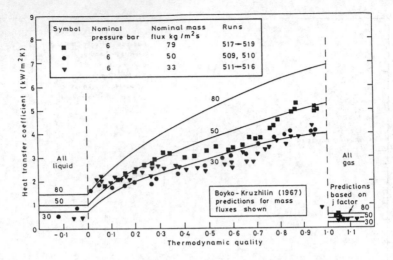

Figure 4: Effect of mass flux on condensing heat transfer coefficients in serrated finning at 6 bar, showing Boyko-Kruzhilin predictions

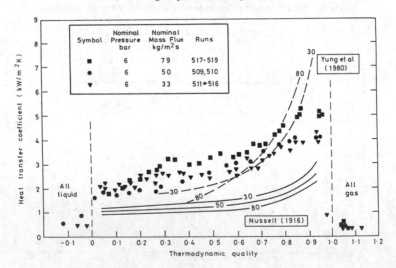

Figure 5: Predicted heat transfer coefficients of Nusselt and Yung et al at 6 bar.

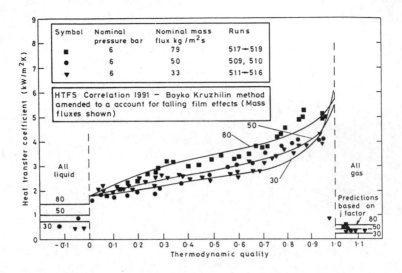

Figure 6: Predicted heat transfer coefficients of HTFS correlation (1991) at 6 bar.

INDEX

O